Time, Quantum and Information

Springer
*Berlin
Heidelberg
New York
Hong Kong
London
Milan
Paris
Tokyo*

Carl Friedrich von Weizsäcker © Ingrid von Kruse

Lutz Castell   Otfried Ischebeck   (Eds.)

Time, Quantum and Information

Springer

Lutz Castell
Max-Planck-Institut für Physik
Föhringer Ring 6
80805 München, Germany

Otfried Ischebeck
ZAE Bayern
Walter-Meissner-Strasse 6
85748 Garching, Germany

Cataloging-in-Publication Data applied for
A catalog record for this book is available from the Library of Congress.

Bibliographic information published by Die Deutsche Bibliothek.
Die Deutsche Bibliothek lists this publication
in the Deutsche Nationalbibliografie; detailed bibliographic data is available in the
Internet at http://dnb.ddb.de.

ISBN 3-540-44033-X Springer-Verlag Berlin Heidelberg New York

This work is subject to copyright. All rights are reserved, whether the whole or part of the material is concerned, specifically the rights of translation, reprinting, reuse of illustrations, recitation, broadcasting, reproduction on microfilm or in any other way, and storage in data banks. Duplication of this publication or parts thereof is permitted only under the provisions of the German Copyright Law of September 9, 1965, in its current version, and permission for use must always be obtained from Springer-Verlag. Violations are liable for prosecution under the German Copyright Law.

Springer-Verlag Berlin Heidelberg New York
a member of BertelsmannSpringer Science+Business Media GmbH

http://www.springer.de

© Springer-Verlag Berlin Heidelberg 2003
Printed in Germany

The use of designations, trademarks, etc. in this publication does not imply, even in the absence of a specific statement, that such names are exempt from the relevant protective laws and regulations and therefore free for general use.

Typesetting by authors/editors
Data conversion: LE-TEX Jelonek, Schmidt & Vöckler GbR, Leipzig
Cover Design: *design & production* GmbH, Heidelberg

Printed on acid-free paper   55/3141/YL   5 4 3 2 1 0

Preface

This publication centers on the extraordinary ideas in and concepts of physics of Carl Friedrich von Weizsäcker. At the time of his 90th birthday on June 28, 2002, it seems the right moment to try such a survey. The themes of two Festschrifts for Carl Friedrich von Weizsäcker on the occasion of his 60th and 70th birthdays (E. Scheibe and G. Süssman (eds.): *Einheit und Vielheit*, and K. Meyer-Abich (ed.): *Physik, Philosophie und Politik*) were his unique capability to encompass physics, philosophy and politics. He may be more known publicly today for his efforts for containment of the Cold War nuclear threat, for the abolition of war as an instrument of international politics, for the social responsibility of scientists, and for the Conciliar Process of the Churches for Justice, Peace and the Integrity of Creation. But physics has been his primary professional vocation and has always remained in the center of his thought and life.

But even in light of the physics focus of this book, it would not do justice to Carl Friedrich von Weizsäcker to restrict his achievements in physics to efforts only accessible to professionals. The contributions in Part I show how his very concentration on physics has led him to take an active part in problems of politics, social change, philosophy and religion.

From his doctorate in 1933 under Werner Heisenberg to the 1950s, Carl Friedrich von Weizsäcker built a successful career as a theoretical physicist in both nuclear physics and astrophysics. The theory of nuclear structure and forces, energy production in stars, the formation of planetary systems from a dust cloud, and the structure of turbulence are the fields where his contributions are still felt today. Part II reports on these achievements and their impact on modern science.

Carl Friedrich von Weizsäcker is one of the rare persons who can creatively link philosophy and natural sciences. Backed with this gift and with his success in the physics of atomic nuclei and stars, he set out, beginning in the mid-fifties, to reconstruct quantum theory as the center of modern natural philosophy. Parts III to V of the book contain contributions to this program. The way into this program is indicated by the concept of time, which is essentially linked to the concept of experience, the factual nature of the past and the indeterminacy of the future. The indeterminacy of the future leads to the central theory of modern physics: quantum theory, been understood as a theory which expresses the essential conditions for experience of the physical world. Its predictions provide information on the outcome of measurements in the form of probabilities. Weizsäcker called the elementary unit of information in quantum theory an *ur*. As an all encompassing theory of physics, quantum theory should contain the possible fundamental forms of matter, elementary particles, and their interaction. It should thus permit the construction of particles and interactions from quantized bits of information. This hypothesis is called the *ur-hypothesis*, which was developed during the 1970s at the *Max Planck Insitut zur Erforschung der Lebensbedingungen der wissenschaftlich-technischen Welt* in Starnberg. It is on this path that Carl Friedrich von Weizsäcker would expect to achieve scientific honesty, eventual scientific discovery and, finally, scientific truth.

Carl Friedrich von Weizsäcker's work in nuclear physics and astrophysics can be considered as the first round on a "circular path", whose following round was marked by two parallel tracks: the scientific program for reconstruction of quantum theory, and his engagement in politics and society. Philosophy provides the common ground of these two tracks: the inquiry into one's own comprehension and understanding, and the moral issue of the engagement of the individual in the society. The person to whom he felt closest on this path has been Niels Bohr.

The driving force in his political engagement has been his involvement with the problems of nuclear power and weapons: under a dictatorship during World War II, as a threat for human existence during the following decades, but also as a source of energy for society. On the scientific track, his starting point is the all-encompassing relevance of quantum theory within science, which is a foundation for our belief in the unity of nature. But the unity of nature includes the history of humankind which, in turn, includes the development of science. So, the two parallel tracks converge again in the search for peace of humanity with nature, and for peace among humankind.

The editors are grateful to Carl Friedrich von Weizsäcker's friends, co-workers and fellow scientists for their contributions in particular for writing texts which can be read by both professional physicists as well as by interested non-physicists. Yet we are aware that a printed book cannot fully present what is to us the most impressive side of Carl Friedrich von Weizsäcker: his ability to listen to others, his power of critical reasoning and the presence of his enormous knowledge in a discussion. He enjoys these capacities up to his present age of ninety.

The editors wish to thank Christoph Castell, Roger Hilton and Michael Rawn for correcting the English expression of translations from German. Christian Oberdorf has cared for the technical work with the edition. We thank the Max Planck Institute for Physics (Werner Heisenberg Insitute), Munich and the Human Science Center of the Ludwig Maximilian University of Munich for financial support. We thank Jörn Behrmann for discussions, and last, but certainly not least, Professor Wolf Beiglböck and the staff of Springer Verlag for their support of this publication.

Munich, May 2003
Lutz Castell
Otfried Ischebeck

Contents

Letters to Carl Friedrich von Weizsäcker on the Occasion of His 90th Birthday

Part I
Carl Friedrich von Weizsäcker and His Role in the Physics of the 20th Century

Carl Friedrich von Weizsäcker and The Kaiser Wilhelm/Max Planck Society
Reimar Lüst 17

The Role of the Physicist in the Nuclear Age
Edward Teller 25

"The Atomic Bomb Reveals the Political Responsibility of Science"
Götz Neuneck 27
1. The Problem of Morality of Science in the Service of Nuclear Weapons Development 27
2. The Uranium Project 1939–1945 29
3. First Attempts to Find a Political Framework to Nuclear Weapons Development 40
4. Coping with the Nuclear Threat During the Cold War 47
5. Paths Through Peril 53
References 55

On the Unfathomableness of Consciousness by Consciousness
Klaus Gottstein 59

Part II
Carl Friedrich von Weizsäcker's Contributions to Nuclear and Astrophysics and Their Impact on the Development of Physics

Introduction: From the Atomic Nucleus to Cosmic Vortex Systems
Helmut Rechenberg 75
1. First Research on Relativistic Quantum Problems (1931–1934) 75
2. The Theory of Nuclear Structure (1934–1937) 76
3. Pioneering the Theory of Nuclear Fusion in Stars (1936–1938).......... 77
4. Nuclear Fission Energy, Turbulence and the Structure of Stellar Systems (1939–1954) 78
References 80

The Origins of Nuclear Physics and Carl Friedrich von Weizsäcker's Semi-Empirical Mass Formula
Karl v. Meyenn 83
1. Theoretical Physics at the Beginning of the 30's 83
2. The Penetration of Quantum Theory into Nuclear Physics 85
3. Gamow's Liquid Drop Model...................................... 87
4. The Neutron and Heisenberg's First Theory of the Nucleus 88
5. Nuclear Physics During the Solvay Conference of 1933................ 91
6. The Discovery of the "American Forces" 92
7. Carl Friedrich von Weizsäcker's Work in Physics During His Study in Leipzig 93
8. A Winter Sojourn in Copenhagen 1933/34: The Discovery of the Weizsäcker–Williams Method 95
9. The Origin of the Semi-Empirical Mass Formula 97
10. As Debye's Assistant at the Kaiser Wilhelm Institute in Berlin 102
11. The Liquid Drop Model and Nuclear Fission 103
References 105

Thermonuclear Processes in Stars and Stellar Neutrinos
Georg Wolschin... 115
1. Introduction 115
2. Energy Evolution in Stars 119
3. Hydrogen Burning.. 123
4. Stellar Neutrinos 127
5. Perspectives 132
References 133

Comets: Fascinating Cosmic Objects
Reimar Lüst.. 135
1. The Nature of Comets 135
2. The Orbits of Comets and Their Origin............................. 137
3. The Physical and Chemical Properties of Comets 138
4. Artificial Cometary Tails – Experiments in Space 138

5	The Preparation of the Space Missions to Halley's Comet and the Spacecrafts	140
6	The Giotto Mission	141

The Spectrum of Turbulence
Siegfried Grossmann 145
1	Introduction	145
2	Historical Remarks	145
3	Turbulent Diffusion	146
4	Scaling Law of Turbulence	147
5	Mean Field Theory of Turbulence	149
6	Temperature Structure Function	153
7	Outlook	154
References		156

From Dust Disks to Planetary Systems
Thomas Henning 159
1	The Scheme – From Fiction to Reality	159
2	Protoplanetary Disks – The First Steps	162
3	Detection of Planets and Planetary Systems	164
4	Interactions – Numerical Simulations	166
5	Tools for the Future	167
References		169

Part III
The Unity of Nature and the Nature of Time

Science and Its Relation to Nature in C.F. von Weizsäcker's Natural Philosophy
Klaus Michael Meyer-Abich 173
1	The Meaning of Knowledge, Time, and History in the Philosophy of Nature	173
2	The Crisis of the Atomic Bomb in the Natural History of Mind	176
3	Understanding Explanation – Weizsäcker's Titanic Project, and Some Suggestions About It	180
References		184

C.F. von Weizsäcker's Philosophy of Science and the Nature of Time
Michael Drieschner 187
1	Philosophy of Time	188
2	Statistical Thermodynamics	191
3	Probability	194
4	Quantum Mechanics	197

X Contents

What Is Missing? – The Fundamental Role of Time in C.F. von Weizsäcker's Conception of Physics and Some Insights from Modern Neuroscience
Eva Ruhnau 203
1 Time and Experience 203
2 Subject/Object Separation and Observation in Physics 205
3 The Now as Transition Point Between Past and Future 207
4 Temporal Logistics of the Brain I 209
5 Separation and Integration in the Brain 210
6 Object Formation in the Brain 212
7 Temporal Logistics of the Brain II 213
8 Experience and Time 214
References 216

Irreversibility via Semichaos
Georg Suessmann 219
1 Stochastic Entropy 219
2 Objections 220
3 Molecular Semichaos 221
4 The Model 223
5 The Equations of Motion 223
6 Distribution of Discs in the Case of a Single Rhombus 225
7 Distribution of Rhomboid-Disc Pairs 226
8 The Time Asymmetric Diffusion Equation 227
9 The Transition to the Continuum 228
10 The H-Theorem 229
11 The Correlations 230
12 Heredity 231
13 Appendix: Boundary Conditions 232

Remarks on Fractional Time
Rudolf Hilfer 235
1 Introduction 235
2 Requirements for time evolution operators 236
3 Consequences from the requirements 238
4 Philosophical remarks 240
References 241

The Dynamics of Modelling
Gérard G. Emch 243
1 Introduction 243
2 Some Models from Thermophysics 244
3 A Longer View 254
References 257

Part IV
The Structure of Quantum Theory and Its Interpretation

An Introduction to Carl Friedrich von Weizsäcker's Program for a Reconstruction of Quantum Theory
Thomas Görnitz and Otfried Ischebeck 263
1   The Motivation for a Reconstruction of Quantum Theory
    and the Elements of the Program 263
2   Quantum Theory and Philosophy 265
3   Time, Probability and Quantization 268
4   Postulates for the Basic Structure of Quantum Theory 269
5   The Ur-Hypothesis 275
6   On the Interpretation of Quantum Theory 276
References 279

Interpreting Quantum Mechanics – in the Light of Quantum Logic
Peter Mittelstaedt 281
1   Classical and Quantum Physics – Their Respective Roles 281
2   Reduction of Ontological Hypotheses 282
3   The Quantum Logic Approach – An a priori Justification? 284
References 290

On the Interpretation of Quantum Theory – from Copenhagen to the Present Day
Claus Kiefer 291
1   Copenhagen Interpretations and Alternatives 291
2   The Emergence of Classical Properties in Quantum Theory 294
3   Quantum Gravity 297
4   Conclusion 298
References 299

Epistemic and Ontic Quantum Realities
Harald Atmanspacher and Hans Primas 301
1   Introduction 301
2   Distinguishing Epistemic and Ontic Perspectives 304
3   Epistemic Descriptions of Quantum Systems 306
4   Ontic Descriptions of Quantum Systems 311
5   Relations Between Epistemic and Ontic Descriptions 315
6   Conclusions 319
References 319

Information and Fundamental Elements of the Structure of Quantum Theory
Časlav Brukner and Anton Zeilinger 323
1   Introduction 323
2   Finiteness of Information, Ur, Elementary System 325

3	Mutually Complementary Propositions	327
4	Measure of Information in a Probabilistic Experiment	330
5	The Catalog of Knowledge of a Quantum System	333
6	Total Information Content of a Quantum System	334
7	Malus' Law in Quantum Physics	337
8	Entanglement – More Information in Joint Properties than in Individuals	343
9	Time Evolution of the Catalog of Knowledge	346
10	Measurement – the Update of Information	349
11	Conclusions	351
	References	353

Multiple Quantization in Fock Space
Dirk Graudenz 357

1	Introduction	357
2	Setting the Stage: Quantization of a Free Scalar Field	358
3	A Framework for Multiple Quantization	361
4	Summary and Open Questions	362
	References	362

Part V
The Ur-Hypothesis and Its Implications for Particle Physics and Cosmology

The Ur-Hypothesis
Lutz Castell 365
References 367

The Momentum Eigenstates and the Lorentz-Invariant State
Lutz Castell 369
References 373

C.F. von Weizsäcker's Reconstruction of Physics:
Yesterday, Today, Tomorrow
Holger Lyre 375

1	The Origin: The Philosophy of Physics	375
2	Yesterday: Urs, Spinors and Spacetime	376
3	Today: Tetrads, Gravity and Gauges	379
4	Tomorrow: Qubits, Holographic Principle and the Ontology of Information	381
	References	384

The Operational Structure of Spacetime
H. Saller 387

1	Ontic or Praxic Spacetime	387
2	Operations and Symmetries	388

3	How Simple Can Be Simple?	390
4	The Binary Alternative, Spin and Space	392
5	The External-Internal Dichotomy	394

Ur Theory and Space-Time structure
David Ritz Finkelstein 399
1 The ur 399
2 The quantum universe 400
3 Quantum logic 401
4 Non-associative logic 402
5 Summation 407
6 Acknowledgements 408
References 408

Weizsäcker's Ur Theory – A Cosmological Point of View
Thomas Görnitz 411
1 Introduction 411
2 The Way to a Connection Between Ur Theory and Cosmology 413
3 The Introduction of Space 414
4 An Effective Energy-Momentum Tensor 417
5 An Estimation for the Entropy of Particles 418
6 Ur Number and Bekenstein–Hawking–Entropy for Particles 419
7 Conclusions 420
References 420

Ur Theory and Cosmological Phase Transition
Jörg D. Becker and Lutz Castell 423
1 Do Large Numbers Have a Meaning? 423
2 From Ur Space to Minkowski Space 423
3 Photon Condensation in an Einstein Universe 425
4 Conclusion 427
References 427

Phase Transitions in an Expanding Universe: Paul Roman's Models and Some Remarks on Entropy
Jörg D. Becker 429
1 Photon Condensation in an Expanding Universe 429
2 Bubble Formation and Entropy 430
3 Linking Ur Cosmology to More Conventional Approaches 430
4 Conclusions and Some Remarks on Entropy 431
References 432

Appendix 435
C. F. von Weizsäcker:
Biographical Data and Selected Bibliography in Physics 435
The Authors 451

Letters to Carl Friedrich von Weizsäcker on the Occasion of His 90th Birthday

Cornell University
Newman Laboratory for Elementary-Particle Physics

Newman Laboratory
Ithaca, NY 14853-5001
(607) 255-4397

Telex: WUI 6713054
Telefax: 607-254-4552
hab11@cornell.edu

27.5.02

Dear Carl Friedrich:

I am happy to congratulate you on your 90th birthday. Our lives in astrophysics have flown much in parallel: You discovered that the proton-proton fusion must be the first nuclear reaction in a star; then Critchfield and I worked it out quantitatively. A few months later we independently found the carbon-nitrogen cycle which is generally accepted as the greatest source of stellar energy.

A few years ago, in Copenhagen and in Ithaca, we had intensive talks on arms control and disarmament, and found to our mutual pleasure that here also our ideas are running very much in parallel. If they only could be fulfilled!

I am wishing you many happy years of further life and thought.

Yours sincerely,

Hans A. Bethe

Hans A. Bethe

HAB/dsh

HOOVER INSTITUTION
On War Revolution and Peace
Stanford, California 94305-6010
FAX: 650-723-1687

Dr. Edward Teller
Senior Research Fellow

Den 4. März 2002

Professor Dr. Carl Friedrich von Weizsäcker
Max-Planck-Institut für Physik
(Werner Heisenberg-Institut)

Lieber Carl Friedrich,

Von diesem Briefe getrennt, werde ich ein Paar Zeilen schicken über die Fragen die in den letzten 70 Jahren nichts an ihrer Drohnung verloren haben. Ich hoffe sehr dass ich über die Beiträge deinem 90. Geburtstag hören kann.

Die Probleme unserer Jugend blieben im ganzen ungelöst, aber die zweite Hälfte des 20. Jahrhundert war sicherlich besser als die erste Hälfte. Es tut mir leid dass wir unsere jugendliche Gespräche nicht wieder von ihren verwirrten Mitte mit erneute Hoffnung neu anfangen können.

Wenn ich Deine religiöse Glauben teilen könnte, würde ich mir wünschen dass Du mich einmal in Purgatorium von einem höheren Himmel besuchen würdest.

Inzwischen, mit besten Wünschen zu Deinem 90. Geburtstag.

Dein

Edward Teller

Dear Carl Friedrich,

apart from this letter, I shall send a few lines addressing those questions that have not lost any of their menacing character during the past 70 years. I hope very much that I can hear the contributions to your 90th birthday.

The problems of our youth remain largely unsolved, but the second half of the 20th century was certainly better than the first half. I feel sorry that we cannot once more take up our youthful discussions from their bewildering middle and engender new hope.

If I could share your religious belief, I would wish that you will one day come from a higher heaven and visit me in purgatory.

Meanwhile, with best wishes on your 90th birthday,

Your

Edward Teller

Jürgen Habermas

Lieber, verehrter Herr von Weizsäcker,

gestatten Sie mir, den glücklichen Umstand Ihres 90. Geburtstages, den Sie bei voller intellektueller Präsenz erleben dürfen, zu nutzen, um aus einem Abstand von zwei bis drei Jahrzehnten auf unsere gemeinsame Zeit am Starnberger Institut zurückzublicken und zu erklären, wofür ich Ihnen dankbar bin.

Damals hat Ihre Einladung, das Institut gemeinsam zu leiten, nicht nur mich überrascht. Daraus haben sich für den einen oder anderen von uns auch Problemen ergeben, die schwer lösbar waren. Aber die persönliche Beziehung zu Ihnen, dem stets wohlwollenden, integren und in seiner hellen analytischen Kraft überlegenen Geist, ist immer intakt geblieben. Das verdanke ich Ihrer wahrhaft platonischen Fähigkeit, die Probleme von der Erdenschwere der schnell beiseite geschobenen organisatorischen Aspekte zu befreien und auf die Höhe intellektuell spannender Themen zu heben. Sie haben von Ihrer einzigartigen Gabe, aus dem Stand über beliebig komplexe Sachverhalte in vollkommener Klarheit zu sprechen, einen zivilen Gebrauch gemacht. Deshalb konnte sich unsere Zusammenarbeit in dem schwebenden Medium einer allenfalls unterbrochenen, jederzeit wieder aufzunehmenden Konversation über die wahrlich interessierenden Dinge vollziehen.

2

Auch in meiner Bibliothek steht natürlich die lange Reihe der gelben Weizsäcker-Leinenbände aus dem Hanser-Verlag. Vermutlich wissen Sie nicht, lieber Herr von Weizsäcker, wie viel und vor allem was ich von Ihnen gelernt habe. Sie würden sich wundern, wenn Sie einen Blick in den ersten dieser Bände aus dem Jahre 1971 werfen würden. In dem zweiten Teil dieses besonders zerlesenen Buches, das mit „Die Einheit der bisherigen Physik" beginnt und mit dem Abschnitt über Quantentheorie und einem „Entwurf der Einheit der Physik" schließt, sind fast alle Seiten mit den Anstreichungen eines ganz inkompetenten, aber offensichtlich lernbegierigen und engagierten Lesers übersät.

In einem der späteren Bände finde ich neben der Widmung einen Zettel mit dem ungeduldigen Hinweis auf drei Stellen: S. 116-121, S. 132-33; S. 474-5. Das war es, worüber Sie mit mir sprechen wollten - Ihre christliche Überzeugung, dass man die Moral der Gerechtigkeit nicht verselbstständigen dürfe: „letzter Grund der Möglichkeit menschlichen Zusammenlebens ist die Liebe und nicht die Moral".

Ich wünsche Ihnen, dass Sie bei guter Gesundheit bleiben und die Anstrengung der zahlreichen verdienten Feiern zu Ihrem Ehrentage nicht nur überstehen, sondern genießen, und bin

mit herzlich Grüßen

Dear, honored Mr. v. Weizsäcker,

please permit me to take the opportunity presented by the happy circumstance of your 90th birthday, which you are able to celebrate with full intellectual presence, to look back two or three decades at our time together at the Institute in Starnberg and explain my gratitude to you.

Back then, your invitation to jointly direct the Institute was a surprise not only to myself. From this, problems arose which, for one or the other of us, were difficult to solve. But my personal relation to you, the ever benevoler, upright and superior spirit stemning from clear analytical capability, has always remained intact. I owe this to your truly Platonic capacity to liberate problems from the earthly weight of organizational mundanities and lift them to intellectually exciting heights. You put at the service of the public your unique gift for impromptu speaking with complete clarity on complex subjects. Therefore our cooperation could manifest itself in the intangible medium of a conversation on truly interesting things, interruptions of years not preventing resumption at any moment.

On my bookshelf too, there are, of course, the yellow linen volumes of Weizsäcker published by Hanser Verlag. Perhaps you are not aware, dear Mr. von Weizsäcker, how much and, above all, what I have learnt from you. You would be astonished if you were to take a look at the first of these volumes from 1971. In the second part of this particularly well worn book, which starts with "The unity of the hitherto existing physics" and ends with the section on quantum theory and an "Essay on the unity of physics", almost all pages are strewn with underlinings by a totally incompetent but apparently curious and enthusiastic reader.

In one of the later volumes, I find next to the dedication a slip of paper with an anxious reference to three passages, pp. 116–121, pp. 132–133, and pp. 474–5, concerning something you wanted to discuss with me—your Christian conviction that the morality of justice should not be held up in isolation: "the ultimate basis of the possibility for humans to live together is love not morality."

I wish you continuing good health and trust that you will not merely bear the numerous festivities which you have merited on the occasion of your day of honor, but that you will genuinely enjoy them, and I remain

with cordial greetings

Your Jürgen Habermas

Ruth Grosse

Juli 2002

Lieber Herr v. Weizsäcker,

ich habe Ihnen hier einen Traum aufgeschrieben, den ich - angeregt durch die diversen Feiern zu Ihrem 90. Geburtstag – kürzlich hatte, und der in ganz unbefangener Weise das widerspiegelt, was für mich wichtig war in den zwei Jahrzehnten der Zusammenarbeit mit Ihnen und den Wissenschaftlern im damaligen Institut.

„Eine Gruppe von Menschen, darunter ich, machte eine Wanderung unter Ihrer Führung. Wir durchquerten eine sehr fremdartige, wilde, immer wieder atemberaubend schöne Landschaft: Eisgraue, abweisende, bizarre Berggipfel im Hintergrund, Geröllhalden mit seltsamen, kostbaren Steinen, ausgetrocknete Flussbetten, gelbe, dünenartige Sandflächen. Das Gehen in weglosem Gelände war sehr mühsam, zumal der Boden nicht sicher war; stellenweise war es sumpfig, dann wieder taten sich Triebsandlöcher auf, oder tiefe Wasserläufe versperrten den Weg. Aber Sie waren unermüdbar und guter Dinge. Sie liefen voraus, um den Weiterweg zu erkunden, Sie kamen zurück gerannt, um uns Instruktionen zu geben und zu ermuntern. Unsere Klagen, wenn z.B. einer bis zur Brust im Morast steckte, ein anderer von einem reissenden Bach mitgerissen wurde, ein dritter sich den Fuss verstaucht hatte, berührten Sie nicht weiter. ‚Kümmert Euch um Euch selbst, ich habe keine Zeit für diese Dinge' riefen Sie und eilten wieder voraus. Sie hatten eine Kinderträgerhose an mit zu kurzen, weiten, flatternden Hosenbeinen, Ihre Stirne war extrem hoch, zu hoch, die Haut gelblich mit vielen Sommersprossen, die Augen waren immer geschlossen. Sie schienen aber alles durch die Augenlieder hindurch zu erkennen, jedenfalls orientierten Sie sich in dem unwegsamen Gelände mit traumwandlerischer Sicherheit und vermieden mit grossem Geschick alle Hindernisse, die uns anderen so zu schaffen machten.

Weit oben am Berg verlief eine Strasse, auf der andere Menschen anscheinend mühelos wanderten, in derselben Richtung wie wir. ‚Warum gehen wir nicht auf der Strasse, da kämen wir besser voran', fragten wir, ‚man müsste nur 200 Meter hinaufsteigen und schon wären wir auf einer sicheren, ebenen Strasse?'
Aber einige unter uns sagten, so eine Strasse sei reizlos, die Wildnis fernab der Zivilisation sei viel aufregender und interessanter. Also blieben wir unten und schlugen uns weiter durch. Unser Führer, Sie selbst, war
 inzwischen nicht mehr bei uns.

Auf unserer Wanderung kamen wir an mehreren Burgen und Schlössern vorbei, die alle von Ihren „Verbündeten" oder „Anhängern" bewohnt waren. Alles schöne, schweigsame Menschen, engelähnliche Gestalten mit hellem Teint, aquamarineblauen Augen und langen, schwarzen Augenwimpern. Sie übernahmen in Ihrer Abwesenheit die Führung der Gruppe und brachten uns in verwunschenes Wüstenland, in vom Sand verwehte frühzeitliche Städte, mit halb verfallenen mächtigen Kirchen und prächtigen, maroden Stadtpalästen. Wir sahen, dass keine Bleibe war in diesem herrlichen, verlassenen Geisterland. Wir sehnten uns nach zu Hause."

Die heutige Traumwissenschaft spricht dem Traum lediglich eine Funktion der Verarbeitung von „Tagesresten" zu. Anscheinend aber können es auch „Lebensreste" sein.

Ruth Goosse

July 2002

Dear Mr. v. Weizsäcker,

I have written down a dream, which I had recently – inspired by the various festivities on the occasion of your 90th birthday. Put simply, it reflects what has been important for me in the two decades of collaboration with you and with the scientists in the former Institute in Starnberg.

"A group of persons, among them myself, made a journey under your leadership. We passed through a breathtakingly beautiful land: forbidden peaks, gray with ice, in the far, boulders with rare, precious stones, dried creeks, yellow dune shaped stretches of sand. Making progress in this terrain was very cumbersome, especially as the ground was not safe; at places it had been swampy, then there were shifting sands, and deep waters that barred the way. But you suffered no fatigue and were of good spirits. You went ahead in order to find out the further path, you came running back to give us instructions and to encourage us. Our complaints, for example when one of us was stuck in the swamp up to his chest, when another had been torn away by a torrent, when a third had sprained his foot, did not distract you. "Care for yourselves, I don't have time for these things" you called and you hurried ahead. You wore a pair of children's trousers with braces and short fluttering trouser-legs, your voice was extremely high, too high, the skin yellowish with freckles, your eyes were always shut. You appeared to recognize everything through your eye lids; in any case you moved in this pathless terrain with sleep-walking security and you avoided with great aptitude all obstacles that caused us so many troubles.

Far up on the mountain, there was a road on which other people apparently walked without difficulties in the same direction as we. "Why don't we go on this road, we could advance more easily there", we asked, "we need only climb two hundred meters and we would be on a safe and even road?" But some of us said, such a road has no charm, the wilderness far from civilization is much more exciting and interesting. So we stayed below and continued to battle our way. Our guide, yourself, had meanwhile vanished from our midst.

On our journey, we passed several castles and palaces, which were all occupied by your "allies" or "adherents", All of them beautiful, silent persons, of angelic appearance, with light complexion, deep blue eyes and long black eye-lashes. In your absence they took over the lead of the group and brought us into an enchanted desert country, to ancient cities covered with the sand, with ruined churches and splendid but derelict municipal palaces. We saw that there was no possibility to stay in this marvelous, deserted land of spirits. We longed to come home."

The modern science of dreams attributes to the dream only the function of digesting the "day's remnants". Apparently, these can also be "life's remnants".

Ruth Grosse

Part I

Carl Friedrich von Weizsäcker and His Role in the Physics of the 20th Century

Carl Friedrich von Weizsäcker and The Kaiser Wilhelm/Max Planck Society *

Reimar Lüst

Carl Friedrich von Weizsäcker has been a member of the Kaiser Wilhelm/Max Planck Society since 1936. The only interruption in his membership during this long span of time was a three year period during the war. The Institute of Physics, the present Werner Heisenberg Institute, has been his scientific home the longest. It is, therefore, quite appropriate that this speech on the occasion of the completion of the 90th year of your life bear the title: Carl Friedrich von Weizsäcker and the Kaiser Wilhelm/Max Planck Society.

As I was president of the Max Planck Society for twelve of the years of Carl Friedrich von Weizsäcker's affiliation, this topic has fallen to me this afternoon. A more fitting title for what I want to present today would perhaps have been: Carl Friedrich von Weizsäcker through the eyes of one of his disciples.

It probably does not happen very often that a single book determines the whole scientific career of a student. This happened with me through von Weizsäcker's book *"Zum Weltbild der Physik"*. I read it in 1945 in a prisoner of war camp in Texas, where I had begun to study mathematics and physics in the camp university. I was captivated by what I read in the first chapter on *"Modern physics and the world picture of physics"*. I do not claim to have understood everything I read. I remember in particular, my dear Mr. von Weizsäcker, your discourse on *"The relation of quantum mechanics to the philosophy of Kant"*, a problem which has occupied you up to the present. At the time I was simply a 22 year old U-boat sailor with a special emergency wartime diploma from secondary school. I knew, however, that I wanted to study under you.

That this was not possible I learned in March, 1946 when I came back from captivity and, two days later, wanted to enroll for study at Göttingen. I was told ex cathedra that the dead line for inscription had been a week earlier. I was not admitted any more and, thus, I could not attend your great course on *"The history of nature"*, which had already become legendary.

---

* Speech given on July 4, 2002 on the occasion of C.F. von Weizsäcker's 90th birthday at the Max Planck Institute for Physics in Munich

I went to Frankfurt, where I met an understanding dean, Erwin Madelung, who was able to by-pass all bureaucratic obstacles and admitted me to study theoretical physics.

I learned more about you during the following three years while I was waiting to be able to move to Göttingen to study with you. You were already a well-known atomic physicist. Your book on *"The Atomic Nuclei"*, which you had written in 1937 in Leipzig, was in the library. I heard about the Bethe–Weizsäcker cycle on energy production in the sun's interior. You had written a paper on this in 1937, which had appeared in the *Physikalische Zeitschrift*.

Otto Hahn gave a public speech in the winter of 1946 in his home town Frankfurt, in which he told us how he had discovered nuclear fission. I did not know then that in 1936 you had worked at the Kaiser Wilhelm Institute for Chemistry, which was directed by Otto Hahn. You have written: "During Christmas 1938, I worked and lived in Dahlem. Otto Hahn, in whose institute I had previously worked for half a year, called me on the telephone and asked if I could imagine a Radium atom behaving like Barium in every chemical reaction. He had discovered the fission of Uranium. Two months later, bewildered and upset, I went to Picht to discuss with him how the world would be affected by the possibility of an atomic bomb".

In the spring of 1936, after your post-doctoral research under Heisenberg was completed, you were appointed as an assistant replacing Max Delbrück and you worked in the department of Lise Meitner at the Kaiser Wilhelm Institute for Chemistry whose director had been Otto Hahn.

Half a year later in the fall of 1936, you changed to the Kaiser Wilhelm Institute for Physics in Berlin, which was under the leadership of Peter Debeye at the time. You became concurrently a reader at the University of Berlin.

At the beginning of the War, the government placed the Kaiser Wilhelm Institute for Physics under the command of the army's office for weaponry. The government had hoped to profit in some way from the technical application of atomic energy. Debeye, a Dutch citizen, did not want to remain in Germany. He left the Institute and went to the United States of America.

You were a member of the so-called "Uranium Club" which investigated the technical conditions for the development of atomic weapons and for the production of nuclear energy. Heisenberg was called in from Leipzig to play a leading role.

Heisenberg gave an address to the annual assembly of the Max Planck Society in 1971, in which he reported on the history of the Institute of Physics and summarized the contents of a meeting held on June 4, 1942 in the Harnack House: "It could be reported to government experts that the construction of an atomic reactor from natural Uranium and heavy water is a viable possibility, but that the construction of atomic bombs would require a technical effort which is beyond all our means. The government subsequently issued an order to us for the construction of a reactor, or, as we then called it, a Uranium burner or atomic pile, but there was no order for the construction of atomic bombs".

Much has been written on this meeting from very different points of view. The same is true regarding your journey to Copenhagen with Werner Heisenberg in

September 1941. Young people today are not able to understand the nature of the situation and what it meant to embark on such a journey.

In the heated discussion in connection with the play "Copenhagen", there is no mention of Niels Bohr coming to Göttingen in 1951 and visiting Werner Heisenberg and the Institute. Heisenberg would in vain use at least fifty matches for lighting his pipe while lecturing. He would always admonish us excitedly, like a little school boy, not to help him light his pipe. This became a ritual at his lectures. There was no sign of discord between Niels Bohr, Heisenberg and you.

The subordination of the Institute under the army's office for weaponry was eventually lifted after another meeting with government representatives. Special mention is due to you and Karl Wirtz, as you succeeded in replacing the director Diebner with Werner Heisenberg. What would have happened to the Kaiser Wilhelm/Max Planck Institute for Physics if Werner Heisenberg had not have taken over the directorship then?

In 1942 Heisenberg managed to rescind your being drafted to the eastern front. You were able to obtain a professorship in theoretical physics at the University of Strasbourg. This marks an end to your first period in the Kaiser Wilhelm Society, which one might describe: "Carl Friedrich von Weizsäcker, nuclear physicist". What you experienced after the end of the War is a much cited history and has been presented in two movies. With Otto Hahn, Werner Heisenberg and others, you were taken prisoner by the Americans and interned by the British until the spring of 1946 at Farm Hall. When you returned to Germany, you together with Karl Wirtz and later Ludwig Biermann, began the reconstruction of the Institute in Göttingen under the new name Max Planck Institute for Physics under the direction of Werner Heisenberg.

In 1949 I finished my diploma examination under Erwin Madelung at the University of Frankfurt and intended to begin a new field of study in Göttingen under your direction, atomic and nuclear physics.

I did not know that your favorite subject was astronomy. In your book *"Der Garten des Menschlichen"* (The Garden of the Human), you wrote: "For my twelfth birthday in June, 1924, I had wished for a rotating stellar map adjustable to days and seconds. Soon afterwards we left Basel, where my father was the German consul, for our summer vacation in the Jura Mountains in the Canton of Bern. We stayed at the lonely pension Mont Crosin. The Swiss national holiday was celebrated there on the evening of 1 August, with traditional bonfires and rockets. A dance started up with a long polonaise. During one of the separations of the queue, I managed to lose my partner, who was about my age. With map in hand, I escaped from the crowd into the warm, wonderful starry night, completely alone."

I was not expected when I arrived at the Max Planck Institute for Physics in Böttingerstrasse during the first days of March 1949. Nevertheless I was greeted casually and told to go to the second floor, where Mr. von Weizsäcker welcomed me warmly, even though he was in a great hurry. He told me that the Institute's colloquium was about to begin and I should come with him right away. We would continue our conversation later.

I was surprised that you were not an abstracted professor, what one might have suspected after reading your books. You were neither aloof nor pompous. You appeared very young, I did not yet know your real age. You spoke in a clear voice with a very slight Swabian accent.

After the seminar lecture you took me to your office, which was opposite Heisenberg's. I saw the name tag of Max von Laue on the door to one side of your office and Ludwig Biermann on the door to the other side. Only then did I fully realize where I was. You explained that you would accept me as a doctoral student, but that as your work here in Göttingen was on astrophysical problems, you could only give me a topic on astrophysics and not – as I had imagined – on quantum and atomic physics. A fundamental paper of yours on the further development of Kant's theory on the origin of the planetary system had just appeared in the *Zeitschrift für Naturforschung*, in which you applied the hydrodynamic equations with turbulence. You proposed that I should investigate the problem of transport of angular momentum in a rotating gaseous disc. The solution of this problem was important, since the Sun rotates relatively slowly nowadays, although the rotation of the central mass should actually have increased during its contraction.

From this first conversation in your office, I remember especially your desk. It was covered with papers and books. A book by Karl Barth, probably his dogmatics, caught my attention. In "Der Garten des Menschlichen", you report on the only conversation you had with Karl Barth (in the early 1950s): "I see the direct path from Galileo to the atomic bomb and I am concerned whether I can continue to pursue physics, an enterprise that I cherish so much. He replied: Mr. von Weizsäcker, if you believe what all Christians confess and almost none believes – the second coming of Christ – then you can, you must continue to do physics; otherwise you should not."

The 1950s were a marvelous time at the Institute in Göttingen for us students. We could pursue our scientific work freely and did not care much about our future. Nevertheless, I could not have survived the first years in Göttingen without financial help from Carl Friedrich von Weizsäcker. This financial aid came from a fund established and sustained with the honoraria he received for his numerous lectures.

I remember that sometimes, after discussions in your office, you invited me on the spur of the moment to have lunch with your family at your home. Your wife took good care of a somewhat bashful student. I recall evenings among a larger circle of acquaintances at your house, when we played charades.

We met once a week on the top floor of the Institute in a small working group, and discussed questions concerning the genesis of stars and galaxies, problems of turbulence and shock waves, even the theory of the crack of a whip.

During this period in Göttingen, from 1946 to 1958, you contributed decisively to the standing and character of the Institute. It is a time that can be designated: Carl Friedrich von Weizsäcker, astrophysicist.

During this time you were occupied also with philosophical questions. This became evident to us students from your philosophical seminar, which took place every Thursday evening until late at night. Our doorkeeper, Mr. Czierpka, disrespectfully called it the bible class. I still recall one of the lectures by Picht. These seminars allowed us to learn something of Greek philosophy and Kant.

When the Institute moved from Göttingen to Munich in 1958, you accepted a professorship in philosophy at the University of Hamburg. Nevertheless, you remained a scientific member of the Institute and spent most of your summer vacations at the Institute in Munich, staying here on the campus.

In 1957, before the time in Göttingen drew to an end, you became the initiator of the Göttingen Declaration of eighteen atomic physicists. The central principle of this group aroused much attention: "We do not feel competent to make concrete proposals for the politics of the superpowers. For a small country like the Federal Republic, we believe that the best course to protect it and to promote world peace is if Germany renounces explicitly the possession of nuclear weapons of any kind. None of the signers of this declaration is prepared to participate in the production, testing or use of atomic weapons."

In 1967, the foundation of an autonomous institute was proposed to you from several different sides. The proposed institute should deal with all questions pertaining directly or indirectly to dangerous potential conflicts in the world. Werner Heisenberg spoke about it in his address to the annual assembly in 1971: "Since Carl Friedrich von Weizsäcker first took the initiative for the Göttingen Declaration in 1957, he has labored tirelessly to reflect on the philosophical and political consequences arising from atomic physics. In order to give him and his collaborators the necessary context for pursuing this work, the Max Planck Institute for the Research on the Conditions of Life in the Scientific-Technical World has been founded. Von Weizsäcker had chosen this complicated name for the Institute instead of a more familiar name such as peace research or research on the future, because he knew how much one has to beware of clichés, especially in this field. In the relatively near future the world will experience rapid and thoroughgoing changes; the beginnings are already visible. It is good that people with training in the natural sciences and philosophy reflect together sincerely and responsibly on these problems."

When the time of your retirement from active service drew closer, divergent opinions and controversy arose regarding whether your institute should be continued. I do not want to pass silently over this conflict on the present occasion. You accepted obediently but with regret the decision of the Senate of the Max Planck Society to close your institute. This was one of the most painful personal decisions of my life. At the beginning of the deliberations the aim had not been to close the Institute. A successor appeared to have been found in the person of Ralf Dahrendorf. The Institute in Starnberg was to be named the Max Planck Institute for Social Sciences and Ralf Dahrendorf, Franz Emanuel Weinert and Jürgen Habermas were to direct the work. But Dahrendorf decided to remain in England.

The third period of your activity in the Max Planck Society, from 1970 to 1980, can be characterized: Carl Friedrich von Weizsäcker, cosmopolitan scientist. You formulated your task clearly in your book, *"Der Mensch in seiner Geschichte"*: "Natural sciences, especially theoretical physics, is the profession in which I have been educated and in which I still work today. Philosophy is the attempt to understand what we think and what we do. Politics is the dire obligation of the physicist in the age of the atomic bomb".

It is not up to me to present an appraisal of the very comprehensive and rich work of Carl Friedrich von Weizsäcker. Indeed, I am not capable of doing it. Today, I am touched by very similar sentiments to those on your 70th birthday, when I said: "In spite of my joy on this occasion, I stand here with a certain uneasiness. After – all these rich and impressive displays of appreciation, what can I add? Unfortunately, I can not even use the timid excuse of the first president of the Max Planck Society, Otto Hahn. At the celebration of Otto Hahn's 80th birthday the first President of the German Federal Republic, Theodor Heuss, told a story about how he had almost given a kiss to Otto Hahn. This happened after an address by you, Mr. von Weizsäcker, at the assembly of the Max Planck society in 1950 in Cologne on the subject: "Where does science lead us?" When seeing out Heuss, Hahn admitted that he felt uneasy about the task which he was now performing. When Heuss asked why, Hahn confessed: "I am only an *Oberrealschüler* (a person with a mere modern secondary school diploma)." I cannot use this excuse, as I graduated from a secondary school that was grounded in the classics. Nevertheless, I think many of us have occasionally felt as '*Oberrealschüler*' in front of you."

Allow me to return again to Otto Hahn: In his memoirs he describes the book that you and he jointly authored during the interment in England: "Von Weizsäcker is apparently a polymath. In addition to physics he reads political and historical essays, Shakespeare and other English poets. To me, his knowledge is incredibly broad."

Two things I want to remark at the end: your gift for speaking in public and your gift for listening. In your speeches, you show a mastery of presentation, you arrive at explaining even the most difficult problems. Your insights and your use of language have been and are always impressive in their clarity. It is not only I who admire you in this regard. In your speeches you have affected scientists as well as non-scientists. Your speeches continue the tradition of the famous Cosmos Lectures by Alexander von Humboldt. As with Humboldt, a captive audience always fill the lecture hall or church.

Of particular importance have been your lectures in the former GDR. Since 1959, you have been a member of the German Academy of Leopoldina Natural Scientists and you have participated regularly at the biennial assembly in Halle. Your lectures have always had overwhelming audiences, both those on the official program of Leopoldina and those in the Market Church of Halle, which had not been publicly announced, but were advertised by word of mouth only.

You are also a master in the art of listening. Nothing better can happen to a speaker than to see you in the front row. With your agreeing nods you reassure the lecturer over and over again. In addition, you understand marvelously how to summarize and explain a complicated matter. By questioning the speaker or through some remarks, you always succeed in making a lecture clearer, even on subjects which are hard to understand, and you are able to draw conclusions which the lecturer himself did not see.

I experienced in my very first lecture at the colloquium of the Institute in Göttingen how much you assist a speaker. The topic concerned the origin of the planetary system; I was reporting on a paper by the Danish physicist ter Haar. Heisenberg, von Laue and Eucken sat in the front row. Of course, I was terribly excited. I had hardly begun when Eucken interrupted me and said: "First tell us where this paper on which you are reporting has been published." I was so disturbed that I answered: in Svenska Dagbladed. This answer earned me the loudest laughs of my career. Laue laughed so hard that he could not calm down during the rest of the lecture. You, Mr. von Weizsäcker, stood up and said: Mr. Lüst is almost right, it wasn't Svenska Dagbladed but the publication of the Swedish Academy of Sciences, the *Svenska Vetenes Kapsakademica*. At the end of the lecture, Mr. von Weizsäcker defended me against the attacks of the physical chemist, Arnold Eucken, who did not appreciate the theory I presented on the origin of the planetary system.

A birthday is above all an occasion for thanks. Besides research and discoveries of basic importance in physics and astrophysics, besides a scientific and philosophical *oeuvre* of extraordinary breadth and diversity, you have stood up to the task of making the public aware of the relation between progress in science and the basic questions of our human and social existence, by appealing to your vast knowledge of natural history, philosophy and religion.

Many consider Alexander von Humboldt as the last universal scholar. I think that the Max Planck Society can be proud and thankful for having you as a member, a universal scholar of our time – a scholar of natural sciences and humanities.

Ten years ago, Hans Zacher, the former president of the Max Planck Society said it clearly: "In your person you combine the most varied ways for grasping matters related to human being and human sense, for understanding and interpreting and ultimately transforming them into moral obligations. You link the most diverse disciplines in science with insights of religion and aesthetics. This is a remarkably broad and unified way to understand the world and life".

Let me return once more to my perspective as your student. In the name of all your students, I thank you warmly for what you have given us at the start and throughout our careers. All your students are marked by the experience of having a teacher who took science so seriously that he felt responsible for its consequences, and made this clear to the public.

Let me close with a very personal remark. At the occasion of my 60th birthday, you wrote in the paper, *Recollections on the theory of planets*: "After the War the young physicist Reimar Lüst came to me in Gottingen. He asked me for a topic for a doctoral thesis. He was interested in quantum theory and nuclear physics. But I inflicted on him what I had once imagined as the worst demand a thesis supervisor could make; I made him work on the theory of the planetary system. He had to calculate the transport of angular momentum, a calculation that had been too difficult to me. I believe that he has solved that problem." Dear Mr. von Weizsäcker, I still thank you for making me work on this problem. My entire scientific orientation, as well as my entire personal life, has been determined by it.

The Role of the Physicist in the Nuclear Age

Edward Teller

The topic on "The Role of the Physicist in the Nuclear Age" is both particularly appropriate and entirely absurd. It is appropriate because at the very beginning of our carriers, about 70 years ago, Carl Friedrich and I had long and explicit conversations, which in subsequent years have never been repeated. In sharp contrast to this statement, I must acknowledge that in the intervening period, the problems have assumed in my mind an increasingly insoluble character.

This is due to two sharply contrasting factors. One is that the discoveries connected with physics have established more conflicts with common sense than the discoveries in any other science. The other sign of the difficulty is that any solution to the problems created in the 20th century affects the interaction of common people throughout the globe, and the demand that the underlying problems be even understood by the common people appears to become even more remote.

Let me start by mentioning the solid facts of modern physics which – to say the least – are difficult to grasp. They are statements made by Einstein's relativity, Bohr's and Heisenberg's quantum mechanics, and finally, recent developments in astronomy.

In Einstein's relativity theory, the relativity of time is the most striking. It is almost unimaginable to rethink the concepts of past, present and future, as depending on the questions on whose past, present or future we are talking about. In quantum mechanics, a sharp contrast is found between past and future events. The past is considered fixed, while the future is uncertain and subject only to probabilistic predictions. Finally, in astronomy, we are not only observing the Milky Way System over a million times the volume of the habitation of earth, but an expanding universe containing a billion Milky Way Systems. Moreover, this enormous volume contains a great multiplicity of black holes whose interior is beyond any possibility of exploration. Somehow, it seems more acceptable to call the universe infinite than to consider the immense complications of what may be only a tiny part of this infinity.

It may well be that in the near future, biology may become even more amazing than physics but at any rate, it is not yet so in the present situation.

The practical consequences of physics have led to the knowledge of the use of atomic energy. If it is a relatively novel knowledge of the 20th century which requires novel solutions to problems, then one must acknowledge that in order to arrive at the required solutions, one needs popular understanding. The lack of it has led to the incredibly dangerous "green movement" which attempts to stop progress of new knowledge as leading only to more dangers. The age-old tension between science and religion is about to escalate in this tension between scientific knowledge and any picture of the world with which the common man can be comfortable.

I have tried to put the problem into a few words. Having done so, I must reiterate that whatever else is said, my addiction to knowledge remains unchanged. In simple words, our task is and remains to find knowledge and establish this knowledge in the understanding of everyone. I believe that in spite of the obvious fact that in the last 70 years of our life, this job has become even more difficult.

During one of our Copenhagen evenings, Carl Friedrich and I separately put down a modern version of the Ten Commandments. I believe I remember one of Carl Friedrich's commandments: "Not to believe anything that you have understood unless you can say it so simply that everyone can understand it." I believe that this commandment contains the essential task of the physicist in the atomic age.

"The Atomic Bomb Reveals the Political Responsibility of Science"*

Götz Neuneck

1 The Problem of Morality of Science in the Service of Nuclear Weapons Development

The Oppenheimer hearings have demonstrated the complex and conflicting motivations of scientists in the service of nuclear weapons development. They have also given an example of how these conflicting motivations can become falsified in the public debate. In a successful 1950's theater play by the German author Heinar Kipphardt ("*In Sachen Robert Oppenheimer*"), Oppenheimer's political problems are linked together with an alleged moral conflict in his work on nuclear weapons. But Oppenheimer never felt any moral qualms for his participation in the Manhattan Project and even asked C.F. von Weizsäcker in 1958 about the possibility of banning this play in Germany ([86], p. 370).

The work of German nuclear scientists during the Second World War was complicated even more as they worked under (or for) a dictatorial and inhumane regime. While today, a level of transparency has been achieved, when looking back at the scientific and technical work in the Uranium Project, questions are still being posed on the motivations and intentions of the scientists implicated, such as:

- Did the nuclear scientists participate in the Uranium Project in order to put the brakes on it, to ensure the failure of creating a nuclear weapon, and if so, should this be considered as moral success rather than a professional failure?
- Can there be an ethical implication in the fact that the historical evolution of the Uranium Project oriented it initially to the development of a reactor, thus subordinating the construction of a bomb to a subsequent secondary goal?
- Were these scientist saved from the guilt of putting a weapon of mass destruction into Hitler's hands only by the luck destiny?

---

* Zeit und Wissen, 1992, p. 28. I am grateful for comments and supplements to: Klaus Gottstein, Otfried Ischebeck, Wolfgang Liebert and Jürgen Altmann. I would like to thank Jeffrey Hathaway, Kim Bennett and André Rothkirch for their editing and translation assistance. The quotations in this text were translated by the author.

– Did they purposely create a myth after the war by stating that their participation in the Uranium Project had posed moral problems for them?

There is also a debate on the criteria for assessing the answers to these issues, which can be characterized by the following questions: Should the war-time efforts of scientists involved in military programs be weighted fully in calculating their life's achievements in science or are these special efforts in a special time?

The most disputed book on the motives and intentions of the members in the Uranium Club was written by Robert Jungk and came to the conclusion that these physicists purposely delayed the development of a nuclear weapon on moral grounds [45]. Several books and papers took up the issue again in recent years, most of them focusing on Werner Heisenberg as the leading scientist in the Uranium Project [4, 10, 57, 58, 61, 71]. While some authors carefully develop and balance their arguments, others arrive at more direct conclusions. In Rose's book, Heisenberg is depicted as a liar, fool and mediocre physicist, in Power's view, Heisenberg comes close to an anti-fascist member of the resistance. These opposite pictures may indicate that the analyses do not take full account of the complexity of the situation.

From today's outlook, we can see that the problem of the political responsibility of science which the bomb has revealed must be traced from the discovery of fission and chain reaction, over war time efforts for reactors and bombs to the efforts for containment of the nuclear threat during the Cold War. C.F. von Weizsäcker's experiences with the German nuclear program during World War II shaped his approach to problems lying in the triangle of political power, public interest and personal integrity. With respect to science, Weizsäcker's lifelong conclusion was: *"The atomic bomb reveals the political responsibility of science"* [88, p. 28]. The ambivalenve of scientific knowledge and discoveries became clear and modern science lost its innocence. Much of Weizsäcker's attitude, activities and proposals must be seen in this context.

In the following, we will give a short description of the course of the German nuclear program during World War II and intend to show how the complex situation led to complex structures in the motivations, intentions and action of its participants. Nuclear weapons could only be controlled in the course of the Cold War by a combined effort of scientists, politicians and engineers from East and West.

The idea that a few split-off neutrons might perhaps cause a chain reaction, thus setting free large amounts of energy occurred immediately after Hahn and Strassmann's discovery of nuclear fission in December 1938 [29]. Weizsäcker recalls: *"During the Christmas days of 1938, I worked and lived in Dahlem. Otto Hahn, in whose institute I had earlier worked for half a year, called me on the phone and asked me if I could conceive of radium behaving in every chemical reaction as barium does. He had discovered nuclear fission."* ([88], p. 567). The news of Hahn's discovery at the Kaiser Wilhelm Institute (KWI) for Chemistry reached America in January 1939 [68]. In February of 1939, Niels Bohr and John Wheeler published the "Bohr–Wheeler-theory" of fission, which proved that the two uranium isotopes U^{235} and U^{238} behave entirely differently in the neutron-induced fission process: U^{235} is fissionable by slow thermal neutrons, while U^{238} is not. Bohr mistakenly thought that this made a nuclear weapon only an academic possibility . [7] Otto Frisch and Lise

Meitner described the fission mechanism in the journal *Nature* in March 1939 [52]. Within a few months it became clear that neutrons could induce chain reactions, which was confirmed by Halban, Joliot and Kowarski [32]. The US group of Fermi, Szilard and Zinn kept secret that the fission process emits on average 2.42 neutrons and hence makes a chain reaction possible. But the Joliot group published their results in March 1939 in *Nature* [33]. This article prompted some German scientists to contact the German army ([4], p. 1; [64]). Siegfried Flügge, the theoretician at the KWI for Chemistry, found the correct theoretical interpretation as well [17].

When Joliot and his collaborators had discovered that additional neutrons are emitted in a fission process, C.F. von Weizsäcker was invited by Otto Hahn to a seminar in February or March 1939 to discuss this finding. On the evening of that day, he spoke with his friend, the philosopher Georg Picht, and stated *"that one could probably make a bomb which could destroy the entire city of London* [85]." *"We considered the atomic weapon as a real possibility which could be constructed within humanity as we know it."* ([86], p. 398) And both concluded *"that from now on, the scientist carries a political responsibility which he cannot relinquish in any way."* ([81], p. 204) Weizsäcker said he wanted to show *"how physicists stumbled into this affair without knowing how."* ([86], p. 326.) But *"this discovery would not fail to radically change the political structure of the world."* ([81], p. 203)

Looking back to these discussions Weizsäcker wrote: *"To a person finding himself at the beginning of an era, its simple fundamental structures may become visible like a distant landscape in the flash of a single stroke of lightning. But the path toward this in the dark is long and confusing. At that time [i.e. 1939] we were faced with a very simple logic. Wars waged with atom bombs as regularly recurring events, that is to say, nuclear wars as institutions, do not seem reconcilable with the survival of the participating nations. But the atom bomb exists. It exists in the minds of some men. According to the historically known logic of armaments and power systems, it will soon make its physical appearance. If that is so, then the participating nations and ultimately mankind itself can only survive if war as an institution is abolished."* ([81], p. 203, quoted after Rhodes 1986, p. 312)

2 The Uranium Project 1939–1945

2.1 The Uranium Club

With increasing political tension in Europe, the discovery of fission entered the realm of weapons development in England, the United States, Germany and in the Soviet Union. The story is told in various books, see: [4, 10, 27, 39, 44, 45, 58, 71]. In Germany, several scientists and persons in the industry wrote letters to the Ministry for Culture and Education, as well as to the Army [46, 64]. M. Walker wrote: "These scientists were probably motivated in varying degrees by nationalism, patriotism, and ambition, professional as well as personal." ([71], p. 18) In April 1939, the Army Weapons Bureau (*Heereswaffenamt*) organized a secret meeting in Berlin to discuss the possibility of producing a nuclear explosive. Siegfried Flügge recalls a discussion among the young assistants at the KWI for Physics in January 1939 on why geological uranium deposits had so far never exploded through a chain reaction [20]. In a survey article of June 1939, he discusses the control of the chain reaction in order to avoid explosion [17]. He added a popular publication in a large daily newspaper two weeks before the outbreak of the war where he explained that the energy stored in one cubic meter of Uranium is enough to lift the water content of the Wann Lake in Berlin into the stratosphere [18].

In the United States, Albert Einstein wrote the two famous letters to President Roosevelt. On August 2, 1939 he writes: "I understand that Germany has actually stopped the sale of uranium from the Czechoslovakian mines which she has taken over. That she should have taken such early action might perhaps be understood on the ground that the son of the German Under-Secretary of State, von Weizsäcker, is associated with the KWI in Berlin, where some of the American work on uranium is now being repeated."([90], p. 12) In a letter of March 7, 1940 he again mentions C.F. von Weizsäcker: "The latter [the Kaiser Wilhelm Institute for Physics] has been taken over by the government and a group of physicists, under the leadership of C.F. von Weizsäcker, who is now working there on uranium in collaboration with the Institute of Chemistry. The former director [Peter Debye[2]] was sent away on leave of absence, apparently for the duration of the war." ([75], p. 120)

C.F. von Weizsäcker agrees with Einstein's description of his occupation and plans, although he insists that his father did not play any role in it. He formulates his personal motivation for participation in the nuclear program during the decisive days at the turn of 1938 to 1939: *"Of course, there were physicists who looked for success in technical realizations. As chance had it, not me. I was not made like that. But many others wanted that."* ([86], p. 326) Weizsäcker explained his motivation: *"I gladly took part in it and I persuaded Werner Heisenberg, who was Professor in Leipzig, to join in. I said, 'we must face the results of this issue and after that, we must decide independently what we do with them'"*. ([81], p. 362)

"I felt at that time that, by being somebody who was able to work on such an important assignment, I might gain access to those responsible for the policy-making, which I though I might thus be able to influence rather than simply being at the receiving end." [87]

However, C.F. von Weizsäcker would find that his search for a new world order and the quest for political influence through the Uranium Project almost turned into a catastrophe from which he was only saved by the ultimate failure of the program. The search for a new world order imposed by the development of nuclear weapons remained, however, on the international agenda.

Soon after the outbreak of WWII in September 1939, the Department for Nuclear Research was formed in the *Heereswaffenamt* under Kurt Diebner. Diebner's assistant was Erich Bagge, who had obtained his doctorate in Leipzig under Werner Heisenberg. On September 16, the Army Weapons Bureau assembled scientists for a wartime

---

[2] The Dutch physicist and Nobel laureate Peter Debye left Germany and joined Cornell University.

uranium project and a second meeting on uranium fission was convened on September 26 to form the *Uranverein (Uranium Club)*. Among the participants were Walter Bothe (KWI for Medical Research, Heidelberg), Siegfried Flügge, Hans Geiger, Wilhelm Hanle, Paul Harteck, Otto Hahn (KWI for Chemistry), Werner Heisenberg (University of Leipzig) and Carl Friedrich von Weizsäcker (KWI for Physics). During the meeting, Heisenberg discussed a possible uranium reactor (*Uranbrenner* or *Uranmaschine*), noting that with sufficiently enriched uranium it would explode. Paul Harteck proposed a heavy water reactor with a layered structure of uranium, thereby avoiding resonance absorption in U^{238}.

Fission research was made a secret of the State, and publication of scientific results was suppressed. Instead, all scientific results had to be reported to the Army. The fission program was to be headquartered in the KWI of Physics in Berlin-Dahlem and the Army Weapons Bureau took control of the Institute on October 5, 1939. Diebner became the administrative head to coordinate research throughout Germany. Several decentralized groups were to work on specific questions: Heisenberg in Leipzig became scientific adviser and was responsible for theory, Harteck (Hamburg) for isotope separation and Bagge for the measurement of collision cross sections of neutrons and deuterium (for use of heavy water as a moderator). Heisenberg remained the leading person in the Uranium Project, acting as an advisor to the Berlin group and as director of a smaller group for reactor experiments in Leipzig. But he never obtained administrative authority comparable to that of Oppenheimer's. The Heidelberg group under Walther Bothe established detailed accounts on the nuclear processes in a reactor and on their energy balance. But the overall picture presents the Uranium Project as a lose association of groups which, furthermore, was constantly beset with internal rivalries on competence as well as animosities. Otto Hahn had a kind of honorary membership in the *Uranverein*. Throughout the war, he continued his radiochemical research and published its results [30]. Independent of the program under Heisenberg, Fritz Houtermans set up a third laboratory for fission research in Manfred von Ardenne's research institute in Berlin-Lichterfelde. This laboratory was financed by the Post Ministry.

Over the years, not more than 70 to 100 physicists and chemists worked full time or part time on the Uranium Project, which employed a total of about 1,000 persons. The Manhattan Project (see for example [26]), on the other hand, employed over 180,000 people and systematically entered the industrial phase, while the Uranium Project remained mostly confined to scientific institutes. While some of the scientists involved pushed for industrial development, others remained more interested in the scientific aspects of fission and chain reactions. The institutional weakness of the program was further demonstrated by the fact that essential scientific instruments and production facilities were located in occupied countries: the only cyclotron available was in Paris[3] and heavy water was produced in Norway.

---

[3] The cyclotron in Paris was operational by the winter of 1941–42 and became a very useful scientific instrument. However, the equipment frequently failed at critical stages of experiments due to subtle sabotage by the French technicians. Bothe and Gentner later built Germany's first cyclotron in 1944.

In retrospect, the lack of an industrial phase in the Uranium Project should not be regretted. The production of V1 and V2 missiles and Jumo jet engines for the Luftwaffe by using forced labor in mines during the later years of the war demonstrates the conditions under which this kind of industrial production would have been organized.

2.2 Reactor Designs and Experiments

On December 6, 1939, Heisenberg submitted the first part of a basic report entitled "On the Possibility of Technical Energy Production from Uranium Splitting I", in which he stated that a controlled fission reactor was technically feasible and that enriched uranium U235 would be a powerful explosive. He pointed out that isotope separation was the "surest method" of obtaining a reactor and that this was "the only method for producing explosives [37]." Part II of the report followed on February 29, 1940. Heisenberg considered heavy water and very pure graphite as the only feasible moderator substances. He discusses enrichment: the more the nuclear fuel is enriched with U235, the smaller the reactor can be built, eventually allowing ones smaller than 1 m^3.[4,5]

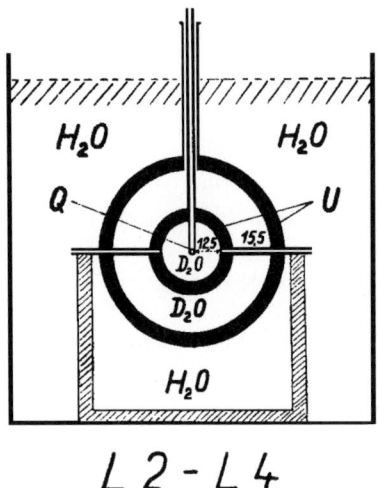
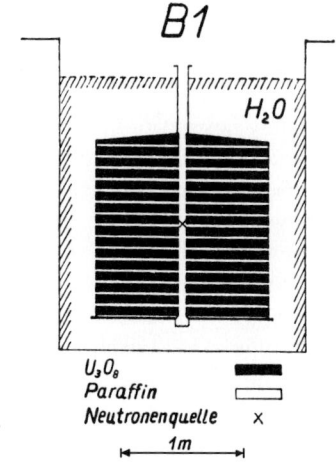

Fig. 1. Schematic Diagram of the Leipzig Experiments L2–L4. [38, p. 50]

Fig. 2. Schematic Diagram of the Berlin B1-reactor [38, p. 52]

---

[4] Bernstein points to the fact that Heisenberg's reports also include a set of important technical errors which hindered any progress in the German program and which he only became aware of years later at Farm Hall after hearing of the first bomb at Hiroshima [4].

[5] Heisenberg and Wirtz described the German nuclear experiments between 1939 to 1946 in the "FIAT Reviews" edited by W. Bothe and S. Flügge.

From July 1940, Heisenberg and Weizsäcker began to work on a reactor design. An alternating layer design was produced by Weizsäcker's graduate students Karl-Heinz Höcker and Paul Müller [78]. Heisenberg and Weizsäcker's group in Berlin used a cylindrical or spherical reactor casing filled with parallel layers of uranium oxide UO_2 and paraffin as a moderator immersed in ordinary water. Neutrons were emitted from the center by a radium-beryllium source. Multiplication of neutrons could not be found. The experiments were conducted in the so-called "Virushaus", thus named because it was set up on the land of the KWI for Biology and Virus Research. It contained a 2-meter-deep pit, surrounded by a brick wall and filled with water, for reactor experiments. The water was used to act as a neutron reflector and for protection against radiation.

In December 1940, Heisenberg and von Weizsäcker, working with Fritz Bopp and Karl Wirtz, started on experiments for neutron multiplication. For this research, von Weizsäcker was able to rely on his previous work in three areas: the passage of radiation through matter, which had been the field of his dissertation in Leipzig, the structure of nuclei, to which he had contributed the formula for the masses of nuclei, and thermonuclear energy generation in stars. On January 6, 1942, Heisenberg, von Weizsäcker and collaborators reported to the Army Weapons Bureau on reactor experiment B3 (B standing for Berlin).[6]

In August 1940, Paul Harteck reported on experiments in Hamburg with a uranium reactor using 15 tons of carbon in the form of dry-ice (CO_2) as a moderator, but its 185 kilograms of uranium oxide proved inadequate for neutron multiplication. This experiment was the first attempt to build a neutron-multiplying pile [64]. Eventually, graphite was ruled out and heavy water remained the only candidate for a moderator with natural uranium as a nuclear fuel. In Heidelberg, Walter Bothe and Peter Jensen measured neutron absorption by carbon nuclei in a ball of electrographite with a diameter of 110 cm, which had been produced by Siemens. They calculated an absorption cross section of 6.4×10^{-27} cm^2, twice the value which Fermi had obtained. They reported on their measurements in January 1941 and graphite was excluded as a moderator. But the high value of the absorption cross section was due to traces of boron stemming from electrodes used in the production of graphite

---

[6] Documents on the Uranium Project are held in various institutions, among them the Atomic Museum of Haigerloch, the Heisenberg Archives in the Max Planck Institute for Physics in Munich, the Forschungszentrum Karlsruhe, the Niels Bohr Library of the American Institute of Physics/New York City and the Deutsches Museum in Munich. E. Bagge collected 14 volumes of secret reports on German fission research from 1939 to 1945, where reports of the Berlin group are contained in Vol. 2: F. Bopp, E. Fischer, W. Heisenberg, C.F. von Weizsäcker, K. Wirtz: Vorläufiger Bericht über die Ergebnisse an einer Schichtenkugel aus 38-Metall und Paraffin (1.6.42), C.F. von Weizsäcker: Über den Temperatureffekt der Schichtenmaschine (1.9.41), Vol. 3: F. Bopp, E. Fischer, W. Heisenberg, C.F. von Weizsäcker, K. Wirtz: Untersuchungen mit neuen Schichtenanordnungen aus U-Metall und Paraffin, Vol. 5: W. Heisenberg, F. Bopp, E. Fischer, C.F. von Weizsäcker, K. Wirtz: Messungen an Schichtenanordnungen aus 38-Metall und Paraffin. Heisenberg's reports are contained in his Collected Works Part AII.

 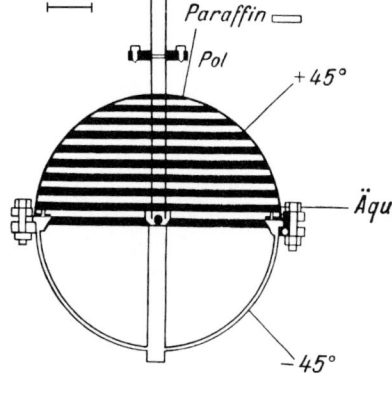

Fig. 3. Berlin Experiments B3. [38, p. 53]

Fig. 4. Schematic Diagram of the Berlin Experiment B3. [43, p. 54]

through electrolysis. In addition, Heisenberg had calculated that a graphite-based reactor would require much more material than a heavy water reactor.

While graphite was thus ruled out in the German program, Fermi's group in Chicago achieved the first chain reaction on December 2, 1942 using 350 tons of graphite, 36 tons of uranium oxide and 5.6 tons of metallic uranium. Two years later, three power reactors became operational at Hanford, Washington, producing the plutonium for the Trinity test and the Fat Man bomb.

Bothe and Jensen later published their measurements in 1944, which shows how far the Uranium Project was behind the Manhattan Project [8]. It is commonly accepted that Bothe is responsible for the mistake in the experiments which became crucial for the failure of the Uranium Project. But the controversy on this point went on for a long time. In 1980, Gentner, in the last letter of his life to colleagues, claimed that Bothe had been aware of the trace boron and its implications, but that Bothe simply did not know where or how to obtain a more pure sample of graphite. Bothe and Jensen's measurements were repeated in 1980 and found to be correct, and it was discovered that Heisenberg had not respected the error analysis correctly [49].

H.A. Bethe points to the essential difference between the Uranium Project, which remained essentially confined to scientific institutes, and the Manhattan Project, which became an industrial program: "In America, Enrico Fermi did a similar experiment and concluded that graphite was marginal. He suspected that an impurity in the graphite was responsible for the problem. Leo Szilard [51], who was working alongside Fermi, had studied chemical engineering before going into physics. He remembered that electrodes of boron carbide were commonly used in the manufacture of graphite. It was known that one atom of boron absorbs about as many slow neutrons as 100,000 atoms of carbon. Very small boron impurities would "poison" the graphite for use as a nuclear reaction moderator. Szilard therefore went around to the American graphite manufacturers and convinced one of them to make boron-free graphite." [5]

In an interview with *Der Spiegel* in 1967, to the question of whether by using ultra-pure graphite, German physicists could have during the war, succeeded in producing a chain reaction in a reactor, the prerequisite for building an atomic bomb, Heisenberg replied: "Probably, but in the big picture of atomic energy developments, it would still have changed only very little. Let's assume we had in 1943/44 – anything earlier would have been highly unlikely – managed to finish a critical reactor. Then every agency in charge of armament that wanted to even think about atomic bombs would have had to argue as follows: If one now wants to produce plutonium this way, one would have to increase expenses by at least a factor of a hundred, to invest a lot of manpower, take it away from other weapons productions, and so on. And that was really impossible in the war situation then." [35]

In the year 1942, it looked increasingly as if moderation by D_2O would work. In April 1942, Döpel and Heisenberg observed neutron multiplication of 13% in the L4 reactor (L standing for Leipzig), which contained about 140 kilograms of heavy water and 750 kilograms of uranium oxide powder arranged in concentric shells. Thus up to this time, the German efforts for obtaining a chain reaction were ahead of the American program. Heisenberg predicted that this design could be critical with 5 tons of heavy water and 10 tons of uranium metal. The L4 reactor was destroyed in June 1942 by an explosion of oxyhydrogen gas, which was formed by the contact of uranium oxide with heavy water.

Natural uranium remained the only reactor fuel for the German nuclear program, as uranium enrichment was considered too difficult in all processes studied. The process most thoroughly studied was thermodiffusion, while diffusion through a porous wall was not considered. Paul Harteck reported in March 1941 to the Army Weapons Bureau that the physicists agreed that a separation of uranium isotopes for reactors was not economically feasible ([60], pp. 344–346).

The possibility for creating transuranium elements in a uranium reactor, which could then be extracted and used for weapons, was recognized.[7] Already in July 1940, C.F. von Weizsäcker [77] suggested in a five-page report to the Army Weapons Bureau that neptunium, then called Eka-Rhenium, formed by neutron absorption in

---

[7] In June 1940, the Princeton physicist L. Turner published an article "The Missing Heavy Nuclei" in The Physical Review, June 1, 1940, p. 950 but suppressed a subsequent article "Atomic Energy from 238U" (later published in The Physical Review 69, p. 366, 1946), where he postulated that neutrons captured by U238 create two new transuranic elements 93 (neptunium) and 94 (plutonium). In Berkeley, P. Anderson and E. McMillan discovered both elements in their cyclotron. See Bernstein 2001, p. 29.

a uranium reactor, could be used as the explosive material in a fission bomb.[8] A year later, in August 1941, Fritz Houtermans reported the possibility of using plutonium in a bomb [42]. He discussed his findings with Heisenberg and Weizsäcker, but his report remained in the safe of the Post Ministry ([60], pp. 372–373). German scientists reported in February 1942 to the Army Weapons Bureau on the status of fission work, describing the possibility of generating plutonium in reactors, to be tested "once we have a functioning atomic reactor". The German scientists never found traces of Plutonium, so their concept of breeding Plutonium remained hypothetical [67].

2.3 Supply of Uranium and Heavy Water

To begin with, there were no uranium deposits in Germany. Annexation of Czechoslovakia in 1938 solved part of this problem when uranium could be acquired from the Jochimsthal mines. The first ton of highly purified uranium oxide was delivered in January 1940 to the Army Weapons Bureau by the Auer Company located in Oranienburg near Berlin. Starting in May 1940, 1,200 tons of uranium ore of Congolese origin could be seized in Belgium. About a thousand tons of this were later secured by the invading American troops in Stassfurt, south of Magdeburg. 31 tons had been found in Toulouse and this material was then used in the calutrons of Oak Ridge for Little Boy.

The supply of heavy water, however, turned out to be a greater problem. Estimates from the Leipzig reactor experiments indicated that 4–5 tons of heavy water were needed for a reactor with natural uranium for it to become critical. The only available major source of heavy water was the Norsk Hydro plant Vermok near Rjukan, 60 km west of Oslo. After the German invasion of Norway in April 1940, Norsk Hydro was put under control of I.G. Farben and the annual production was raised from 120 to 3,600 liters. The State of Norway was charged with the costs for the new catalytic plant.

Paul Harteck and Klaus Clusius had suggested in 1941 that a new plant for heavy water production be built in Germany, but the Army Weapons Bureau preferred Norway as it was the cheapest solution. This decision proved to be fatal to the Uranium Project. A first Allied commando raid on the Vermok plant on November 19, 1942 turned into disaster. But in a second commando operation, on February 28, 1943, Norwegian saboteurs destroyed the electrolysis plant, with the loss of 500 kg of heavy water. The plant was, however, quickly repaired. Massive air bombardment followed on November 16, 1943, damaging the electrolysis building and thus interrupting the production of heavy water for good. On February 20, 1944, the last shipment of heavy water from Rjukan to Germany was lost when the ferry carrying it across Lake Tinnsjö was sabotaged. In all, about 120 persons (commandos, guards, civilians) died for the cause of destroying the facility at Vermok. The end of heavy water production caused severe shortcomings to the experimental groups and air dilution

---

[8] The "German reports" are listed in Walker 1989, p. 268. They are available in the Forschungszentrum Karlsruhe and the Niels Bohr Library of the American Institute of Physics/New York City.

further downgraded existing stocks [12]. Heavy water production provided another indication that turning the Uranium Project into an industrial program had failed. In 1943, Karl-Hermann Geib developed the "dual temperature exchange sulphide process" in Leuna, which is now regarded as the most cost-effective process for producing heavy water. The process has never been used in Germany, but it was developed in parallel by J.S. Spevack at Columbia University and became the basis of the post-war American plants [74].

2.4 The Slow-Down of the Uranium Project in 1942

In the wake of the total mobilization in Germany, the head of Army Weapons Research, Erich Schumann, ordered a review of the uranium project. Declaring that the army could no longer support projects that would not yield results in the foreseeable future, he considered cancellation of support for fission research. The German project slowed down significantly when the Manhattan project started to become more active. In February 1942, the Army Weapons Bureau decided to withdraw almost entirely from fission research and to relinquish physics at the KWI. On February 26, Heisenberg, Hahn and other scientists delivered a series of lectures on nuclear research to a restricted audience ([4], pp. 337–347). These afternoon sessions were sponsored by the Reich Education Ministry in Berlin and marked the beginning of the ministry's backing for the project under the Reich Research Council. This event is in stark contrast to procedures of the Manhattan Project, in which secrecy always remained at the utmost level.

Heisenberg became acting head of the KWI for Physics on July 1, 1942 and reactor research ceased in Leipzig. In Berlin, Heisenberg began planning a series of large reactor experiments involving as much of Germany's uranium and heavy water as possible. The uranium metal plates of his design proved both difficult to manufacture and ineffective, and while awaiting their delivery, Heisenberg turned to the theory of elementary particles. C.F. von Weizsäcker left Berlin to become professor for theoretical physics in Strasbourg. He kept close ties with the nuclear program at the KWI of physics, but he published work on the formation of planets and on the philosophy of science, as well. He returned to the KWI in 1944.

In March 1942, Albert Speer, who was appointed Minister for Armament in February, placed the German economy on a war footing. Projects that did not promise results within two years were eliminated or downgraded in priority. On June 4, 1942, Speer and the directors for weapons development of the three sectors of the Wehrmacht, Field Marshal Milch, Senior General Fromm and General-Admiral Witzell, met with nuclear physicists, among them Otto Hahn and Werner Heisenberg, in the conference and social center of the Kaiser Wilhelm Society in Berlin. In the meeting, Heisenberg discussed information on the large American nuclear program. He explained how a successful reactor might be used in submarines, that U^{235} can be used to make a bomb and that a reactor could generate plutonium. He gave no figure for a critical mass. He estimated the time required for the production of an atomic bomb by enrichment as being at least two years.

It should be understood that, at this time, the Ministry for Armament under Albert Speer was ready to take risks and efforts in the development of a new *"Wunderwaffe"* ("wonder weapon"). A week after the meeting in the Harnack House, on June 13, 1942, Speer, Milch, Fromm and Witzell were present at a corresponding meeting in Peenemünde where they witnessed a test flight of the V2 missile. Although the guidance failed and the missile hit the ground a kilometer away, the spectators nonetheless marveled at its possibilities [55]. Although technical development had not yet been perfected, Hitler signed the order for series production of the V2 on December 22, 1942; the first V2 missiles were launched on England only in September 1944 [66].

Speer had basically been prepared to support further research on nuclear energy and nuclear weapons. When he asked the nuclear scientists to submit a proposal to continue the program, he himself increased the requested sum of some one hundred thousand Reichsmark to one or two million. The program was subsequently oriented to the development of a reactor for the propulsion of submarines. But when tungsten for tank grenades became scarce in the summer of 1943, the national stock of 1,200 tons of uranium was allotted to this purpose. This was a clear indication that the Ministry for Armament had by then abandoned prospects for nuclear power and nuclear weapons [66].

2.5 Final Efforts for a Critical Reactor in Haigerloch

Due to allied bombing raids, in autumn of 1943, Berlin research institutes began moving to small towns in central and southern Germany for safety. The KWI for Physics was divided between Berlin and the neighboring towns of Hechingen and Haigerloch on the Schwäbische Alb. On January 1, 1944, Walther Gerlach, then a professor of physics in Munich, was appointed "plenipotentiary" of fission research sponsored by the Reich Research Council. He then ordered the remainder of the Heisenberg team to move to Hechingen and Haigerloch. Otto Hahn's group continued its work in nearby Tailfingen.

In March, the Heisenberg team in Haigerloch began war-time Germany's last attempt to achieve a critical level through the B8 reactor. The reactor design had meanwhile been changed from alternating layers of uranium oxide and heavy water to an array of 664 metallic uranium cubes suspended from 78 metal chains in a vessel with 1.5 tons of heavy water. The neutron multiplication factor obtained was 70%. Heisenberg had calculated that an enlargement by 50% would make the reactor critical.

Until the end of the War, the following results for reactor development have been achieved [47].

- The contribution of fast neutrons to the increment of neutron has been determined (Heisenberg, Döpel, Stetter, Lindner).
- The necessity of separating uranium and moderator was recognized (Heisenberg, von Weizsäcker, Höcker, Wirtz, Bothe, Fünfer).

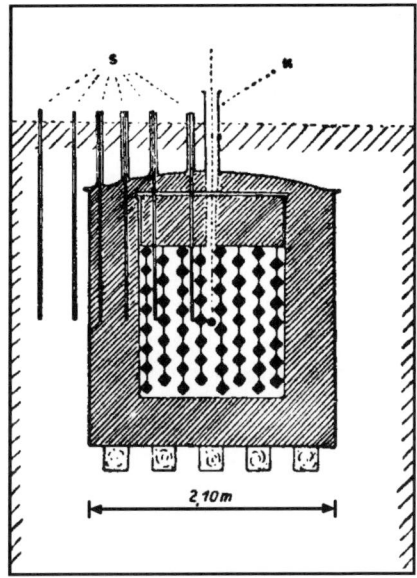 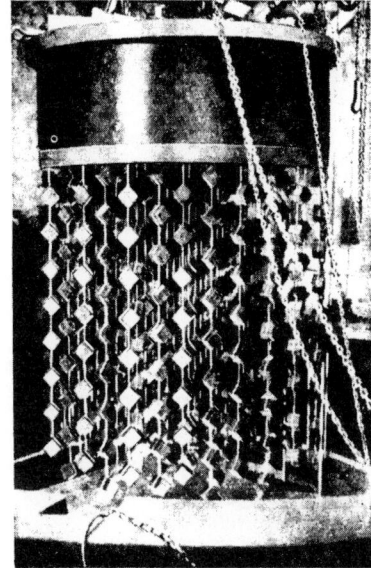

Fig. 5. Schematic diagram of the Haigerloch experiment B8. [43, p. 54]

Fig. 6. Lattice of uranium cubes of the B8 reactor [38, p. 59].

- The resonance absorption during the slowing down of neutrons to thermal energy was theoretically understood and experimentally verified (Sauerwein, Flügge, von Weizsäcker, Höcker).
- A 2-group diffusion theory for rapid and slow neutrons was developed (Bopp).
- The geometry for a critical reactor has been determined from experimentally verified material constants (Bothe, Heisenberg).
- The influence of the reactor envelope (water or graphite) for back scattering was known (Bopp, Fischer).
- The influence of temperature effects has been determined (von Weizsäcker, Müller, Heisenberg).
- The cross sections and the diffusion constant of reactor materials have been measured (von Droste, Flammersfeld, Flügge, Döpel, Heisenberg, Harteck, Höcker, Müller).
- Cadmium was foreseen as material for regulation of the reactor (Flügge, Heisenberg).
- The main fission products with mass numbers from 83 to 140 and their chain of disintegration were known (Seelmann-Eggebert, Götte).

The significance of retarded neutrons for the control of a reactor has not been recognized. As well, xenon poisoning of a reactor has not been foreseen.

One month after the D-day invasion of Europe, the ALSOS Mission, an American science intelligence unit, arrived in Europe in August 1944. In November it became clear that no German bomb existed. When French troops approached the hills of the Schwäbische Alb, the teams prepared for surrender. In a great rush, C.F. von Weizsäcker had tried to bury some of the material in a field before the Alsos Mission arrived on April 23 capturing scientists and equipment in Hechingen and Haigerloch. Diebner, Gerlach and Heisenberg were arrested in Bavaria on May 1–3.

When after the War in 1955 Germany was permitted to work again on nuclear reactors, planning for a natural uranium/heavy water research reactor was taken up in the Max Planck Institute in Göttingen. These plans then led to the FR2 reactor built in 1962 at the Karlsruhe nuclear research center under Karl Wirtz. The reactor remained in operation until 1981.

3 First Attempts to Find a Political Framework to Nuclear Weapons Development

3.1 Physics During World War II Under the Nazi Regime – a Faustian Pact?

While Diebner and Bagge represented the Party line in the Uranverein, Heisenberg and his fellow scientists were neither politically naive nor were they resistance fighters, nor ardent supporters of the Party line. They did indeed put up measures of solidary among physicists and deceney. Otto Hahn once found his name on a list of dismissed professors presented in the exposition *"The Eternal Jew"* ([30], p. 146f). He was personally and professionally loyal to his former Jewish director Fritz Haber and to Lise Meitner. Nazi circles were aware of Heisenberg's and Bothe's struggles with the spokesmen of *"Aryan Physics"* during the 1930s. Heisenberg was publicly accused of being a *"White Jew"*. It was Heinrich Himmler himself who dismissed Heisenberg from these pressures. Bothe has lost his chair at the University of Heidelberg for his opposition against the Nazi regime. It should also be mentioned that the son of Max Planck was sentenced to death for his participation in the failed *coup d'état* of July 20, 1944, although Planck was not a member of the *Uranverein*. Wolfgang Gentner came into close contact with the French Resistance when he was in charge of the German research efforts at Frédéric Joliot-Curie's laboratory in Paris [53]. He intervened several times in freeing Joliot-Curie and also Paul Langevin from arrest. Documents discovered after the war show that the Gestapo had long questioned Gentner's loyalty, while on the other hand, Joliot-Curie could explain only very little about the Uranium Project when he was questioned by the Alsos Mission after the liberation of Paris.

Nevertheless, all members of the Uranium Club considered themselves as patriots, rooted in the traditions of loyalty to the state, patriotism and professional duty. None of the members in the Uranium Club ever withheld scientific results or refused to carry out research tasks. No attempt of sabotage, of espionage, or even voluntary delay of experiments by a member of the *Uranverein* has been reported. But it should be mentioned as well that this sense of patriotic duty had its limits. For example, early during the war, Bothe's assistant Heinz Maier-Leibnitz had volunteered for the Luftwaffe. When he was released from duty and returned to Heidelberg, Bothe told him that he was not obliged to do war-related research, as he had already served his country. Indeed, in spite of his expertise, Maier-Leibnitz never participated in the Uranium Project. In other cases, work in the Uranium Project served to keep scientists from having to perform military service. Otto Hahn reports that when a collaborator of his was summoned to go into forced labor because his wife was Jewish, Hahn was able to prevent this by testifying that his colleague's year long exposure to uranium in the laboratory had rendered him unable to do physical work ([30], p. 160).

Few of the German scientists raised moral concerns on the use of science in the service of the development of weapons of mass destruction. Otto Hahn felt deeply troubled by the destructive potential of his discovery, perhaps being made sensitive to the issue by his involvement as a chemist in the preparation of poison gas attacks during World War I. But he rejects Robert Jungk's statement that he later regretted having published his discovery of nuclear fission ([30], p. 166).

So while the Uranverein was neither a network of active members of the resistance, nor beset with widespread and permanent moral conflict, it was to some extent a circle for safeguarding the independence of science on a basis of give and take between fundamental research and research for technical and military application. The structure of the Uranium Project as a rather loose network of institutes, many of which only worked for the Project to a certain extent, favored this. Heisenberg used his position as acting director of the KWI for Physics for work in elementary particle physics. Paul Harteck argued in an interview after the war: *"In those days in Germany we [had] no support for pure science ... So we had to go to an agency to get funding. I was always realistic about such things. The War Office had the money and so we went to them."* [13] Bothe felt no restraint in applying for financial support from the military budget to construct a cyclotron at the physics department of the KWI for medical research and in his proposal to Albert Speer he emphasized the "critical" importance of his research to the military. Yet, Bothe knew well that this was a hollow claim and Speer was not fooled either. When Speer was later shown the cyclotron in Heidelberg, he inquired on the uses of the instrument and Bothe answered that it is useful for progress in medicine and biology. Speer did not ask further questions ([66], p. 242).

At the beginning of World War II, the general feeling in the *Uranverein* was that the war would be short. Although the development of atomic weapons was then considered to be attainable in a matter of a few years, it was uncertain whether the development for nuclear weapons was feasible. The Uranverein was asked to find out. Around 1941, there was a feeling that nuclear weapons could become operable in wartime if the war continued for several more years. But in the years that World War II continued, the progress in reactor development dragged behind previous predictions and financial support was reduced. It again became the conviction of scientists and military leaders that an atomic bomb would not play a role in the course of war. So the moral pressure on the physicists posed by their work on a nuclear weapon decreased as the war continued. Most of the members of the *Uranverein* remained ambivalent about the project, apparently choosing to keep their concerns to themselves – at least until after the war.

Weizsäcker's dream that atomic physicists could, through their capability of building nuclear weapons, gain influence on Hitler, or whoever became his successor, never became real. None of the scientists of the Uranium Project was given an audience with Hitler. In contrast to this, Hitler had an open ear to Wernher von Braun, whose presentation of V2 missiles on July 7, 1943 at Hitler's headquarters made a big impression on the "Führer". Atomic weapons had nourished his fantasy of mass destruction. He would certainly not have refrained from using them in an attack on London or Moscow. On the other hand, he always remained suspicious of nuclear physics labeling it "Jewish physics" and pointing out Philipp Lenard who had been one of his very few early supporters from the scientific community ([66], p. 242).

History kept a benign eye on the members of the *Uranverein* by not letting them finish the development of an atomic bomb.

3.2 The Visit of Copenhagen in September 1941: First Steps in the Search for a New World Order?

In a 1941 letter to a friend, Heisenberg cautiously expressed his apprehension of the perspectives of a nuclear holocaust: "Perhaps we humans will recognize some day that we really have the power to destroy the Earth completely so that we can easily create a "last judgment" through our own guilt or something close to it. But it is perhaps still fantasy to imagine this [34]." As to Heisenberg and von Weizsäcker, it was felt by most physicists involved in the nuclear program that both the destructiveness of nuclear weapons and the promises of nuclear power would require a new kind of international regime for safeguarding peace, which ought to be established after World War II had come to an end. Weizsäcker had reached the conclusion that atomic physicists were to become the carriers of the idea of the necessity for peace.

At the height of the German military power, in September 1941, Weizsäcker pressed Heisenberg to take the issue to Bohr. In these months, the German (and Nazi) domination of Europe by a victory in World War II appeared possible. Germany had then occupied all surrounding countries it (with the exception of Switzerland) in short campaigns and its military strength was still growing. Since June 1941, it had waged war against the Soviet Union and a victory seemed likely.

Bohr's institute, in which Heisenberg and Weizsäcker had themselves worked, had become synonymous with the internationalization of physics, and Bohr had become the undisputed "scientific and moral father" to theoretical physicists of many nations, in particular to Heisenberg and also to von Weizsäcker.[9] Formally, the journey had been organized in the frame of a seminar at the German Cultural Institute.[10] Bohr, however, refused to attend this seminar, as he boycotted every Danish-German event as part of the popular resistance to the occupation of Denmark in April 1940. In addition, Bohr was deeply embittered by the systematic persecution of Jewish colleagues and of Jews in general by the Nazi regime.

---

[9] Victor Weisskopf mentions in his memoires ([76], p. 119): One evening, labeled *Comic Physics*, Weizsäcker and others performed a parody on Goethe's Faust with Bohr as God the Father, Ehrenfest Faust and Pauli Mephisto.

[10] Heisenberg and Weizsäcker were used as "Goodwill Ambassadors" for the German Cultural Propaganda after the beginning of the War in 1939. See chapter 6 and 7 in Walker 1995.

Bohr was, however, ready to welcome Heisenberg and they had a meeting perhaps on September 16.[11] As Bernstein put it, "the intent and content of this meeting remain one of the most controversial events in history" ([4], p. 43; [60], p. 383). Heisenberg, being torn between the need for secrecy and the desire for communication, apparently used very cautious words. He indicated to Bohr the progress that had been achieved in the German nuclear program. The conversation failed when Bohr became deeply disturbed by Heisenberg's explanations ([57], p. 483). However, today there are several interpretations on the meaning of this conversation. According to one interpretation, Bohr felt that Heisenberg was trying to suggest he collaborate in the German nuclear weapons program. Possibly, Bohr had gotten the impression that his best student Heisenberg had turned into the leader of a program for the development of a weapon of mass destruction in order to promote Hitler's victory over Europe. His respect for Heisenberg's capabilities in theoretical physics caused him the utmost concern.

Bohr's post-war letters to Heisenberg inquiring as to the intentions of his visit to Copenhagen and on the possible governmental agencies backing this journey show that even to him these questions had not been settled even long after the War [54]. Yet, when Bohr was exposed to nuclear weapons development in 1943 in Los Alamos, he began to think along the same "political lines" as his two visitors had in Copenhagen in the year 1941 ([24], p. 288 and Chapt. V).

Bohr never published a summary of his meetings in wartime Copenhagen, but Heisenberg wrote in his memoirs on this subject ([36] and [31], pp. 1984–86). Margarete Bohr saw this visit as hostile and Bohr's son Aage reported that the meeting had upset Bohr. There are many speculations about Heisenberg's intentions. Some think that one reason was an effort to renew a collegial relationship between scientists; others claimed that Heisenberg pumped Bohr for more information on the allied nuclear program or tried to find some agreement on not making the atomic bomb. After Heisenberg's death in 1976 his wife Elisabeth claimed that her husband tried to convince Bohr to find some international agreement among the acting nuclear scientists not to make nuclear weapons.[12] Others believe that Heisenberg could have been convinced that Germany was going to win the war and that Bohr should terminate his uncooperative stance towards Germany.

Michael Frayn's play *Copenhagen* has renewed the public interest in Heisenberg's visit to German-occupied Copenhagen in 1941. The drama portrays several versions of the meeting with Niels Bohr and his former student Werner Heisenberg, who represented the occupying force. Frayn does not see his play as "an attempt to adjudicate between these differing views of Heisenberg's personality, or these differing accounts of his activity" ([22], p. 112). On the contrary, the message is that "thoughts and intentions, even one's own – perhaps one's own most of all – remain shifting and elusive" ([22], p. 99).

---

[11] The place and time they met is unclear. It is said that they took a walk to the *Langelinie*, as they had used to walk and discuss there in former years ([4], p. 43).

[12] Further voices can be found in [60], p. 383.

3.3 The Farm Hall Tapes: Revelation or New Ambiguities?

From July 1945 to January 1946, Carl Friedrich von Weizsäcker was interned in Farm Hall near Cambridge together with his colleagues from the Uranium Project: Erich Bagge, Kurt Diebner, Walther Gerlach, Otto Hahn, Paul Harteck, Werner Heisenberg, Horst Korsching, Max von Laue and Karl Wirtz. These were the leading members of the *Uranverein*, with the exception of von Laue who had not been part of the Uranium Project, and not including Walter Bothe.

The group was separated from their families and not free to move. But otherwise daily life was not difficult, contrary to the life of many other Germans in destroyed cities or in prisoner of war camps. The group was served by five German prisoners of war including a cook. Starting in November 1945, the detained scientists could even send small parcels to their families in Germany with coffee, tea, cocoa and some cigarettes.

The detained scientists held regular seminars on physics, and some were taken on excursions to London or received visits from British colleagues. In July 1945, Max Planck proposed that Otto Hahn take over the presidency of the Kaiser Wilhelm Society (which later became the Max Planck Society) and on November 15, 1945, the Swedish Academy awarded Otto Hahn the Nobel Prize for chemistry for the year 1944 [31].

After all, one of the reasons for their internment was to prevent them from leaving for the Soviet Union or France, either by being kidnapped or through voluntary emigration. The second reason was to assess the achievements and the organization of the German wartime Uranium Project. The third purpose of their internment was to single out those scientists who ought to receive key positions in science in Germany. Biographical sketches of the ten detainees were drafted by the Allied Administrators. Weizsäcker is characterized as follows: "Outwardly very friendly and appears to be genuinely cooperative. He has stated, both directly and in monitored conversations, that he was sincerely opposed to the Nazi regime and anxious not to work on an atomic bomb. Being the son of a diplomat he is a bit of one himself" ([4], p. 144).

But apart from conditions of life and the reasons for their internment, there was the problem that all members of the group, with the exception of von Laue, had to come to terms with an explanation of their work in the Uranium Project whose final objective had been a workable atomic bomb in Hitler's hands, even if this goal had become more and more distant as the Project went on. The two bombs on Hiroshima and Nagasaki suddenly and unexpectedly pulled atomic bombs from a distant possibility into reality. The group's discussions were clandestinely recorded by the British secret service. After the war, it had long been known that the recordings existed, but their publication was delayed for 50 years [4, 41].

When the group was informed in the afternoon of August 6 by their liaison officer, Major Rittner, of the Hiroshima atomic explosion, the first reaction of the interned physicists was bewilderment and disbelief [4]

"HEISENBERG: All I can suggest is that some dilettante in America who knows very little about it has bluffed them by saying: "If you drop this it has the equivalent of

20,000 tons of high explosive" and in reality it doesn't work at all.
HAHN: At any rate, Heisenberg, you're just second-raters and you might as well pack up.
HEISENBERG: I quite agree.
HAHN: They are 50 years further advanced than we. . . .
WEIZSÄCKER: I think it is dreadful of the Americans to have done it. I think it is madness on their part.
HEISENBERG: One can't say that. One could equally well say: "That's the quickest way of ending the war."
HAHN: That's what consoles me."

When the news was confirmed on the evening broadcast of the BBC, the group was completely stunned. Their recorded discussion turned to the question of why the German program was unable to achieve building a bomb ([4], p. 122 and [2]):
"HEISENBERG: We wouldn't have had the moral courage to recommend to the government in the spring of 1942 that they should employ 120,000 men just for building the thing up.
WEIZSÄCKER: I believe the reason we didn't do it was because all the physicists didn't want to do it, on principle. If we had all wanted Germany to win the war we would have succeeded.
HAHN: I don't believe that, but I am thankful we didn't succeed . . .
HEISENBERG: The point is that the whole structure of the relationship between the scientist and the state in Germany was such that although we were not 100% anxious to do it, on the other hand we were so little trusted by the state that even if we had wanted to do it, it would not have been easy to get it through.
DIEBNER: Because the officials were only interested in immediate results. They didn't want to work on a long-term policy as America did.
WEIZSÄCKER: Even if we had gotten everything that we wanted, it is by no means certain whether we would have gotten as far as the Americans and English have now. There is no question that we were very nearly as far as they were, but it is a fact that we were all convinced that the thing could not have been completed during the war.
HEISENBERG: Well, that's not quite right. I would say that I was absolutely convinced of the possibility of our making a uranium engine, but I never thought we would make a bomb, and at the bottom of my heart I was really glad that it was to be an engine and not a bomb. I must admit that."

As interesting as these statements are, the tape recordings of the discussion taken secretly should not be overly interpreted as the ultimate expression of the members of the *Uranverein* on their motivations and moral standards in the Uranium Project. Weizsäcker was described by their guards as "the son of a diplomat" and much of a diplomat himself. As a diplomat he apparently took into account that their conversation was secretly recorded and his sentences appear to be phrased accordingly. During a tape-recorded conversation with Diebner, Heisenberg, however, did believe that their guards would not use "the real Gestapo methods" such as recording conversations ([4], p. 78). The members of the Uranium Club had become accustomed to life in an ambiguous professional and ethical environment. The end of the war had changed the environment, but their situation remained ambiguous. They had the desire and possibly the chance to the keep their position in the national scientific establishment and to reintegrate into the international world of physics while just coming out of a military nuclear program under the Nazi dictatorship. On the other hand, they had to explain the reasons for not having succeeded at the scientific and technical objective of building a reactor and a bomb while the American program had succeeded in both.

We may thus understand the recorded discussion as a diplomatic attempt to make the best of these ambiguities. What diplomats do in such cases is to phrase sentences which permit several, even contradictory interpretations but on which all interested parties can agree, as long as the different interpretations do not contradict essential facts. Today, we no longer have to present the history of nuclear weapons development during World War II in this diplomatic manner. But we must develop an understanding of the particular situation of the life of scientists working in Germany during World War II.

The private notes of the interned scientists were clearer. In August 1945, Weizsäcker writes in his personal notes: *"Today we carry, and this refers to everybody who has helped to promote the knowledge of the atomic nucleus, part of the guilt for the death of 90,000 men, women and children, for the wounds and homelessness of hundreds of thousands. And none of us can get around the question of whether because of this work to which we have devoted our lives that still in our lifetime not 90,000 but 90 millions will suffer the same death."* ([82], p. 17).

Let us contrast this with some quotations about the personal motives in building the bomb by American scientists. At a speech in Los Alamos in November 1945, Oppenheimer said [56]: *But when you come right down to it the reason that we did this job is because it was an organic necessity. If you are a scientist you cannot stop such a thing. If you are a scientist you believe that it is good to find out how the world works; that it is good to find out what the realities are; that it is good to turn over to mankind at large the greatest possible power to control the world and to deal with it according to its lights and its values.*

Victor Weisskopf argues in the same direction ([76], p. 128): *Today I'm not quite sure whether my decision to participate in this awesome – and awful enterprise was solely based on the fear of the Nazis beating us to it. It may have been more simply an urge to participate in the important work my friends and colleagues were doing. There was certainly a feeling of pride in being part of a unique and sensational enterprise. Also, this was a chance to show the world how powerful, important, and pragmatic the esoteric science of nuclear physics could be.*

In June 1945 a group under the leadership of James Franck and Leo Szilard submitted a written proposal to the Secretary of War, the so-called Franck-Report, urging the government not to use the bomb over an inhabited region [21]. In particular, the report argued that: "If the United States were the first to release this new means of indiscriminate destruction upon mankind, she would ... precipitate the race for armaments, and prejudice the possibility of reaching an international agreement on the future control of such weapons." This was the first time that scientists have publicly expressed their opinion about the awesome consequences of their work.

4 Coping with the Nuclear Threat During the Cold War

4.1 The Russell–Einstein Manifesto 1955

When the American scientists left the secret weapons laboratories after the end of the war, some of them began to contribute their knowledge to the public and political discussion on the evolving nuclear threat and its implication to world affairs [69]. In the 1950s, the international community became increasingly worried and frightened about the development, the testing and the production of nuclear and thermonuclear weapons and with the growth of intercontinental bomber fleets both in the United States and, with some delays, in the Soviet Union. The nuclear arms race came into motion. The Korean War and the Hungarian insurrection demonstrated the instability of East-West relations. The shock of Sputnik on October 5, 1957 made intercontinental missiles with nuclear warheads a reality. But nuclear power also held the promise that everybody would be provided with energy for lighting, heating, water desalination, for industrial processes and transport and communication. The bomber development and wide area radar surveillance became the technical basis for the intercontinental air travel of the masses. Missile development became the basis for global communication and other space applications such as weather forecasting and warnings of natural disasters. The post-war world thus held both enormous promises and threats in one hand.

Upon the initiative of Bertrand Russell, a manifesto was published on July 9, 1955 which demands the renouncement of nuclear weapons and culminates in the sentence: *"There lies before us, if we choose, continual progress in happiness, knowledge, and wisdom. Shall we, instead, choose death, because we cannot forget our quarrels? We appeal as human beings to human beings: Remember your humanity, and forget the rest. If you can do so, the way lies open to a new Paradise; if you cannot, there lies before you the risk of universal death."* [62] Albert Einstein, who died in April prior to the publication of the manifesto, and ten other scientists, signed the declaration, among them ten Nobel Prize winners. This document has since been called the Russell–Einstein Manifesto [63].

Three days after the Russell–Einstein Manifesto, the "Mainau Declaration" was introduced by Otto Hahn to the periodical meeting of Nobel Prize winners in Lindau on Lake Constance. Born, Heisenberg and Weizsäcker had written draft proposals which formed the basis for the declaration [46, 50]. Its final sentences include Weizsäcker's experience during the Third Reich and the lesson he learned from the presence of nuclear weapons: *"All nations must come to the decision to renounce the use of force voluntarily as a last resort in politics. If they are not ready for this, they will cease to exist"*. All 16 Nobel Prize winners present in Lindau signed the declaration as well as Compton and Yukawa who were not present. A year later, 51 signatures had been obtained. Other declarations by scientists followed. In a speech broadcast worldwide on April 23, 1957 by more than 100 radio stations, Albert Schweitzer asked for a nuclear test ban, as radioactive isotopes were accumulating in organisms to ever higher doses. On June 2 of the same year, in a declaration initiated by Linus Pauling, 2,000 scientists asked the government of the United States to end nuclear tests.

4.2 Tactical Nuclear Weapons in Europe and the Göttingen Declaration

Two political themes occupied West German society during the mid 1950s: the unexpected economic recovery (*Wirtschaftswunder*) on the one hand and the threat of the ultimate destruction of Central Europe as a habitable region by a nuclear war on the other. The apprehension of a nuclear war in Europe grew when increasing numbers and varieties of tactical nuclear weapons were deployed.

The first American tactical nuclear weapons were shipped to Europe around 1952 and by 1959 their number had grown to about 2,500 ([14], Chapt. 4). Between 1968 to 1979 this number reached its highest at about 7,000 weapons. About two thirds of these weapons were deployed in West Germany. Soviet tactical nuclear weapons for the European theatre were produced some years later but their number grew steadily. "Small" nuclear weapons were developed for use with artillery. Nuclear artillery of 280 mm with warheads with a yield of 12 kt was first sent to Europe in 1953. These very immobile systems were replaced starting in 1956 by 203 mm howitzers (12 kt). Nuclear ammunition with a 0,1 kt yield for self-propelled 155 mm guns have been deployed since 1969.

Even the infantry in Europe was equipped with nuclear weapons. The Davy Crockett was the smallest and lightest of these battlefield nuclear weapons deployed by the U.S. Army from 1961 to 1971. Between 1956 and 1963, 2,100 were produced. This weapon consisted of a projectile launched from either a 120-millimeter or 155-millimeter recoilless rifle. It had a maximum range of 1.24 miles (120 mm) to 2.49 miles (155 mm). The W54 warhead used on the Davy Crockett weighed just 51 pounds, with a variable explosive yield of 0.01, 0.02 and 1 kiloton [11]. W54 was also used as atomic demolition munition. On the other end of the yield spectrum, medium-range missiles and cruise missiles deployed in Europe carried thermonuclear warheads in the megaton range. This high yield was considered necessary due to the inaccuracy of these missiles whose CEP was in the range of 1 to 3 miles. Nike-Hercules air defense missiles also carried a nuclear warhead.

Nuclear tactics required a break with the traditional notions of warfare and military organization. Military exercises with nuclear weapons were held in Nevada in 1951 and 1952. Allied divisions in Europe were to become reorganized into "pentomic divisions" for fighting on the nuclear battlefield. This division was to canalize enemy attacking columns and strike them with nuclear missiles, such as "Honest John" or "Davy Crockett", and nuclear artillery. Whether these tactics would have been effective is debatable. Casualties in the tactical nuclear environment were expected to be extremely heavy. Creation of "unintended obstructions" like tree blow-down, residual radiation, fires, and other collateral damage, implied that tactical nuclear weapons would complicate military operations for the side that used them.

The "Carte Blanche" NATO exercise was held in June 1955 to evaluate the air force alert plans for the event of an atomic war. The exercise involved the territory of the Federal Republic, Holland and Eastern France. Within 48 hours, 335 theoretical atomic bombs had been dropped. The German public was thus exposed for the first time to a demonstration of what had heretofore been an abstract doctrine [48]. What could "defense" mean in a conflict that would bring total devastation to the territory to be defended?

In the Soviet Army, a new service, the Strategic Rocket Forces, was created in 1959. A group of authors at the Voroshilov General Staff Academy wrote the first study of nuclear battlefield tactics. In 1962, a new edition of this work was published under the title *Military Strategy* and under the editorship of Marshal V.D. Sokolovsky, who had been chief of the General Staff when the text was drafted [65]. This manuscript summed up the General Staff's assumptions regarding the revolutionary impact of the nuclear-rocket revolution upon military affairs: *"Military strategy under conditions of modern war has become the strategy of deep nuclear rocket strikes in conjunction with the operations of all services of the armed forces in order to effect the simultaneous defeat and destruction of the economic potential and armed forces throughout the entire depth of the opponent's territory in order to accomplish the aims of war in a short period of time."* Only about 25 years later did military experts in the Soviet Union inquire what these "aims of war" could possibly mean – if there was a meaning at all.

German generals and politicians were appalled as were most civilians. But Konrad Adenauer gave West Germany's political, economic, social and military integration into the Western world utmost priority and regarded the risk of nuclear war as a necessary attribute of this integration. The political opposition to nuclear armament of the Bundeswehr was confined to the parties which were not in power like the Social Democratic Party. The Communist Party had been outlawed in 1956. Opposition among military staff which came up with proposals for robust conventional defense was eliminated on the administrative level [9].

However, the German public did in fact react. From September 1956 to July 1958, the movement *"Kampf dem Atomtod"* ("Fight Atomic Death") united scientists, theologians, intellectuals and workers in an effort to cut down or prevent nuclear armament of the Federal Armed Forces through public protest. The movement received support from the Social Democratic Party, the Protestant Church and from the trade unions. Opinion polls revealed that 80 – 95% of the population was hostile to the government's military plans. Yet, public apprehension on the danger of nuclear war did not bring Adenauer's government into political peril. In the election to the Bundestag in 1953, the conservative CDU/CSU coalition had obtained the majority by one seat and the election of September 1957 extended its majority to a comfortable 43 seats. The movement *"Kampf dem Atomtod"* eventually slackened in mid 1958 and Khrushchev's Berlin ultimatum of November 27 pushed the politics of détente and the public anti-nuclear movement further into the defensive.

C.F. von Weizsäcker recalls: *"In the fall of 1956 we, the atomic scientists, realized that the first preparations had been taken to equip the Federal Armed Forces with atomic weapons"* ([82], p. 34). In September 1956, the Federal Minister of Defense, Theodor Blank, stated that the Federal Armed Forces would be equipped in such a way that it would be able to fulfill its tasks in a nuclear war [16]. On October 16, Blank was replaced by Franz Josef Strauss who had previously been Minister for Atomic Affairs, and on November 19, the members of the group for nuclear physics in the German Atomic Commission wrote to Strauss asking him to declare publicly that the Federal Republic would neither produce nor store atomic weapons. If he were to refuse, they would publish their letter. Franz Josef Strauss became very upset about this *"unreasonable demand"* but spoke twice with Hahn and others on January 29 and February 23, 1957. He explained that: *"the Germans cannot stand up to the Russians with bow and arrow"*. Chancellor Adenauer declared on April 4 in a press conference that *"tactical nuclear weapons are in principle only a continuation of the artillery and it should be self-evident that the German forces become equipped with the most modern models."*

In 1957, C.F. von Weizsäcker explained the responsibility of science in the nuclear age quite clearly and unambiguously [79]: *"The personal responsibility of the natural scientist corresponds to the practical relevance of his professional field. I would like to use a comparison. Each natural scientist learns how to execute experiments carefully, otherwise his science would degenerate into humbug. I believe that, as long as the examination of the repercussions of our inventions to human life is not made with care equal to the care taken in experiments, we are not ready for life in the technical age. A Hippocratic oath for natural scientists and engineers has been considered. (...) The responsibility (of scientists in the nuclear age) is concrete. Only our action in each individual case shows whether we can live up to it. (...) Responsibility of man in the technical world thus means at least: in the midst of planning and technical devices, he has to learn to remain a human being. Perhaps he must still learn in decisive points to become a human being."*

The natural scientist should take responsibility for the consequences of his discoveries: *"But this is the difference between immaturity and maturity. An adult person assumes the responsibility also for things which he could not have foreseen in case somebody has to assume it."* ([86], p. 332) Accordingly, Weizsäcker came to the conclusion that an appeal for disarmament would not be sufficient. *"Therefore we had in particular to state in public that none of us would be personally ready to make bombs, to test them or use them"* ([86], p. 39). He drafted a new text which was published on April 12, 1957 and which became known as the Göttingen Declaration. The declaration was signed by 18 nuclear physicists, among them some of the leading scientists of the former Uranium Project, the president of the Max Planck Society and vice-chairman of the German Atomic Commission (Hahn), four winners of the Nobel Prize (Born, Hahn, Heisenberg, von Laue)[13], ten out of twelve members of the task group on nuclear physics in the German Atomic Commission (Bopp, Haxel, Heisenberg, Kopfermann, Mattauch, Maier-Leibnitz, Paul, Riezler, Walcher,

---

[13] A fifth Nobel Prize to members of this group was awarded in 1989 to Wolfgang Paul.

Göttingen Declaration of the German Nuclear Physicists

The plans of the German Army of acquiring atomic weapons has caused deep concern to the undersigned nuclear physicists. Some of them have already communicated their concern to the appropriate members of the government. Today, the debate has become public, and the undersigned feel compelled to point out publicly certain facts known to the experts, but seemingly not sufficiently known to the public.

1. *Tactical atomic bombs have the same destructive effects as normal atomic bombs.* The designation "tactical" is used in order to express that they are to be used not only against human settlements, but also against troops in surface combat. Every single tactical atomic bomb or atomic grenade has similar effects as the first atomic bomb which destroyed Hiroshima. Since tactical atomic weapons are today available in large numbers, their destructive effect would be on the whole much larger. These bombs are designated as "small" only in comparison to the recently developed "strategic" bombs and mainly to the hydrogen bombs.

2. *There is no limit known to the possibility of increasing the destructive effect on life and property of strategic atomic weapons.* Today a tactical atomic bomb can destroy a small city; a hydrogen bomb can make uninhabitable a region the size of the industrial district of the Ruhr. The whole population of the German Federal Republic could be exterminated today by means of the spreading radioactivity of hydrogen bombs. We do not know any practical possibility to protect large populations from this danger.

We know that it is very difficult to draw political consequences from these facts. Since we are not politicians, one might deny us the right to judge these questions; however, our activity in pure science and its applications, which brings us into contact with many young people in this field, has bestowed upon us a responsibility for the possible consequences of this activity. This is why we cannot keep silent in these political matters. We support wholeheartedly the idea of freedom as represented today by the Western world against the ideas of communism. We do not deny that the mutual fear of the hydrogen bombs represents today an essential contribution to the maintenance of peace in the whole world and of freedom in a part of the world. However, we consider, in the long run, this way to maintain peace and freedom as completely unreliable and the danger in the case of its failure as deadly.

We do not feel competent to make concrete propositions for the policies of the great powers. We think that today a small country such as the Federal Republic can protect itself best and promote world peace by renouncing explicitly and voluntarily the possession of atomic weapons of any kind. Be that as it may, none of the undersigned would be ready in any way to take part in the production, the tests, or the application of atomic weapons.

At the same time, we emphasize the utmost importance of the peaceful application of atomic energy, which should be supported with all means, and we will participate in this task as we did before.

April 13. 1957
Prof. Fritz BOPP, Chairman of the Institute for Theoretical Physics, University of Munich
Prof. Max BORN, Bad Pyrmont
Prof. Rudolf FLEISCHMANN, Physical Institute of the University of Erlangen
Prof. Walter GERLACH, Chairman of the Institute of Physics, University of Munich
Prof. Otto HAHN, President, Max Planck Society, Göttingen
Prof. Otto HAXEL, Director of the Second Physical Institute, University of Heidelberg
Prof. Werner HEISENBERG, Director of the Max Planck Institute of Physics, Göttingen
Prof. Hans KOPFERMANN, Director of the First Physical Institute, University of Heidelberg
Prof. Max von LAUE, Director of the Fritz Haber Institute of the Max Planck Society, Berlin-Dahlem
Prof. Heinz MAIER-LEIBNITZ, Laboratory for Technical Physics of the Technical University of Munich
Prof. Josef MATTAUCH, Director of the Max Planck Institute for Chemistry, Mainz
Prof. Friedrich-Adolf PANETH, Director of the Max Planck Institute for Chemistry, Mainz
Prof. Wolfgang PAUL, Prof. of Experimental Physics, University of Bonn
Prof. Wolfgang RIEZLER, Prof. of Radiation and Nuclear Physics, University of Bonn
Prof. Fritz STRASSMANN, Director of the Institute for Inorganic Chemistry, University of Mainz
Prof. Wilhelm WALCHER, Director of the Physical Institute, University of Marburg
Prof. Carl Friedrich von WEIZSÄCKER, Division Director of the Max Planck Institute for Physics, Göttingen
Prof. Karl WIRTZ, Division Director of the Max Planck Institute for Physics, Göttingen

Source: Bulletin of the Atomic Scientists, Vol. XIII, No. 6, pp. 228, Chicago, Ill., June 1957.

Weizsäcker) and the head of a section for nuclear research and reactor design in Karlsruhe (Wirtz).

An analysis of the political implications of the Göttingen Declaration indicates three levels. The declaration:

– was an attenuation of Adenauer's current attempts to publicly downplay the effects of tactical nuclear warfare.

- pointed to the fact that as a result of the Second World War, Germany was fixed to a minor role in international politics, especially in the context of nuclear armament.
- made it clear that even an option for nuclear weapons development was not possible any longer.

In their immediate reaction, Konrad Adenauer and Franz Josef Strauss were infuriated and deeply embarrassed and they dismissed the Declaration in public as a "thoughtless experiment". Strauss spoke of Otto Hahn as "an old idiot who cannot retain his tears and cannot sleep at night when he thinks of Hiroshima". A discussion was arranged by the Chancellor's Office in Bonn on April 17, where Adenauer and Strauss with deputy ministers Hallstein, Rust and Globke took part on behalf of the government with the Inspector General of the Federal Armed Forces, Heusinger and the commander of NATO forces in Central Europe, General Speidel, and Gerlach, Hahn, Laue, Riezler and von Weizsäcker as signatories of the declaration. In the joint communiqué of the meeting, the scientists were able to maintain their position and the government had to give in [30].

The Federal Republic had agreed by international law at the London Conference on October 3, 1954, not to produce atomic, chemical and biological weapons [6]. However, especially Franz Josef Strauss maintained the argument that Germany should keep alive the technical capability for nuclear weapons development for the case that, some day, the American engagement in Europe would end. Nuclear weapons were then widely considered as a necessary requisite for political sovereignty, demonstrated most conspicuously by Charles de Gaulle's build up of the Force de Frappe and the withdrawal of France from military integration into NATO after the first few Mirage IV with atomic bombs were in service. The Göttingen Declaration rendered any German aspiration for a possession of nuclear weapons illusionary [50]. Through the Göttingen Declaration, the scientists made it clear to the government that German scientific establishments would not lend their capacities for nuclear weapons development.

The Declaration points to the limitation for sovereignty of Germany in the East-West conflict and to the danger of nuclear proliferation if the desire for sovereignty becomes excessive. Weizsäcker explains: *"At first it has appeared and still appears to us for political reasons that a nuclear armament of individual national states, such as France, Germany, Sweden, would be a misfortune to the world and for the respective nations as well. Seriously, where will these bombs fall if not on one's own country? Is there not a better way to draw the bombs of the big powers on to oneself than by possessing one's own nuclear weapons? Furthermore: how long would it take until Syria, Israel, and Egypt would also possess nuclear weapons?"* ([82], pp. 35–36).

On the other hand, beyond this dispute between the government and the leading nuclear physicists, the threat of nuclear war in Europe was commonly felt by the government as well as by the commanders of the Bundeswehr and the German population. But there was profound disagreement on the question of how to deal with this threat. In the end, the refusal of the German scientists to work for nuclear weapons development brought West Germany even more closely under American protection, and this was in line with Adenauer's principal political guideline of integrating West Germany into the Western alliance.

Konrad Adenauer's government continued its attempts to obtain a hold on nuclear weapons. After Sputnik was launched on October 4, 1957, Adenauer on November 16 even concluded an agreement with the French government for the joint development of nuclear weapons, which Charles de Gaulle immediately annulled in June 1958 after he came to power [3]. A further attempt was the creation of a NATO Multilateral Force composed of ships armed with missiles and nuclear warheads and with a crew from several NATO member states [23]. Over the years, a kind of double key procedure was installed, where the Americans had the nuclear weapons and the Germans had delivery systems: artillery, short and medium range missiles and aircraft. The nuclear warheads remained under strict American custody and were to be released to West German troops in the case of an alert by order of the US-President based on a demand by the Supreme Allied Commander in Europe, who was to be a US citizen. A demand for release was never made.

The fact that Germany does not possess nuclear weapons has become a rule in international politics and was probably a major underlying factor for the Soviet government's willingness under Gorbatchev to end the Cold War. It is codified as an international law in the treaty on German reunification of 1990. It shapes Germany's demand for permanent membership in the United Nations Security Council, while the present five permanent members are, at the same time, the recognized nuclear powers. Thus, the Göttingen Declaration of the 18 nuclear physicists was a step towards the eventually peaceful resolution of the East-West conflict and of Germany's position in today's world affairs.

5 Paths Through Peril

The Russel–Einstein Manifesto created the momentum for the foundation of the *Pugwash Conferences on Science and World Affairs* where nuclear physicists from East and West have met since 1957 [63]. In the deepening bipolar partition of the world, science, and physics in particular, was still one of the rare fields where communication between East and West has remained possible and which could possibly have an influence on politics. In this function, scientists could benefit from their international relations, their obligation to the objective nature of science, and from their understanding of scientific-technological developments [59].

The group of the 18 signatories of the Göttingen Declaration became aware of the Pugwash Conferences and began taking part in this movement ([50], pp. 66 and 311). C.F. von Weizsäcker, in March and April 1958, participated in the second Pugwash Conference in Lac Beauport, Quebec and traveled to the USA and England. On these journeys he learned of the debate on concepts for nuclear arms control which had emerged during that time. These concepts provided analytical access to the nuclear arms race which formed a heretofore intractable dynamic complex of mutual threat perception, military research and industry, administration, military planning and political steering. Mutual threat perception was analysed by mathematical theory which permitted finding strategies out of the dilemma of the arms race. Using methods from systems analysis, operational procedures for the reduction of the number of weapons systems, for a nuclear test stop, non-proliferation, and verification of arms control agreements could be discussed. The Pughwash Conferences permitted an exchange of this research between East and West [15].

Weizsäcker published his new experiences in May 1958 in the weekly DIE ZEIT under the title: "Living with the Bomb", which introduced the discussion on nuclear arms control in Germany [92]. The title indicates that a modus vivendi of civilian life in the presence of nuclear weapons and nuclear threat was sought which should permit reducing the danger of nuclear war step by step and eventually eliminating it. In order to carry out the necessary scientific investigations, the "*Vereinigung Deutscher Wissenschaftler*" (VDW, Association of German Scientists) was founded on October 1, 1959 upon the initiative of Born, Hahn, von Laue and Weizsäcker and according to the model of the "Federation of American Scientists" ([50, 70]). The association established a research unit in Hamburg, where Carl Friedrich von Weizsäcker had held a chair in philosophy since 1957.

Upon Weizsäcker's initiative, this unit became engaged in research and the organization of a commission for civilian shelters in the event of a nuclear war. This study laid the ground for a comprehensive investigation on the sequels to nuclear warfare in Central Europe, which was published in 1970 [80]. The study showed in detail how the defense of West Germany with TNWs would destroy the very country that was to be defended. It had a deep impact on the German debate over the consequences of the use of tactical nuclear weapons [40]. For this work, Weizsäcker was awarded the prestigious "Peace Price" in 1963. In his address he introduced the peace problem, global governance and the technical dimension of modern war into the intellectual life of West Germany thus building the momentum to establish peace research in Germany. Three theses of his speech in Frankfurt became legendary: 1. World peace is necessary, 2. World peace is not the Golden Age, 3. World peace demands an extraordinary moral effort ([82], pp. 125–138).

The VDW research unit then tackled the problem of modern conventional armament and planning of warfare in Central Europe, as the conventional superiority of the Warsaw Pact provided the rationale for the deployment of thousands of TNWs. The essential problem of conventional armament was identified in the dual capability of mobile forces for both attack and defense. Mobile armored units had become the creed of military strategists based on the experiences gained from the Blitzkrieg and defensive battles during World War II. Armored vehicles were now supplemented with aircraft and attack helicopters. The dual capability of these units for attack and defense gave rise to an arms race in the conventional domain by build up on both sides in numbers and technical modernization. Based on this analysis, Horst Afheldt designed strategic and tactical concepts to decouple offensive and defensive capabilities. His proposal for a new defense concept in Central Europe relied only on dispersed capabilities with reduced mobility, like anti-tank weapons for the infantry and rocket artillery, not unlike the Swiss defense system [1]. Concepts of this type should make the use of TNW obsolete, both for the attacker and for the defender. Most military experts in Germany did not agree with these concepts of "non-offensive defense" as they called for discarding the armored mobile units in which the German army took so much pride. But whether the experts liked these concepts or not, they had to admit that these concepts demonstrated the essential weaknesses of NATO's defense plans. The research on non-offensive conventional defense was continued from 1970 in Starnberg at the "Max Planck Institute for Research on Living Conditions of the Scientific-Technological World", where it continued after the closure of the Institute until October 1989.

Carl Friedrich von Weizsäcker has always maintained that the concepts for arms control and defense which he developed in Hamburg and Starnberg together with his collaborators have been paths *in the face of* peril [82]. One could not claim they have indicated sure paths *out* of peril, but they represented the best propositions which he and his fellow scientists could propose in the given situation. The instabilities and danger in the East-West military systems were again felt in the early years of the 1980s and were expressed by the peace movement in Germany as well as in other countries. In retrospect, the eventually peaceful settlement of the East-West conflict at the end of this decade shows that these paths *in the face of* peril contributed to finding ways *out* of the peril of nuclear war between NATO and the Warsaw Pact.

At the end of Zeit und Wissen, Weizsäcker's "opus magnum" is the following legacy: *"One has been unable to overcome the institution of war for centuries now. But since the invention of modern weapons this will be an inalienable right. It would be life-threatening for all of us, to consider the collapse of one of the competing empires as the solution to the problem. War is even more probable today than it was ten years ago".* ([88], p. 1158)

References

1. H. Afheldt: *Defensive Verteidigung*. Rowohlt, Hamburg 1983
2. [www.aip.org./history/heisenberg/p11a.htm]
3. U. Albrecht: *Das deutsch-französische Atomwaffenprojekt*. VDW info. Vereinigung Deutscher Wissenschaftler Nr. 3, October 1996
4. J. Bernstein: *Hitler's Uranium Club. The Secret Recordings at Farm Hall*. The American Institute of Physics, 1996; Paperback Edition Springer-Verlag, 2001
5. H.A. Bethe: The German Uranium Project. *Physics Today* 53, no. 7 (July 2000)
6. Bundesminister der Verteidigung: *Zur Sicherheit der Bundesrepublik Deutschland*. Weissbuch, Bonn 1983
7. N. Bohr, J.A. Wheeler: The Mechanism of Nuclear Fission. *Physical Review* 56, 426–450 (1939), see the paper reproduced in: Wohlfahrt 1979, pp. 141–190 and French/Kennedy 1985
8. W. Bothe, P. Jensen: Die Absorption thermischer Neutronen in Kohlenstoff. *Zeitschrift für Physik* 122, 749–755 (1944)
9. D. Brill: *Bogislaw von Bonin im Spannungsfeld zwischen Wiederbewaffnung-Westintegration-Wiedervereinigung – ein Beitrag zur Entstehungsgeschichte der Bundeswehr 1952–1955*. Nomos, Baden Baden 1987
10. D.C. Cassidy: *Uncertainty: The Life and Times of Werner Heisenberg*. H. Freeman, New York 1992

11. T.B. Cochran, W.M. Arkin and M.M. Hoenig: *Nuclear Weapons Databook, Vol I: U.S. Nuclear Forces and Capabilities.* Ballinger, Cambridge MA 1983
12. P.F. Dahl: *Heavy Water and the Wartime Race for Nuclear Energy.* IOP, Philadelphia 1999
13. Interview in: *Atomic Bomb Scientists.* Memoirs, 1939–1945, J.J. Ermenc (ed.), Westport 1989, pp. 77–160
14. M. Evangelista: *Innovation and the Arms Race.* Ithaca 1988
15. M. Evangelista: *Unarmed Forces: The Transnational Movement to End the Cold War.* Cornell University Press, Ithaca N.Y. 1999
16. *Frankfurter Allgemeine Zeitung*, 24 September (1956)
17. S. Flügge, G.v. Droste: Energetische Betrachtungen zu der Entstehung von Barium bei der Neutronenbestrahlung von Uran. *Zeitschrift für Physikalische Chemie B 4* 1, 11–15 (1939)
18. S. Flügge: *Deutsche Allgemeine Zeitung* of 15.8.1939, reprinted in [31, p. 23, pp. 26–29]
19. S. Flügge: Kann der Energieinhalt der Atomkerne technisch nutzbar gemacht werden? *Die Naturwissenschaften* 27 23/24, 402–410, (1939) see Wohlfahrt 1979, pp. 119–140
20. S. Flügge: How fission was discovered. In: J.W. Behrens and A.D. Carlson (eds.): *50 Years of Nuclear Fission.* American Nuclear Society, Inc., vol. 1, 1989, pp. 26–29
21. J. Franck, D.J. Hughes, J.J. Nickson, E. Rabinowitch, G.T. Seaborg, J.C. Stearns and L. Szilard: Bull. At. Sci. 1 (10), 2 (1946); see the text at: [http://www.nuclearfiles.org/docs/1945/450611-franck-report.html]
22. Michael Frayn: Copenhagen, Methuen Drama, London 1998
23. L. Freedman: *The Evolution of Nuclear Strategy.* New York 1981, p. 327
24. A. P. French, P.J. Kennedy: *Niels Bohr. A Centenary Volume.* Cambridge/Mass. 1985
25. R. Garwin, G. Charpak: *Megawatts and Megatons. A Turning Point in the Nuclear Age?* Alfred Knopf, New York 2001
26. S. Goldberg: Groves and the Scientists: Compartmentalization and the Building of the Bomb. *Physics Today*, 38–43 (August 1995)
27. S. A. Goudsmit: *ALSOS*, Woodbury, New York 1947
28. D. Hafemeister: *Physics and Nuclear Arms Today.* Readings from Physics Today, American Institute of Physics, New York 1991
29. O. Hahn, F. Strassmann: Über den Nachweis und das Verhalten der bei Bestrahlung des Urans mittels Neutronen entstehenden Erdalkalimetalle. *Die Naturwissenschaften* 27, 1, 11–15 (1939); see for the following original papers: Wohlfahrt 1979, pp. 65–77
30. O. Hahn: *Mein Leben.* Bruckmann, Munich 1968
31. Atom-Museum Haigerloch: *Geschichte deutscher Atomforschung. Der erste Atommeiler: Originalberichte deutscher Wissenschaftler.* Haigerloch 1982
32. H. von Halban, F. Joliot and L. Kowarski: Liberation of Neutrons in the Nuclear Explosion of Uranium. *Nature* 142, 470–471 (1939); see Wohlfahrt 1979, pp. 111–114
33. H. van Halban, F. Joliot, L. Kowarski: Number of Neutrons Liberated in the Nuclear Fission of Uranium. *Nature* 143, 680 (1939); see Wohlfahrt 1979, pp. 115–118
34. Letter to H. Heimpel, 1st October 1941 (Werner Heisenberg Archive, Max Planck Institute for Physics, Munich)
35. W. Heisenberg: *Der Spiegel* No. 28, 3rd July 1967
36. W. Heisenberg: *Der Teil und das Ganze.* Munich 1969
37. W. Heisenberg: Die Möglichkeit der technischen Energiegewinnung aus der Uranspaltung. In: W. Blum, H.-P. Dürr and H. Rechenberg (eds.): *Werner Heisenberg.* Gesammelte Werke/Collected Works Part AII. Springer-Verlag, Berlin 1989, pp. 378–396
38. W. Heisenberg, K. Wirtz: Grossversuche zur Vorbereitung der Konstruktion eines Uranbrenners. In: W. Bothe, S. Flügge (Eds.): *Fiat Review of German Science (Naturforschung und Medizin in Deutschland) 1939–1946*, reprinted in [31, p. 50]

39. J. Herbig: *Kettenreaktionen. Das Drama der Atomphysiker.* München 1976
40. A comparable study from 1988: F. von Hippel, B.G. Levi, T.A. Postol and W.H. Daugherty: Civilian Casualties from Counterforce Attacks. In: F. von Hippel (ed.): *Scientific American.* September 1988, Citizen Scientist, New York 1991, pp. 137–149
41. D. Hoffmann: *Operation Epsilon. Die Farm-Hall-Protokolle oder die Angst der Alliierten vor der deutschen Bombe.* Rowohlt-Verlag, Berlin 1993
42. F. Houtermans: *Zur Frage der Auslösung von Kernreaktionen.* G-94, unpublished report (August 1941)
43. See on the Internet: [www.aip.org./history/heisenberg/p11a.htm]
44. D. Irving: *Der Traum von der deutschen Atombombe.* Gütersloh 1967
45. R. Jungk: *Brighter than a Thousand Suns.* New York 1958
46. H. Kant: *Werner Heisenberg and the German Uranium Project.* Preprint 203, Max-Planck-Institut für Wissenschaftsgeschichte, Berlin 2002 [http://www.mpiwg-berlin.mpg.de/PREPRINT.HTM]
47. G. Kessler, K. Wirtz: Von der Entdeckung der Kernspaltung zur heutigen Reaktortechnik. In: KfK-Nachrichten 4/88: *50 Jahre Kernspaltung.* Forschungszentrum Karlsruhe, 1988
48. H. Kissinger: *Nuclear Weapons and Foreign Policy.* Harper & Brothers, New York 1957
49. L. Koester: Zum unvollendeten ersten deutschen Kernreaktor 1942/44. *Die Naturwissenschaften* 67, 573–575 (1980)
50. E. Kraus: *Von der Uranspaltung zur Göttinger Erklärung. Otto Hahn, Werner Heisenberg, Carl Friedrich von Weizsäcker und die Verantwortung des Wissenschaftlers.* Würzburg 2001
51. W. Lanouette: *Genius in the Shadow.* A Biography of Leo Szilard, Chicago 1994
52. L. Meitner, O. Frisch: Disintegration of Uranium by Neutrons: a New Type of Nuclear Reaction. *Nature* 143, (1939, February 11) pp. 239–240; see Wohlfahrt 1979, pp. 97–100
53. Max Planck-Institute for Medical Research: *Walther Bothe and the Uranverein: KWImf Nuclear Physics during the War.* [http://sun0.mpimf-heidelberg.mpg.de/shared/docs/institute/index.php3?LANG=ge]
54. These letters were published in 2002 in the internet by the Niels Bohr Archive, Copenhagen, [www.nbi.dk/NBA/papers/introduction.htm.]; see also W. Sweet: *The Bohr Letters. No more Uncertainty.* Bulletin of the Atomic Scientists May/June 2002, 20–27, and a reply by K. Gottstein (ed.): Bulletin of the Atomic Scientists September/Oktober 2002, 4–5 and 72
55. M.J. Neufeld: *The Rocket and the Reich. Peenemünde and the Coming of the Ballistic Missile Era.* Cambridge/Mass 1995
56. R. Oppenheimer: Speech to the Association of Los Alamos Scientists, 2.11.1945; see: A. Kimball, S. and C. Weiner (eds.): *Robert Oppenheimer: Letters and Recollections.* Stanford/Ca., p. 317
57. A. Pais: *Niels Bohr's Times.* Oxford University Press 1991
58. T. Powers: *Heisenberg's War. The Secret History of the German Bomb.* Little, Brown and Company; Boston, New York, Toronto, London 1993
59. [www.pugwash.org]
60. R. Rhodes: *The Making of the Atomic Bomb.* Simon and Schuster, New York 1986
61. P.L. Rose: *Heisenberg and the Nazi Atomic Bomb Project – A Study in German Culture.* University of California Press, Berkeley 1998
62. The Russell–Einstein Manifesto of July 9, 1955
63. J. Rotblat: The Early Days of Pugwash. *Physics Today,* June 2001 [http://www.physicstoday.org/pt/vol-54/iss-6/p50.html]
64. M. Schaaf: *Der Physikochemiker Paul Harteck (1902–1985).* CENSIS-Report-33–99 Hamburg 1999

65. V.D. Sokolovsky et al: *Voennnia Strategiia.* Voyenizdat, Moscow 2d ed. 1963
66. A. Speer: *Erinnerungen.* (Inside the Third Reich: Memoirs translated by Richard and Clara Winston, New York 1970) Berlin 1969
67. K. Starke: Transuranium research in Germany 19349 to 1945. *Atomkernenergie/Kerntechnik* 41, 264–266 (1982)
68. Stuewer, Roger A.: Bringing the News of Fission to America. *Physics Today*, October 1985. In: Hafemeister 1991, pp. 261–268
69. See for example the work of L. Szilard and E. Rabinowitch In: L.M. Rieser: *Leo Szilard and the Early Days of the Bulletin of the Atomic Scientists*, In: G. Marx (ed.): Leo Szilárd Centenary Volume, Budapest 1998, pp. 51–58. A.K. Smith, A. Peril and A. Hope (eds.): The Scientists' Movement in America, 1945–1947, Chicago 1965
70. VDW: Forschen in Freiheit und Verantwortung, 1984
71. M. Walker: *German National Socialism and the Quest for Nuclear Power. 1939–1949.* Cambridge University Press, 1989
72. M. Walker: *Die Uranmaschine. Mythos und Wirklichkeit der deutschen Atombombe.* Berlin 1990
73. M. Walker: *Nazi Science: Myth, Truth, and the German Atomic Bomb.* New York 1995
74. Ch. Waltham: *Early History of Heavy Water.* Department of Physics and Astronomy, University of British Columbia, Vancouver B.C., Canada
V6T 1Z1, [www.physics.ubc.ca/ ~waltham/d2098/paper/node10.html]
75. S.R. Weart, G.W. Szilard: *Leo Szilard: His Version of the Facts.* Selected Recollections and Correspondence, Cambridge/Mass. 1978
76. V. Weisskopf: *The Joy of Insight.* Basic Books, New York 1991
77. C.F. von Weizsäcker: *Eine Möglichkeit der Energiegewinnung aus* U238. G-59, unpublished report (July 17, 1940)
78. C.F. von Weizsäcker, P. Müller and K.-H. Höcker: *Berechnung der Energieerzeugung in der Uranmaschine.* G-60, unpublished (26 February 1940)
79. C.F. von Weizsäcker: *Die Verantwortung der Wissenschaft im Atomzeitalter.* (Kleine Vandenhoeck-Reihe 42). Göttingen 1957
80. C.F. von Weizsäcker: *Kriegsfolgen und Kriegsverhütung.* Hanser-Verlag, München 1970
81. C.F. von Weizsäcker: *Wege in der Gefahr. Eine Studie über Wirtschaft, Gesellschaft und Kriegsverhütung.* Hanser-Verlag, München 1976 (The Politics of Peril, Seabury Press 1978)
82. C.F. von Weizsäcker: *Der bedrohte Friede. Politische Aufsätze.* Hanser-Verlag, München, Wien 1981
83. C.F. von Weizsäcker: *Wahrnehmung der Neuzeit.* Hanser-Verlag, München, Wien 1983
84. C.F. von Weizsäcker: *Der Garten des Menschlichen. Beiträge zur geschichtlichen Anthropologie.* Hanser-Verlag, München 1984
85. Interview with the Magazine Stern 1984, reprinted in: Bewusstseinswandel, 1988, pp. 362–376
86. C.F. von Weizsäcker: *Bewusstseinswandel.* Hanser-Verlag, München, Wien 1988
87. C.F. von Weizsäcker: *The Political and Moral Consequences of Science...* Transcriptions of Lectures held at CERN, Geneva, January 1988, quoted in Bernstein 2001 p. 30
88. C.F. von Weizsäcker: *Zeit und Wissen* Hanser-Verlag, München, Wien 1992
89. C.F. von Weizsäcker: *Lieber Freund! Lieber Gegner!* Hanser-Verlag, München, Wien 2002
90. R.C. Williams, P.L. Cantelon: *The American Atom.* Philadelphia 1984
91. H. Wohlfarth: *40 Jahre Kernspaltung.* Wissenschaftliche Buchgesellschaft, Darmstadt 1979
92. DIE ZEIT, 15 May 1958, p. 4; 22 May 1958; p. 3; 29 May, p. 3; 5 June 1958, p. 3. Reprinted in: Der bedrohte Friede, 1981, pp. 42–87

On the Unfathomableness of Consciousness by Consciousness
Why do physicists widely agree on the assured extent of their professional knowledge, but not so philosophers? – *An exchange of letters and a discussion with Carl Friedrich von Weizsäcker*

Klaus Gottstein

On the occasion of Carl Friedrich von Weizsäcker's 81st birthday, a colloquium was held on July 3, 1993 in Niederpöcking on the Starnberg Lake. One of the topics was "The epistemological foundation of physics from Kant to von Weizsäcker". Walter Schindler and Thomas Görnitz gave the introductory speeches. I took part in the discussion which followed. Afterwards I got the impression that I had not succeeded in making my points clear. Therefore, on July 5, I wrote an explanatory letter to Prof. von Weizsäcker. This was the beginning of a correspondence on the questions intimated by the title above. In addition, a private discussion between Prof. von Weizsäcker and myself took place. Some readers may be interested in the answers which Carl Friedrich von Weizsäcker has given to the naïve questions and views of a former physicist with hardly any education in philosophy. The letters and my notes on our discussion follow in chronological order. (I am, of course, responsible for any errors in my representation of von Weizsäcker's remarks in my notes.)

July 5, 1993

Dear Mr. von Weizsäcker,

I am afraid that my improvised remarks in the discussion after the speeches of Messrs. Schindler and Görnitz were not sufficiently clear. My first comment concerned the remarkable circumstance that the physicists of a given historical period agree to a large extent on the observable phenomena of nature and on their underlying laws, while in philosophy there are always different schools with different conceptions of truth and which accuse each other of error. To me, this appears to be due to the fact that physics is concerned with phenomena which present themselves in the same way to all human beings possessing the same brains and the same state of consciousness as conditioned by the historical era under consideration. In philosophy, however, the brain seeks to discover which possibilities of cognition exist for the brain. Isn't that so as if a man wants to take measure of himself? This can never lead to unambiguous results, I should think! One cannot help remembering Münchhausen who tore himself out of the swamp by his own pigtail. There is, perhaps, a connection between this difficulty of the human brain in grasping the conditions of its capacity for gaining knowledge and in representing these conditions unambiguously and free of contradiction, and Gödel's Theorem on the impossibility of proving the consistency of a formal system by its own means (see, for example: Alfred Gierer: Physics, Life and Mind, Piper Verlag 1985). It seems that the realm of religions, that means of faith, begins at this point.

My second remark concerned the often cited connection between Kant's inclusion of the subject into objective reality in the course of its recognition, and quantum theory where one cannot abstract from the observer (Mr. Schindler spoke about this). It appears to me that this is not a real connection, but rather a kind of parallelism. After all, in quantum theory it is only shown that the experimenter with his apparatus, in observing atomic processes, interferes with the process to be observed (by emitting light quanta and by other means), so that the process cannot be observed objectively. But this is of no consequences under macroscopic conditions. Also according to quantum mechanics, we can still speak of macroscopic objects observed independently of the subject, the experimenter; in astronomy, for example. Kant, however, had in mind the macroscopic world when setting out his epistemology, which seeks to determine what can be known of reality to the human subject. He knew nothing yet about atomic processes. Therefore, quantum theory and the Critique of Pure Reason deal with different matters, which nevertheless have in common that the observing subject becomes entangled with the observed phenomena – in the case of Kant in a very general way as the human subject which is dependent on its sensory experiences and for which the 'thing-in-itself' is inaccessible anyway and, therefore, meaningless; in quantum theory, however, the observing subject is entangled only for observations of processes in domains where classical physics is no longer valid.

Would you agree?

Cordially,

Klaus Gottstein

July 12, 1993

Dear Mr. Gottstein,

I thank you very much for having more fully explained the sense of your contribution to the discussion. I must confess, however, that I see the problem in a slightly different way. It is a particularity of life that living beings are capable of performing many activities, without being able to explain what they are really performing, or even without reflecting on what they are performing. With animals this is plainly so, but it is also largely so with humans. Now, science is one of the great cultural discoveries in which one has seen that man is capable of very complex activities when certain methods are applied; for example, empirical and mathematical methods. Anyone who then reflects more closely on what was really done here encounters the difficulty that he acted like animals do who are able to do something without being able to provide the information why that is so.

This is already implied in your parable of the brain which wants to understand the brain itself. The question asked here, on the other hand, is an entirely meaningful question, and particularly so when it does not attempt to solve all great problems at once, but toils on the problems of understanding which arise from science anyway and which prove to be important. Thomas Kuhn has differentiated between normal science as puzzle-solving under a given paradigm, that is, under a given example of solving riddles, and scientific revolutions in which one is forced to change paradigms. Scientific revolutions are mostly made possible through a reflection on what one has really done, a reflection which one may very well call philosophical. In this sense, for example, the philosophical reflection of Ernst Mach on the meaning of classical mechanics contributed significantly to Einstein's being able to discover the theory of relativity and Heisenberg's being able to discover quantum mechanics. Only when normal science resumes fully its activity under a new paradigm, it is no longer necessary to tackle the more difficult question which I have called the philosophical one.

Moreover, I believe that the connection between Kant and quantum theory can play an explanatory role after all though, as you correctly say, it is not identical at all with these problems. Kant's question was how experience was actually possible. Thereby he clarified the important role of the judgement of the scientific subject in understanding scientific objects. Precisely this question came up again in quantum theory in the context of special examples. That is, methodically Kant asks similar questions, the concrete contents, on the other hand, are different. I might add that, in the context of quantum theory, classical physics has been cast into a new light as a sort of limiting case. Kant's problem which, of course, referred only to classical physics, is indeed cast into a new light by quantum theory. But I shall not try to detail these matters in this letter.

Unfortunately, I am a little short of time and I cannot provide a more complete account. I have written on this issue in some of my books; perhaps one day we shall be able to find an opportunity for a quiet discussion.

Cordially,
CF Weizsäcker

October 5, 1993

Dear Mr. von Weizsäcker,
...

It is certainly correct that humans and other living beings are capable of many things without knowing how they have done it. But this cannot mean that something can be done which is logically impossible. When the brain wants to understand the brain – and not that in a medical-physiological sense in which the human brain is an object as other objects of natural science, but in the philosophical sense as an organ that experiences the world – then it is true that this desire is legitimate. But, to choose another analogy, isn't this like the eye desiring to look at itself without the help of a mirror? So it is not really surprising that in philosophy there are as many answers to one question as there are philosophical schools.

Accidentally, I found a description of Karl Jaspers' philosophy in the book by Jeanne Hersh: "The Philosophical Marvel".

"... *Objectivity, therefore, is much more than just a scientific requirement: it is of an ontological nature. Just because of this, Man strives to grasp the world objectively as a totality. But he cannot escape the fundamental situation of all thinking: that he, as the thinking subject, faces an object-reality which he wants to raise to a totality. And here he is astonished: just this is impossible. He can never be included in this presumed totality insofar as he thinks himself as a subject. The world will never combine for him into a totality ... The thinker, therefore, always remains in the Kantian subject-object split and in a world which can never become for him a unity and a totality. We cannot reach the knowledge of totality. It is true that we can always proceed toward a horizon which all the time surmounts what we already know; but we can never proclaim a knowledge claiming that it refers to the totality. Totality is denied to us by our situation as recognizing subjects ... "*

This quotation seems to express in a much more perfect way what I have tried to say with my simple examples of the yardstick which attempts to measure itself, the brain which wants to understand itself, and the eye, which strives to see itself without a mirror.

Cordially
Klaus Gottstein

November 3, 1993

Dear Mr. Gottstein,

...

The topic which you raise again in your letter is basically too difficult for a brief discussion by an exchange of letters. We should talk quietly about it, the two of us, which I will be glad to do.

For the time being let me say that even Jaspers' very careful formulation does not convince me at all. Perhaps it is consistent in the tradition of a thinker greater than Jaspers, namely Kant. On the other hand, I know that there have been thinkers in the history of human thought of the same greatness as Kant who look at these matters very differently. From antiquity I would name Plato; from Asia both the Vedanta philosophy and Buddha. All of these do not see in knowledge the contraposition of a subject with precise knowledge of itself with a recognized object in such a way that, as Jaspers means, the subject can never become the object of cognition. Definitely not. This distinction, in all its rigor, is only due to Descartes, if I understand correctly, and I would say that already Fichte, Schelling and Hegel after Kant sensed that a transgression is needed here, and a different view. But, as I said, this is a wide field and we should discuss these matters orally.

Cordially,
Carl Friedrich Weizsäcker

December 6, 1993

Dear Mr. von Weizsäcker,

At the outset, with regard to your letter of November 3, perhaps already this remark: The fact that great thinkers have seen the possibilities of human self-knowledge in very different ways was exactly the starting point of my remarks at the colloquium discussion on the occasion of your last birthday. If the human brain would be able to analyze its own essence in an irrefutable manner, all serious philosophers would have to agree on the possibilities of human self-knowledge, just as all serious physicists basically agree on the theory of relativity and on quantum mechanics. The fact that such an agreement does not exist among the great philosophers, appears to me to confirm that man is overburdened with the task of self-analysis in a similar way as the yardstick with the task of self-measurement. It would require an external yardstick. Without such an external yardstick, "arbitrary" results are obtainable. The yardstick external to man, however, would not belong to the realm of science any more, but to that of religion. After all, the thinkers whom you have named in your letter, Plato, Buddha and the Vedanta philosophers, may perhaps be attributed to this realm.

With the help of the "religious yardstick" man can recognize himself and his individual relation to divinity, and this in a dimension which is no longer directly accessible to reason, to science. Religious revelation has taken different forms in different cultural areas. This non-unambigiousness religion and philosophy have in common. It would be an interesting question what this commonality means. For non-religious persons an answer readily at hand is that religion is just as imagined by man as is philosophy. But what answer should believers give? Again, a ready-made answer is: all religions except one's own are imagined by men; only one's own religion is founded by God. For the adherents of the major world religions this is, no doubt, the "correct" answer in each case. The question remains what consequences we have to draw from this. In one of your books and during one of our discussions you said that you would have become, of course, a Buddhist or a Muslim if you had not happened to be born in Europe, but had been born in East Asia or in the Middle East. For this reason, and because of the wisdom of the great Asian religions, you could not adhere to the claim of Christianity to be the only salvation-bringing religion, as Karl Barth has done. This answer has never satisfied me from a logical point of view, since you and I were not born and raised in East Asia or in the Middle East but in Europe in the same spiritual world as Karl Barth. Therefore, to us the commandments of our religion should be valid, including the "Go ye therefore, and teach all nations and baptize them ...". This does not exclude that, seen from a higher level, the representatives of other religions have received the order to convert *us*. If this were so, it would not be our concern. For us, there would remain only one explanation: God's ways are not our ways. We have a different message.

Cordially,
Klaus Gottstein

March 9, 1994

Dear Mr. von Weizsäcker,

Since our last discussion and my letter of December 6, 1993, I have continued to search for philosophical statements regarding the possibility of man to know himself – not only as an object, but in the totality of one's own self. The search is not yet finished, the literature is too extensive. But I want to give you an intermediate report as this may save time during our upcoming conversation.

I do not intend to return to Karl Jaspers who thought that the perceiver cannot attain the knowledge of totality, as there is always only a horizon which discloses itself. In your letter of November 3 you have termed Jaspers' statement as Kantian but not convincing. I have come across the following statements, which I can give here only as a kind of anthology, for which I beg your pardon:

Bertrand Russell: "The value of philosophy consists, on the contrary, essentially in its uncertainty, which is inherent in her."
(quoted from Wilhelm Weischedel, 34 große Philosophen in Alltag und Denken).

Karl Popper and John Eccles: "We have to face the fact that we live in a world in which almost everything which is truly important remains essentially unexplained" *(quoted from Volker Spierling, Kleine Geschichte der Philosophie).*

Maurice Merleau-Ponty considers as hopeless the endeavor to penetrate to a theoretically fixed structural entity of basic human experiences. ....Philosophy must be inherent in the original, indissoluble union ("mixture") of man and the world which no analysis can circumvent. Starting from a finiteness of man, newly understood and positively grasped in this way, philosophy consummates itself as hermeneutics of experience or interpretation of existence, respectively. It renounces the concept that the world, things and consciousness could be generally determined and become adequately fixed by thinking. *(Margot Fleischer in: Philosophen des 20. Jahrhunderts, Wissenschaftliche Buchgesellschaft, 3. Auflage 1992).*

Merleau-Ponty speaks of the "paradox of time": *"...I cannot really take possession of my time before I understand myself completely, and this moment can never come, because, if it came, it would still be just a moment surrounded by a horizon of the future, which itself would stand in need of unfolding in order to be understood....The reflection... clarifies everything except its own role."* (quoted according to Erich Christian Schröder, in: Margot Fleischer (Hrsg.), Philosophen des 20. Jahrhunderts).

Michel Foucault emphasizes the impossibility to reach behind our entanglement with the world through a reflection on the subject, let alone with the intention of providing an ultimate substantiation. No longer is a reason common to all held as the governor for knowledge and truth. *(quoted from: Erich Christian Schröder, in: Margot Fleischer (Hrsg.), Philosophen des 20. Jahrhunderts).*

According to Michel Foucault, human activity, life and language develop their own, quasi-transcendental structure of laws which man struggles in vain to dominate. The attempt to think the unthought and to catch up with the continuously escaping origin leads to an endless duplication in empiricism and transcendentalism which does not help us advance at all.
(Bernhard Waldenfels in: Margot Fleischer (Hrsg.), Philosophen des 20. Jahrhunderts).

For Theodor Adorno, the patient acceptance of the heterogeneous, dispersed and refractory given is essential to philosophy. Philosophy interprets it, seeks for traces of hope in it, tries to unravel its ciphers, but without entertaining the illusion that world could be discoverable or moldable into a unity of meaning, or could open a view to a "world behind".
(Margot Fleischer, loc. cit.).

The thesis of W.V.O. Quine is: There exist total theories of the world which deeply differ from each other, yet are empirically equivalent in the sense that they imply exactly the same observation-categorical principles. *(Felix Mühlhölzer in: Margot Fleischer, loc. cit.).*

With Ernst Bloch, the starting point is "being", that existence from time immemorial which we always are already and which we can never get hold of because as the existing one it remains unfathomably ahead of thinking though itself pushing and driving us. ... Bloch has circumscribed the non-constructibility of the absolute question which we are to ourselves, in the well-known formula: "darkness of the presently lived moment." We live, I am, but just this directness of life, of the "being" which carries us and from which everything wells up, cannot be caught up with, neither experiencing nor understanding. *(Wolfdietrich Schmied-Kowarzik, in: Margot Fleischer, loc. cit.).*

This is as far as I have got. But perhaps these samples are already sufficient to show that the limits of self-knowledge are felt also by modern, not at all religious, thinkers. The philosopher Jeanne Hersch, a disciple of Jaspers, I must admit, has expressed this in the following words: "Existence knows that it has not created itself" (Jeanne Hersch, The Philosophical Marvel). This is very close to statements such as the one by Gregor of Nazianz (330–390), one of the Early Fathers of the Church: God in his being is inaccessible for human thought. *(quoted from Wolf-Dieter Hauschild, Gregor von Nazianz, in: Klassiker der Theologie, Beck-Verlag)*

...

Cordially,
Klaus Gottstein

K. Gottstein
Notes on the conversation with Carl Friedrich von Weizsäcker on March 16, 1994

Just to finish this subject and then proceed to questions more important to me I recalled the second topic of my letter of July 5, 1993 (the parallel between Kant's inclusion of the subject into the objective reality in the process of perceiving it, and the statements of quantum mechanics) and mentioned that Kant's statement, of course, referred to the macroscopic world, whereas quantum mechanics leads to differences from classical physics only in the microscopic world, so that there is no real connection between Kant and quantum mechanics. C.F. von Weizsäcker, however, stayed with this topic for a while, it is especially close to him. He mentioned the dispute between Boltzmann and Mach on the existence of atoms, a dispute based on the fact that even the behavior of macroscopic bodies (energy, temperature, the second law of thermodynamics) cannot be understood without contradictions if one supposes on the basis of classical physics that macroscopic bodies are composed of atoms. Since Mach realized this, he denied Boltzmann the right to assert the existence of atoms. In this context I stated that the necessity to apply quantum mechanics here to macroscopic bodies is due to the inclusion of the concept of microscopic atoms into the consideration. – C.F. von Weizsäcker agreed with this and added that the chemists of the 19th century had such great success with the notion of atoms because, in their happy naiveté, they did not see the associated complications. Moreover, C.F. von Weizsäcker maintained that the study of Kant could definitely contribute to the understanding of quantum mechanics. He, C.F. von Weizsäcker, had just written a paper on Kant and quantum mechanics. – Regarding the relation of science and philosophy, C.F. von Weizsäcker pointed to Heidegger whom he had known well and who had remarked that scientists do not think. They don't reflect on why their methods are successful, what they do with these methods and what the consequences are of their procedures. It is not self evident, said C.F. von Weizsäcker, that the application of mathematics describes nature "correctly". But scientists prefer not to think about this, because they can obtain a university chair more easily by producing scientific results then by writing poor philosophical treatises. I objected that one could just as well write good philosophical treatises and then obtain a chair in philosophy. C.F. von Weizsäcker replied to my remark, which obviously referred to his own career, that this was very difficult.

The discussion then turned to the first topic of my letter of March 9, 1994: the question whether total self-knowledge is possible. Jaspers and the philosophers I had cited in my letter had denied this possibility. C.F. von Weizsäcker indicated that he was not at all convinced or impressed by these citations. One had to know, he said, that these are philosophers of the 20th century. Jaspers shows philosophical arrogance when he says that he knows that the ego cannot know itself. Other centuries came to totally different conclusions. He pointed especially to Descartes who had shown that everything can be doubted except that one doubts, and therefore exists. From this point Descartes started in constructing his philosophy. He even derived a proof for the existence of God which, however, is the weakest point in his system of thought. He, C.F. von Weizsäcker, said that he did not share Descartes' philosophy but he could defend Descartes as an attorney, so to speak. He added that he developed his own philosophy from a dispute with Descartes. In any case, Descartes showed that one cannot doubt one's own consciousness.

I raised two objections:

1. Jaspers and the other philosophers which I had cited did not doubt their own existence but the possibility of grasping the totality of their own selves.

2. The centuries named by C.F. von Weizsäcker, in which philosophers could supposedly understand their own selves, are only the few centuries of the Age of Enlightenment. In the many centuries before, philosophers were of a different opinion. I mentioned Thomas Aquinas and the Early Fathers of the Church who understood man as a creature of God and God as inaccessible to reason. Faith in revelation had priority over philosophical wisdom which does not contribute anything to salvation.

C.F. von Weizsäcker replied that with the Gentile Christians (i.e. Christians of non-Jewish origin, of which Thomas Aquinas was one), two different cultures encountered each other: the Greek thinking of Aristotle, and Christianity which arose from Jewish sources. In order to reconcile both, Thomas made the compromise to give precedence to Christian revelation. At his time, C.F. von Weizsäcker said, Thomas could not do more. For having a unified view of the world without compromises he had been born either 700 years too late or 900 years too early. Even today religion and enlightenment are still incomplete. The next steps must lead both sides to the realization that each of them needs the other one for its own perfection (religion the enlightenment as well as enlightenment the religion).

In this context C.F. von Weizsäcker mentioned that Niels Bohr with his philosophy of complementarity is probably the most important thinker of the 20th century. He is the Socrates of our time who, like Socrates, proclaims that he knows nothing in order to show to others who pretend to know that they, too, know nothing. Then, when a solution is asked for, the reply is: "We shall talk more tomorrow". Bohr's philosophy of complementarity allows to view an object under different – even seemingly incompatible – aspects. Here is an indication of how the philosophy of the future could look.

I then mentioned that the present epoch, in any case, lacks a world view determining the ethical behavior of men. Although many wise books have been written – among them those by C.F. von Weizsäcker – none of these are evaluated in the sense that they would become sufficiently effective in politics. Christianity once played the role of the dominant world view in Europe, today it is no longer binding for the majority of people. Marxism which had been the dominant world interpretation in some countries for several decades has collapsed. The peaceful coexistence of men requires a common ethical basis. I asked if one could "invent" an ethics because it is needed for practical purposes, as Hans Küng has tried to do. Isn't ethics a result of a firmly rooted world interpretation? Can one generate ethics without this foundation?

C.F. von Weizsäcker objected that Küng sets out from the statement: "No world peace without peace between religions!" – I asked if this does not represent an overestimation of religion. Are religions still taken so seriously that a conclusion of peace between the leaders of different religions can still affect world politics? But I admitted that religious differences can provide the pretext for expulsions, civil wars and other belligerent actions. But wouldn't politicians find other reasons when religious pretexts were no longer available? (Remark at the writing of these notes: For a long time now there has been peace between the Catholic and Protestant leaderships, but which effect does this have in Northern Ireland?)

C.F. von Weizsäcker defended Küng's "Project World Ethos." It did not represent an ethics on command. After all, Küng worked out the common ethical foundations of the world religions by detailed investigations of which the results are laid down in several books. In this research he found, i.a., that the "Golden Rule" is valid everywhere: What you do not want anybody to do to yourself, do not do to anybody! In addition, there are common elements in mysticism and meditation.

At this point, C.F. von Weizsäcker inserted a biographical note: At the age of twenty, on the occasion of a Protestant church service, he noticed the preacher's great pathos. He then realized that this pathos served to drown his own disbelief: The preacher did not believe what he said. At a Catholic service, the Latin mass and the solemn singing had given him, C.F. von Weizsäcker, the impression: This is the tradition of two thousand years, this I can bear. He had thought of converting to the Catholic faith. But there was the dogma of infallibility of the Pope when he spoke ex cathedra. Had he already been a Catholic, then he would not have left because of this. But to submit to this dogma voluntarily through conversion would have been too great a sacrifice.

Here I, too, made a biographical remark: I was molded by the Protestant youth circle in the parish of pastor Martin Niemöller in Dahlem at the time of the struggle of the Church with the Nazi state. I have met preachers who believed what they said. Today, their belief would perhaps be termed fundamentalist, but it enabled believers to risk their lives in offering resistance.

C.F. von Weizsäcker remarked that he admired these believers and their actions. But he had not taken part in this. Niemöller and Gollwitzer he had known very well. He had met Niemöller in 1958 at the synod in Spandau in which he had participated as an invited expert. The topic was atomic armament. Opinion was divided among the members of the synod, the Church was tried to the breaking point. Niemöller proved to be the preserver of unity. Although he himself was a radical pacifist, he had learned as church president to respect the opinions of differently minded persons. He called on the quarreling parties to reassemble under the gospel in spite of their divergences. This appeal had its effect.

After this excursion, we came back to Hans Küng and the problem of cooperation among the religions in the elaboration of a common ethics. C.F. von Weizsäcker said that he supports Küng and only regrets that Küng is not on good terms with the Pope. Catholics, of course, should be included in this cooperation. To him as a Protestant school boy sitting in class with Catholics and Jews, it had already seemed strange that the contingency of birth decides on one's faith. If his parents had been Catholics or Buddhists, he would have become a Catholic or a Buddhist. It could not be true that God had given the right faith only to one religion and had left all others in error.

I replied that with such reflections he, C.F. von Weizsäcker, placed himself in the role of God. The Christian has a clear mandate: "Go ye therefore, and teach all nations and baptize them!" (compare the letter of December 6, 1993)

C.F. von Weizsäcker responded that lack of tolerance is the main shortcoming of Christians. A Buddhist could be Christian simultaneously without any problem. Why is the inverse impossible? C.F. von Weizsäcker's Japanese Christian acquaintances (in Japan a small minority) had told him that it is not their aim to convert all Japanese to Christianity; rather it is their aim to make the Japanese people Buddhists again. To him the wording of the dogmas is not essential, C.F. von Weizsäcker said.

I objected that then he could not really speak the Christian Confession of Faith. C.F. von Weizsäcker replied that in this matter he followed his great grandfather who had been a liberal professor of protestant theology. Someone had told his great grandfather that the Confession of Faith must be changed since one knows now that several statements therein are not correct. His great grandfather had replied that he was against a modification of the Confession of Faith but that he would propose to allow anyone when speaking the Confession to think what he considers to be correct. The Bible contains several contradictions, C.F. von Weizsäcker said. He had discovered the following contradiction in the Old Testament: At one point (Second Book of Samuel, Chapter 24), David's census is said to be God's order, and the observance of this order is subsequently punished by God. This seems to be the original version. The later chronicler (First Book of Chronicles, Chapter 21) has apparently modified the text in order to erase this contradiction. Now Satan inspires King David with the idea of the census, and God punishes.

For the history of creation, C.F. von Weizsäcker said, he had constructed for himself the following anecdote which he thought had a high degree of veracity. To the Rabbi who around 500 B.C. in Babylon writes down the history of creation comes a young man and asks him if he really believes that God has created the world in six days. The Rabbi answers: "Don't you see that this story is a parable?" The texts of the Old Testament which partly originate around 900 B.C., were apparently, according to recent findings, edited and compiled in an "editorial conference" around 500 B.C.

C.F. von Weizsäcker returned again to Kant. Logically, the existence of God could not be deduced from Kant's system. But Kant presupposed the existence of God for moral reasons and restricted his system so that there remained room for the belief in God.

C.F. von Weizsäcker quoted what many of his Buddhist acquaintances had told him: Europeans do not realize what damage they cause with their Aristotelian logic: they ruin the world.

I replied that people like the Babylonians, Assyrians and Huns who had known nothing about Aristotle, had also inflicted terrible damages. To this C.F. von Weizsäcker answered that the Europeans with their Aristotelian method of thinking had finally defeated these people. Moreover, the reproach of Buddhists against Europeans is not that of cruelty; rather it concerns the thoughtlessness of their actions with respect to the consequences.

To this I said that grim consequences are in most cases collateral effects or long term consequences of measures with good intentions. In order to be able to prevent such undesirable collateral effects and long term consequences, the possibility of their occurrence must be forecast, and for this interdisciplinary and often international cooperation is necessary. What must we do today? I stressed that, regardless of the unclarified philosophical and theological questions, what now matters is to avoid catastrophes. I mentioned the early efforts by C.F. von Weizsäcker within the the Federation of German Scientists and the Starnberg Institute, and the Federation of American Scientists and the Pughwash Movement. C.F. von Weizsäcker said that the atomic bomb was the alarm signal, indicating to us that we cannot proceed in the same way. The message has been heard, the Third World War has so far been avoided.

I mentioned once again my hope that a social theory and a world view for the modern era could be found – perhaps in the form of C.F. von Weizsäcker's anticipated mutual approach of religion and science.

March 22, 1994

Dear Mr. von Weizsäcker,

In the meantime I have read with great interest your essay: "On the Structure of Physics: Kant and Quantum Theory", which you mentioned to me during our conversation on March 16. To my surprise, you have not only discussed there the question of whether there is a genuine connection between Kant's thoughts on the role of the subject in the observation of "objective reality" and the concepts resulting from quantum mechanics, or whether this is only a parallelism. But on the last page of the essay you make also statements regarding the second topic of our discussion, namely the question whether it is possible for consciousness to grasp itself in its totality. In our conversation, you designated as philosophy of the 20th century the collection of statements by modern philosophers from my letter of March 9 (expressing the view that man cannot reach ultimate clarity on himself) which should be confronted with the quite different findings of earlier centuries. In particular, you mentioned Descartes who had inferred his existence from the indubitable fact of his doubt. I had remarked in this context that also the cited philosophers of the 20th century did not dispute the certitude of their own existence. After all, the knowledge of one's own being cannot be equated to grasping the totality of one's own consciousness.

Now I see on page 8 of your essay on Kant and quantum theory that, in speaking of Descartes, you characterize a certitude of self-consciousness which goes beyond the absolute certitude of self-existence, with the words "that may be" which express a lower level of certainty. At the end of the relevant paragraph, you write: "I find myself in the thinking and observing consciousness and interpret this in the frame of evolution." After all, when one finds oneself somewhere, then one does not know how one got there and has to rely on interpretations.

In the last paragraph of page 8 of your essay you write: "The problem (of mind and form) cannot be solved in a rational-speculative way. It presupposes, according to my conviction, religious experience: ethical practice and meditative perception." Do I interpret correctly that – much in the sense of the statements in my letters of July 5, October 5 and December 6 of 1993 – you come to the conclusion that there are insurmountable barriers for the rational knowledge of the brain by the brain, therefore of the totality of the self by the self, and that behind these barriers lies the domain of religion, in which the last questions can be answered by faith?

...

Cordially,
Klaus Gottstein

June 29, 1994

Dear Mr. Gottstein

...

Your letter of March 22 is still on my desk, unanswered. To blame for this is the fact that I receive vast amounts of mail these days...

The pecularity is, as it appears to me, that both of us, you and I, have basically a very similar conception of human consciousness in the context of nature, but that again each of us spontaneously chooses ways of expression in which the other cannot join and which provokes the other to contradiction.

From my point of view, for example, my main stumbling-stone with respect to your manner of expression has been when you say that there are limits to the rational cognition of the brain by the brain. First, it should be said that in the tradition, as it exists since Descartes, nobody could have spoken like that, because for Descartes consciousness, the res cogitans, was of a totally different substance than matter, the res extensa, to which the brain belongs. In my understanding of Descartes, he held the opinion that consciousness fundamentally knows itself. The question how this relates to the brain was difficult for him. According to his conviction, animals which also have a brain after all, were automatic machines without any consciousness. On the other hand, he could not deny the link between matter and consciousness for humans. He placed it into a causal mutual influencing in the only unpaired organ in the human head, the pineal gland. For Descartes it would have been self-evident that the brain cannot recognize itself, because the brain is just matter and not thinking substance.

Now, this is not my personal opinion at all. I merely wanted to say that the history of the philosophy of consciousness up to Kant is not understandable if this division by Descartes is not taken into consideration and that the mind-body problem, which still confronts biologists today, is a consequence of this Cartesian separation.

When one attempts to remove this separation, when one simply denies its existence, and this would be my interpretation especially also of quantum theory, then indeed it becomes almost self-evident that consciousness cannot know itself completely; I like to express this by the formula: "consciousness is an unconscious act". This, however, I mean at first again in a descriptive way, and for this I do not necessarily need the recourse to the brain. It is sufficient that one knows, for example, psychoanalysis since Freud. Fundamentally, also earlier thinkers, for example Goethe, have known this quite well. When I then tried to describe the role of the brain in this context, then I first had to start from the fact that according to quantum theory, at least as I interpret it, already the atom or the elementary particle is part of something virtually conscious. 'Virtually' means that it may take several billions of years before a consciousness arises which can even reflect on itself. Schelling said nature is the mind which does not recognize itself as mind. In this way of speaking I may perhaps say of man that the brain is not the carrier of consciousness, but the brain is an aspect of consciousness which one gets to see when one approaches it from the outside, that is spatially.

Whether or not this solves our controversy is not clear to me, but I wanted to make at least these few remarks.
Cordially
Carl Friedrich Weizsäcker

August 1, 1994

Dear Mr. von Weizsäcker,
I thank you very much for your letter of June 29, 1994 which I found in my mail after longer journeys. I am very glad to take it from your lines that, strictly speaking, we agree in substance and that you have only disapproved of my manner of expression, which did not sufficiently take into account the tradition since Descartes, although you, too, dissociate yourself from Descartes' separation of thinking and material substance. Your statement that, after removing this separation, it becomes almost self-evident that consciousness cannot know itself completely, and my statement that the brain cannot know itself completely for logical reasons (in my letter of July 5, 1993 I had suspected a link with Gödel's theorem of incompleteness) show to me satisfactory agreement. This is so because, actually, I meant human consciousness when I spoke of the "brain". Strictly speaking, I only wanted to say that *man* cannot understand himself completely. You seem to agree with this. When I spoke of the brain, this was a concession to natural science which apparently has identified the brain as the seat of consciousness. When the brain is destroyed, there is no more consciousness, even if the heart continues to beat for a long time. But recourse to the brain is not important for me either.

As you will remember, the starting point of our discussion was my remark that the conspicuous inability of philosophers to agree on the deepest elements of human existence including the basic questions of ethics – which is tantamount to the occurrence of different schools of philosophy feuding with one another – could perhaps be traced back to the fundamental inability of man to attain self-knowledge. Physicists, on the other hand, can reach consent, in an equally remarkable way, on the results of their measurements and on whether or not these results agree with the ideas of theory. They just have to deal with res extensa. Here the yardstick does not intend to measure itself, but objects which are external to it, and this is possible.

There remains only the question what consequences can be drawn from this realization. Is philosophy, in as much as it gives statements on the totality of human existence, only an expression of the prevailing social circumstances and views, interesting enough as such, but not a gateway to truth? I am inclined to give an affirmative reply to this question.

...

Cordially,
Klaus Gottstein

Part II

Carl Friedrich von Weizsäcker's Contributions to Nuclear and Astrophysics and Their Impact on the Development of Physics

Introduction: From the Atomic Nucleus to Cosmic Vortex Systems

Helmut Rechenberg

1 First Research on Relativistic Quantum Problems (1931–1934)

In early 1929, shortly before graduating from the *Gymnasium* in Berlin, Carl Friedrich von Weizsäcker wrote to Werner Heisenberg and mentioned his intention to begin a university study of philosophy. The Leipzig professor of theoretical physics, whom he knew personally since a couple of years, replied: "One should work on physics while being young, and in our century it is necessary to explain nature with the tools of physics. Philosophy, as you know from Plato, requires a mature person, say about 50 years of age. Hence you should first study physics" [1]. The young man followed this advice and began his scientific education in the summer semester 1929 at the University of Berlin – as Heisenberg was on a world tour then – changing in the winter semester 1929/30 to Leipzig, where he was admitted soon to Heisenberg's seminar. Already in spring 1931 he finished his apprentice piece of research, a paper on the "Determination of the position of an electron by a microscope" [2]. The investigation dealt with a proof of Heisenberg's indeterminacy (or uncertainty) relation on the basis of a detailed application of the theory of optical image formation, using the relativistic quantum electrodynamics recently developed by Heisenberg and Pauli.

In the following year the Ph.D. student began to labor – as he admitted later – on his thesis problem, a topic emerging from the interest of his advisor in the high-energy phenomena observed in cosmic radiation, i.e., "The passage of fast corpuscular rays through a ferromagnetic substance" [3]. At times pushed or assisted by Heisenberg, von Weizsäcker obtained several useful results for the energy determination of fast charged particles: thus he demonstrated that it was the strength of magnetic induction B (and not of the magnetic field H), which played a role in the energy measurement, and that the stopping power for electrons and protons depended on the electric conductivity of the materials contained in the measuring device. Heisenberg recommended the acceptance of the thesis in his report of 3rd June 1933 with the words: "The work has been formulated in a clear and condensed way; several nice and *anschauliche* considerations exhibit a clear physical understanding of the quantum-mechanical methods", and he graded the work as being "very good" [4].

Also the third publication of the young scholar, composed in early 1934 as a research fellow at the famous Copenhagen institute of Niels Bohr, concerned a question then important for the interpretation of cosmic-ray phenomena, namely the "Radiation emitted from collisions of very fast electrons" [5]. In those days the validity of relativistic quantum electrodynamics for describing processes involving highest energies was seriously doubted; yet von Weizsäcker proved, by presenting an *anschauliche* consideration and applying a suitable Lorentz transformation in the calculation, that normally in the collisions of highly energetic charged particles only *small* energy exchanges occur, hence the available quantum electrodynamics should still be applicable to the problem treated. His procedure was later called the "Weizsäcker–Williams method" [6].

2 The Theory of Nuclear Structure (1934–1937)

In spring of 1934 von Weizsäcker returned to Leipzig as the main assistant at Heisenberg's institute – a position which he kept until spring of 1936. In these years nuclear theory became the central topic at the institute, altough the professor concentrated his interest on the more fundamental problem of nuclear forces and left the detailed description of the constitution of atomic nuclei to his collaborators and students [7]. Weizsäcker concerned himself particularly with the task to calculate the binding energies – or mass defects – of atomic nuclei, which were obtained from mass spectroscopic measurements. In his first paper on the subject, which he submitted in July 1935, he started from a reasonable asumption for Heisenberg's nuclear exchange potential, i.e. $J = \exp(-br)$, with r denoting the radial distance and b a constant, and further took into account two empirical facts: first, the fast increase of the mass defects with rising atomic number (A), and second, the stronger binding of nuclei consisting of even numbers of protons (Z) and neutrons (N). In [8] dealing with the many-body problem of nuclei, he applied the so-called Thomas–Fermi method, and he further added the Coulomb energy in the energy calculation. Thus he obtained an expression for the total energy of a nucleus, that is,

$$E(Z, N) = E_V + E_O + E_Z, \tag{1}$$

the terms of which he visualized easily in the following manner: the term E_V (proportional to $Z + N$) should represent the volume energy, the term E_O (proportional to $(N + Z)^{2/3}$) the surface energy, and the term E_Z (proportional to Z^2) the Coulomb energy. Equation (1) has been called "Bethe–Weizsäcker mass formula", since Hans Bethe and Robert Bacher derived essentially the same expression in a later published report on nuclear physics [9].

Several members of the Leipzig institute applied this description in their further investigations: e.g., the Chinese student Foh-san Wang (who had already assisted von Weizsäcker in checking his calculations) considered nuclei with different numbers of protons and neutrons and extended (1) to these cases [10]. In September 1935, von Weizsäcker gave the main review on the "Structure of atomic nuclei" at the annual conference of the German Physical Society in Stuttgart, where he announced further results obtained by introducing relativistic, spin-dependent forces. These he worked out in his *Habilitation* thesis, which he submitted for publication in July 1936, although the investigations were completed already in January of that year [11]. The author started from the standard fundamental theory of nuclear forces of those days, the "Fermi-field theory" – proposed in January 1934 by Heisenberg when extending Enrico Fermi's description of beta-decay – and assumed, in addition, that protons and neutrons have to be described by the relativistic Dirac equation – which was not generally accepted then because of the observed anomalous magnetic moments of the proton and the neutron. It should be mentioned that the results of Weizsäcker's *Habilitation* thesis and of the simultaneous Bethe–Bacher report [9] coincided largely, except that the American authors displayed more detailed calculations for specific nuclei. Before von Weizsäcker left Leipzig and joined in summer 1936 the *Kaiser-Wilhelm-Institut (KWI) fürPhysik* (directed by Peter Debye, formerly professor for experimental physics at the University of Leipzig), he began to write a book, indeed the first special monograph devoted to the new nuclear theory, which he had enriched by his own investigations. It appeared in print in early 1937 and carried the reputation of the young author as a leading expert of this topic into the scientific world [12].

3 Pioneering the Theory of Nuclear Fusion in Stars (1936–1938)

In Berlin von Weizsäcker immediately turned his interest to a completely new scientific problem, the energy production in stars and its relation to the creation of chemical elements heavier than the original hydrogen. He included the results of these considerations in an extended two-part review, entitled in English translation "On the transmutation of chemical elements in stars" [13]. In Part 1, submitted for publication in January 1937, he started from Arthur S. Eddington's "build-up hypothesis" which he even generalized as: "The temperature in the interior of the stars takes on values that allow the transmutation of the lightest nuclei by the action of hydrogen. The energy liberated in these transmutations (or fusion processes, as we call them today!) accounts for the radiation emitted from the star due to its (surface) temperature." (l.c., p. 178) That is, both the energy production in stars and the creation of heavier elements should be coupled closely, because in the stars together with the radiation energy also the neutrons are produced that can be used for building up the heavier from light nuclei. The author then looked at various examples of thermally excited transmutations of lightest nuclei (proton, neutron, triton), applying the "droplet model" of atomic nuclei proposed recently by Niels Bohr, and he studied particular reaction cycles that would result in a continuous energy production. He arrived at the conclusion that, on the one hand, the temperatures existing in the interior of stars might indeed be high enough to ignite the corresponding transmutation processes (nuclear reactions) – which start from hydrogen as the primary star material – and that, on the other hand, the neutron production resulting in these reactions is also sufficient to build up the observed amounts of heavy elements in stars.

In Part 2, which he submitted in August 1938, von Weizsäcker withdrew from an essential aspect of his previous, strong build-up hypothesis. Especially, he now did not "exclude the possibility that the chemical elements (existing in normal stars) have been created already before the formation of stars by a different process". The reason was the observation that, if one connected the energy production in stars too tightly with the build-up process of heavy nuclei, this would "lead, when on tries to carry out the calculation quantitatively, to a series of hardly surmountable difficulties" (l.c., p. 633). Hence he proposed instead a weaker build-up principle, namely one which does not couple the two processes in *normal* stars but only admits them in very massive ones. At the same time, von Weizsäcker also gave up the reaction cycle, favored in 1937 for the stellar energy production, i.e., the processes leading from the H-nucleus via those of D and Li to He 4, because this cycle required the existence of an intermediate nucleus of mass 5, meanwhile been shown to be an unstable object. He now proposed rather a "carbon cycle" involving the isotope C-14, later called the "Bethe–Weizsäcker cycle", and he concluded: "The energy source of stars would therefore consist first of the decomposition of the elements lighter than carbon and then in the process (i.e., the carbon cycle!) indicated." (l.c., p. 639) However, if in a star the carbon-12 isotope had been destroyed already by other nuclear reactions – hence it was not available anymore for energy production – an oxygen cycle might be used alternatively.

At the same time as von Weizsäcker in Germany considered these problems in cosmology, he received a strong competition from colleagues in the USA, notably by immigrants from Germany and Europe, such as Hans Bethe, George Gamow und Edward Teller. Again it was particularly Bethe, who carried out rather detailed calculations on the carbon cycle in an extended essay on energy production in stars, submitted in September 1938 and published in 1939 [14]. These investigations were also quoted, when he received in 1967 the Nobel prize in physics. In an interview conducted in 1966, Bethe judged quite critically about the earlier work of his competitor in Germany; especially he rejected the strong build-up hypothesis in Part 1 of 1937. Obviously he had forgotten entirely about Part 2, where von Weizsäcker withdrew from this hypothesis. A careful historian should therefore – in spite of admitting the more detailed calculations by Bethe – not forget about the previous, pioneering ideas and results of von Weizsäcker on this fundamental issue of astrophysics: they definitely appeared in print first [15].

4 Nuclear Fission Energy, Turbulence and the Structure of Stellar Systems (1939–1954)

In his lecture course on "The history of nature", delivered in the summer semester 1946 before a general audience at the University of Göttingen, von Weizsäcker presented the recent scientific knowledge about the history of the universe, the earth and life, and about the nature of the human species [16]. Thus he spoke on the origin of nuclear energy and the problems of its application for peaceful and military purposes, or on the recent knowledge, obtained especially with the help of the methods derived from atomic physics, to determine the age and the structures in the universe and on earth. These lectures clearly revealed a shift of his scientific interests during the eight years.

The whole development began in December 1938 with the discovery of nuclear fission by Otto Hahn and Fritz Strassman from the Berlin *KWI für Chemie*. This result had not been anticipated by any expert of nuclear physics, including the great old master Niels Bohr and the young von Weizsäcker; but once the chemists had found the fission process of the uranium nuclei, the physicists were able immediately to explain the new phenomenon. Thus von Weizsäcker used a model of Wilfried Wefelmeier, a colleague at the institute: this model easily accounted for an elongation of the shape of heavy nuclei and their possible breaking up by neutrons into two nuclei of approximately equal mass [17]. As a consequence of the liberation of extra free energetic neutrons in the fission process the technical creation of nuclear energy in a nuclear chain reaction now appeared to be possible, which might be applied both for peaceful and military purposes. When the war broke out in September 1939, von Weizsäcker was soon drafted into the secret German nuclear-energy project of the *Heereswaffenamt* (Army Weapons Office) to explore the chances for nuclear energy. Together with some students he studied the theory of experimental set-ups preparing "uranium machines" (nuclear reactors) devices at his *KWI für Physik* – they consisted of layers of uranium and moderator substances to slow down the fast fission-neutrons for stimulating further fission reactions with the rare uranium isotope U-235. Perhaps his most prominent result within this military project was contained in a theoretical study of July 1940 showing that the frequent uranium isotope U-238 can be transmuted by neutron capture into a transuranium element, a substance having the same excellent fission properties as U-235 – hence, if it were produced in greater amounts (for example, in a functioning reactor), it might be used as material in nuclear explosives [18].

In the fall of 1942 von Weizsäcker left the German nuclear-energy project (which was heading then slowly towards achieving a critical reactor) and went back to occupy an academic position and to work on purely scientific problems. Being appointed associate professor of theoretical physics at the (German) University of Strasbourg, he continued studying fundamental astrophysical problems. So he submitted in August 1943 a detailed paper dealing with the origin of the planetary system [19]. He assumed, in particular, that the planets originated from a flat gaseous cloud, rotating about the sun and consisting of hydrogen and a 1% admixture of heavier elements, while exhibiting turbulent inner motions; in this cloud he then derived the existence of convection currents, "which distinguish certain positions for the formation of planets (rule of Titius and Bode)" (l.c., p. 319). The same convection currents, he concluded, would also explain the senses of the moon's orbital motion and of the rotation of planets. He continued cosmogonic research – in the footsteps of the great Immanuel Kant (one of his favorite philosophers) – in the following decade. Further, he discussed with Werner Heisenberg, during their post-war internment at Farm Hall (1945/46), the fundamental physical problem of turbulence. Like Heisenberg, he published afterwards important papers on the theory of statistical turbulence [20], which he applied then to a number of astrophysical problems, e.g. the motion of cosmic gas clouds or the size of spherical stellar clusters [21].

When Heisenberg after his release to Germany in early 1946 renewed the Berlin *KWI für Physik* (of which he had become a director in 1942) in Göttingen as *Max-Planck-Institut für Physik*, von Weizsäcker became head of its theory divison. There he cooperated closely with Ludwig Biermann, the head of the new astrophyics division within the institute, e.g., on theoretical plasma physics, which assumed an important role in the research on problems of cosmology and further in the attempts (starting in the early 1950s) to perhaps obtain technically enormous amounts of energy from processes of nuclear fusion on earth. Thus von Weizsäcker taught and inspired a generation of German post-war students, who became leaders in stellar and plasma physics, before he finally fulfilled for himself the dream of his youth: in 1957 he left Göttingen and took over the chair philosophy at the University of Hamburg.

References

1. C.F. von Weizsäcker: Erinnerungen. Manuscript 2000, p. 18
2. C.F. von Weizsäcker: Ortsbestimmung enes Elektrons durch ein Mikroskop. *Z. Physik* 70, 114–130 (1931)
3. C.F. von Weizsäcker: Durchgang schneller Korpuskularstrahlemn durch ein Ferromagnetikum. *Annalen der Physik* 17, 869–896 (1933)
4. W. Heisenberg: Gutachten, 2 June 1933. In H. Rechenberg and G. Wiemers (eds.): Werner Heisenberg: *Gutachten- und Prüfungsprotokolle für Promotionen und Habilitationen* (1929–1942). ERS-Verlag, Berlin 2001, p. 69
5. C.F. von Weizsäcker: Ausstrahlung bei Stößen sehr schneller Elektronen. *Z. Physik* 88, 612–625 (1934)
6. See E.J. Williams: Nature of high energy particles of penetrating radiation and status of ionization and radiation formulae. *Physical Review* 45, 729–730 (1934)
7. See, e.g, the historical review of H. Rechenberg: Die Theorie der Atomkerne in Leipzig. In Ch. Kleint and G. Wiemers (eds.): *Werner Heisenberg in Leipzig* (1927–1942). Akademie Verlag, Berlin 1993, pp. 30–52
8. C.F. von Weizsäcker: Zur Theorie der Kernmassen. *Z. Physik* 96, 431–458 (1935)
9. H.A. Bethe and R.F. Bacher: Nuclear physics. A. Stationary states of nuclei. *Reviews of Modern Physics* 8, 81–221 (1936)
10. F.S. Wang: Über die erweiterte Thomas-Fermi-Methode bei Atomkernen. *Z. Physik* 100, 734–741 (1936)
11. C.F. von Weizsäcker: Über die Spinabhängigkeit der Kernkräfte. *Z. Physik* 102, 572–602 (1936). See also Heisenberg's recommendation of 23 January 1936, in Ref. 4, pp. 218–219
12. C.F. von Weizsäcker: *Die Atomkerne. Grundlagen und Anwendungen ihrer Theorie.* AVG, Leipzig 1937. Perhaps it is a pity that the author did not Refer to the most recent experiments showing the charge independence of nuclear forces; however, this result became only substantiated later in 1936 and entered the theoretical description after he had completed his manuscript
13. C.F. von Weizsäcker: Über Elementumwandlungen im Innern der Sterne. I; II. *Physikalische Zeitschrift* 38, 176–191 (1937); 39, 633–646 (1938)

14. H.A. Bethe: Energy production in stars. *Physical Review* 55, 434–456 (1939)
15. For an account of the whole sitaution between 1937 and 1939, see in J. Mehra and H. Rechenberg: *The Historical Development of Quantum Theory*. Vol. 6/2, Springer-Verlag, New York, Berlin, etc. 2001, pp. 991–998
16. C.F. von Weizsäcker: *Die Geschichte der Natur.* S. Hirzel, Zürich 1948
17. C.F. von Weizsäcker: Zum Wefelmeierschen Modell der Transurane. *Naturwissenschaften* 27, 133 (1939)
18. C.F. von Weizsäcker: Eine Möglichkeit der Energiegewinnung aus U238. Geheimbericht, 17. Juli 1940
19. C.F. von Weizsäcker: Über die Entstehung des Planetensystems. *Z. Astrophysik* 22, 319–355 (1943)
20. C.F. von Weizsäcker: Das Spektrum der Turbulenz bei großen Reynoldschen Zahlen. *Z. Physik* 124, 614–627 (1948)
21. C.F. von Weizsäcker: Die obere Grenze der Größe der Spiralnebel. *Naturwissenschaften* 35, 188–189 (1948); Eine Bemerkung über die Gestalt kugelförmiger Sternhaufen. *Z. Astrophysik* 35, 252–254 (1955)

The Origins of Nuclear Physics and Carl Friedrich von Weizsäcker's Semi-Empirical Mass Formula

Karl v. Meyenn

1 Theoretical Physics at the Beginning of the 30's

When Carl Friedrich von Weizsäcker began his studies of physics in the summer of 1929 at the University of Berlin, quantum mechanics had been completed to a large extent. Victor Weisskopf who was four years older describes this unique state of physics in his memoirs:[1] "I was aware that I was living in one of the most exciting periods in the history of science. Quantum mechanics had been conceived and was developed from 1925 to 1927 by Niels Bohr, Werner Heisenberg, Erwin Schrödinger, Max Born, Wolfgang Pauli, P.A.M. Dirac, and others. They paved the way for an understanding of most atomic and molecular phenomena and laid the foundation of our knowledge on the properties of matter."

Most physicists spent the bulk of their time with applications of the newly gained theory to problems of atomic, molecular, or solid state physics, which so far had resisted any treatment by classical physics. Some mourned, however, that they were born too late and thus had no chance to participate in the exciting development of fundamental physics. All the more they investigated the new, at first still incomprehensible phenomena, which resulted from the research on nuclei and cosmic rays which were being carried out on an increasing scale.

The enormous energies, which were transferred in the radioactive decays, had astonished scientists since their discovery. The separation of nuclear processes from what happens in the electron cloud of atoms has permitted an independent development of a closed theory of atoms. This made a closer study of nuclei more difficult at first. In the early editions of his influential treatise *Atombau und Spektrallinien*, Arnold Sommerfeld attempts to make readers aware of this strange world of the nuclei of atoms which is "generally closed from the exterior world",[2] which is only governed "by the law of probability, of the spontaneous uncontrollable chance".[3]

---

[1] [207, p. 25].

[2] In his inaugural address of the Sommerfeld exhibition in the *Deutsches Museum* in Munich, Weizsäcker admitted in December 1984, that he also had eagerly read "when I had started my studies" the "totally disintegrated double volume".

[3] [179, p. 53].

"Only exceptionally a door opens, which leads out of the inner world of an atom into the outer world, the α- or β-rays which are emitted in such processes are messengers from a world which is otherwise closed to us."

Nobody then had any doubt that these particles which leave the nuclei had to be for sure also building blocks of the atomic nuclei. Since 1919, besides α- and β-particles, protons also belonged to this group of particles, which had become artificially liberated by a transformation of the nucleus, induced by irradiation with α-particles.[4] However, while the motion of these α-particles and protons inside the nucleus could be described by non-relativistic quantum theory,[5] this was not possible any more for the light electrons.[6] Their restriction to the small nuclear volume, which required extremely relativistic velocities by the Heisenberg uncertainty relations, referred them to the domain of validity of Dirac's theory of electrons. The paradoxical phenomena associated with this theory, especially the reversal of charge in the case of a penetration of a steep potential barrier, which was first discovered by Oskar Klein, had become clearly apparent with these *nuclear electrons*.

The widely discussed enigma of the continuous β-spectrum has also been attributed to theses *donkey electrons*[7], as well as the mysterious anomalies in the absorption and scattering of short wave γ-rays[8] and the violation of the conservation laws of energy, momentum, spin and statistics which was observed in some nuclear processes.[9] All these problems seemed to point to the necessity of a further revolution of the conceptual frame of physics. "So many spoke about the conjecture," von Weizsäcker explained,[10] "that a step similar to the one from classical mechanics to quantum mechanics, which the theory of the electrons in the atom had made necessary, would again be necessary for the understanding of the atomic nucleus ... Just as in the case of classical mechanics, which has proved its validity for macroscopic objects, did this theory only retain its validity as a limiting case of quantum mechanics. Now quantum mechanics, which was adapted to the electron cloud, seemed again to become dissolved as the limiting case of a more general *nuclear mechanics*."

---

[4] See for example the report by Kasimir Fajans published in *Die Naturwissenschaften* [61].

[5] Protons and α-particles move in the interior of the nucleus with about $1/10$ of the speed of light. Therefore, the forces can be supposed to be static (i.e. non-retarded).

[6] Compare also [182].

[7] In October 1931, Niels Bohr had seen during the first large international meeting on nuclear physics no other way out of these difficulties as to assume a violation of energy conservation in nuclear processes. See in particular also the remarks on the failure of the conservation laws in the *Handbuch der Radiologie* by Guido Beck, Heisenberg's first assistant in Leipzig [9, pp. 446–449].

[8] The effect in question is the so called Meitner–Hupfeld effect, which could not be represented by the otherwise very successful Klein–Nishina formula and which could be explained only by the appearance of pair production processes (compare [42] and [147, pp. XXXVff.]).

[9] With regard to some inconsistencies in the band structure of nitrogen, which were observed by Ralph Kronig, Walter Heitler and Gerhard Herzberg, the term of a *nitrogen anomaly* was used (see for example [147, pp. XXVff.]).

[10] [218, p. 209].

Already, calls for such a "super quantum mechanics" or "hyper wave mechanics" appeared in various papers on nuclear physics of this time.[11]

A report written in 1932 by Heisenberg's former doctoral student Edward Teller[12] on the *state and limitations of the present theories on the structure of nuclei* counts "the questions of nuclear physics to the presently most interesting ones" and which are not to be regarded as hopeless in spite of all difficulties.[13] It was because of these problems in nuclear physics, that the elite of theoretical physics of these days turned to the extension of relativistic quantum and field theory.

2 The Penetration of Quantum Theory into Nuclear Physics

The fact that quantum theory was also the proper theory for the structure and the processes in the interior of atomic nuclei was shown in the successful treatment of α-decay by the Russian phycisist Georg Gamow, who was then 24 years old. In the middle of June 1928, he came to Göttingen together with Vladimir Fock as one of the "three musketeers from Leningrad"[14] to complete his training under the direction of Max Born. Within only a few weeks, he succeeded in explaining the escape of α-particles from the atomic nucleus as a quantum mechanical tunneling phenomenon.[15] This gave justification to the connection between the mean lifetime of α-emitters and the velocity of the particles sent out from such a nucleus which had already been found in 1911 by Hans Geiger and John Michael Nuttall.[16] The new conception rapidly paved its way and aroused high expectations among physicists for obtaining new insights into the structure and processes of nuclei. When shortly after, Ernest Rutherford, James Chadwick and Charles Drummond Ellis published their classic standard treatise on nuclear physics, which provided an excellent overview of the state of knowledge in nuclear physics shortly before the discovery of the

---

[11] For example [112, p. 124, 181].

[12] Teller had obtained his doctorate in March 1930 in Leipzig under Heisenberg and became friend with the young von Weizsäcker. Since 1931 he worked in Göttingen with Arnold Eucken at the *Institut für Physikalische Chemie* (see also the biography of E. Teller by Blumenberg and Panos [26]).

[13] "Stand und Grenzen der heutigen Theorien über den Kernaufbau", [190, p. 547]. This report still largely relies on the concept of nuclear electrons, although the discovery of the neutron is already mentioned.

[14] This was the name of the triumvirate formed there together with Lev Landau and Dimitri Iwanenko (see [97, p. 48]).

[15] See [49].

[16] See also Fritz Houtermans' comprehensive paper [112] in the *Ergebnisse der exakten Naturwissenschaften* and R. Stuewers historical presentation [184] of this discovery which occurred simultaneously by Ronald W. Gurney and Edward U. Condon. – A relation corresponding to the law of Geiger and Nuttall between decay energy and decay probability for the β-decay was found only in 1933 by Bernice W. Sargent (see [171]).

neutron, Gamow's interpretation already belonged to the core of the theory.[17] In the sequel it influenced the experiments with accelerated protons by John Cockcroft and Ernest T.S. Walton in Cambridge, in which they succeeded in 1932 the first artificial transformation of a Lithium nucleus.[18]

A transfer of the same considerations to β-decay turned out to be much more problematic because of the particularities of nuclear electrons, which we have already mentioned.[19] Already Gamow's elder colleague in Leningrad, Jakov Frenkel, had previously tackled with this problem.[20] In an investigation submitted on June 14, 1928, he showed that the "Pauli–Fermi theory of the electron gas", when applied to the inner structure of atomic nuclei, gave a much too big value of the mean kinetic energy of the nuclear electrons.[21] "We thus see", he concluded, "that the electrostatic forces are not capable to contain the over-compressed electron gas."

But the knowledge of the nuclei, which was available to the theoreticians, was still meager. Except for the results on the natural radioactive decays of the heavy elements, some scattering experiments and artificially caused "demolitions"[22] of light nuclei including their energy balance, this knowledge consisted, above all, of the masses of all known isotopes which have been measured spectrographically with great precision, in particular during the 20's by Francis William Aston. Initially, the distribution of "mass defects" which was deduced from these measurements as a function of the entire mass number constituted one of the few facts on which theory building could rely. It gave a mean binding energy per nucleon of about 8 MeV.

---

[17] [170, pp. 328ff.]. This excellent book belonged, together with the book published shortly after by Gamow, to the last expositions of nuclear physics which was based on the concept of nuclear electrons before this concept was made obsolete by the discovery of the neutron.

[18] Already in May 1932, the German nuclear physicists were informed on these exciting experimental results by Rutherford and Chadwick during the Bunsen Conference in Münster (see [57]).

[19] Heisenberg and Pauli have applied their just newly developed "quantum dynamics of wave fields" also to the theory of Gamow, Gurney and Condon of radioactive nuclear decays, as they still hoped to be able to explain the spread of the primary β-spectra by radiative effects. – Opposing this concept, the Hungarian physicist Johann Kudar, who worked in Berlin with Schrödinger, had published numerous investigations on the β-decay on the basis of the concept of nuclear electrons and the violation of the conservation of energy as proposed by Bohr.

[20] When Gamow left Leningrad in the beginning of June 1928 (see [97, pp. 52f.]), he must have been informed on the work of his Leningrad teacher, before he began in Göttingen to tackle the problem of α-decay. Frenkel has participated later with the elaboration of the compound nuclear model.

[21] [80, p. 241]. The kinetic energy became about 1000 times greater than the rest energy m_0c^2 of the electrons.

[22] The term "atomic demolition" (*Atomzertrümmerung*) has been propagated especially by the textbook of Hans Pettersson and Gerhard Kirsch which appeared in 1926, although this book has dealt primarily with processes of the formation of nuclei. For example, at Rutherford's first artificial process of transformation of nuclei in 1919, a high energy α-particle penetrated the nitrogen nucleus N^{14} and transformed it – under liberation of a proton – into an oxygen nucleus O^{17} with additional three units of mass.

In addition, some nuclei made themselves noted by their electromagnetic interaction with the atomic electrons in provoking a *hyper fine structure* of their spectra. In particular, these lines were carefully measured and analyzed by Hermann Schüler and his collaborators at the *Astrophysical Laboratory* of the *Einstein-Foundation* in Potsdam and they provided important insights in the angular momenta and magnetic momenta of nuclei.[23]

3 Gamow's Liquid Drop Model

During a stay in Copenhagen in the winter semester 1928/29, Gamow has designed his *liquid drop model* in connection with his decay theory.[24] We shall take a closer look at it, as it would later become the basis of Weizsäcker's mass formula. The binding energy per nucleon is called the packing fraction or the mass defect. Their contributions had been determined by Aston's measurements of mass for a great number of isotopes and were represented as a function of the mass number A and atomic number Z. For the light nuclei, James Chadwick, Ernest Rutherford and Etienne Samuel Bieler had shown deviations from pure Coulomb scattering. This permitted an estimate of the range of nuclear forces, but it proved to be impossible to determine the precise form of the proton–neutron interaction. In addition, the question if, besides Coulomb repulsion between protons, their should exist a notable proton–proton or neutron–neutron interaction remained without answer (until the summer of 1936).

The experimental results permitted only to conclude with certainty, that – as with the molecules of a liquid drop – an interaction takes place only between immediately neighboring nuclear particles. The relation $R = R_0 A^{1/3}$ between the nuclear radius R and the mass number A (the proportionality constant R_0 is about $2 \cdot 10^{-13}$ cm) proved to be an important sign post in the search for a theoretical model of nuclei.

Rutherford had remarked that protons can be ejected only from nuclei with atomic weights different from $4 N_\alpha$ (N_α stands for the number of α-particles contained in the nucleus). Therefore Gamow supposed "for sure" in his model that the α-particle must be "a very important element of nuclear structure", which "does not lose its individuality also inside the nucleus". Apart from this, he supposed that these α-particles are not assembled in shells as the electrons of the atoms, but behave as in the ground state of a Bose gas.[25] "Such an assembly of particles with rapidly varying attractive forces can be regarded as a *liquid drop*. The surface tension holds this conglomerate of α-particles together and compensates the inner pressure, which is caused by the zero point energy."

---

[23] This field had then been treated theoretically by means of Dirac's electron theory in particular by Enrico Fermi [69] as well as Pauli's former assistants Ralf de Laer Kronig [131] and Hendrik Casimir [44, p. 319–328]. See also Hermann Schüler's address during the *Züricher Vortragswoche* of May 20–24, 1931.

[24] See for this and the following in particular R.H. Stuewer's study [186] on the historical origins of the liquid drop model.

[25] [87, p. 717].

In a further investigation, which Gamow completed in January 1930 with Rutherford in Cambridge as a Rockefeller fellow, he determined the nuclear binding energy which resulted from his model. After taking into account the influence of the Coulomb repulsion between the α-particles, he obtained the desired minimum and the rising behavior of the curve of mass defect at higher mass numbers which explained the instability of the heavy nuclei.[26] But the model gave a qualitatively sensible result only for light nuclei. The nuclear electrons were held responsible also for this failure of the theory.[27]

When Gamow, in the spring of 1931, collected his results in a book, he expressed his growing skepticism towards the nuclear electrons by marking the respective paragraphs with warning stars. This has been the first book which contained the term *nucleus* also in its title and which thus replaced the designation *radioactivity* which until then had been dominant in the literature. Nuclear physics had clearly established itself as a new scientific discipline.[28]

In Zurich as well as in Rome, the first international conferences on nuclear physics took place in the same year of 1931, at which Gamow's theory played a central role, before the great turn of the year 1932 should place nuclear physics on a new foundation.[29]

4 The Neutron and Heisenberg's First Theory of the Nucleus

This development was inaugurated in the beginning of 1932 by the discovery of the neutron.[30] With this discovery, the right of existence of the nuclear electrons disappeared, and all difficulties which had been associated with them as well. The impact of this discovery for the development of the theory of nuclei is perhaps comparable with impact of the discovery of the electron for the development of the theory of atoms. Hendrik Casimir, who by recommendation of his teacher Paul Ehrenfest "from the beginning of June to the beginning of August 1932 had to serve as the personal theoretical slave" in Lise Meitner's institute in Dahlem, "became instantly very excited, ... especially as the spell of neutron fascinates him much (of course)." Because of that generally prevailing euphoria Hans Bethe has called the

---

[26] For N_α α-particles enclosed in the nuclear volume (R = nuclear radius, R_α = radius of an α-particle) we obtain by Heisenberg's uncertainty the mean energy $E_{\text{Vol}} = N_\alpha E_\alpha = -h^2/(4m_\alpha R_\alpha^2) \, N_\alpha^{1/3}$ ($E_\alpha = -h^2/4m_\alpha R^2$, $R = R_\alpha N_\alpha^{1/3}$; by the virial theorem $|V_{\text{pot}}| = 2E_{\text{kin}}$; $E_{\text{kin}} = p^2/2m_\alpha$). In account of the energy of Coulomb repulsion $E_{\text{Coul}} = (2eN_\alpha)^2/R = (4e^2/R_\alpha)N_\alpha^{5/3}$, the total energy $E = E_{\text{Vol}} + E_{\text{Coul}}$ has an expression of the form $-aN_\alpha^{1/3} + bN_\alpha^{5/3}$.

[27] Compare [182].

[28] [89]. Until then, radioactivity had been in the forefront of all books on nuclear physics.

[29] The participants of the symposium held in May 1977 by Roger Stuewer in Minnesota report on the impact of the discovery of the deuteron, the neutron, the positron and of the first artificial transformation of nuclei. See also Brown and Rechenberg [43, pp. 27ff.].

[30] See [46, 47].

now beginning decade of nuclear physics *the happy thirties* – in spite of all political turmoil, which would then increasingly burden not only his own, but also the lives of many other physicists:[31] "The neutron was discovered, and as far as theory was concerned, it was now possible for the first time to think of a rational quantum mechanics of the nucleus."

Werner Heisenberg belonged to the first who created a new basis for a quantum theory of nuclei, based on their composition by neutrons and protons. Although he had not yet made up his mind on the nature of the neutron as an elementary or as a composed particle, he had decided – as he communicated to Bohr on June 20, 1932 – "to devolve all principle difficulties on the neutron, and to do basically quantum mechanics." Nevertheless, the hitherto supposed free electrons in the nucleus should continue to play a role for some time as a mediator of the nuclear force, before they disappeared completely from the theory of nuclei.

In his first paper on nuclei Heisenberg assumed an interaction between proton and neutron caused by an exchange of electrons, which he imagined similar to the bond in the H_2^+-molecular ion which was familiar to him.[32] In October 1935, von Weizsäcker explained in a review during the physics congress in Stuttgart:[33] "The exchange force distinguishes itself from the ordinary force by the fact that it does not leave the state of the interacting particles unchanged, but it is connected with a change of place of the electric charge. This exchange force obeys to laws very similar as the chemical molecular forces; it leads to a saturation of the effect of the force, which means that it does not allow a single proton to bind to an arbitrary large number of neutrons and vice versa and thus leads automatically to a behavior of nuclei which resembles a liquid."

Besides this, Heisenberg assumed a comparatively much weaker van der Waals-like interaction between neutrons; protons should continue to act on each other only through the pure Coulomb force. After he had established a corresponding hamiltonian, using the isospin formalism which was introduced at this occasion (called "ρ-spin"), and had verified its validity with several special cases, he began to discuss the problem of nuclear stability, as further insights into the nature of nuclear forces could be expected mainly from stability problems.

With some modifications, Gamow's liquid drop model could be taken over in the new theory: the place of the "conglomerate", composed by α-particles and some surplus protons and free electrons, was now taken by protons and neutrons. This new model had, in particular, to take into account that the neutrons and protons obeyed Fermi statistics. According to the laws of quantum theory such a collection of identical particles should show a completely different behavior from a corresponding classical system. While the α-*particles* were understood by Gamow as a Bose system,

---

[31] This is the title of one of Bethe's [18] essays.

[32] This significance of the exchange interaction for nuclear physics could have caused Heisenberg in the winter of 1929/30 to have his doctoral student Teller calculate once more the wave functions and energy levels of the hydrogen molecular ion according to wave mechanics.

[33] [211, p. 780]. See also the detailed treatment in von Weizsäcker's book *Die Atomkerne* [215, § 14].

which produced an *inner pressure* by their *zero point energy*, the neutrons and protons were dominated by the Pauli principle and should behave as a completely degenerate Fermi gas, in which all states are occupied up to the Fermi level. Thereby, the constant density of the nuclear matter (up to small variations due to the Coulomb interaction) could be explained. In these considerations, all spin–spin and spin–orbit interactions were left aside for the moment.

In his second paper, which was completed in July 1932 during a stay in Ann Arbor, Heisenberg tried to justify his assumption more thoroughly "that the neutron can be understood as a elementary constituent of the nucleus".[34] In the third and last part of this nuclear trilogy, submitted shortly before Christmas 1932, he eventually calculated according to the method of Thomas and Fermi the total energy resulting from the complete hamiltonian by treating the nucleus "as a gas of free particles ... which obey the laws of Fermi statistics."[35] He obtained the following dependency of the energy of heavy nuclei from the mass number A:

$$E = -aA + bA^{5/3} + c \, . \tag{1}$$

The first term on the right was produced by a proton-neutron exchange force $J(r)$, which rapidly decreases with the distance r, the second by Coulomb interaction among protons.

Heisenberg revised his conception of nuclear forces under the influence of Fermi's student Ettore Majorana, who had visited him early in 1933 in his Leipzig institute. Already on Februar 23, 1933 he could communicate to Bohr, that "Majorana (jr.) has written a very nice paper" on nuclear physics. Instead of arriving at a the constant density of nuclear matter by repulsive forces, what had been the usual way up to this time, Majorana achieved the same result by using a negative exchange force. He chose for $J(r)$ the form which Heisenberg later preferred, too.[36] In addition, Majorana showed by means of the Hartree approximation, which was known from atomic physics that the nucleons of nuclei with many constituents move almost freely inside the nuclei, and are only reflected at the nuclear boundary back to the interior. This was the condition that the Thomas-Fermi method, which had proved so successful for the quantum mechanical treatment of many-body systems, could be applied also to nuclei.[37]

By means of a quantum mechanical argument, Eugene P. Wigner could show how the strong increase of the nuclear bounds from H^3 to He^3 and He^4.[38] He had "supposed for the explanation of this fact that in the deuteron the potential energy of the bound and the kinetic zero point energy are nearly balancing out. If a further deuteron is added, and an α-particle is thus built up, additional attracting forces occur which push the protons and neutrons still a bit closer together."[39] When Eugene Feenberg repeated

---

[34] [102, II, p. 163].
[35] [102, III, p. 591].
[36] See [103, p. 308].
[37] See [216, p. 72ff.].
[38] [235].
[39] Cited according to [211, p. 781].

Wigner's calculations in 1935 by using an exchange force,[40] he could achieve results in correspondence with experience only by n–p, p–p and n–n interactions of approximately equal strength. At first, the significance of this result was not taken seriously, as it was considered as the result of an inappropriate approximation.[41] For the complicated nuclei one still had to rely on the statistical method of Thomas and Fermi and on the procedure of Hartree which, meanwhile, had been improved by Vladimir Fock.[42]

In this way, the particular stability of α-particles could now be understood. Majorana's Ansatz has led, in addition, to spin-independent nuclear forces, while Heisenberg had obtained for parallel or anti-parallel position of particle spins an attractive and respectively a repulsive interaction.[43] Although Majorana's forces were given preference for some time, it was finally shown that they alone are not sufficient for the description of scattering experiments with deuterons.[44]

5 Nuclear Physics During the Solvay Conference of 1933

Already in June of 1932, it has been decided in Brussels to choose nuclear physics as the topic of the next Solvay conference, which was planned for October 1933.[45] Heisenberg and Bohr should present the main speeches on the whole subject of nuclear physics. Heisenberg has completed a first draft already on June 30, 1933. "It contains a short section on principal questions," as he reported to Bohr, "then a somewhat more extensive on Gamow's liquid drop model and its extension through the neutron hypothesis, and finally applications to the curve of mass defects and alike. As soon as I have completed the report, I shall send a copy."

Following his habits, Heisenberg sent his meanwhile finished report on nuclei to Pauli. Pauli had just finished reading the proofs of his well known 'Handbuch' article on wave mechanics and looked forward to new tasks when he came back from a restful summer vacation in southern France. "Although I have not achieved anything in the field of nuclear physics," as he remarked with self-criticism, he had been invited by the Solvay Committee, as well.

In a detailed letter of July 14, Pauli first attempts to find an agreement with Heisenberg on an "order of the facts". By all means, he wanted to retain the conservation of "all discrete quantized quantities" in all nuclear processes. This requirement was

---

[40] Feenberg [63,64] has used for $J(r)$ the Ansatz $A\,e^{-\alpha r^2}$ which made calculations easier, and he calculated the binding energies of the deuteron and the α-particle on the basis of the theories of Wigner and Majorana (he ignored Heisenberg's Ansatz, as it did not lead to a stable α-particle). Comparison with the experimental binding energies of the deuteron and the α-particle showed that the two constants A and α were identical for both theories.

[41] In the beginning of 1936, Heimo Dolch, Heisenberg's doctoral student in Leipzig, tried to disprove Feenberg's result in his dissertation using a variational method.

[42] See [78].

[43] See Jordan's survey reports [118, 120].

[44] See especially the survey report by G. Wentzel [229, p. 7].

[45] See [186].

especially intended for Heisenberg's "favorite idea, which I hated, that the neutron can be decomposed into electron and proton."

"If it is true," he concluded, "that the neutron has spin 1/2 and Fermi statistics, it then follows *that a neutron can never become disintegrated by external fields (or otherwise) into an electron and a proton*; but well in a more complicated way, for example into a proton, an electron and a neutrino." At this occasion, Pauli also doubted the existence of the *"exchange forces between protons and neutrons which are postulated by you and Majorana"*. He therefore did not exclude that these forces could have "a completely different origin than the *exchange*".

In the definitive form of his speech, which was handed out at the beginning of the congress to all of the 41 participants, Heisenberg presented his liquid drop model which was built on the assumption of an exchange force of the Majorana type $J(r) = a e^{-br}$. "By this Ansatz," von Weizsäcker explained later,[46] "the binding energies of the nuclei were estimated by quantum theory and the parameters were then determined in such a way that the theoretical binding energies coincided with the experimental values as precisely as possible."

Pauli tried instead above all by his remarks, which he prepared for the discussion in Brussels, to attract attention to his neutrino and the ensuing consequences for the problem of nuclear forces. Returning from Brussels, Enrico Fermi instantly set down to work out this idea into a theory of β-decay. Heisenberg's former assistant Felix Bloch, who had left Leipzig due to the racial laws, and had gone to Rome working with Fermi, became a witness to these events. On February 10, 1934 he informed Bohr: "Fermi has made now, by the way, a quite sympathetic theory of β-decay, in which he has tried to incorporate quantitatively Pauli's idea of the neutrino. As far as the experiments suffice for a quantitative comparison, the β-spectrum seems to follow quite nicely."[47] But the immediate attempt by Heisenberg to apply Fermi's theory also to an explanation of the nuclear forces gave an interaction which was too small by a factor of 10^{12}!

6 The Discovery of the "American Forces"

But all further attempts to determine the precise form of the nuclear interaction by theoretical means remained at first fruitless. Still in May 1936 in a report published in *Forschungen und Fortschritte*, von Weizsäcker has explained,[48] that "the forces between protons and neutrons, which cause the cohesion of nuclei, are a priori completely unknown. The present task of nuclear theory is therefore to seek a preliminary order the nuclear properties by simple assumptions on these forces, and to obtain from this a stepwise improvement of the 'Ansatz' for the nuclear forces. By this way the result was obtained that the force between proton and neutron is stronger than the force between two protons or two neutrons".

Only two months later, American physicists, Merle A. Tuve, Norman P. Heydenburg and Lawrence R. Hafstad, came to the surprising conclusion that all nuclear

---

[46] [211, p. 780].
[47] See [147, p. XLVII].
[48] [212, p. 172].

particles are subject to a *single* universal law of interaction. By the analysis of their proton-proton scattering experiments, Gregory Breit, Edward U. Condon and Richard D. Present have made the discovery that the forces between proton-neutron and proton-proton are almost of the same strength.[49] "The scattering of protons by protons is primarily due to Coulomb forces," Jordan explained this new matter of facts.[50] "This contribution can be calculated mathematically with precision: essentially Rutherford's famous scattering formula is valid, although with a modification which is due to the fact that the scattering and the scattered particle are identical in this case. For a determination of the ensuing modification it is essential, according to Mott, that the protons are subject to Fermi statistics and have spin 1/2. – If there exists a non-electric *nuclear force* between protons, the Coulomb scattering will be superimposed by an anomaly. This is now revealed by *central* collisions (zero angular momentum) and also then only in the case when a sufficiently high velocity permits a close encounter of both protons in spite of the Coulomb repulsion. Such an anomaly has now effectively been found by Tuve, Heydenburg and Hafstad." To this result "the bold conjecture has been linked that a universal law is valid for all non-electric nuclear forces between the heavy particles, which contains the three cases neutron-proton, proton-proton, neutron-neutron."

While many theoreticians have still tried to find more general field theoretic formalisms for the description of these charge independent "American forces",[51] Carl David Anderson and Seth N. Neddermeyer discovered, in the spring of 1937, a new particle in the cosmic rays. These cosmic ray mesons, which were then termed as *heavy electrons, mesotrons or mesons*, caused still further confusion, as they were for a long time mistakenly regarded as the carriers of the nuclear force field which had already been postulated by Hideki Yukawa in 1935.[52] Only the π-mesons, discovered in 1947 by Cecil F. Powell and his collaborators, enabled a final clarification of the problem of nuclear forces.[53]

7 Carl Friedrich von Weizsäcker's Work in Physics During His Study in Leipzig

As Carl Friedrich von Weizsäcker has reported several times, his approach to physics was motivated by an early aroused interest in philosophical and cosmological problems. As a fourteen year old boy he had the opportunity on February 3, 1927 in

---

[49] During a physics conference held in September 1937 in Bad Kreuznach, the consequences of these new discoveries were extensively discussed in a lecture [113] by Friedrich Hund.

[50] [120, p. 277].

[51] The phenomenological description of the nuclei by static forces should now have become supplemented by a field theoretic description, in which the nuclear interactions are provided by exchange of corresponding field quanta in accordance with quantum theory.

[52] See [43], [113] and [228]. See also the report by von Weizsäcker [222] on the theory of mesons, which he had given in the course of series of lectures on cosmic radiation which was held at the Kaiser Wilhelm-Institut für Physik in Berlin.

[53] See for example [41, p. 139ff.].

Copenhagen to get to know Werner Heisenberg, who had already been famous as the discoverer of quantum mechanics, when Heisenberg visited his parents. This was just the time when Heisenberg developed his ideas on the uncertainty principle.[54] When, shortly after, the Weizsäckers moved from Copenhagen to Berlin, Carl Friedrich visited the *Bismarck-Gymnasium* in Wilmersdorf. Also there in Berlin he seems to have used, on several occasions, the opportunity for meetings with Heisenberg, which drew him ever more into the sacred circle of physics.

Still at the age of seventeen, immediately after graduation, he studied mathematics and physics at the University of Berlin. "In the summer of 1929 I studied in Berlin, winter 1929/30 until winter 1930/31 in Leipzig, summer 1931 in Göttingen, winter 1931/32 until summer 1933 again in Leipzig physics and mathematics, from the onset with a preference for theoretical physics," as is written in his curriculum vitae on the occasion of his doctor's degree.[55] Already at the start of April 1931, he had finished, following a suggestion by Heisenberg, a high level analysis on the "determination of the position of an electron by a microscope" using the quantum electrodynamics of Heisenberg and Pauli. These questions stood in close connection with Bohr's efforts to extend Heisenberg's uncertainty relations to electromagnetic field variables. Bohr had again and again delayed the publication of his own considerations on these questions in spite of Heisenberg's insistence – "excuse me if I play somewhat 'Pauli' " – which were then much discussed in his institute together with Rudolf Peierls and Lev Landau.[56]

As Heisenberg intended to travel to the United States in the summer of 1931, he recommended to his new pupil to spend a semester with Max Born in Göttingen.[57] Back in Leipzig, Weizsäcker now began to prepare his dissertation: he should investigate the problem of the magnetic deviation of charged particles at a passage of charged particles through a ferromagnet, which has been important for the research on cosmic rays, a field which was actively pursued in Leipzig.[58] This investigation, which apparently did not mean much to him, has been accepted by the faculty of philosophy of Leipzig on June 29, 1933 on the basis of a recommendation by Heisenberg and Friedrich Hund.

When Heisenberg visited Bohr in Copenhagen in December 1931 with his newly bred doctoral student returning from a skiing holiday in Norway, Weizsäcker was introduced to this great Dane. Heisenberg wrote in his letter of thanks on February 3, 1932: "The day in Copenhagen on Christmas was very nice, I have learnt a lot and the young Weizsäcker was very happy over the evening with you." An invitation to Bohr

---

[54] Heisenberg's famous paper reached the journal editor on March 23, 1927.

[55] See also the interview held in January 1991 in Starnberg by Konrad Lindner [135].

[56] Bohr published these, but without reference to Weizsäcker's preparatory work, finally in April 1933 together with Léon Rosenfeld (see Rosenfeld's later report in "A journey to Laplacia").

[57] Max Born said on October 13, 1963 in his address on the occasion of the award of the peace prize of the German book trade to von Weizsäcker: "I can not call him my student even if he has once attended a course of mine in Göttingen. I really got to know him only after the War in Edinburgh, where I then was a professor for natural philosophy."

[58] See [208].

followed, to join them, from 10 to 20 March 1932 on the skiing hut close to Kufstein. There "we would, if you came, be four: you, Bloch, Weizsäcker and myself."

At this occasion the discovery of the neutron and its consequences for nuclear physics was probably already in the center of their conversations, which then caused Heisenberg to move in the direction of this subject. Weizsäcker remembers this time: "In May 1932 I sat with Heisenberg in Brotterode in the Thuringian Forest. For the Pentecost holidays, Heisenberg always looked for a region where he couldn't have a hay fever, somewhere high up or on the sea side, in Helgoland. There he had the idea that atomic nuclei should consist of protons and neutrons and that this would give a consistent theory of atomic nuclei . . . With that Heisenberg had entered nuclear physics. And I said to myself about at the same time that, having finished my dissertation which was a little boring, I will enter nuclear physics."[59]

8 A Winter Sojourn in Copenhagen 1933/34: The Discovery of the Weizsäcker–Williams Method

When his doctoral dissertation was finished, von Weizsäcker traveled to Copenhagen in the fall of 1933. Now he too wanted to learn more of Bohr's way of working. From September 14 to 20, Bohr had convened a small meeting of physicists – still before the beginning of Solvay Conference mentioned above. The meeting was, however, overshadowed by the political events in the German Reich and the dismissals of many physicists. Therefore, in a letter to Heisenberg, Bohr had pointed to his great desire "to speak on all pleasant and sad things." To the pleasant things apparently also belonged the message of the Nobel prize, which Heisenberg should now receive retrospectively for 1932. On the other hand, a great insecurity reigned among German physicists how one should behave in face of the political inference of the NS regime. Schrödinger for example, protesting against the dismissal of his assistant Fritz London, had abandoned his position in Berlin. Samuel Goudsmit, who was then touring in Europe, even expressed the expectation "that Heisenberg has sense and courage enough to make a similar move like Schrödinger in protest to what has been done to his teacher and outstanding colleagues."

Even travelling now became more difficult. "The future is totally uncertain," Heisenberg reported to Bohr on June 30. "How my travel plans will come out in the fall, I cannot say yet. I hope I will be allowed to come to Denmark and Belgium, but one cannot predict anything for sure. – Weizsäcker looks forward to study with you in Copenhagen, he has passed his examination with great success." Coming back again to his visit in Copenhagen, Heisenberg wrote on August 31 how "tremendously excited" he was "to learn from you on the positive electrons, to get to know your opinion on their connection with Dirac's holes, etc." But in particular he felt glad "that Dirac and Fermi come to Copenhagen."

Although there are no published reports on this September meeting in Copenhagen, its course can be reconstructed indirectly. On a photography, which had been

---

[59] Weizsäcker's interview with Konrad Lindner [135, p. 127].

taken as usual during the meeting,[60] 47 participants can be seen, among them, next to Bohr, Dirac and Heisenberg also Paul Ehrenfest, Max Delbrück and Lise Meitner in the first row; behind them von Weizsäcker sitting next to Teller, as well as Viktor Weisskopf, Walter Heitler, Hendrik Casimir, Felix Bloch, Rudolf Peierls and the Rockefeller fellow Evan James Williams, who had arrived from Manchester on September 4. Fermi, however, was not present.

Especially Christian Møller and his collaborators in Copenhagen had, during the past years, dealt with the scattering theory of relativistic particles. The curvature of particle trajectories, which were observed by Lise Meitner and Kurt Philipp in the laboratory by artificial excitation of atomic nuclei with neutrons and by Patrick Blackett and Giuseppe Occhialini in cloud chambers which were exposed to cosmic radiation, had now impressively confirmed the existence of the positive electrons which were found by Anderson in 1933.[61] There has been great excitement among theoreticians and a general interest in learning more on the interaction of fast electrons with a radiation field (bremsstrahlung and pair production). The question if quantum electrodynamics can still be employed at energies which are 137 times greater than the rest energy of the electron was particularly disputed.[62] This was contradicted by a formula for electron deceleration, which had been deduced by Hans Bethe still at Sommerfeld's institute,[63] and which could not be brought into agreement with the capacity for penetration of the known particles in the cosmic rays.

As the newly arrived Williams had been intensively occupied during the past years with the problems of deceleration of β-rays, he was asked at this occasion to report on his research.

Apparently von Weizsäcker took also part in the ensuing discussion. The problem of deceleration was familiar to him from his doctoral dissertation, which he had just finished: "When I had enlarged and written down on Bohr's request a lecture by Williams," he explained on a physics conference in 1985 in Munich, "I gave the paper to Bohr for appraisal. We then spoke on it for three hours. Bohr was very tired, almost distracted, and I could for the first time answer to all of his questions. But in the last hour I was driven into the corner, and Bohr said triumphantly and without any malice: 'Now I understand the point. The point is that everything is just the complete inverse of what you have said. That is the point!' ".

Indeed, the clue has been to use as a reference system in this scattering process the system of the fast moving particle, instead of the laboratory system (for example a lead nucleus). In the case of an electron then energies were at most of the order of an electron mass which thus remained within the domain of validity of quantum electrodynamics. The distance of an electron slowed down in lead, as calculated by this method, should be less than one centimeter. But the observed penetration capability of the particles in cosmic rays was much higher. Therefore, according to theory, these

---

[60] [33, p. 224].
[61] Compare the reports by Meitner and Philipp [144], W. Bothe [36] and especially the historical essay of Roqué [164].
[62] Compare [83].
[63] [10].

could not be electrons. This realization has then become an essential contribution to already mentioned discovery of the mesotrons (resp. muons) by Anderson and Neddermeyer.[64]

Weizsäcker elaborated his investigation while still being in Copenhagen and submitted it for publication in February 1934.[65] The American physicist David Jackson has emphasized the significance of this contribution to the understanding of processes of bremsstahlung:[66] "Conventionally, bremsstrahlung is described in the laboratory as the acceleration of a swiftly moving, light charged particle by the Coulomb field of an essentially stationary heavy nucleus, accompanied by the emission of radiation. In the Weizsäcker–Williams method of virtual quanta, one instead views the process in the rest frame of the incident particle. In this frame the heavy nucleus is incident at high speed upon the stationary light particle. The nucleus' Coulomb field appears, because of the Lorentz transformation, very much like a pulse of electromagnetic radiation, with transverse electric and magnetic fields having a broad frequency spectrum with a number of virtual quanta per unit frequency interval inversely proportional to frequency."

Von Weizsäcker could show that also these virtual light quanta of frequency ω' did satisfy the conservation of energy: "There is, in fact, an apparent dilemma, and a neat explanation due to Weizsäcker. The spectrum of virtual quanta of the heavy nucleus contains energies $\hbar\omega$ far above the rest energy of the light particle. Such quanta, when scattered and transformed to the laboratory frame, would have energies on the order of $\gamma\hbar\omega'$, far greater than γmc^2, violating conservation of energy – the largest possible photon energy is $(\gamma - 1)mc^{2\star}$. As Weizsäcker showed in detail, however, this violation does not occur, because of the scattering of quanta with energies $\hbar\omega'$ greater than mc^2 in the incident particle's rest frame is described by the Compton cross section, not the simple Thomson (recoilless) cross section ... All of these features combine to conserve energy back in the laboratory and give all the details of the less transparent calculations."

9 The Origin of the Semi-Empirical Mass Formula

Von Weizsäcker's semi-empirical mass formula has become even more well known than this extremely helpful method for the analysis of inelastic collisions between particles with high energies. A theoretical representation of mass defects as a function of the mass number was initially thought of being useful in the search for a suitable Ansatz for the nuclear forces. The slow increase of the curve with rising mass number could already be satisfactorily explained by Gamow's liquid drop model, but difficulties remained with the strong increase of mass defects with the light nuclei, in which surface effects play a dominant role.

---

[64] Compare [228].
[65] [209]. Compare also [153].
[66] [116, p. 40].
$\star$ $\gamma = (1 - \beta^2)^{(-1/2)}$; $\beta = v/c$

In his Solvay report[67] on the state of nuclear physics Heisenberg had already indicated how the precise form of the nuclear interaction can be determined. Pascual Jordan has tried to explain the difficulty of finding an analytical expression for deriving the expression of the nuclear forces from the mass by an analogous example from atomic physics:[68] "Also the laws governing the electron clouds of atoms would not be easily detectable if the only empirical data at hand would be just the total binding energies of the *ground states* of all atoms (relatively to the completely ionized nucleus)."

Already on January 7, 1934, Pauli communicated to the physicists in Leipzig the exciting news that Fermi had succeeded in "formulating a theory of the β-decay with neutrinos ... Fermi claims that the dependence on energy and lifetime as well as the shape of the velocity spectrum follows quite well." After Heisenberg had "thought a little on Fermi's work", he came to the conclusion on January 18 that "Fermi's matrix elements for the creation of a pair: electron + neutrino ... – in a similar way which leads to the Coulomb force by the possibility for the creation of light quanta with the atomic electrons – give rise in second approximation to a force between neutron and proton." In the calculation it turns "out that an *exchange interaction* of neutron and proton results, which has according to the specific Ansatz ... the form given by Majorana or mine." But the initial excitement gave soon way to a disenchantment. What Heisenberg had first only attributed to a "sloppiness" of his own calculations, turned out to be the fact: the calculated value of the exchange integral was not sufficient – as already mentioned – for the explanation of the nuclear forces.

When Heisenberg prepared, in the spring of 1934, a lecture on nuclear physics for Cambridge, he noted in a letter to Bohr on April 17 "that I am basically very happy with the overall development of nuclear physics. I have the feeling that the old analogy [between atomic physics and nuclear physics], which we have, as I remember, discussed on a walk on the Langelinie,[69] ... contains a very big part of the truth."

It now became part of the duties in Weizsäcker's position as a new assistant to care for the ever growing number of collaborators in the Leipzig institute. Work in Leipzig was predominantly on problems of nuclear physics.[70] "At that time this was pure basic research," he noted in his comments on this work from the outlook of later developments.[71] During this time, Weizsäcker has probably started with the study of the ever growing literature on nuclear physics, as is indicated by numerous book

---

[67] [103]. See also [187].

[68] [118, p. 212].

[69] Von Weizsäcker has reproduced the scheme of this formal analogy, which follows in Heisenberg's letter to Bohr, also in his book *Die Atomkerne* [215, p. 29].

[70] In particular, the Chinese Wang Fo-San belonged to this group, who collaborated with von Weizsäcker during this time, and he also supported him with the numerical analysis of his results (see [210, p. 437] and [201]). Further investigations, which were closely related with Weizsäcker's work in nuclear physics, were those by Siegfried Flügge [75, 76], Heimo Dolch [52], Hans Euler [58, 59] and Helmut Volz [197, 198].

[71] See [226, p. 101].

reviews and review lectures of this time.[72] These activities later became beneficial for his book *Die Atomkerne*, published in 1937.

At this time, the basic references were Heisenberg's essay "Über den Bau der Atomkerne" (On the structure of atomic nuclei) of the year 1932, the so-called Solvay-Report and a speech, which Heisenberg has given in September 1934 in Copenhagen. When the latter was published in the *Zeeman Festschrift*,[73] Casimir termed it "Heisenberg's dream" because of the "vague thoughts" in it. A further investigation by Heisenberg on the structure of light nuclei ("*Die Struktur der leichten Kerne*") was finished in the summer of 1935 at the same time as von Weizsäcker's discussion of the nuclear masses.

A determination of the binding energies from first principles proved to be impossible and the calculation of complicated nuclei by means of various approximation methods has, up to this time, not provided "much more than an arbitrarily compiled heap of formulas for approximation".[74] Therefore, von Weizsäcker decided to determine the nuclear energies first by a semi-empirical procedure. As a starting point of his considerations, he choose Heisenberg's mass formula (1) of the Solvay Report. But in the derivation given there, just the effect of the nuclear surfaces has been left out which is essential for light nuclei.[75]

The energy content of a nucleus can be expressed as a sum of three terms, a volume term E_V proportional to the nuclear volume, a surface energy E_O and a Coulomb energy E_C. The main part is, of course, made up by the volume term which is caused by the strongly attractive action by the nuclear particles. This part is enhanced by the surface, respectively Coulomb energy, which becomes effective especially with small resp. large mass numbers and which thus lead to a minimum in the region of the iron nucleus ($A = 52$).[76] In addition, each such mass formula had to single out the stable nuclei, represented by their values of N and Z, which occur in nature for each given mass number $A = N + Z$.

The part of the Coulomb and volume term had already been estimated by Heisenberg and Majorana by means of the Thomas-Fermi method.[77] In accordance with this method, independent particle states for the neutrons and protons were assumed

---

[72] See, for example, [211] and [212].

[73] See [155, vol. II, p. 353f., 405].

[74] [218, p. 210].

[75] Already in his first presentation of the liquid drop model, Gamow [96, p. 634f.] had taken into account a contribution of the boundary to the binding energy. This idea has then been taken up by Fermi's collaborator Gian Carlo Wick, who has frequently visited Heisenberg's institute in Leipzig at that time (see [2, p. 128]). Wick pointed out in 1934 in a study on the behavior of nuclear matter that the particles at the boundary can only contribute to the energy of the nucleus with about one half of their bindings. The quantitative elaboration of this consideration has then been carried out by von Weizsäcker. See also the presentation by Bethe and Bacher [19, p. 163–168].

[76] The particular stability of the iron nucleus $^{52}\text{Fe}_{26}$ could later be explained in the frame of Wefelmeyer's geometric nuclear models as the tightest spherical packing of an α-particle surrounded by 12 others (see [202, 203] and [218, p. 227]).

[77] See Heisenberg's Solvay-Report [178].

as well as a constant particle density $\rho_N = N/V$, $\rho_Z = Z/V$ over the nuclear volume V. The eigenfunctions can be represented in this approximation by plane waves $\Psi = (1/\sqrt{V})e^{\frac{i}{\hbar}p\cdot r}$. On account of the Pauli principle, all states in such a *nuclear liquid* are then occupied by two particles up to a maximum momentum P. ρ_N, respectively ρ_Z, can be expressed through P by an integration over the phase space

$$\rho = \frac{2}{h^3}\int_0^P V\psi\psi^* \, d\mathbf{p} = \frac{8\pi}{3}\frac{P^3}{h^3} \tag{2}$$

and the kinetic energy becomes (M = mass of the nucleons)

$$E_{\text{kin}} = \frac{2}{h^3}\int_0^P V\frac{\hbar^2}{2M}|\nabla\psi|^2 \, d\mathbf{p} = \frac{4\pi}{5}\frac{P^5}{Mh^3}. \tag{3}$$

Inserting the respective particle densities from (2) gives the mean kinetic energy

$$\bar{E}_{\text{kin}} = \frac{4\pi h^2}{5M}\left(\frac{3}{8\pi}\right)^{5/3}\int (\rho_N^{5/3} + \rho_Z^{5/3}) \, d\mathbf{r}. \tag{4}$$

If an exchange interaction $J(r)$ between proton and neutron is supposed, the corresponding expression for the potential energy is

$$\bar{E}_{\text{pot}} = -\int f\left[\rho_N(r), \rho_Z(r)\right] d\mathbf{r}, \tag{5}$$

where f is supposed to be a symmetrical function in ρ_N and ρ_Z. For the total volume energy one obtains

$$E_V = \int\left[\frac{4\pi h^2}{5M}\left(\frac{3}{8\pi}\right)^{5/3}(\rho_N^{5/3} + \rho_Z^{5/3}) - f\left[\rho_N(r), \rho_Z(r)\right]\right]d\mathbf{r}$$

$$\approx \left[\frac{4\pi h^2}{5M}\left(\frac{3}{8\pi}\right)^{5/3}\left[(Z/V)^{5/3} + (N/V)^{5/3}\right] - f\left[\frac{Z}{V}, \frac{N}{V}\right]\right]V. \tag{6}$$

The Coulomb energy is determined correspondingly[78]

$$E_{\text{Coul}} = -\frac{1}{2}\iint \frac{e^2}{|r - r'|}\rho_Z(r)\rho_Z(r') \, d\mathbf{r} \, d\mathbf{r}'$$

$$\approx -\frac{1}{2}\frac{3}{5}(Ze)^2\left(\frac{3}{4\pi}V\right)^{-1/3} \approx \text{const} \cdot Z^{5/3}, \tag{7}$$

where, in a first approximation, a variation of the particle densities by the Coulomb interaction is not taken into account.

The lighter the nuclei, the more the surface effects become important. These should, according to von Weizsäcker, consist of a *kinetic* and a *potential* part which

---
[78] [103, p. 309f.].

result from a different distribution of energy and density in the surface layer of the nucleus. The latter is caused already by the fact, which had been mentioned, that the binding forces of particles at the surface are only half saturated and thus contribute only by half to the binding energy. Otherwise, a discontinuous or too steep decrease of density at the nuclear rim would cause, by virtue of the uncertainty relations, a prohibitively big kinetic energy.[79] Von Weizsäcker took account of this by an extension of the Thomas-Fermi method, inserting an additional term which balances this steep decrease of the density.

The calculation, using an exchange integral of the Majorana type $J(r) = a\mathrm{e}^{-br}$, and the requirement, that both the mass defect and the nuclear radius of O_8^{16} should be correctly reproduced, gave a "qualitatively correct representation of the dependence of the mass defect on the atomic weight" and "permits to conclude that the relative order of magnitude of ϵ_V, ϵ_O and ϵ_C as well as the width of the surface ($d \approx 0{,}65$ electron radii) corresponds already closely to reality."[80] On the other hand, the values of the exchange integral obtained from a and b did not correspond to Wigner's much more exact calculations which he had found by his quantum theoretical analysis of the He_2^4 nucleus. Von Weizsäcker saw the cause of this discrepancy in an improper use of the Thomas–Fermi method.

By this negative result, the actual purpose of the investigation, to obtain more precise information on the interaction between proton and neutron, has failed for the time being. Yet, von Weizsäcker recognized the possibility "to use the qualitative agreement of our results with experience as a kind of phenomenological presentation of the nuclear masses."

"For facilitating the ordering of the experimental materials", he remarked in explaining the use of such a representation,[81] "and in order that the data present an immediately interpretable form to a future quantitative theory, we now attempt to express the curve of the nuclear energies by an interpolating formula, whose form is theoretically plausible, while a number of constants remain theoretically undetermined and are determined by a best fit to the experience."

By recourse to the then available mass defects, decay energies and criteria for stability, he now constructed the following expression for the total energy:

$$E(Z, N) = \left[-\sqrt{\alpha^2 + \beta^2} + \sqrt{\alpha^2 + \beta^2 \frac{(Z-N)^2}{(Z+N)^2}} \right]$$
$$\times \left[(Z+N-1) - \gamma(Z+N-1)^{2/3} \right]$$
$$+ \frac{3e^2}{r_0(Z+N)^{1/3}} \left(1 - \delta \frac{|Z-N|}{Z+N} \right) \left[\frac{Z^2}{5} - \left(\frac{Z}{2}\right)^{4/3} \right]. \quad (8)$$

---

[79] As a simple consideration shows, the decrease in density can not be on an interval shorter than $d = \hbar/\sqrt{2ME_{\text{kin}}}$. For the investigation of this boundary effect, Bethe [13] has proposed a scattering experiment with slow neutrons.

[80] [210, p. 442f.]. ϵ_V, ϵ_O and ϵ_C are the mass defects per particle corresponding to E_V, E_O and E_C; the calculation was made for the case $N = Z = A/2$.

[81] [210, p. 449].

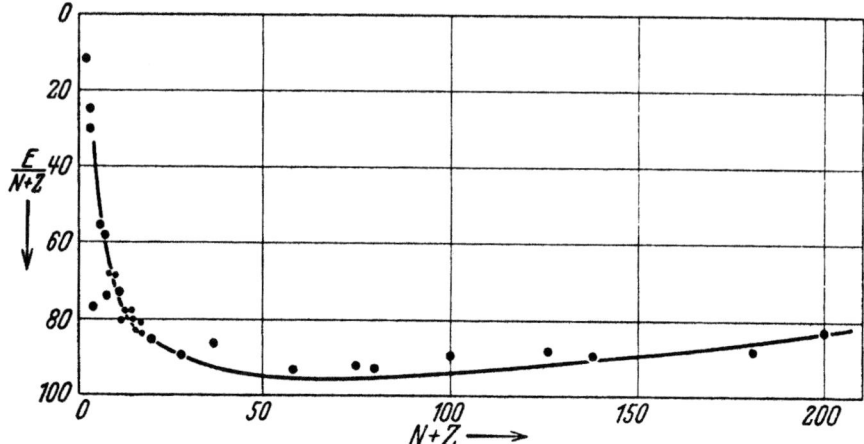

Fig. 1. Package fraction of nuclei ([210])

The constants $\alpha = 2{,}6$, $\beta = 18{,}4$, $\gamma = 1{,}07$, $\delta = 1{,}1$ and $r_0 = 0{,}42$ were determined by comparison with empirical quantities (using two different methods). Figure 1 gives the theoretical curve and the experimental values of the package fraction as a function of mass number. Von Weizsäcker supposed that the "not yet completely satisfactory" agreement with the experience could be due to the phenomenon of shell closure, which had already been indicated by Gamow, K. Guggenheimer and Walter Elsasser.[82]

When shortly afterwards Hans Bethe and Robert Bacher published their well known report on nuclei in the *Reviews of Modern Physics*,[83] they devoted a long section to Weizsäcker's mass formula. They introduced several simplifications which facilitated its use. In this form the formula entered the literature of textbooks on nuclear physics, where it is mostly cited as the Bethe–Weizsäcker formula.[84]

10 As Debye's Assistant at the Kaiser Wilhelm Institute in Berlin

Following his work on the mass formula, von Weizsäcker began to investigate the spin dependence of the neutron-proton interaction which up to then had not been taken into account. By the requirement for relativistic invariance he could show among other things that the exchange forces chosen by Heisenberg and by Majorana are

---

[82] Gamow [92] as well as the two emigrants K. Guggenheimer [99, 100] and W. Elsasser [53–55] in Paris had noticed particularly stable configurations at mass numbers 50, 82 and 126 and have therefore concluded a shell type arrangement of nuclear particles. See [56, p. 187f.].

[83] [19, p. 165–168]. The mass formula was updated then by Bethe [15].

[84] See the remarks by Stuewer [186, p. 97].

only special cases of a more general form of the interaction. At the start of 1936, the result has been submitted as a thesis for habilitation to the Faculty of Philosophy of the University of Leipzig.[85] On May 6, 1936, he reported in his habilitation lecture on the "Aims of new investigations on the atomic nucleus".

This put an end to the apprenticeship in Leipzig. When a new conference of theoreticians convened in Copenhagen in June 1936, at which "almost everybody took part who had a name in theoretical circles", a favorable occasion was presented for many physicists to get information on the embattled market for employment.[86] Pauli too has then looked for a new collaborator, as Weisskopf's position as an assistant drew to an end in the fall. By Heisenberg he had been informed on Weizsäcker's capabilities and he was therefore very interested to have him now as an assistant. But also the director of the just finished new building of the *Kaiser Wilhelm Institute for Physics*, Peter Debye, had to offer an assistantship. Debye had been called to Berlin in order to take the direction of this excellently furnished institute,[87] which should focus on nuclear and low temperature physics. Von Weizsäcker decided for the attractive offer from Berlin.

There was a second applicant for Pauli's assistant position who was fitted with similar attributes. The person in case was the physicist Markus Fierz, born in Basle, who obtained his doctor's degree in January 1936 under Gregor Wentzel at the *University of Zurich* by a thesis on the problem of the artificial transformation of protons into neutrons and who spent the summer semester of 1936 in Leipzig. After Pauli had once more inquired from Heisenberg on the position as an assistant, he made his offer to Fierz during the conference in Copenhagen.[88] That this has not been an easy decision for him is revealed from a letter of October 7, 1936 to Weisskopf: "Debye has caught Weizsäcker", he writes, "now I will try it with Fierz. It isn't for the eternity."

When Weisskopf could then state "with grim enjoyment" that von Weizsäcker had made numerous mistakes in the calculation of magnetic moments of heavy nuclei in his thesis for habilitation, the critical Pauli revised his hitherto so positive judgement: "The measure of sloppiness in Weizsäcker's work exceeds altogether and by far the tolerable measure, and my pain of not having had him as an assistant has been alleviated by this. Up to now I am getting along well with Fierz." Such an incident did, however, not hinder Pauli to continue for many years a cordial relation and exchanges with von Weizsäcker.

11 The Liquid Drop Model and Nuclear Fission

When during a seminar in Copenhagen in the spring of 1935, Bohr was informed by Christian Møller of Fermi's successful transformation of nuclei by neutron bom-

---

[85] [213].

[86] As a consequence of the numerous dismissals of "non-Aryan" scientists there has been a great lack of available positions for theoretical physicists outside of the German Reich.

[87] Compare Debye's description [51] of the equipment of the new institute.

[88] See [71, p. 135].

bardment, he immediately developed his conception of these processes. He imagined that the neutron, when penetrating the nucleus, at first transfers his total energy to the other nuclear particles and forms an intermediate transitory nucleus. As such a state could not be attributed to a single but to all particles of the nucleus it was called a *collective state* or a *compound nucleus*. Only afterwards – under emission of protons, neutrons and/or γ-radiation – the transition to an eventually stable nucleus should be possible.

While the preceding liquid drop model could only account for the equilibrium properties of a nucleus, the model of the compound nucleus was particularly suited to the description of dynamic nuclear processes. Bohr made his model, which was also called the sand bag model, known to the public in February 1936 through the journal *Nature*. Gregory Breit and Eugen Wigner, starting from similar concepts, have then developed a theory of nuclear processes which fitted to this picture and which permitted a detailed description of nuclear reactions.[89]

Bohr now tried together with his collaborator Fritz Kalckar – using Weizsäcker's mass formula – to determine the spectrum of excitations of such a nucleus. Bohr's model has led to a flood of further publications which have then found their echo in the reports on nuclei by Bethe [17] and by Livingston and Bethe [136].[90]

But also in Berlin and Leipzig efforts were made to refine the hitherto existing liquid drop model by accounting for group formation of several nuclear particles. In particular, Heisenberg's new collaborator, Hans Euler, developed a theory for heavy nuclei in which the α-particles originate by a "formation of lumps" as a kind of *molecules* in the *nuclear liquid*.[91]

The liquid drop model should once more prove its power with the explanation of nuclear fission which became known in the beginning of 1939: "It is well known, that very small drops of a liquid," von Weizsäcker explained in an essay in the journal *Forschungen und Fortschritte*,[92] "are less stable due to the surface tension than big ones; therefore the tendency of liquids to unite several small drops to one bigger drop. In the same way, the smallest atomic nuclei are less stable than the bigger ones. However, the atomic nuclei must, because of their electrically positive protons, be compared to electrically charged liquid drops. When the drop becomes too big, the electric repulsion gains predominance and has the tendency to disintegrate again into smaller drops ... While the atomic nuclei were earlier regarded as always spherical, Schüler and Schmidt[93] have shown by experiment and Wefelmeier by theory,[94] that they can assume non-spherical shape in certain cases ... But even a liquid drop must have the tendency to assume an elongated cigar-like form, when it becomes strongly charged electrically. This elongation works in an opposite way to the surface tension which has the tendency to keep the surface as small as possible; on the other hand, the mutually repulsive electric charges, which are contained in the liquid,

---

[89] See especially the contribution by Stuewer [183] to the Bohr-Festschrift.
[90] See also [81].
[91] See [218].
[92] [221, p. 11]. See also the comprehensive historical report by Eduardo Amaldi [1].
[93] [176].
[94] See [219].

become, on average, farther separated through the elongation. Bohr and Wheeler have investigated this situation more closely by numerical calculation. This reveals that an atomic nucleus, which could disintegrate almost by himself, can be transferred from a shape with spherical symmetry by the addition of a small amount of energy into a strongly elongated shape, which then decays spontaneously into two halfs. This amount of energy is provided in the case of nuclear fission by the neutron which serves as a projectile."

At this time, von Weizsäcker had already begun to turn to his favorite topic of astrophysics.[95] The intensive work on nuclear physics has brought him on detours closer to the realization of the dreams of his youth – probably contrary to his own expectation. In his autobiographical notes he remarks: [96] "In the writing of the book [on atomic nuclei] I noticed, that the knowledge of nuclear reactions now permitted to tackle *Eddington's* problem of the energy sources of the stars. I figured out the carbon cycle, which *Bethe* has found at the same time, and which he has worked out more completely. This has led to the question of the history of the development of the stars and there to the inspiration of an enlightening sketch of theory on the origin of the planetary system, which I later recognized as a reappraisal of *Kant's* theory by modern means."

References

1. E. Amaldi: From the discovery of the neutron to the discovery of nuclear fission. *Physics Reports* 111, 1–331 (1984)
2. E. Amaldi: *Gian Carlo Wick during the thirties*. In [7, p. 128–140]
3. F.W. Aston: A new mass-spectrograph and the whole number rule. Bakerian lecture. *Proceedings of the Royal Society* A115, 487–514 (1927)
4. F.W. Aston: *Mass spectra and isotopes*. London, 1933
5. F.W. Aston: Masses of some light atoms determined by a new method. *Nature* 135, 541 (1935)
6. E. Bagge: Beiträge zur Theorie der schweren Atomkerne. I: Zur Frage des Neutronenüberschusses in den schweren Atomkernen; - II: Über die Abflachung des Gamowbergs bei einer Anregung der Atomkerne. *Annalen der Physik* 33, 359–388; 389–403 (1938)
7. G. Battimeli and G. Paoloni (eds.): 20th *century physics: Essays and Recollections. A selection of historical writings by Edoardo Amaldi*, Singapur, 1998
8. G. Beck: Über die Systematik der Isotopen. I und II. *Zeitschrift für Physik* 47, 407–416; 50, 548–554 (1928)
9. G. Beck: Kernbau und Quantenmechanik. In: *Handbuch der Radiologie*, Band IV: *Quantenmechanik der Materie und Strahlung*, Teil 1: *Atome und Elektronen*. Leipzig, 1933, pp. 279–450
10. H.A. Bethe: Zur Theorie des Durchgangs schneller Korpuskularstrahlen durch Materie. *Annalen der Physik* 5, 325–400 (1930)
11. H.A. Bethe: Masses of light atoms from transmutation data. *Physical Review* 47, 633–634 (1935)

---

[95] See [220].
[96] [225, p. 558].

12. H.A. Bethe: The capture and scattering of neutrons. *Physical Review* 47, 640 (1935)
13. H.A. Bethe: Theory of disintegration of nuclei by neutrons. *Physical Review* 47, 747–759 (1935)
14. H.A. Bethe: Ionization power of a neutrino with magnetic moment. *Proceedings of the Cambridge Philosophical Society* 31, 108–115 (1935)
15. H.A. Bethe: An attempt to calculate the number of energy levels of a heavy nucleus. *Physical Review* 50, 332–341 (1936)
16. H.A. Bethe: Nuclear radius and many-body problem. *Physical Review* 50, 977–979 (1936)
17. H.A. Bethe: Nuclear Physics. B. Nuclear dynamics, theoretical. *Reviews of Modern Physics* 9, 69–244 (1937)
18. H.A. Bethe: *The happy thirties.* In: [181, pp. 11–26]
19. H.A. Bethe and R.F. Bacher: Nuclear Physics. A. Stationary states of nuclei. *Reviews of Modern Physics* 8, 82–229 (1936)
20. H.A. Bethe and H. Bethe: Enrico Fermi in Rome, 1931–32. *Physics Today*, Juni 2002, pp. 28–29
21. H.A. Bethe and W. Heitler: On the stopping of fast particles and on the creation of positive electrons. *Proceedings of the Royal Society* A149, 176–183 (1934)
22. H.A. Bethe and R. Peierls: The scattering of neutrons by protons. *Proceedings of the Royal Society* A149, 176–183 (1935)
23. J.M. Blatt and V.F. Weisskopf: *Theoretical nuclear physics.* New York, London, 1952
24. F. Bloch: Bemerkung zur Elektronentheorie des Ferromagnetismus und der elektrischen Leitfähigkeit. *Zeitschrift für Physik* 57, 545–555 (1929)
25. F. Bloch and G. Gamow: On the probability of γ-ray emission. *Physical Review* 50, 260 (1936)
26. S.A. Blumenberg and L.G. Panos: *Edward Teller. Giant of the golden age of physics.* New York, 1990
27. N. Bohr: Chemistry and the quantum theory of atomic constitution. *Journal of the Chemical Society* 1932, pp. 349–384
28. N. Bohr: Properties and constitution of atomic nuclei. *Nature* 138, 695 (1936)
29. N. Bohr: Neutron capture and nuclear constitution. *Nature* 137, 344–348 (1936) – Deutsche Fassung: Neutroneneinfang und Bau der Atomkerne. *Die Naturwissenschaften* 24, 241–245 (1936)
30. N. Bohr: Wirkungsquantum und Atomkern. *Annalen der Physik* 32, 5–19 (1938)
31. N. Bohr: The Rutherford memorial lecture 1958: Reminiscences of the founder of nuclear science and of some developments based on his work. *Proceedings of the Physical Society* 78, 1083–1115 (1961)
32. N. Bohr: *Collected works.* Vol. 9: *Nuclear physics (1929–1952).* Amsterdam, 1986
33. N. Bohr: *Collected works.* Vol. 8: *The penetration of charged particles through matter (1912–1954).* Amsterdam, 1987
34. N. Bohr and F. Kalckar: On the transmutation of atomic nuclei by impact of material particles. I. General theoretical remarks. *Det Kongelige Danske Videnskabernes Selskab. Mathematisk-fysiske Meddelelser* 14 (10), 1–40 (1937)
35. M. Born: Carl Friedrich Freiherr von Weizsäcker, Bindeglied zwischen zwei Kulturen. *Physikalische Blätter* 20, 518–520 (1964)
36. W. Bothe: Das Neutron und das Positron. *Die Naturwissenschaften* 21, 825–831 (1933)
37. G. Breit: Some recent progress in the understanding of atomic nuclei. *Review of Scientific Instruments* 9, 63–74 (1938)
38. G. Breit, E.U. Condon and R.D. Present: Theory of scattering of protons by protons. *Physical Review* 50, 825–845 (1936)

39. G. Breit and E.P. Wigner: Capture of slow neutrons. *Physical Review* 49, 519–531 (1936)
40. E. Bretscher: *Kernphysik*. Vorträge gehalten am Physikalischen Institut der Eidgenössischen Technischen Hochschule Zürich im Sommer 1936 (30. Juni–4. Juli). Berlin, 1936
41. D.M. Brink: *Nuclear forces*. Oxford, 1965. Deutsche Übersetzung: *Kernkräfte*. Einführung und Originaltexte. Braunschweig, 1971
42. L.M. Brown and D.F. Moyer: Lady or tiger? the Meitner–Hupfeld effect and Heisenberg's neutron theory. *American Journal of Physics* 52, 130–136 (1984)
43. L.M. Brown and H. Rechenberg: *Origin of the concept of nuclear forces*. Bristol and Philadelphia, 1996
44. H.B.G. Casimir: *Haphazard reality. Half a century of science*. New York, 1983
45. B. Cassen and E.U. Condon: On nuclear forces. *Physical Review* 50, 846–849 (1936)
46. J. Chadwick: Possible existence of a neutron. *Nature* 129, 312 (1932)
47. J. Chadwick: The existence of a neutron. *Proceedings of the Royal Society* A136, 692–708 (1932)
48. J. Chadwick and E.S. Bieler: Collisions of alpha particles with hydrogen nuclei. *Philosophical Magazine* 42, 923–940 (1921)
49. E.U. Condon: Tunneling – How it all started. *American Journal of Physics* 46, 319–323 (1978) – E. Merzbacher: The early history of quantum tunneling. *Physics Today*, August 2002, 44–49
50. A.S. Dawydow: *Theorie des Atomkerns*. Berlin, 1963
51. P. Debye: Das Kaiser Wilhelm-Institut für Physik. *Die Naturwissenschaften* 25, 257–260 (1937)
52. H. Dolch: Zur Theorie der leichtesten Kerne (Dissertation). *Zeitschrift für Physik* 100, 401–439 (1936)
53. W. Elsasser: Sur le principe de Pauli dans les noyaux. *Journal de physique et le radium* 4, 549–556 (1933)
54. W. Elsasser: Sur le principe de Pauli dans les noyaux. II. *Journal de physique et le radium* 5, 389–397 (1934)
55. W. Elsasser: Sur le principe de Pauli dans les noyaux. III. *Journal de physique et le radium* 5, 636–639 (1934)
56. W. Elsasser: *Memoirs of a physicist in the atomic age*. New York und Bristol, 1978
57. O. Erbacher and K. Philipp: Die Radioaktivität auf der Bunsentagung in Münster. *Die Naturwissenschaften* 20, 586–589 (1932)
58. H. Euler: Über die Wechselwirkung in den schweren Atomkernen. *Die Naturwissenschaften* 25, 201 (1937)
59. H. Euler: Über die Art der Wechselwirkung in den schweren Atomkernen. *Zeitschrift für Physik* 105, 553–575 (1937)
60. R.D. Evans: *The atomic nucleus*. New York, 1955
61. K. Fajans: Die künstliche Zerlegung des Stickstoffatoms. *Die Naturwissenschaften* 7, 729–730 (1919)
62. H. Falkenhagen (ed.): *Quantentheorie und Chemie*. Leipziger Vorträge. Leipzig, 1928
63. E. Feenberg: Neutron-proton interaction. Part I: The binding energies of the hydrogen and helium isotopes. *Physical Review* 47, 850–856 (1935)
64. E. Feenberg: Neutron-proton interaction. Part II: The scattering of neutrons by protons. *Physical Review* 47, 857–859, (1935)
65. E. Feenberg: Semiempirical theory of the nuclear energy surface. *Reviews of Modern Physics* 19, 239–258 (1947)

66. E. Fermi: Eine statistische Methode zur Bestimmung einiger Eigenschaften des Atoms und ihre Anwendung auf die Theorie des periodischen Systems der Elemente. *Zeitschrift für Physik* 48, 73–79 (1928)
67. E. Fermi: Statistische Berechnung der Rydbergkorrektion der s-Terme. *Zeitschrift für Physik* 49, 550–554 (1928)
68. E. Fermi: *Über die Anwendung der statistischen Methode auf die Probleme des Atombaues.* In: [62, pp. 95–111]
69. E. Fermi: Über die magnetischen Momente der Atomkerne. *Zeitschrift für Physik* 60, 320–333 (1930)
70. E. Fermi: Versuch einer Theorie der β-Strahlen. *Zeitschrift für Physik* 88, 161–177 (1934)
71. M. Fierz: Physik in den dreißiger Jahren – ein Rückblick. *Physikalische Blätter* 36, 133–136 (1980)
72. S. Flügge: Ein wellenmechanisches Modell des Neutrons? *Zeitschrift für Physik* 81, 491–495 (1933)
73. S. Flügge: Der Einfluß der Neutronen auf den inneren Aufbau der Sterne. Göttinger Dissertation. *Zeitschrift für Astrophysik* 6, 272–292 (1933)
74. S. Flügge: Gibt es ein Neutron der Masse 2? *Zeitschrift für Physik* 95, 312–318 (1935)
75. S. Flügge: Zum Aufbau der leichten Atomkerne. *Zeitschrift für Physik* 96, 459–472 (1935)
76. S. Flügge: Die Massendefekte der leichtesten Atomkerne auf Grund der neuen Annahmen über die Kräfte. *Zeitschrift für Physik* 105, 522–536 (1937)
77. S. Flügge and A. Krebs: Kernphysik. *Physikalische Zeitschrift* 38, 13–36 (1937)
78. V. Fock: Näherungsmethode zur Lösung des quantenmechanischen Mehrkörperproblems. *Zeitschrift für Physik* 61, 126–148 (1930)
79. A.P. French and P. Kennedy: *Niels Bohr: A centenary volume.* Cambridge, Mass., 1985
80. J. Frenkel: Anwendung der Pauli-Fermischen Elektronengastheorie auf das Problem der Kohäsionskräfte. *Zeitschrift für Physik* 50, 234–248 (1928)
81. F.L. Friedman and V.F. Weisskopf: The compound nucleus. In [154, pp. 134–162]
82. O.R. Frisch, F.A. Paneth, F. Laves und P. Rosbaud (eds.): *Beiträge zur Physik und Chemie des 20. Jahrhunderts.* Lise Meitner, Otto Hahn, Max von Laue zum 80. Geburtstag. Braunschweig, 1959
83. P. Galison: The discovery of the muon and the failed revolution against quantum electrodynamics. *Centaurus* 26, 262–316 (1983)
84. G. Gamow: Zur Quantentheorie des Atomkernes. *Zeitschrift für Physik* 51, 204–212 (1928)
85. G. Gamow: Zur Quantentheorie der Atomzertrümmerung. *Zeitschrift für Physik* 52, 510–515 (1928)
86. G. Gamow: Discussion on the structure of atomic nuclei. *Proceedings of the Royal Society* A123, 386 (1929)
87. G. Gamow: Über die Struktur des Atomkerns. *Physikalische Zeitschrift* 30, 717–720 (1929)
88. G. Gamow: Mass defect curve and nuclear constitution. *Proceedings of the Royal Society* A126, 632–644 (1930)
89. G. Gamow: Quantum theory of nuclear structure. In *Convegno di Fisica Nucleare Ottobre 1931- IX.* Rom, 1932, pp. 65–81
90. G. Gamow: *Constitution of atomic nuclei and radioactivity.* Oxford, 1931. Ins Deutsche übertragen von C. und F. Houtermans: *Der Bau der Atomkerne und die Radioaktivität.* Leipzig, 1932
91. G. Gamow: Teoria quantica della struttura nucleare. *Nuovo Cimento* 9, XXXII–XXXV (1932)

92. G. Gamow: Empirische Stabilitätsgrenzen von Atomkernen. *Zeitschrift für Physik* 89, 592–596 (1934)
93. G. Gamow: Modern ideas of nuclear constitution. *Nature* 133, 744–747 (1934)
94. G. Gamow: Les noyaux atomiques. *Annales de l'Institut Henri Poincaré* 5, 89–114 (1935)
95. G. Gamow: Zusammenfassender Bericht. Kernumwandlungen als Energiequelle der Sterne. *Zeitschrift für Astrophysik* 16, 113–160 (1938)
96. G. Gamow: The energy-producing-reaction in the sun. *Astrophysical Journal* 89, 130–133 (1939)
97. G. Gamow: *My world line*. An informal autobiography. New York, 1970
98. A.E.S. Green: *Nuclear physics*. New York, Toronto, London, 1955
99. K. Guggenheimer: Remarques sur la constution des noyaux atomiques. I. *Journal de physique et le radium* 5, 253–256 (1934)
100. K. Guggenheimer: Remarques sur la constution des noyaux. II. *Journal de physique et le radium* 5, 475–485 (1934)
101. D.R. Hartree: The wave mechanics of an atom with a non-Coulomb central field.I: Theory and methods; II: Some results and discussion; III: Term values and intensities in series in optical spectra. *Proceedings of the Cambridge Philosophical Society* 24, 89–110; 111–133; 426–437 (1928)
102. W. Heisenberg: Über den Bau der Atomkerne. I, II, und III. *Zeitschrift für Physik* 77, 1–11 (1932); 78, 156–164; (1932), 80, 587–596 (1933)
103. W. Heisenberg: Considérations théoriques générales sur la structure du noyau. In [178, pp. 289–344]
104. W. Heisenberg: Die Struktur der leichten Atomkerne. *Zeitschrift für Physik* 96, 473–484 (1935)
105. W. Heisenberg: Bemerkungen zur Theorie des Atomkerns. In *Pieter Zeeman. 1865–25. Mai 1935*, pp. 108–116
106. W. Heisenberg: *Die Physik der Atomkerne*. 8 lectures given in spring of 1942 at the Technische Hochschule Charlottenburg. Braunschweig, [1]1943, [2]1947
107. W. Heisenberg (ed.): *Kosmische Strahlung*. Berlin, 1943
108. W. Heisenberg: C.F. von Weizsäcker zum 60. Geburtstag (28. Juni 1972). *Physikalische Blätter* 28, 319–321 (1972)
109. W. Heitler: Über die bei sehr schnellen Stößen emittierte Strahlung. *Zeitschrift für Physik* 84, 145–167 (1933)
110. W. Heitler and F. Sauter: Stopping of fast particles with emission of radiation and the birth of positive electrons. *Nature* 132, 892 (1933)
111. G. Hermann: Die naturphilosophischen Grundlagen der Quantenmechanik. *Die Naturwissenschaften* 23, 718–721 (1935)
112. F.G. Houtermans: Neuere Arbeiten über die Quantentheorie des Atomkerns. *Ergebnisse der exakten Naturwissenschaften* 9, 123–221 (1930)
113. F. Hund: Theoretische Erforschung der Kernkräfte. *Physikalische Zeitschrift* 38, 929–935 (1937)
114. F. Hund: Symmetrieeigenschaften der Kräfte in Atomkernen und Folgen für deren Zustände. *Zeitschrift für Physik* 105, 202–228 (1937)
115. D. Iwanenko: Interaction of neutrons and protons. *Nature* 133, 981–982 (1934)
116. J.D. Jackson: The impact of special relativity on theoretical physics. *Physics Today*, Mai 1987, 34–42
117. J.H.D. Jensen: Über den Austausch im Thomas-Fermi-Atom. *Zeitschrift für Physik* 89, 713–719 (1934) (1963) Zur Geschichte der Theorie des Atomkerns. Nobel-Vortrag, gehalten am 12. Dezember 1963 in Stockholm. In *Les Prix Nobel en 1963*. Stockholm, 1964, pp. 153–164

118. P. Jordan: Fortschritte der Theorie der Atomkerne. *Die Naturwissenschaften* 24, 209–216 (1936)
119. P. Jordan: *Anschauliche Quantentheorie*. Berlin, 1936
120. P. Jordan: Kernkräfte. *Die Naturwissenschaften* 25, 273–279 (1937)
121. P. Jordan: Fortschritte der Theorie der Atomkerne. *Ergebnisse der exakten Naturwissenschaften* 16, 47–103 (1937)
122. H. Kallmann: *Einführung in die Kernphysik*. Leipzig und Wien, 1938
123. H. Kallmann and H. Schüler: Hyperfeinstruktur und Atomkern. *Ergebnisse der exakten Naturwissenschaften* 11, 134–175 (1932)
124. P. Kirchberger: Amerikanische Forschungen über den Kern des Atoms. *Die Umschau* 40, 586–588 (1936)
125. G. Kirsch: Atomzertrümmerung. *Ergebnisse der exakten Naturwissenschaften* 5, 165–191 (1926)
126. G. Kirsch and E. Teller: Vom Bau der Atomkerne. In *Müller-Pouillets Lehrbuch der Physik*, 4. Band, 3. Teil: *Elektrische Eigenschaften und Wirkungen der Elementarteilchen der Materie*. Braunschweig [11] 1933, pp. 541–578
127. Ch. Kleint and G. Wiemers (eds.): *Werner Heisenberg in Leipzig 1927–1942*. Berlin, 1993
128. E.J. Konopinski and G.E. Uhlenbeck: On the Fermi theory of β-radioactivity. *Physical Review* 48, 7–12 (1935)
129. E.J. Konopinski and G.E. Uhlenbeck: Higher order derivatives in the interaction "Ansatz" of the Fermi theory. *Physical Review* 48, 107–108 (1935)
130. A.J. Kox and D.M. Siegel (eds.): *No truth except in the details*. Amsterdam, 1995
131. R. de L. Kronig and S. Frisch: Kernmomente. *Physikalische Zeitschrift* 32, 457–472 (1931)
132. A. Landé: Neutrons in the nucleus. I. *Physical Review* 43, 620–626 (1933)
133. A. Landé: Neutrons in the nucleus. II. *Physical Review* 43, 624–626 (1933)
134. W. Lenz: Über die Anwendbarkeit der statistischen Methode auf Ionengitter. *Zeitschrift für Physik* 77, 713–721 (1932)
135. K. Lindner: *Es fragt Konrad Lindner: Carl Friedrich von Weizsäcker über sein Studium in Leipzig*. In [127, pp. 123–135]
136. M.S. Livingston and H.A. Bethe: Nuclear physics. C. Nuclear dynamics, experimental. *Reviews of Modern Physics* 9, 245–390 (1937)
137. E. Majorana: Über die Kerntheorie. *Zeitschrift für Physik* 82, 137–145 (1933)
138. J. Mattauch: Erkenntnisgewinn in der Kernphysik durch Fortschritte der Apparaturen und Methoden. *Die Naturwissenschaften* 27, 185–195; 201–205 (1939)
139. J. Mattauch: *Kernphysikalische Tabellen*. Mit einer Einführung in die Kernphysik von S. Flügge. Berlin, 1942
140. L. Meitner: Über die von I. Curie und F. Joliot entdeckte künstliche Radioaktivität. *Die Naturwissenschaften* 22, 172–174 (1934)
141. L. Meitner: Atomkern und periodisches System der Elemente. *Die Naturwissenschaften* 22, 733–739 (1934)
142. L. Meitner and M. Delbrück: *Der Aufbau der Atomkerne. Natürliche und künstliche Kernumwandlungen*. Berlin 1935
143. L. Meitner and O.R. Frisch: Disintegration of uranium by neutrons. A new type of nuclear reaction. *Nature* 143, 239–240 (1939)
144. L. Meitner and K. Philipp: Die bei Neutronenanregung auftretenden Teilchenbahnen. *Die Naturwissenschaften* 21, 286–287 (1933)
145. K. v. Meyenn: Theoretische Physik in den dreißiger Jahren. Die Entwicklung einer Wissenschaft unter ideologischen Zwangsbedingungen. *Gesnerus* 39, 417–435 (1982)

146. K. v. Meyenn: Physics in the making in Pauli's Zürich. In [172, pp. 93–130]
147. K. v. Meyenn: Die Vor- und Frühgeschichte des Neutrinos. In [155, Band IV/3, S. VII-LXV]
148. K. v. Meyenn, K. Stoltzenburg and R.U. Sexl: *Niels Bohr 1885–1962. Der Kopenhagener Geist in der Physik.* Braunschweig, Wiesbaden, 1985
149. St. Meyer: Bemerkung über Atomgewichte und Packungseffekte. *Die Naturwissenschaften* 15, 623–625 (1927)
150. A.I. Miller: Werner Heisenberg and the beginning of nuclear physics. *Physics Today* 38 (11), 60–68 (1985)
151. N.F. Mott: Wellenmechanik und Kernphysik. In *Handbuch der Physik*. Band 24. Berlin, 1933, pp. 785–841
152. S.A. Moszkowski: Models of nuclear structure. In *Handbuch der Physik*, Band 39: *Bau der Atomkerne*. Berlin, Göttingen, Heidelberg, 1957, p. 411–550
153. H. Olsen and H. Wergeland: Bremsstrahlung. In [82, pp. 66–73]
154. W. Pauli (ed.): *Niels Bohr and the development of physics*. Essays dedicated to Niels Bohr on the occasion of hid seventieth birthday. London, 1955
155. W. Pauli: *Wolfgang Pauli. Wissenschaftlicher Briefwechsel mit Bohr, Einstein, Heisenberg u.a.* Band I, II, III und IV/1, 2, 3. Berlin, Heidelberg, New York, 1979, 1985, 1993, 1996, 1999 und 2001
156. R. Peierls: Quantum theory of atomic nuclei. *Report of Progress in Physics* 2, 27–38 (1935)
157. R. Peierls: The development of our ideas on the nuclear forces. In [181, pp. 183–202]
158. R. Peierls: Introduction. In Niels Bohr, *Collected Works*. Vol. 9: *Nuclear Physics (1929–1952)*. Amsterdam, 1986, pp. 3–83
159. H. Pettersson and G. Kirsch: *Atomzertrümmerung. Verwandlung der Elemente durch Bestrahlung mit α-Teilchen*. Leipzig, 1926
160. F. Rasetti: *Elements of nuclear physics*. London, Glasgow and Bombay, 1937
161. H. Rechenberg: Die Theorie der Atomkerne in Leipzig (1930–1942). In [127, pp. 30–52]
162. F. Reines (ed.): *Cosmology, fusion and other matters*. George Gamow Memorial Volume. Colorado, 1972
163. W. Riezler: *Einführung in die Kernphysik*. Leipzig, 1937
164. X. Roqué: The manufacture of the positron. Studies in History and Philosophy of Modern Physics. 28 (1997) p. 73
165. L. Rosenfeld: Nuclear reminiscences. In [162, pp. 289–299]
166. E. Rutherford: Discussion on the structure of atomic nuclei. *Proceedings of the Royal Society* A123, 373–382 (1929)
167. E. Rutherford: Discussion on the structure of atomic nuclei. *Proceedings of the Royal Society* A136, 735–762 (1932)
168. E. Rutherford: Radioaktivität und Atomtheorie. *Die Naturwissenschaften* 24, 673–680 (1936)
169. E. Rutherford and J. Chadwick: Scattering of alpha particles by helium. *Philosophical Magazine* 4, 605–620 (1927)
170. E. Rutherford, J. Chadwick and Ch.D. Ellis: *Radiations from radioactive substances*. Cambridge 1930
171. B.W. Sargent: Nuclear physics in Canada in the 1930s. In [177, pp. 221–240]
172. A. Sarlemijn and M.J. Sparnaay (eds.): *Physics in the making. Essays on the developments in 20th Century Physics in Honour of H.B.G. Casimir on the occasion of his 80th Birthday.* Amsterdam, 1989
173. E. Scheibe and G. Süssmann (eds.): *Einheit und Vielheit*. Festschrift für Carl Friedrich von Weizsäcker zum 60. Geburtstag. Göttingen, 1973

174. Th. Schmidt: Über die magnetischen Momente der Atomkerne. *Zeitschrift für Physik* 106, 358–361 (1937)
175. H. Schüler: Hyperfeinstrukturen und Kernmomente. *Physikalische Zeitschrift* 32, 667–670 (1931)
176. H. Schüler and Th. Schmidt: Bemerkungen zu den elektrischen Quadrupolmomenten einiger Atomkerne und dem magnetischen Moment des Protons. *Zeitschrift für Physik* 98, 430–436 (1936)
177. W.R. Shea (ed.): *Otto Hahn and the rise of nuclear physics*. Dordrecht, 1983
178. *Solvay-Report 1933: Structure et propriétés des noyaux atomiques*. Rapports et discussions du septième conseil de physique tenu à Bruxelles, du 22 au 29 octobre 1933 sous les auspices de l'Institut International de Physique Solvay. Paris, 1934
179. A. Sommerfeld: *Atombau und Spektrallinien*. Braunschweig, ^{2}1921
180. A. Sommerfeld: Über den Packungseffekt. In *Convegno di Fisica Nucleare Ottobre 1931- IX*. Rome, 1932, pp. 155–157
181. R.H. Stuewer (ed.): *Nuclear physics in retrospect: Proceedings of a Symposium on the 1930s*. Minneapolis, 1979
182. R.H. Stuewer: *The nuclear electron hypothesis*. In [177, pp. 19–67]
183. R.H. Stuewer: *Niels Bohr and nuclear physics*. In [79, pp. 197–220]
184. R.H. Stuewer: *Gamow's theory of α-decay*. In [196, pp. 147–186]
185. R.H. Stuewer: Mass-energy and the neutron in the early thirties. *Science in Context* 6, 195–238 (1993)
186. R.H. Stuewer: The origins of the liquid drop model and the interpretation of nuclear fission. *Perspectives in Science* 2, 76–129 (1994)
187. R.H. Stuewer: The seventh Solvay Conference: Nuclear physics at the crossroads. In [130, pp. 333–362]
188. L. Tamm: Exchange forces between neutrons and protons, and Fermi's theory. *Nature* 133, 981 (1934)
189. E. Teller: Das Wasserstoffmolekülion. *Zeitschrift für Physik* 61, 458–480 (1930)
190. E. Teller: Stand und Grenzen der heutigen Theorien über den Kernaufbau. In *Müller-Pouillets Lehrbuch der Physik*. Band 4, 3. Teil, pp. 546–553. Braunschweig, 1933
191. E. Teller: *Erinnerungen und Bilder: Heisenberg in Leipzig*. In [127, 1993, pp. 108–109]
192. E. Teller and J.A. Wheeler: On the rotation of the atomic nucleus. *Physical Review* 53, 778–789 (1938)
193. L.H. Thomas: The calculation of atomic fields. *Proceedings of the Cambridge Philosophical Society* 23, 542–548 (1927)
194. M.A. Tuve: The forces which govern the atomic nucleus. *Scientific Monthly* 47, 344–363 (1938)
195. M.A. Tuve, N.P. Heydenburg and L.R. Hafstad: The scattering of protons by protons. *Physical Review* 50, 806–825 (1936)
196. E. Ullmann-Margalit (ed.): *The kaleidoskope of science*. Dordrecht, 1986
197. H. Volz: Über die Größe der Kernkräfte. *Zeitschrift für Physik* 105, 537–552 (1937)
198. H. Volz: Über die Größe der neuen Kernkräfte. *Die Naturwissenschaften* 25, 200–201 (1937)
199. H. Volz: Zusammenhang der neuen Kräfte mit dem Massendefekt der Kerne. *Verhandlungen der Deutschen Physikalischen Gesellschaft* 18, 38 (1937)
200. Wang Foh-san: Über die erweiterte Thomas-Fermi-Methode bei Atomkernen. *Zeitschrift für Physik* 100, 734–741 (1936)
201. Wang Foh-san: *Einige Erinnerungen an meine Studienzeit in Deutschland*. In [127, pp. 110–117]

202. W. Wefelmeier: Ein geometrisches Modell des Atomkerns. *Die Naturwissenschaften* 25, 525 (1937)
203. W. Wefelmeier: Ein geometrisches Modell des Atomkerns. *Zeitschrift für Physik* 107, 332–346 (1935)
204. M. Wein: *Die Weizsäckers. Geschichte einer Familie.* Stuttgart, 1988
205. Ch. Weiner and E. Hart (eds.): *Exploring the history of nuclear physics.* New York, 1972
206. V.F. Weisskopf: Statistics and nuclear reactions. *Physical Review* 52, 295–303 (1937)
207. V.F. Weisskopf: *The joy of insight. Passions of a physicist.* Basic Books, 1991
208. C.F. von Weizsäcker: Durchgang schneller Korpuskularstrahlen durch ein Ferromagnetikum. *Annalen der Physik* 17, 869–896 (1933). Dissertationsschrift vom 29. Juni 1933, mit beigefügtem Lebenslauf
209. C.F. von Weizsäcker: Ausstrahlung bei Stößen sehr schneller Elektronen. *Zeitschrift für Physik* 88, 612–625 (1934)
210. C.F. von Weizsäcker: Zur Theorie der Kernmassen. *Zeitschrift für Physik* 96, 431–458 (1935)
211. C.F. von Weizsäcker: Die für den Bau der Atomkerne maßgebenden Kräfte. *Zeitschrift für technische Physik* 16, 385–391 (1935)
212. C.F. von Weizsäcker: Fortschritte der Theorie des Atomkerns. *Forschungen und Fortschritte* 12, 171–172 (1936)
213. C.F. von Weizsäcker: Über die Spinabhängigkeit der Kernkräfte. *Zeitschrift für Physik* 102, 572–602 (1936)
214. C.F. von Weizsäcker: Metastabile Zustände der Atomkerne. *Die Naturwissenschaften* 24, 813–814 (1936)
215. C.F. von Weizsäcker: *Die Atomkerne: Grundlagen und Anwendungen ihrer Theorie.* Leipzig, 1937
216. C.F. von Weizsäcker: Über die Möglichkeit eines dualen β-Zerfalls von Kalium. *Physikalische Zeitschrift* 38, 623–624 (1937)
217. C.F. von Weizsäcker: Besprechung von Kallmann 122. *Die Naturwissenschaften* 26, 201 (1938)
218. C.F. von Weizsäcker: Neuere Modellvorstellungen über den Bau der Atomkerne. *Die Naturwissenschaften* 26, 209–217; 225–230 (1938)
219. C.F. von Weizsäcker: Zum Wefelmeierschen Modell der Transurane. *Die Naturwissenschaften* 27, 133 (1939)
220. C.F. von Weizsäcker: Umwandlung chemischer Elemente im Inneren der Sterne. *Forschungen und Fortschritte* 15, 159 (1939)
221. C.F. von Weizsäcker: Die theoretische Deutung der Spaltung von Atomkernen. *Forschungen und Fortschritte* 17, 10–11 (1941)
222. C.F. von Weizsäcker: *Theorie des Mesons.* In [107, pp. 90–110]
223. C.F. von Weizsäcker: *Atomenergie und Atomzeitalter: Zwölf Vorlesungen.* Frankfurt a. M., 1957
224. C.F. von Weizsäcker: Eine Anwendung der Theorie der Kristallgitter auf Fragen der Bevölkerungsstatistik. *Physikalische Blätter* 18, 553 (1962)
225. C.F. von Weizsäcker: *Selbstdarstellung.* In [226, pp. 553–597]
226. C.F. von Weizsäcker: *Der Garten des Menschlichen. Beiträge zur geschichtlichen Anthropologie.* München und Wien, 1977
227. C.F. von Weizsäcker: *Reminiscence from 1932.* In [79, pp. 183–190]
228. G. Wentzel: Schwere Elektronen und Theorien der Kernvorgänge. *Die Naturwissenschaften* 26, 273–279 (1938)
229. G. Wentzel: Probleme der Kraftwirkungen im Atomkern. In: *Neue Wege exakter Naturerkenntnis.* Fünf Wiener Vorträge. Vierter Zyklus. Wien, 1939

230. J.A. Wheeler: Niels Bohr and nuclear physics. *Physics Today*, October 1963, 36–45. Also contained in [148, pp. 226–246]
231. J.A. Wheeler: *Some men and moments in nuclear physics*. In [181, pp. 217–322]
232. G.C. Wick: Über die Wechselwirkung zwischen Neutronen und Protonen. *Zeitschrift für Physik* 84, 799–800 (1933)
233. G.C. Wick: Sulla proprietà della materia nucleare. *Il Nuovo Cimento* 11, 227–234 (1934)
234. E.P. Wigner: Über die Streuung von Neutronen an Protonen. *Zeitschrift für Physik* 83, 253–258 (1933)
235. E.P. Wigner: On the mass defect of helium. *Physical Review* 43, 252–257 (1933)
236. E.J. Williams: Application of the method of impact parameter in collisions. *Proceedings of the Royal Society* A139, 163–186 (1933)
237. E.J. Williams: General survey of the theory and experiment for high energy electrons. In [40, pp. 139–141]

Thermonuclear Processes in Stars and Stellar Neutrinos

Georg Wolschin

1 Introduction

In the 1920s Eddington formulated the hypothesis that fusion reactions between light elements are the energy source of the stars – a proposition that may be considered as the birth of the field of nuclear astrophysics [1]. It was accompanied by his pioneering work on stellar structure and radiative transfer, the relation between stellar mass and luminosity, and many other astrophysical topics. Atkinson and Houtermans [3] showed in more detail in 1929 – after Gamow [2] had proposed the tunnel effect – that thermonuclear reactions can indeed provide the energy source of the stars: they calculated the probability for a nuclear reaction in a gas with a Maxwellian velocity distribution. In particular, they considered the penetration probability of protons through the Coulomb barrier into light nuclei at stellar temperatures of $4 \cdot 10^7$ K. From the high penetration probabilities for the lightest elements they concluded that the build-up of alpha-particles by sequential fusion of protons could provide the energy source of stars. An improved formula was provided by Gamow and Teller [4].

Hence, hydrogen and helium (which were later – in the 1950s – identified as the dominant remnants of the big bang) form the basis for the synthesis of heavier elements in stars – but details of the delicate chain reactions that mediate these processes remained unknown until 1938. This is in spite of the fact that rather precise models of the late stages of stellar evolution existed or were soon developed. At that time, white dwarfs were generally considered to be the endpoints of stellar evolution, but in the 1930s Oppenheimer, Volkoff and Chandrasekhar [5] added neutron stars and black holes for sufficiently massive progenitors as final stages. Although this idea was strongly rejected by Eddington, it proved to be true when the first rotating neutron star (pulsar) was detected in 1967 by Bell and Hewish.

Probably the most important breakthrough regarding the recognition of fusion cycles occured in 1937/8 when Weizsäcker [6,7] and Bethe [9] found the CNO-cycle – which was later named after their discoverers Bethe-Weizsäcker-cycle (Fig. 1) – in completely independent works, and Bethe and Critchfield [8] first outlined the proton-proton chain (Fig. 2). After a brief review of thermonuclear reactions in Sect. 2, these nucleosynthesis mechanisms are reconsidered in Sect. 3.

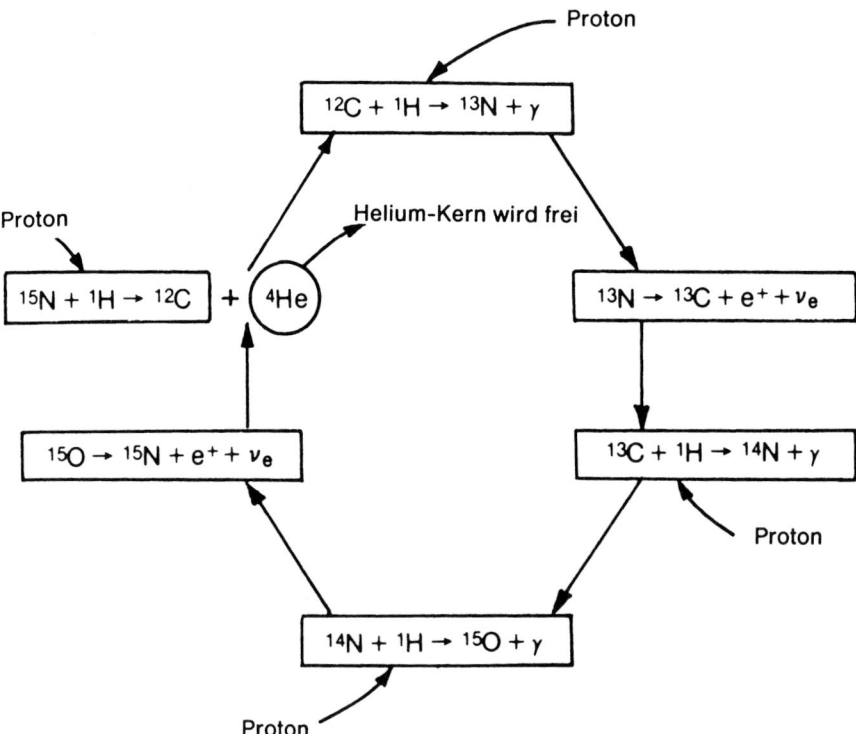

Fig. 1. At temperatures above 20 Million Kelvin corresponding to stars of more than 1.5 solar masses the Bethe–Weizsäcker-cycle dominates the proton-proton chain because its reaction rate rises faster with temperature. This CNO-cycle was first proposed by Weizsäcker [7] and Bethe [9]. Here, carbon, oxygen and nitrogen act as catalysts. From: H. Karttunen et al., Fundamental Astronomy. Springer (1987)

After the war, stellar nucleosynthesis was studied further by Fermi, Teller, Gamow, Peierls and others, but it turned out to be difficult to understand the formation of elements heavier than lithium-7 because there are no stable nuclei with mass numbers 5 or 8. In 1946 Hoyle interpreted the iron-56 peak in the relative abundances of heavier elements vs. mass (Fig. 3) as being due to an equilibrium process inside stars at a temperature of $3 \cdot 10^9$ K. Later Salpeter showed that three helium nuclei could form carbon-12 in stars, but the process appeared to be extremely unlikely. To produce the observed abundances, Hoyle predicted an energy level at about 7 MeV excitation energy in carbon-12, which was indeed discovered experimentally, generating considerable excitement and progress in the world of astrophysics. Burbidge, Burbidge, Fowler and Hoyle then systematically worked out the nuclear reactions inside stars that are the basis of the observed abundances and summarized the field in 1957 [10].

The role of stellar neutrinos was considered by Bethe in [9]. Neutrinos had already been postulated by Pauli in 1930 to interpret the continuous beta-decay spectra,

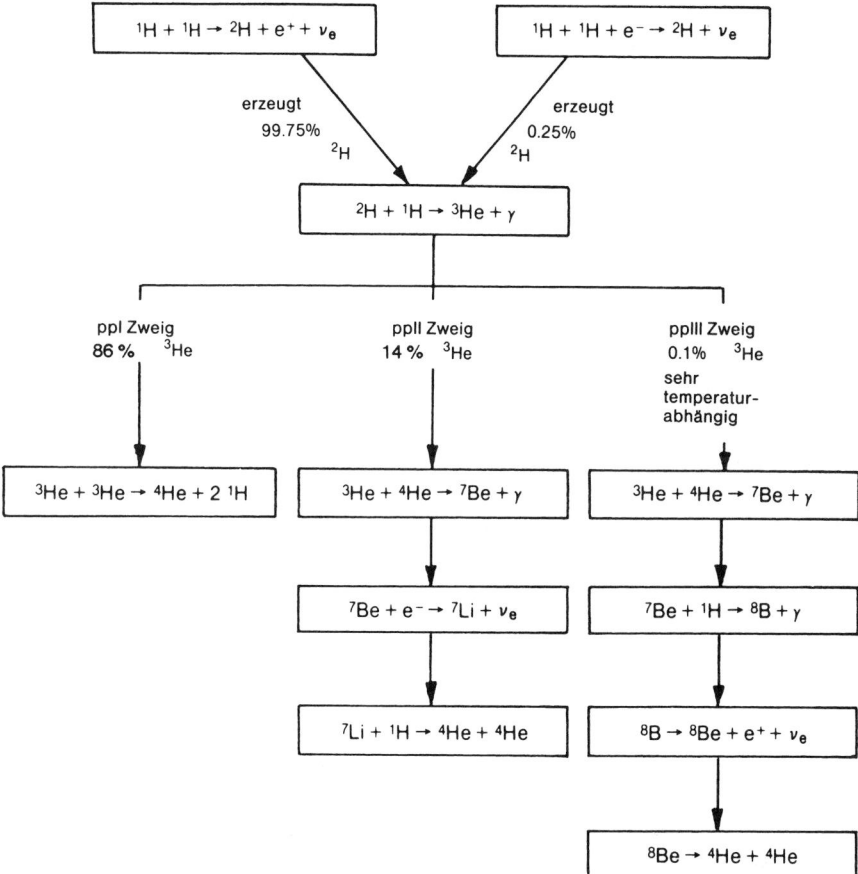

Fig. 2. Proton-proton reactions are the main source of stellar energy in stars with masses close to or below the solar value. They were already briefly considered by Weizsäcker [6] and discussed in more detail by Critchfield and Bethe [8]. Today it is known that the dominant ppI branch is supplemented by the ppII branch, and the small, very temperature-dependent ppIII branch. In the latter two branches, additional electron neutrinos of fairly high energy are produced. The approximate partitions refer to the sun. From: H. Karttunen et al., Fundamental Astronomy. Springer (1987)

but could not be confirmed experimentally until 1952 by Cowan and Reines. Bethe argued in 1938 that fast neutrinos emitted from beta-decay of lithium-4 (which would result from proton capture by helium-3) above an energy threshold of 1.9 MeV might produce neutrons in the outer layers of a star. However, this required the assumption of long-lived lithium-4, which turned out to be wrong. Bethe did not pursue stellar neutrinos further in his early works, and he or Weizsäcker also did not explicitly consider the role of neutrinos in the initial p–p reaction, or in the CNO-cycle at that time.

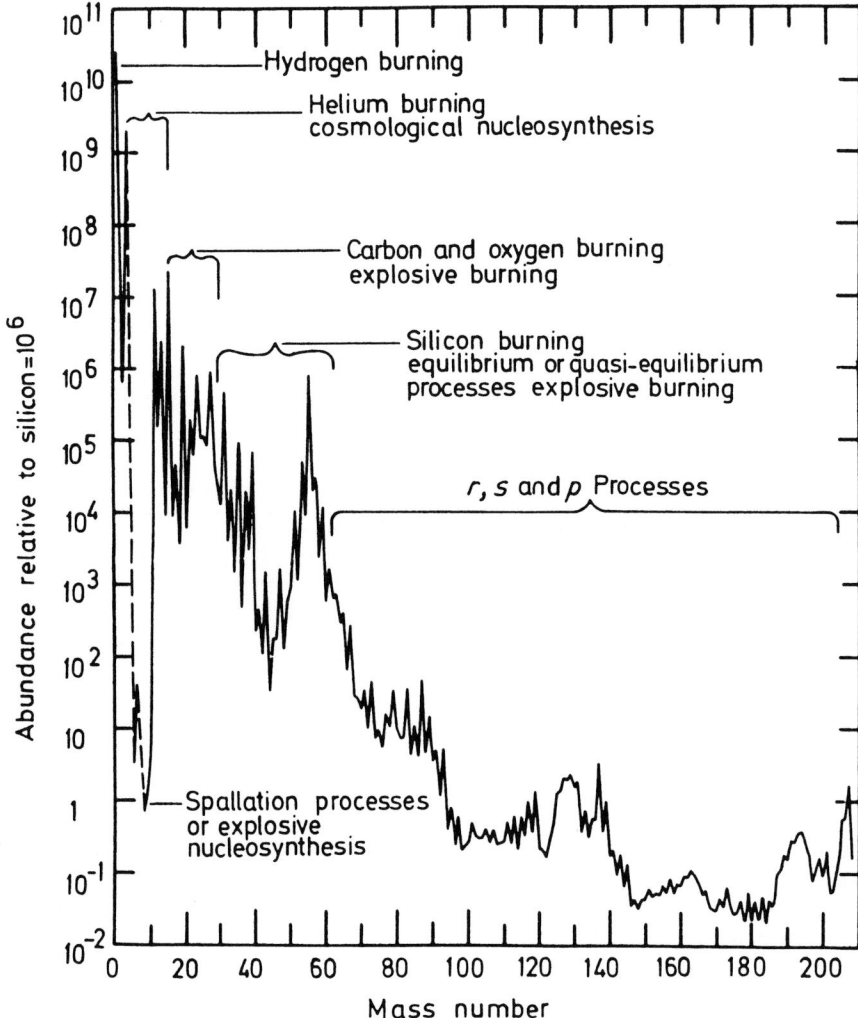

Fig. 3. Solar system abundances of the nuclides relative to silicon (10^6) plotted as function of mass number. The stellar nuclear processes which produce the characteristic features are outlined. The p–p-chain and the CNO-cycle of hydrogen burning are discussed in the text. Elements up to $A \leq 60$ are produced in subsequent burnings at higher temperatures, beyond $A \geq 60$ in supernovae through the r-, s- and p-processes. Source of the data: A.G.W. Cameron, Space Sci. Rev. 15, 121 (1973). From: K.R. Lang, Astrophysical Formulae, p. 419. Springer (1980)

In a normal star, electron neutrinos that are generated in the central region usually leave the star without interactions that modify their energy. Hence, the neutrino energy is treated separately from the thermonuclear energy released by reactions, which undergoes a diffusive transport through the stellar material that is governed by the temperature gradient in the star. Stellar neutrinos are generated not only in nuclear burning and electron capture, but also by purely leptonic processes such as pair annihilation or Bremsstrahlung.

Neutrinos from nuclear processes in the interior of the sun should produce a flux of 10^{11} neutrinos per cm^2 second on the earth. In 1967 Davis et al. – following suggestions by Pontecorvo, and by Bahcall and Davis to use neutrinos " ... to see into the interior of a star and thus verify directly the hypothesis of nuclear-energy generation in stars" – indeed succeeded to measure solar neutrinos with a detector based on 390,000 liters of perchloroethylene. When electron-neutrinos travelling from the sun hit the chlorine-37 nuclei, they occasionally produced argon-37 nuclei, which were extracted and counted by their radioactive decay. The results were published in 1968 [11]. However, these were neutrinos with higher energies which are not produced in the main branch of the proton-proton cycle. More than 90 per cent of the neutrinos are generated in the initial p–p reaction, and these were observed first by the Gallex collaboration in 1992 [12] using gallium-71 nuclei as target in a radiochemical detector. Together with the corresponding result of the Sage collaboration [35], this confirmed experimentally the early suggestions by Weizsäcker and Bethe that p-p fusion is the source of solar energy.

The measurements [11,12] showed that less than 50 per cent of the solar neutrinos that are expected to arrive on earth are actually detected. The subsequent controversy whether this is due to deficiencies in the solar models, or caused by flavor oscillations was resolved at the beginning of the 21st century by combined efforts of the SuperKamiokande [14] and SNO [15]-collaborations in favor of the particle-physics explanation: Neutrinos have a small, but finite mass, and hence, they can oscillate and therefore escape detection, causing the "solar neutrino deficit" – with the size of the discrepancy depending on energy. The identification of oscillating solar neutrinos [15] was actually preceded by evidence for oscillations of atmospheric muon-neutrinos – most likely to tau-neutrinos [13]. Origin of atmospheric neutrinos are interactions of cosmic rays with particles in the earth's upper atmosphere that produce pions and muons, which subsequently decay and emit electron- and muon-neutrinos (or antineutrinos).

Oscillation experiments are, however, only sensitive to differences of squared masses. Hence, the actual value of the neutrino mass is still an open issue, and presently only the upper limit of the mass of the antielectron- neutrino can be deduced from tritium beta decay to be $2.2\,\text{eV}/c^2$ [16]. From neutrinoless double beta decay [17], a lower limit of 0.05–$0.2\,\text{eV}/c^2$ has been deduced in 2002, but this result is only valid if the neutrino is its own antiparticle, which is not certain.

Solar neutrinos are considered in Sect. 4, and a brief outline of some of the perspectives of the field is given in Sect. 5.

2 Energy Evolution in Stars

Stellar and in particular, solar energy is due to fusion of lighter nuclei to heavier ones, which is induced by thermal motion in the star. According to the mass formula that was derived by Weizsäcker in 1935 [18], and by Bethe and Bacher independently in

1936 [19], the difference in binding energies before and after the reaction – the mass defect – is converted to energy via Einstein's $E = mc^2$ [20], and is then added to the star's energy balance. The binding energy per nucleon rises with mass number starting steeply from hydrogen because the fraction of the surface nucleons decreases, then it flattens and reaches a maximum at iron-56, the most tightly bound nucleus; afterwards it drops slowly towards large masses. Although this smooth behavior of the fractional binding energy per nucleon is modified by pairing and shell effects, the overall shape of the curve ensures that energy can be released either by fission of heavy nuclei or by fusion of light nuclei, as it occurs in stars, thus providing our solar energy.

In case of main-sequence stars such as the sun there are no rapid changes in the star that could compete with the time-scale of the nuclear reactions and hence, the energy evolution occurs through equilibrium nuclear burning. Most important in the solar case is hydrogen burning, where the transformation of four hydrogen nuclei into one helium-4 nucleus is accompanied by a mass loss of 0.71 per cent of the initial masses, or 0.029 u. It is converted into an energy of about 26.2 MeV, including the annihilation energy of the two positrons that are produced, and the energy that is carried away by two electron neutrinos. From the known luminosity of the sun, one can calculate a total mass loss rate of $4.25 \cdot 10^9$ kg/s. At such a rate, the hydrogen equivalent of one solar mass could sustain radiation for almost 10^{11} years.

The reactions between nuclei inside stars are due to the thermal motion, and are therefore called *thermonuclear*. Before stars reach an explosive final (supernova) stage, the energy release due to these reactions is rather slow. From the hydrostatic equilibrium condition in the sun one derives the central temperature as

$$T_\odot \leq \frac{8}{3} \frac{G\mu}{R} \frac{M_\odot}{R_\odot} . \quad (1)$$

With the gas constant R, the average number of atomic mass units per molecule μ (0.5 for ionized hydrogen), the gravitational constant G, the solar mass $M_\odot = 1.99 \cdot 10^{30}$ kg and the solar radius $R_\odot = 6.96 \cdot 10^8$ m one finds the central solar temperature

$$T_c \leq 3 \cdot 10^7 \text{ K} . \quad (2)$$

Numerical solutions by Bahcall et al. [22] yield a central temperature $T_c = 1.57 \cdot 10^7$ K and a central pressure $P_c = 2.34 \cdot 10^{16}$ Pa, with a central solar density of $\rho_c = 1.53 \cdot 10^5$ kg/m^3. For these large values of temperature, the assumption of an ideal gas is indeed justified. The reaction rates are strongly dependent on temperature and therefore, massive stars have much greater luminosities with only slightly higher central temperatures. As was noted by Bethe already in 1938, Y. Cygni has $T = 3.2 \cdot 10^7$ K and a luminosity per mass unit of 0.12 W/kg, whereas the sun's luminosity per mass unit is only about $2 \cdot 10^{-4}$ W/kg (the most recent best-estimate value [21] of the total solar luminosity being $3.842 \cdot 10^{26}$ W).

Expressed in units of energy, however, the central solar temperature is only about 1.35 keV. This has to be compared with the height of the Coulomb barrier

$$E_{coul} = \frac{Z_1 Z_2 e^2}{R} \quad (3)$$

with the interaction radius R and the proton numbers Z_1, Z_2 of the nuclei that tend to fuse in order to release energy. Since $E_{\text{coul}}(R) \sim Z_1 Z_2$ MeV, more than a factor of 10^3 in thermal energy is missing in order to overcome the Coulomb barrier.

Thermonuclear reactions in stars can therefore only occur due to the quantum-mechanical tunneling that was established by Gamow [2]. The tunneling probability is

$$P = p_0 E^{-1/2} \exp(-2G) \tag{4}$$

with the Gamow-factor

$$G = \sqrt{\frac{m}{2}} \frac{2\pi Z_1 Z_2 e^2}{\hbar E^{1/2}} \ . \tag{5}$$

Here m is the reduced mass and Z_1, Z_2 are the respective charges of the fusing nuclei, and E is the energy. The factor p_0 depends only on properties of the colliding system. For the pp-reaction at an average energy and at solar temperature, P is of the order of 10^{-20}. It steeply increases with energy and decreases with the product of the charges. Hence, at solar temperatures only systems with small product of the charges may fuse, and for systems with larger $Z_1 Z_2$ the temperature has to be larger to provide a sizeable penetration probability. As a consequence, clearly separated stages of different nuclear burnings occur during the evolution of a star in time.

Once the Coulomb barrier has been penetrated, an excited compound nucleus is formed, which can afterwards decay with different probabilities into the channels that are allowed from the conservation laws. The energy of outgoing particles and gamma-rays is shared with the surroundings except for neutrinos, which leave the star without interactions.

Energy levels of the decaying compound nucleus above or below the ionization energy can be of different types, stationary levels of small width which decay via gamma-emission, and short-lived quasi-stationary levels above the nucleon removal energy which can also (and more rapidly) decay via particle emission. Their width becomes larger with increasing energy and eventually also larger than the distance between neighboring levels.

Due to the existence of quasi-stationary levels above the removal energy, a compound nucleus may also be formed in a resonance when the initial energy matches the one of an energy level in the compound nucleus. At a resonance, the cross-section can become very large, sometimes close to the geometrical value. Astrophysical resonant or non-resonant cross-sections are usually written as

$$\sigma(E) = S E^{-1} \exp(-2G) \tag{6}$$

with the astrophysical cross-section factor S that contains the properties of the corresponding reaction. Although it can be computed in principle, laboratory measurements are a better option. However, because of the small cross-sections, these measurements are difficult at low energies. Extrapolations to these energies are fairly reliable for non-resonant reactions where $S(E)$ is a slowly varying function of E, but this is not true in the case of resonances, which may (or may not) be hidden in the region of extrapolation. The present state of the art for measurements of $S(E)$ in an underground laboratory to shield cosmic rays is shown in Fig. 4 for the reaction

$$^3\text{He}(^3\text{He}, 2p)^4\text{He} \tag{7}$$

that is very important in the stellar pp-chain, cf. next section. The solid line is a fit with a screening potential that accounts for a partial shielding of the Coulomb potential of the nuclei due to neighboring electrons. Data from the LUNA collaboration [23] extend down to 21 keV, where the Gamow peak at the solar central temperature is shown in arbitrary units. The peak arises from the product of the Maxwell distribution at a given temperature T and the penetration probability. Its maximum is at an energy

$$E_G = \left[\sqrt{\frac{m}{2}}\pi\frac{2\pi Z_1 Z_2 e^2 kT}{h}\right]^{2/3}. \tag{8}$$

At E_G, the S-factor for the He-3 + He-3 reaction becomes 5.3 MeV b. The average reaction probability per pair and second is given by

$$\langle \sigma v \rangle = \int_0^\infty \sigma(E) v f(E) dE \tag{9}$$

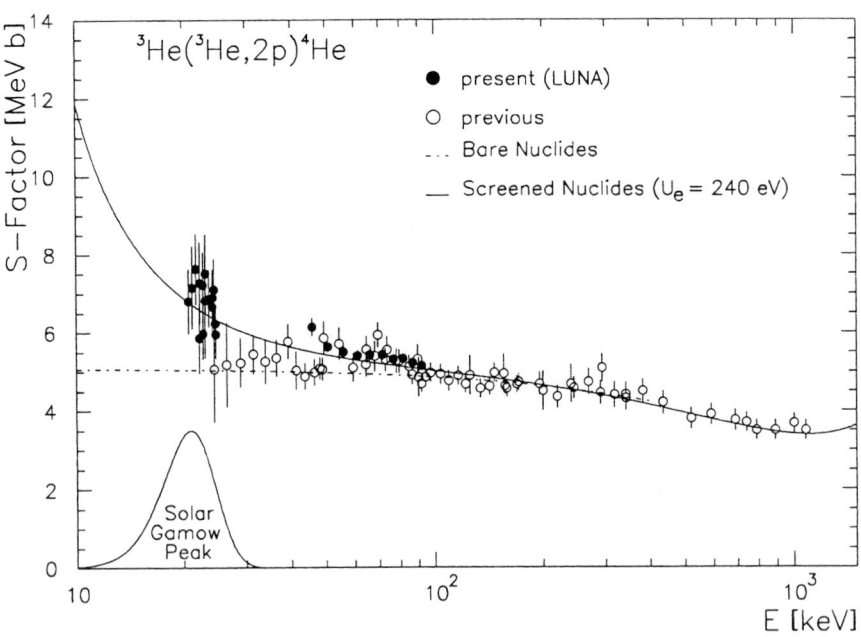

Fig. 4. The astrophysical cross-section factor $S(E)$ for the reaction $^3\text{He}(^3\text{He}, 2p)^4\text{He}$. The solid line is a fit with a screening potential. Data from the LUNA collaboration [23] extend down to 21 keV, where the Gamow peak at the solar central temperature is shown in arbitrary units. From: E.C. Adelberger et al., *Rev. Mod. Phys.* **70**, 1265 (1998)

where $f(E)$ can be expressed in a series expansion near the maximum. Keeping only the quadratic terms, the reaction probability becomes [24]

$$\langle \sigma v \rangle = \frac{4}{3} \left(\frac{2}{m}\right)^{1/2} \frac{1}{(kT)^{1/2}} S_G \cdot \tau^{1/2} \exp(-\tau) \tag{10}$$

with S_G the S-factor at the Gamow peak and

$$\tau = 3E_G/(kT) . \tag{11}$$

The temperature dependence of $\langle \sigma v \rangle$ may be expressed as

$$\frac{\partial \ln \langle \sigma v \rangle}{\partial \ln T} = \frac{\tau}{3} - \frac{2}{3} , \tag{12}$$

which can attain values near or above 20. As a consequence of such large values for the exponent of T, the thermonuclear reaction rates become extremely strongly dependent on temperature, and small fluctuations in T may cause dramatic changes in the energy (and neutrino) production of a star. The corresponding uncertainty in stellar models created the long-standing controversy about the origin of the solar neutrino deficit, which has only recently been decided in favor of the particle-physics explanation, cf. Sect. 4.

3 Hydrogen Burning

Due to the properties of the thermonuclear reaction rates, different fusion reactions in a star are separated by sizeable temperature differences and during a certain phase of stellar evolution, only few reactions occur with appreciable rates. Stellar models account in network-calculations for all simultaneously occuring reactions. Often the rate of the fusion process is determined by the slowest in a chain of subsequent reactions, such as in case of the nitrogen-14 reaction of the CNO-cycle.

In hydrogen burning, four hydrogen nuclei are fused into one helium-4 nucleus, and the mass defect of 0.71 per cent is converted into energy (including the annihilation energy of the two positrons, and the energy carried away by the neutrinos):

$$4 \cdot {}^1\text{H} \rightarrow {}^4\text{He} + 2e^+ + 2\nu_e + 26.2 \text{ MeV} . \tag{13}$$

As net result, two protons are converted into neutrons through positron emission ($beta^+ - decay$), and because of lepton number conservation, two electron neutrinos are emitted. Depending on the reaction which produces the neutrinos, they can carry between 2 and 30 per cent of the energy. Helium synthesis in stars proceeds through different reaction chains which occur simultaneously. The main series of reactions are the proton-proton chain, Fig. 2, and the CNO-cycle, Fig. 1.

In the present epoch, the pp-chain turns out to be most important for the sun – the CNO-cycle produces only 1.5 per cent of the luminosity [22]. The pp-chain starts with two protons that form a deuterium nucleus, releasing a positron and an electron neutrino. (With much smaller probability it may also start with the p-e-p process, Fig. 2). This reaction has a very small cross section, because the beta-decay is governed by the weak interaction. At central solar temperature and density, the mean reaction time is 10^{10} years, and it is to a certain extent due to this huge time constant that the sun is still shining. With another proton, deuterium then reacts to form helium-3. This process is comparably fast and hence, the abundance of deuterons in stars is low.

To complete the chain to helium-4 three branches are possible. The first – in the sun with 85 per cent most frequent – chain (ppI) requires two helium-3 nuclei and hence, the first reaction has to occur twice, with two positrons and two electron neutrinos being emitted. The other two branches (ppII, ppIII) need helium-4 to be produced already (in previous burnings, or primordially). In the subsequent reactions between helium-3 and helium-4, the additional branching occurs because the product beryllium-7 can react either with an electron to form lithium-7 plus neutrino (ppII), or with hydrogen to form boron-8 (ppIII). The energy released by the three chains differs because the neutrinos carry different amounts of energy with them, and the relative frequency of the different branches depend on temperature, density, and chemical composition. The per centages in Fig. 2 refer to the standard solar model at the present epoch [22]. Details of the various parts of the chain including the corresponding energy release, the energies carried away by the neutrinos and the reaction rate constants have been discussed by Parker et al. [26] and Fowler et al. [27].

The other main reaction chain in hydrogen burning is the CNO-cycle, Fig. 1. Here, the carbon, nitrogen and oxygen isotopes serve as catalysts, their presence is required for the cycle to proceed. The main cycle is completed once the initial carbon-12 is reproduced by nitrogen-15 + hydrogen. There is also a secondary cycle (not shown in Fig. 1 since it is 10^4 times less probable). It causes oxygen-16 nuclei which are present in the stellar matter to take part in the CNO-cycle through a transformation into nitrogen-14. The CNO-cycle produces probably most of the nitrogen-14 found in nature. For sufficiently high temperatures, the nuclei attain their equilibrium abundances and hence, the slowest reaction determines the time to complete the whole circle, which is nitrogen-14 + hydrogen (bottom of Fig. 1).

The CNO-cycle dominates in stars with masses above 1.5 times the solar value because its reaction rates rise faster with temperature as compared to pp. Details of the Bethe–Weizsäcker cycle have been discussed by Caughlan and Fowler [28]. The cycle had first been proposed by Weizsäcker in [7]. In this work, he abandoned the main reaction path that he had considered in [6], namely, from hydrogen via deuterium and lithium to helium, because the intermediate nuclei of mass number 5 that were supposed to be part of the scheme had turned out to be unstable.

In the first paper of the series [6], he had considered various reaction chains that allow for a continuous generation of energy from the mass defect, and also of neutrons for the buildup of heavy elements. He had confirmed that the temperatures in the interior of stars are sufficient to induce nuclear reactions starting from hydrogen. In the second paper he modified the results; in particular, he discussed the possibility that some of the elements might have been produced before star formation by another process.

The link between energy evolution in stars and the formation of heavy elements as considered in [6] turned out to end up in difficulties when calculated quantitatively. Hence, he modified his version of the so-called "Aufbauhypothese", according to which the neutrons necessary for the production of heavy elements should be generated together with the energy, and decoupled the generation of energy from the production of heavy elements. He then concluded that stellar energy production should essentially be due to reactions between light nuclei, with the corresponding abundances being in agreement with observations. The CNO-cycle was considered to be the most probable path.

In his independent and parallel development of the CNO-cycle that was published somewhat later [9] and contained detailed calculations, Bethe showed that " ... there will be no appreciable change in the abundance of elements heavier than helium during the evolution of the star but only a transmutation of hydrogen into helium. This result ... is in contrast to the commonly accepted 'Aufbauhypothese' ". Here,

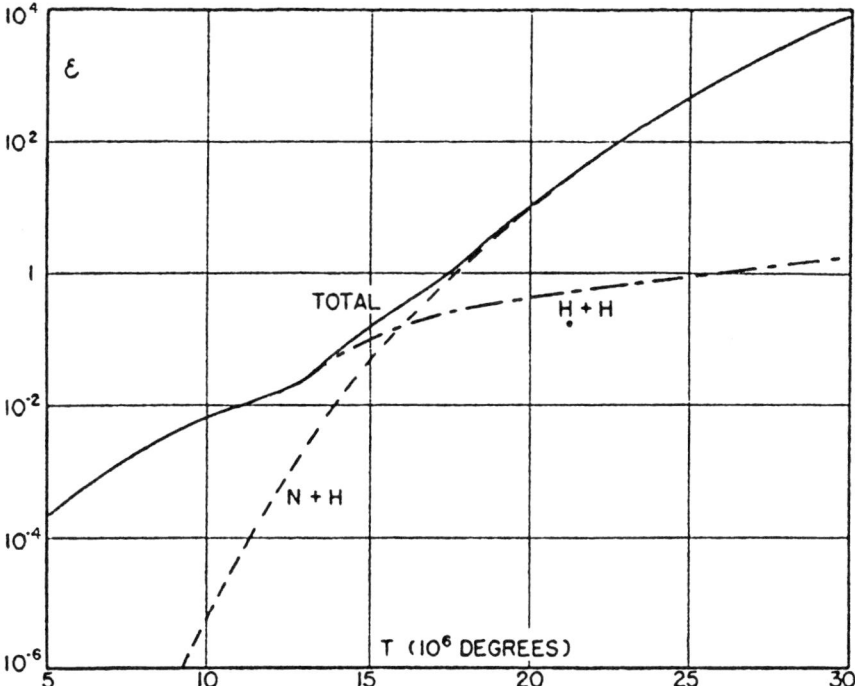

Fig. 5. Stellar energy production in 10^{-4} J/(kg · s) due to the proton-proton chain (curve H + H) and the CNO-cycle (N + H), and total energy production (solid curve) caused by both chains. According to this calculation by Bethe in 1939 [9], the CNO-cycle dominates at higher than solar temperatures. Its role at and below solar temperatures as compared to pp is, however, overestimated, cf. Fig. 6. From: H.A. Bethe, *Phys. Rev.* **55**, 434 (1939)

he referred to Weizsäcker's first hypothesis [6] which had, however, already been modified [7].

Together with Critchfield [8], Bethe also investigated essential parts of the pp-chain (which Weizsäcker also mentioned) – in particular, deuteron formation by proton combination as the first step – and came to the conclusion that it " ... gives an energy evolution of the right order of magnitude for the sun". Details of the pp-chain were developed much later in the 1950s by Salpeter [29] and others. In 1938/9, however, Bethe was convinced that " ... the reaction between two protons, while possible, is rather slow and will therefore be much less important in ordinary stars than the cycle (1)" namely, the CNO-cycle.

In a calculation of the energy production by pp-chain versus CNO-cycle (Fig. 5), Bethe obtained qualitatively the preponderance of H + H at low and N + H at high temperatures. However, the result had to be modified in the course of time as it became evident that the pp-chain dominates the CNO-cycle at solar conditions, although the Bethe–Weizsäcker-fraction will increase considerably in the coming 4 billion years, and eventually supersede the contribution from the ppII-chain (Fig. 6).

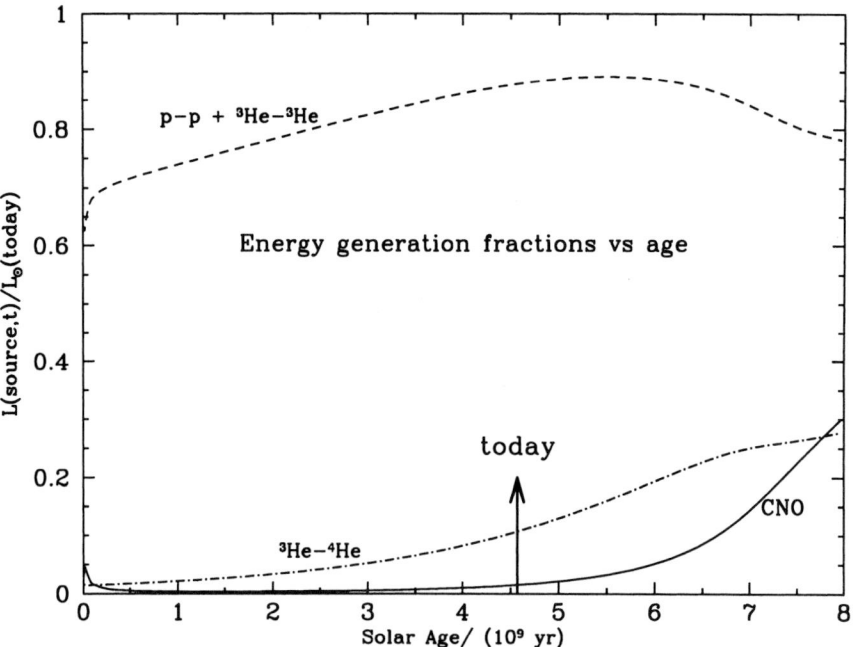

Fig. 6. Fractions of the solar luminosity produced by different nuclear fusion reactions versus solar age, with the present age marked by an arrow [22]. The proton-proton chain is seen to dominate the luminosity fractions – in particular, through the branch that is terminated by the ^{3}He–^{3}He reaction. The solid curve shows the luminosity generated by the CNO-cycle, which increases with time, but is only a small contribution today. From: J.N. Bahcall, M.H. Pinsonneault and S. Basu, *Astrophys. J.* 555, 990 (2001)

Today, detailed solar models allow to calculate the fractions of the solar luminosity that are produced by different nuclear fusion reactions very precisely [22]. The model results not only agree with one another – in the neutrino flux predictions to within about 1 per cent – they are also consistent with precise p-mode helioseismological observations of the sun's outer radiative zone and convective zone [30]. Moreover, the production of heavier elements up to iron in subsequent burnings at higher temperatures [27], as well as beyond iron in the r- and s-process is rather well-understood [31].

4 Stellar Neutrinos

In stellar interiors, only electron neutrinos play a role. The interaction of neutrinos with matter is extremely small, with a cross-section of

$$\sigma_\nu \simeq \left(E_\nu/m_e c^2\right)^2 \cdot 10^{-17} \text{ mb}. \tag{14}$$

Hence, the cross-section for neutrinos with $E_\nu \simeq 1$ MeV is $\sigma_\nu \simeq 3.8 \cdot 10^{-17}$ mb, which is smaller than the cross-section for the electromagnetic interaction between photons and matter by a factor of about 10^{-18}. Associated with the cross-section is a mean free path

$$\lambda_\nu = \frac{u}{\rho \cdot \sigma_\nu} \simeq \frac{4 \cdot 10^{20}}{\rho} \text{ m} \tag{15}$$

with the atomic mass unit $u = 1.66 \cdot 10^{-27}$ kg and ρ in kg/m³. In stellar matter with $\rho \simeq 1.5 \cdot 10^3$ kg/m³, the mean free path of neutrinos is therefore approximately

$$\lambda_\nu \simeq 3 \cdot 10^{17} \text{ m} \simeq 10 pc \simeq 4 \cdot 10^9 R_\odot \tag{16}$$

and hence, neutrinos leave normal stars without interactions that modify their energy. This is different during the collapse and supernova explosion in the final stages of the evolution of a star where nuclear density can be reached, $\rho \simeq 2.7 \cdot 10^{17}$ kg/m³ such that the mean free path for neutrinos is only several kilometers, and a transport equation for neutrino energy has to be applied.

Here only the neutrinos from nuclear reactions in a normal main-sequence star like the sun are considered; their energies are (to some extent, since the continuous distributions overlap) characteristic for specific nuclear burnings. The pp-chain which provides most of the sun's thermonuclear energy produces continuum neutrinos in the reactions ([32]; cf. Fig. 2)

^{1}H + ^{1}H → ^{2}H + e^+ + ν_e (0.420 MeV)
^{8}B → ^{8}Be* + e^+ + ν_e (14.06 MeV)

and the Bethe–Weizsäcker cycle generates

^{13}N → ^{13}C + e^+ + ν_e (1.20 MeV)
^{15}O → ^{15}N + e^+ + ν_e (1.74 MeV)

where the numbers are the maximum neutrino energies for the corresponding reaction. In addition to these continuum neutrinos, there are neutrinos at discrete energies from the pp-chain

$$^1H + {^1H} + e^- \rightarrow {^2H} + \nu_e \quad (1.44\,\text{MeV})$$
$$^7Be + e^- \rightarrow {^7Li^*} + \nu_e \quad (0.861\,\text{MeV} - 90\,\text{per cent})$$
$$\quad\quad\quad\quad\quad\quad\quad\quad\quad\quad (0.383\,\text{MeV} - 10\,\text{per cent})$$

(depending on whether lithium-7 is in the ground state, or in an exited state)

$$^8B + e^- \rightarrow {^8Be} + \nu_e \quad (15.08\,\text{MeV})\,.$$

The CNO-cycle (Fig. 1) which becomes important in stars with masses above 1.5 solar masses, or in later stages of the stellar evolution (Fig. 7) also produces neutrinos at discrete energies

$$^{13}N + e^- \rightarrow {^{13}C} + \nu_e \quad (2.22\,\text{MeV})$$
$$^{15}O + e^- \rightarrow {^{15}N} + \nu_e \quad (2.76\,\text{MeV})\,.$$

For experiments to detect these neutrinos when they arrive on earth 8.3 minutes after their creation the flux at the earth's surface is of interest. Neutrinos from the central

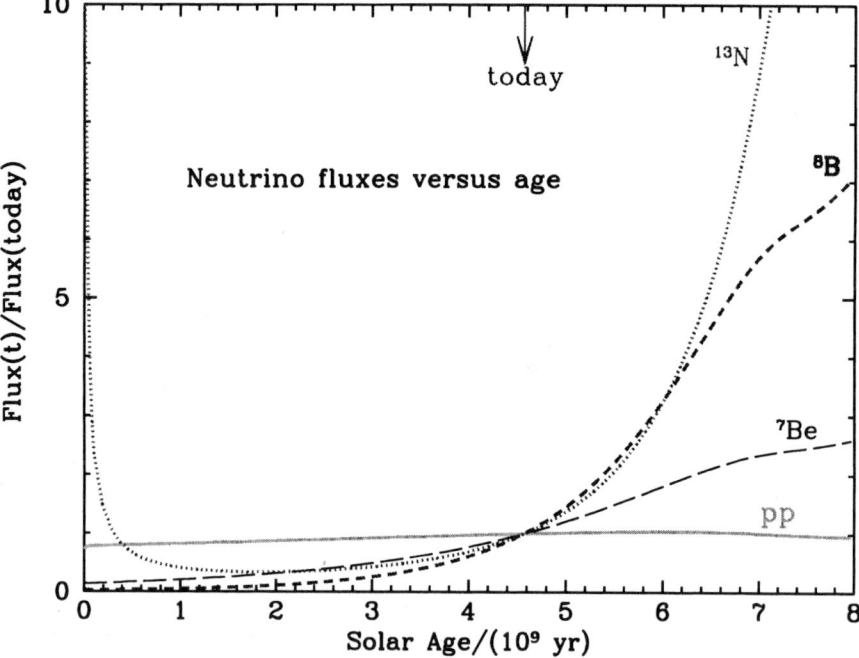

Fig. 7. The proton-proton, beryllium-7, boron-8 and nitrogen-13 neutrino fluxes as functions of solar age, with the present age marked by an arrow [22]. The Standard Solar Model ratios of the fluxes are divided by their values at $4.57 \cdot 10^9$ years, the present solar age. From: J.N. Bahcall, M.H. Pinsonneault and S. Basu, Astrophys. J. 555, 990 (2001).

region of the sun yield a flux of about $10^7/(m^2 \cdot s)$. The precise value as function of the neutrino energy can be calculated from solar models ([25], Fig. 8). Here, solid lines denote the pp-chain and broken lines the CNO-cycle. The low-energy neutrinos from the initial pp-reaction are seen to dominate the flux. However, the first experiment by Davis et al. [11] that detected solar neutrinos on earth with a large-scale underground perchloroethylene tank in 1967/8 – and thus confirmed the theory how the sun shines and stars evolve – made use of the reaction

$$\nu_e + {}^{37}_{17}O \rightarrow e^- + {}^{37}_{18}Ar - 0.814\,\text{MeV}$$

and hence, only neutrinos with energies above 0.814 MeV could be observed through the decay of radioactive argon nuclei – which are mostly the solar boron-8 neutrinos, cf. Fig. 8. The rate of neutrino captures is measured in solar neutrino units, SNU; 1 SNU corresponds to 10^{-36} captures per second and target nucleus. Experimental runs by Davis et al. during the 1970s, 80s and 90s yielded a signal (after subtraction of

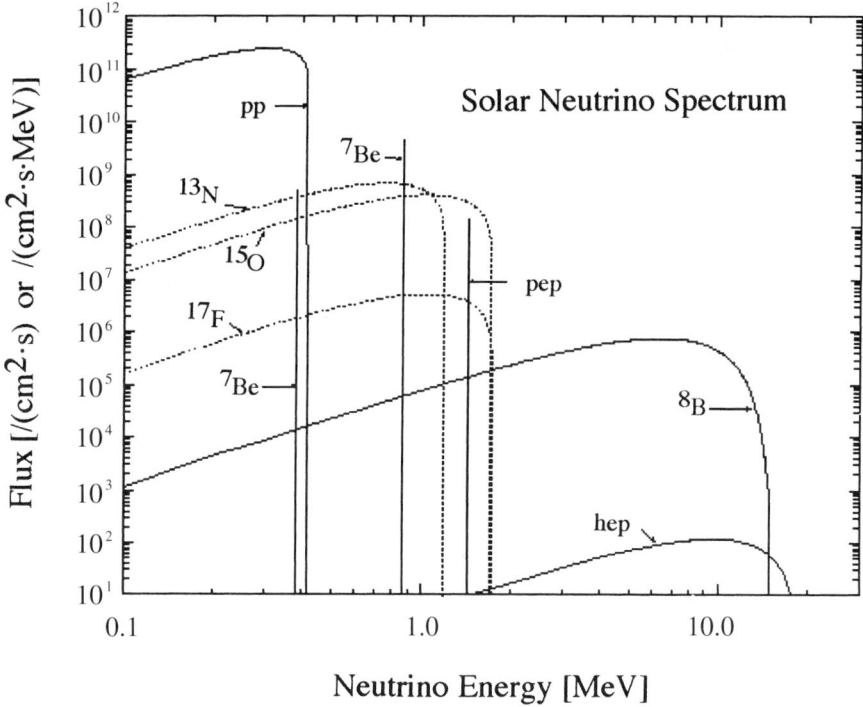

Fig. 8. Spectrum of solar electron neutrinos according to the "Standard Solar Model". The spectrum is dominated by low-energy neutrinos from the p-p chain that have been detected by the Gallex experiment [12, 34]. Solid lines indicate neutrinos from the pp-chain, dashed lines from the CNO-cycle. Neutrinos of higher energies had first been observed by Davis et al. [11]. The calculation is by Bahcall and Pinsonneault [25]. (The hep-neutrinos arise from the ppIV-reaction ${}^3He + p \rightarrow {}^4He + \nu_e + e^+$ which is not shown in Fig. 2). From: J.N. Bahcall and M.H. Pinsonneault, *Rev. Mod. Phys.* **67**, 781 (1995) and astro-ph 0010346

the cosmic-ray background 1.6 kilometers underground) of (2.3 ± 0.3) SNU, whereas the predicted capture rates from a solar model were 0 SNU for pp (because it is below threshold), 5.9 SNU for boron-8 beta decay, 0.2 SNU for the pep-reaction, 1.2 SNU for beryllium-7 electron capture, 0.1 SNU for nitrogen-13 decay and 0.4 SNU for oxygen-15 decay, totally about 8 SNU.

The observation of less than 50 per cent of the expected neutrino flux created a controversy about the origin of the deficit, which was finally – in 2001 – resolved [15] in favor of the particle-physics explanation that had originally been proposed by Pontecorvo in 1968 [33]: on their way from the solar interior to the earth, electron-neutrinos oscillate to different flavors which escape detection, thus creating the deficit. Although deficiencies in the solar models could have been responsible for the discrepancy (in view of the sensitive dependence of the neutrino flux on the central temperature), it could be confirmed [15] that the models are essentially correct, giving the right value of T_c within 1 per cent.

Before this big step in the understanding of stellar evolution and neutrino properties could be taken, there was substantial progress both in experimental and theoretical neutrino physics. In 1992 the Gallex-collaboration succeeded in measuring the pp-neutrinos from the initial fusion reaction, which contributes more than 90 per cent of the integral solar neutrino flux [12]. They used a radiochemical detector with gallium as target, exploiting the reaction

$$\nu_e + {}^{71}_{31}\text{Ga} \rightarrow e^- + {}^{71}_{32}\text{Ge} - 0.23\,\text{MeV}\ .$$

The threshold is below the maximum neutrino energy for pp-neutrinos of 0.42 MeV and a large fraction of the pp-neutrinos can therefore be detected in addition to the pep-, beryllium-7 and boron-8 neutrinos. Gallex – which is sensitive to electron neutrinos only – thus provided the proof that pp-fusion is indeed the main source of solar energy. The result $(69.7 + 7.8/- 8.1)$ SNU was confirmed by the Sage experiment $[(69 \pm 12)$ SNU] in the Caucasus [35]. Again this was substantially below the range that various standard solar models predicted (120–140) SNU, and the solar neutrino deficit persisted. At that time, there were clear indications – but no definite evidence yet – that the flux decreases between sun and earth due to neutrino flavor oscillations – most probably enhanced through the MSW-effect [36] in the sun –, " ... pointing towards a muon-neutrino mass of about 0.003 eV" [34]. The result was later updated to (73.9 ± 6.2) SNU and could be assigned to the fundamental low-energy neutrinos from the pp and pep reactions – but then there remained no room to accomodate the beryllium-7 and boron-8 neutrinos.

Solar neutrinos were also detected in real time with the Kamiokande detector [37] in Japan, a water-Cerenkov detector, and precursor to the famous SuperKamiokande detector. Due to the high threshold of about 7.5 MeV, it could see only the most energetic neutrinos from the decay of boron-8 in the solar center. With the Cerenkov light pattern one could measure for the first time the incident direction of the scattering neutrinos, and prove that they do indeed come from the sun. The result of the boron-8 neutrino flux was

$$flux\ \tfrac{Observed}{Predicted}\ (\nu_e) = 0.54 \pm 0.07\ ,$$

again confirming the deficit. To solve the solar neutrino problems, a larger target volume and a lower energy threshold was needed: the SuperKamiokande detector in the same zinc mine with a threshold of 5 MeV, with 32,000 tons of pure water surrounded by 11,200 photomultiplier tubes for observing electrons scattered by neutrinos (many of the tubes were destroyed in 2001 when one collapsed, emitting an underwater shock wave). The detector was designed to record about 10,000 solar neutrino collisions per year – 80 times the rate of its predecessors –, but also atmospheric neutrinos, and possible signs for proton decay. The result [14] for the boron-8 flux can be expressed as

$$flux \frac{Observed}{Predicted}(\nu_e) = 0.47 \pm 0.02,$$

which was in agreement with the previous findings, but more precise. There was a massive hint that the deficit could be due to neutrino oscillations, since the SuperKamiokande collaboration found evidence in 1998 [13] that muon neutrinos which are produced in the upper atmosphere by pion and muon decays change their type when they travel distances of the order of the earth's radius due to oscillations into another species, most likely into tau neutrinos. The appearance of the tau neutrinos could not yet be detected directly, but oscillations to electron neutrinos in the given parameter range were excluded since the ν_e-flux was unchanged. Accelerator experiments with a long baseline of 700 km between neutrino source and detector are being planned to verify this interpretation [40].

The atmospheric data showed a significant suppression of the observed number of muon neutrinos as compared to the theoretical expectation at large values of x/E_ν, with the travel distance x (large when the neutrinos travel through the earth) and the neutrino energy E_ν which is in the GeV-range for atmospheric neutrinos and hence, much higher than in the solar case. The observed dependence on distance is expected from the theoretical expression for oscillations into another flavor, which yields in the model case of two flavors

$$P = \frac{1}{2}\sin^2(2\theta) \cdot (1 - \cos(2\pi x/L)). \quad (17)$$

Here, θ is the mixing angle between the two flavors considered, and the characteristic oscillation length (the distance at which the initial flavor content appears again) is

$$L = 4\pi E_\nu/\Delta m^2 \simeq 2.48 \left(E_\nu/\text{MeV}\right)/\left(c^4 \Delta m^2/\text{eV}^2\right) [\text{m}] \quad (18)$$

with the difference $\Delta m^2 = | m_2^2 - m_1^2 |$ of the neutrino mass eigenstates. Atmospheric neutrino experiments are thus sensitive to differences in the squared masses of 10^{-4} to 10^{-2} eV^2/c^4 whereas solar neutrinos are sensitive to differences below $2 \cdot 10^{-4}$ eV^2/c^4 due to the lower neutrino energy and the larger distance between source and detector. Whereas such mixings between neutral particles that carry mass had been firmly established many years ago in the case of quarks that build up the K^0 and B^0-mesons and their antiparticles – including the proof that CP is violated [38] for three quark families –, it remained an open question until 1998 whether the corresponding phenomenon [39] exists for leptons.

The atmospheric SuperKamiokande data proved beyond reasonable doubt the existence of oscillations and finite neutrino masses [13, 40], with $\Delta m_{atm}^2 \simeq 2.5 \cdot 10^{-3}$ eV^2 and maximal mixing $sin^2(2\theta_{atm}) \simeq 1$. The corresponding step for solar neutrinos followed in 2001: charged-current results ($v_e + d \rightarrow e + p + p$; sensitive to electron neutrinos only) from the Sudbury Neutrino Observatory (SNO) in Canada with a D$_2$O-Cerenkov detector [15], combined with elastic scattering data from SuperK ($v_\odot + d \rightarrow v + d$; sensitive to all flavors), established oscillations of solar boron-8 neutrinos.

These results were confirmed and improved (5.3σ) in 2002 by neutral current results from SNO ($v_\odot + d \rightarrow v + n + p$); an improved measurement with salt (NaCl; chlorine-35 has a high n-capture efficiency) being underway. The total flux measured with the NC reaction is $(5.09 + 0.64/ - 0.61) \cdot 10^6$ neutrinos per cm$^2 \cdot$ s. This is in excellent agreement with the value from solar models $(5.05 + 1.01/ - 0.81)$, proving that stellar structure and evolution is now well-understood. The currently most-favored mechanism for solar neutrino conversion to myon- and tauon-flavors is the "Large mixing angle" solution, which also implies matter-enhanced (resonant) mixing in the interior of the sun through the MSW-effect [36].

5 Perspectives

Since the early works by Weizsäcker und Bethe, the investigation of thermonuclear processes in stars has developed into considerable detail, and with the advent of stellar neutrino physics an independent confirmation of the origins of solar energy has emerged.

Regarding the physics of stars, improvements in the precise measurements of all the reaction rates at low energies that are involved in the fusion chains may still be expected, as has been outlined in the model case of the ^{3}He +^{3} He system within the energy region of the solar Gamow peak [23].

However, not only a good knowledge of the processes involved in equilibrium burnings at energies far below the Coulomb barrier, but also of explosive burning (with short-lived nuclides at energies near the Coulomb barrier) is of interest, because both contribute to the observed abundances of the elements. This requires new experimental facilities.

An improved understanding of the cross-sections will then put the predictions of the solar neutrino flux and spectrum on a better basis. This entire spectrum will be investigated with high precision in the forthcoming decade. In particular, the new detector Borexino will measure the monoenergetic beryllium-7 neutrinos at 862 keV, which depend very sensitively on the oscillation parameters. The Gallium Neutrino Observatory GNO and the LENS experiment will improve the Gallex measurement of the pp-neutrinos. Together with forthcoming SNO, KamLAND and (after recovery) SuperK results it will be possible to definitely determine all the mixing parameters in a three-family scheme – and verify, or falsify, the LMA solution.

The more detailed knowledge about the physics of stars will thus be supplemented by considerable progress regarding neutrino properties [40]. Questions to be settled are the individual neutrino masses (rather than the difference of their squares); whether neutrinos are their own antiparticles (to be decided from the existence or non-existence of neutrinoless double beta-decay); whether neutrinos violate CP just as quarks do, or maybe in a different manner that opens up a better understanding of the matter-antimatter asymmetry of the universe than has been possible from the investigation of quark systems (to be decided in experiments with strong neutrino beams).

In any case, the Standard Model of particle physics has to accomodate finite neutrino masses, and in future theoretical formulations the relation between quark mixing and neutrino mixing will probably become more transparent.

References

1. A.S. Eddington: *The Internal Constitution of the Stars*. Cambridge University Press, 1926
2. G. Gamow: *Z. Physik* 52, 510 (1929)
3. R.d'E. Atkinson and F.G. Houtermans: *Z. Physik* 54, 656 (1929)
4. G. Gamow and E. Teller: *Phys. Rev.* 53, 608 (1938)
5. S. Chandrasekhar: *An Introduction to the Study of Stellar Structure*. University of Chicago Press, 1939
6. C.F. v. Weizsäcker: *Physik. Zeitschr.* 38, 176 (1937)
7. C.F. v. Weizsäcker: *Physik. Zeitschr.* 39, 633 (1938)
8. H.A. Bethe and C.L. Critchfield: *Phys. Rev.* 54, 248 (1938)
9. H.A. Bethe: *Phys. Rev.* 55, 434 (1938)
10. E.M. Burbidge, G.R. Burbidge, W.A. Fowler and F. Hoyle: *Rev. Mod. Phys.* 29, 547 (1957)
11. R. Davis, D.S. Harmer and K.C. Hoffman: *Phys. Rev. Lett.* 20, 1205 (1968)
12. P. Anselmann et al.: *Phys. Lett. B* 285, 376 (1992); 327, 377 (1994); 342, 440 (1995)
13. S. Fukuda et al.: *Phys. Rev. Lett.* 81, 1562 (1998); 85, 3999 (2000)
14. S. Fukuda et al.: *Phys. Rev. Lett. B* 86, 5651 (2001)
15. Q.R. Ahmad et al.: *Phys. Rev. Lett.* 87, 071 301 (2001); 98, 011 301 (2002)
16. C. Weinheimer et al.: *Phys. Lett. B* 460, 219 (1999)
17. L. Baudis et al.: *Phys. Rev. Lett.* 83, 41 (1999) and *Eur. Phys. J. A* 12, 147 (2001)
18. C.F. v. Weizsäcker: *Z. Physik* 96, 431 (1935)
19. H.A. Bethe and R.F. Bacher: *Rev. Mod. Phys.* 8, 82 (1936)
20. A. Einstein: *Ann. Physik* 18, 639 (1905)
21. C. Fröhlich and J. Lean: *Geophys. Res. Lett.* 25, No. 23, 4377 (1998)
22. J.N. Bahcall, M.H. Pinsonneault and S. Basu: *Astrophys. J.* 555, 990 (2001)
23. E.C. Adelberger et al.: *Rev. Mod. Phys.* 70, 1265 (1998); M. Junker et al.: LUNA collab., *Nucl. Phys. B (Proc. Suppl.)* 70, 382 (1999)
24. R. Kippenhahn and A. Weigert: *Stellar Structure and Evolution*. Springer, 1990
25. J.N. Bahcall and M.H. Pinsonneault: *Rev. Mod. Phys.* 67, 781 (1995); 70, 1265 (1998)
26. P.D. Parker, J.N. Bahcall and W.A. Fowler: *Ap. J.* 139, 602 (1964)
27. W.A. Fowler, G.R. Caughlan and B.A. Zimmerman: *Ann. Rev. Astron. Astrophys.* 5, 525 (1967)
28. G.R. Caughlan and W.A. Fowler: *Ap. J.* 136, 453 (1962)
29. E.E. Salpeter: *Phys. Rev.* 88, 547 (1952)
30. S.A. Bludman and D.C. Kennedy: *Astrophys. J.* 472, 412 (1996)
31. K.R. Lang: *Astrophysical Formulae*, Chapt. 4. Springer, 1980, and references therein
32. J.N. Bahcall: *Phys. Rev.* 135, B 137 (1964)

33. B. Pontecorvo: *Sov. Phys. JETP* 26, 984 (1968); V. Gribov and B. Pontecorvo: *Phys. Lett. B* 28, 493 (1969)
34. T. Kirsten. In: W. Hampel (ed.): *Proc. 4th Int. Solar Neutrino Conf.*, MPI-HD. Heidelberg 1997
35. J. Abdurashitov et al.: *Phys. Lett. B* 328, 234 (1994)
36. S. Mikheyev, A. Smirnov: *Sov. J. Nucl. Phys.* 42, 913 (1985). L. Wolfenstein: *Phys. Rev. D* 17, 2369 (1978)
37. Y. Suzuki et al.: *Nucl. Phys. B (Pro. Suppl.)* 38, 54 (1995)
38. J.H. Christensen, J.W. Cronin, V.L. Fitch and R. Turlay: *Phys. Rev. Lett.* 13, 138 (1964). M. Kobayashi and T. Maskawa: *Prog. Theor. Phys.* 49, 652 (1973). H. Burkhardt et al.: *Phys. Lett. B* 206, 169 (1988)
39. Z. Maki, N. Nakagawa and S. Sakata: *Prog. Theor. Phys.* 28, 870 (1962)
40. F. Feilitzsch et al. (eds.): *Proc. XXth Int. Conf. on Neutrino Physics and Astrophysics.* München, in press (2002)

Comets: Fascinating Cosmic Objects

Reimar Lüst

In spring 1949 I started as a Ph.D. student of Carl Friedrich von Weizsäcker at the Max Planck Institute for Physics in Göttingen. At that time he worked on the theory of origin of the planetary system. This was also the content of my Ph.D. thesis. At the same time and at the same institute Ludwig Biermann studied the theory of cometary tails. For a new understanding of cometary tails I could build up a new Max Planck Institute for Extraterrestrial Physics at the beginning of the sixties. There we developed experiments to create an artificial cometary tail.

Comets have their origin at the beginning of our planetary system. A more detailed study of comets enables us to learn more about the early start of the planetary system. My Ph.D. thesis dealt with the first phases of the planetary system. In this way the circle closes.

1 The Nature of Comets

However not only scientists are interested on comets, but also the broad public is fascinated by the appearance of a comet. Throughout the history the unpredictable appearance of comets has been puzzling mankind and created fear in the hearts of the observers. A comet was regarded an omen, normally presaging disastrous events. Many people shared Aristotle's view that the appearance of a comet signaled adversity or drought.

Reports on the appearances of comets reach back many centuries before Christ. These early records come almost exclusively from China, Japan and Korea. There exists also a Babylonian description interpreted as a reference to the comet of 1140 B.C.

The question was whether comets were celestial objects or phenomena of the atmosphere. The Babylonians and also at first the Greeks were of the opinion that comets were cosmic bodies, like planets. But these ideas changed with Aristotle who ruled out the planetary nature of comets. According to his philosophy comets belonged to the sublunar sphere, being the product of meteorological processes in our atmosphere.

Around the time of Christ's birth, Seneca the Roman, Emperor Nero's teacher, wrote: *"Some day there will arise a man who will demonstrate in what regions of the heavens comets take their way; why they journey so far apart from the other planets, what their size, their nature."*

About 1600 years later, the time was ripe for those men to work out the comet's rightful place in the heaven. Tycho Brahe, generally considered to be the greatest astronomer of his days, made the decisive observation by determining the parallax of the comet of 1577. He showed that the distance of this comet was considerably greater than the distance from the Earth to the Moon. Although he was not able to determine the orbit of the comet, these objects could no longer be placed in sublunar atmospheric space. Since that time comets have been acknowledged as real objects in the solar system.

But still there remained the question of their orbit. Although Kepler was so successful in determining the orbits of the planets, he was wrong in his belief that comets moved along straight lines, but with irregular speed.

Isaac Newton and Edmund Halley, the latter having been eclipsed in History by the more famous Newton, made the final step in describing and calculating the orbits of comets. It was Halley who encouraged the retiring Newton to publish the principles of gravitation he had developed after years of thought, supposedly inspired by the legendary falling apple. Newton recognised that gravity on Earth represented the same law of force as that affecting the motion of planets around the Sun. Halley edited the manuscript and financed the publication in 1687 of Newton's great book, "The Mathematical Principles of Natural Philosophy". But in applying Newton's laws of gravitation in correctly predicting the return of a comet, Halley went an important step further. It turned out to be the first direct confirmation of Newton's theories.

Halley compared the orbits of 24 comets on the basis of observations. He was able to show that these bodies are moving in elliptical orbits like the planets, although some were very elongated, hardly distinguishable from parabolas. From his comparisons he detected that the orbit of the comet 1682 was almost identical with those of earlier years (1607 and 1531). Convinced that these three visits had been made by the same comet, he predicted that it would return in 1759. He knew he would not live to see if his calculations proved to be correct. But he expressed the hope that *"candid posterity will not refuse to acknowledge that this was first discovered by an Englishman"*. Indeed on Christmas night in 1759, almost 17 years after Halley's death, the comet was seen again just as Halley had predicted. This comet now carries his name.

No other comet has been observed over such a long period of time. From the time this comet was mentioned for the first time – by the Chinese in the year 240 B.C. – it has been seen at each periodic visit. The missing record of an observation in the year 164 B.C. has most likely been found just recently on a Babylonian table with cuneiform characters. In 12 B.C. Halley's comet appeared over Rome and was said to have presaged the death of Agrippa. In 66 A.D. it was seen over the city of Jerusalem before the city was destroyed. It appeared again in 451 during the Battle of Châlons when the Roman General Aetius defeated Attila the Hun, and it appeared in 1066 during the Battle of Hastings and was blamed for the defeat of King Harold's armies by William the Conqueror. Also when it returned in 1910, comet Halley caused considerable public concern – especially when it was known that the Earth was actually to pass through the comet's tail. Some worried people went to great lenghts to avoid being harmed by the poisonous cyanogen gas of which the tail was composed.

The English astronomer Halley was thus able to show that comets move around the Sun in an elliptical orbit and therefore belong to our planetary system. However, until 35 years ago it was still an open question as to where to place comets in the solar system and how they originated. Remarkable was the fact that the cometary orbits were not in a plane around the Sun like planetary orbits, but were distributed in a sphere around the Sun.

2 The Orbits of Comets and Their Origin

In 1950 the Dutch astronomer Jan Oort analyzed 19 cometary orbits, which had very elongated ellipses and orbital periods of more than 30,000 years. The cometary orbits had been very accurately determined. Oort's analysis showed that the majority of these comets come from regions situated in the outermost part of our solar system. These regions extend from about 30,000 AU to about 50,000 AU (50,000 AU correspond to 0.8 light years – the distance to the second star, Proxima Centauri, is 4.3 light years). Oort concluded from the analysis of cometary orbits that a cloud of comets must exist in this region.

In the meantime the use of electronic computers had made it possible to analyse a much greater number of cometary orbits, as well as to determine their orbits with greater accuracy. These calculations have confirmed the ideas of Oort. The cloud contains roughly 10^{12} comets with a total mass of 15×10^{18} g or about twice the mass of the planet Earth.

This comet cloud came into being, together with the planets and our own Sun, some four-and-a-half thousand million years ago, through contraction and condensation from an interstellar cloud. The greater part of these comets have since then been far away from the Sun, and because of the very low temperatures out there they have remained practically unchanged as if they were in an astronomical deep-freeze. Consequently, the material in a comet could be representative of the very earliest condensations in our solar system, and observations of its physical and chemical properties could yield information about what the solar system was like in its early stage.

But how does a comet from this cloud come close to the Sun, so that it can be observed from the Earth? The most likely explanation is that a star streaking past the solar system from time to time throws comets in this cloud out of their usual circular orbit into an elliptical one. This elliptical orbit could then be affected so strongly by the large planets, and a comet comes so near to the Sun, that it finally becomes visible. It can be estimated that about once every two million years other stars come close enough to our solar system for their gravitational force to have a detectable effect on the comet cloud.

3 The Physical and Chemical Properties of Comets

So what are comets? The observer on Earth sees a brilliantly shining object, which as its gets closer to the Sun shows the tail that is a characteristic of comets. It was supposed that the brilliance comes from a temporary atmosphere surrounding a solid nucleus. From the Earth, it was impossible to observe this nucleus directly, and one only had clues; in particular it has been possible for only the past few years to receive radar signals from comets, suggesting that they had a radius of about 1 kilometer.

From the observations of the outer envelope the American astronomer, Fred Whipple, arrived at the theory that the core was a dirty snowball, i. e. a conglomerate of ice and dirt. With a density of about 1 g/cm^3 and a diameter of a few kilometres, a core like this would have a mass of 15×10^{18} g, so that the total mass of all the 10^{12} comets comes to only about the order of magnitude of the Earth's mass.

If one of these nuclei – of the kind postulated by Fred Whipple – approaches the Sun, gas is given off by its surface as it heats up. This gas pulls off with its particles of dust, and this builds up a constantly expanding atmosphere, a coma, which can reach dimensions of 10,000 to 100,000 km. The ratio of gas to dust varies from one comet to another, and depends on the distance from the Sun; it can range from ten times more gas than dust to equal proportions. Gas emission begins to develop at about the distance of Jupiter (5 AU) with the lighter gases (such as carbon monoxide). At the distance of Mars, water vapour can also be observed.

As the comet gets closer to the Sun, a tail develops from the core, always pointed away from the Sun. Until the early 1950s there was no theoretical understanding of the observations of comet tails. The German astrophysicist Ludwig Biermann found the answer to this puzzle in the early fifties. A striking fact was that many comets build up not just one, but two tails. One tail consists of particles of dust that are accelerated by the pressure of sunlight and hence blown away in the opposite direction from the Sun. What has been unclear is how the very much smaller gas particles are accelerated, and how the very much longer tails of gas, which sometimes stretch for more than 100 million kilometres, come about. It was Biermann who, from the existence of this electrically-charged tail of gas and the high velocities noted in it (more than 100 km/sec), concluded that this phenomenon could be caused only by a corpuscular radiation from the Sun. Ludwig Biermann thus worked out the theory that a continuous stream of particles must be emitted from the Sun – a wind. This stream of particles was indeed found, in the early 1960s, by the first of the interplanetary space probes, and since then it has been known as the solar wind.

The solar wind is responsible for the comet's ionized tail, and the force it exerts on the gas particles is far stronger than the radiation pressure on the dust particles. Because of this, the dust tails are curved, and do not have the velocities we find in the almost straight, long tails of gas.

4 Artificial Cometary Tails – Experiments in Space

In 1957 and 1958 the Russians and the Americans launched the first artificial satellites. Very soon also space probes had been started with measuring instruments to probe the space where the planets and comets are moving. As Biermann had predicted, the existence of the solar wind could be proven and its strength could be measured for the first time. But how did solar wind interact with the cometary tails? So far this was not possible to observe directly. Also at that time it seems not feasible to send a space probe to a comet to carry out measurements in the neighborhood of a comet in particular within a cometary tail.

For this reason in the Max Planck Instiute for Extraterrestrial Physics the idea was developed to create an artificial cometary tail in space in order to study its interaction with the solar wind.

It would of course have been ideal if it had been possible to inject into Space the same atoms or molecules that had been found in the comet tails. First estimates showed, however, that several tons of carbon monoxide would be needed to produce in Space a cloud of CO+ visible from the Earth. Since there was no hope of having a rocket to carry such a heavy payload, this idea had to be abandoned. On the other hand the element barium was very well suited to the purpose: a few grams of barium would be enough to make a visible cloud, it would be rapidly ionized by sunlight, and the barium ions could – when eluminated by the Sun – be observed from the Earth.

Since at that time there were however only small high-altitude sounding rockets available, we chose to use the upper atmosphere for this experiment, at a hight of between 200 and 250 km. After various, at first unsuccessful experiments in 1962 and 1963, we released the first barium cloud into the ionosphere above the Algerian Sahara in November 1964.

Though the starting point for the barium cloud experiments was our interest in comet physics, this new method has been and still is used for quite different scientific purposes by numerous groups in other countries, for studying the Earth's upper atmosphere. Using barium clouds it is possible to visualize the lines of force in the Earth's magnetic field, just as we make lines of magnetic force visible here on Earth by using iron filings. The observation of barium clouds allowed for the first time reliable measurements of electrical fields and of features of the plasma high up in the Earth's atmosphere. Outside the Earth's atmosphere, but still within its magnetosphere, we carried out only two barium-cloud experiments in the 1960s, using the ESA satellite HEOS and a big Scout rocket made available by the NASA. But it was still not possible to reach interplanetary Space and, consequently, the solar wind. Finally Gerhard Haerendel and his group at the Max-PlanckInstitute in Garching were able to achieve the original aim of the Barium cloud technique. On 27 December 1984, and on 18 July 1985, two Barium clouds were created in the solar wind outside the Earth's magnetosphere at a distance of 17.2 earth radii and 18.6 earth radii.

The opportunity for these experiments arose with the implementation of the AMPTE (Active Magnetospheric Particle Tracer Explorer) mission of the USA, West Germany and the United Kingdom. Although the prime aim of the mission was the long-range tracing of the transport of ions from the solar wind into the magnetosphere and further redistribution therein, it was ideally suited for the study of the initial interaction of the injected plasma with the ambient medium. To this end, a set of plasma and field diagnostic instruments was installed on the German

Ion Release Module (IRM) which carried the Barium and Lithium release containers. Furthermore, a subsatellite developed in the United Kingdom (UKS) was injected into nearly the same orbit as the IRM in order to provide another probe for the plasma and field perturbations created during the release experiments. In addition to the in-situ diagnostics, optical observations from the Earth contributed valuable information about the dynamics of the solar wind interaction with the seeded plasma. 2 kg of Barium were released each time and one was able to observe the interaction of the artificial cloud with the solar wind and the surrounding magnetic field. These measurements can be compared now with the observations in the near neighborhood of Comet Halley from the spacecraft sent to Comet Halley.

5 The Preparation of the Space Missions to Halley's Comet and the Spacecrafts

Four space agencies – the Intercosmos of the USSR Academy of Sciences, the Japanese Institute of Space and Astronautical Science (ISAS), the National Aeronautics and Space Administration (NASA) and the European Space Agency (ESA) – sent spacecraft to Halley's Comet or have been involved with Halley observations from space during the comet's present appearance. Intercosmos launched Vega-1 and Vega-2, ISAS launched Sakigake and Suisei, and ESA launched Giotto, while NASA was able to redirect a spacecraft that was formerly known as the International Sun-Earth Explorer ISEE-3 and was now renamed International Cometary Explorer ICE.

The encounter spacecraft complemented each other in fly by distance, ranging from 600 km to 7 million km, and comet heliocentric distances at the times of encounter ranging from 0.79 to 0.89 AU. The encounters all took place in March 1986, because Halley crossed the ecliptic plane on 10 March and the launch energy required was minimised if the encounter spacecraft could stay close to the ecliptic. The relative fly by speeds, i.e. the speeds in the comet's frame of reference, were all very high. Unfortunately, due to Halley's retrograde orbit, only fast fly by missions were possible.

The whole programmes of the different space agencies were coordinated by an Inter-Agency Consultative Group (IACG) which was formed in 1981 and has met since then very regularly.

Comet Halley was chosen for a number of good reasons:

1. Compared to most of the other comets, the data of its orbit were very well known.
2. The energy requirements for the spacecraft to reach Comet Halley were relatively low.
3. Although Comet Halley often approached the Sun, its dust and gas production rate can be compared with those of "new" comets.
4. Last but not least, the reason for choosing Halley was its name and the role it has played in history.

The Italian painter Giotto di Bondone saw Halley in 1301 and was so impressed by its appearance that he incorporated it a few years later as the "Star of Bethlehem" in one of the frescoes in the Scrovegni Chapel in Padua. The painting shows details of the comet's coma and tail not unlike the drawings made by scientists in the 19th century. This is why ESA has given the name "Giotto" to its comet mission to Halley.

The scientific objectives for the missions have been the following:

1. To measure the production of dust and gases, the size and the shape of the dust particles as well as to define the chemical and isotopic mixing of the gas and dust particles,
2. to investigate the physical processes and chemical reactions in the surroundings of the comet as well as the interaction of the comet with the solar wind plasma, and
3. to discover the nucleus and, if it really exists, to take photos of it with a resolution of up to 50 metres.

The scientific experiments on the various spacecraft complemented and supported each other. The Giotto spacecraft is spin-stabilized, with a nominal spin period of 4 sec. Its diameter is 1.86 m, and the height from the tip of the tripod to the bottom adaptor ring is 2.85 m. At launch Giotto weighed 960 kg.

6 The Giotto Mission

After five years' development work and a great many tests, Giotto was launched on schedule from Kourou in French Guiana on 2 July 1985, by the European Ariane rocket. Its flight was to last nine months. Giotto was put into its orbit so accurately that any course corrections would be needed only a couple of days before the flyby. During its flight, all the instruments were switched on and tested several times.

The nine months of the flight were full of tension for everyone concerned, and were not entirely free of worries. The least of our worries was that Halley's Comet would suddenly disappear. But I shall not easily forget the early morning of 24 January when I was woken up with the news that radio contact with Giotto had been lost. As this was exactly the date on which the American Voyager probe was to fly by the planet Uranus, we did for a while not have the big American reflector telescopes immediately available for sending radio commands to Giotto. At two o'clock the following morning I was relieved to hear at last, from California, that contact with Giotto had been reestablished.

At long last everything was ready for the moment when Giotto was to fly by the core of Halley's Comet, at 3 minutes past midnight on 14 March. It was planned that Giotto would fly by at a distance of 500 kilometres, plus 40 km, on the Sun side of the Comet's nucleus. Final corrections were to be made on 9, 11 and 12 March. To make these we however needed exact data on the Comet's orbit; the normal astronomical observations made from the ground were inadequate. Because of this, there had long before been an international sharing-out of the work with the Intercosmos and NASA as agreed in the Inter-Agency Consultative Group.

The two Russian probes Vega 1 and 2, which flew by Halley on 6 and 9 March 1986 at a distance of 8890 to 8030 kilometres, pinpointed the exact position of the

Comet, and the data was then passed from the Russian control centre in Moscow to the European control centre at Darmstadt. For this one needed however also to know the exact position of the two Russian probes, which could not be determined by the Russians but only with the help of the big American antennas. In this way, finally, Halley's position was plotted to within 40 kilometres. This showed that after almost nine months' flight and travelling 4.44 million kilometres Giotto would, if no course corrections were made from the ground, pass the Comet at a distance of about 800 kilometres, plus or minus 20 km.

Opinions differed among the experimenters as to how near Giotto should come to the Comet, and they argued about this closest distance up to 24 hours before the event. The camera team, especially, wanted to get no closer than 1,000 km, while most of the other experimenters wanted 500 kilometres. They were prepared to take the risk of the probe being totally destroyed by the impact of dust particles, despite its protective shield. In the end, all concerned agreed on a distance of 540 kilometres. Ground control consequently made a slight course correction, 24 hours before the planned time of flyby. The probe was by then 140 million kilometres away from the Earth, and a radio signal needed about 8 minutes to travel this distance.

During the night of 13 to 14 March this stage was reached, after the first comet particles had already been recorded on 12 March at a distance of more than 7 million kilometres from the core. The experimenters were able to follow the instrument data being received at Darmstadt in real time, and these were released to the public a few minutes later. All the experiments worked without a hitch. The closest distance actually reached from the Comet's core was 605 kilometers.

Though Giotto was meant to fly by very close to the Comet's core, it was expected that it would survive. Between eight and sixteen seconds before the closest encounter Giotto was struck by a relatively large dust particle, which made the probe rock. Because of this the antenna was no longer pointing exactly at the Earth, and radio contact was suddenly lost. Such an eventuality had however been foreseen by the engineers, and they had built nutation dampers into Giotto; 32 minutes later these had reduced the spacecraft's nutating movements enough for the variation in antenna pointing to be once more within 1 degree of the required direction, and radio contact was restored.

The two Japanese spacecraft had a fly by-distance of 151,000 km on 8 March and of 7 million km on 11 March, the American Explorer ICE a distance of about 28 million km on 25 March.

Decisive for the success of the space mission was the camera, developed by the Max Planck Institute for Aeronomie in Germany. The quality of the camera had been even better than expectation. The passing velocity of Giotto with respect to the nucleus of comet Halley was 250,000 km/h. The camera had to take pictures from such a fast moving object. One comparison could demonstrate the performance of the camera: One would be able to photograph a pilot in an airplane that passing at a distance of 160 m with sound velocity or equal to 1,200 km/h with a resolution of 4 mm.

For the first time the nucleus of a comet could be photographed. Its surface as well as its size and shape could be observed and measured. As the theory predicted, it is comparable to a dirty snowball and it is one of the darkest bodies in the solar system. The pictures show a lengthy, non-spherical nucleus, comparable to a potato. The long axis has a length of about 15 km, while the smaller one between 7 to 10 km. The surface is irregular and shows spherical structures, quite comparable to craters, valleys and hills. Beside these pictures the other measuring instruments like mass spectrographs and dust counters have given very important results on the nature of this comet.

Giotto suffered some damages during the fly by due to the impact of dust particles. After the encounter the instruments were checked, all worked perfectly with the exception of the camera, which apparently was damaged by the dust impact. It was possible to retarget Giotto to return to the neighborhood of the Earth on July 1990. Using an Earth gravity assist it could be redirected towards another comet, namely to comet Grigg-Skjellerup. On June 10, 1992 Giotto passed his second comet in a distance of about 200 km and again interesting and important results were received on the Earth. At this time Giotto was at a distance of 314 million km and had travelled about 64 billion km from the Earth.

The Spectrum of Turbulence

Siegfried Grossmann

1 Introduction

The clustered distribution of interstellar matter raised C.F. von Weizsäcker's interest in turbulent flow with its nested vortex structures, its intermittent distribution of strongly active, dissipative turbulent bursts amidst more quiet regions, all this strongly fluctuating in time. Evidently many time scales are present, the smaller vortex structures circulating faster, being advected by the slower, larger ones. Also the spatial structures of the turbulent vortices or *eddies* display many scales. The prevailing impression of a turbulent flow field is its structural similarity on the various scales: zoomed smaller parts of the flow field look like larger ones, in a statistical sense. Such systems are properly described by power or scaling laws of the physical quantities of interest.

An observable quantity of prime importance in turbulent flow is the distribution of energy $\langle v^2(r) \rangle$ among the eddies of various sizes r, with $v(r, t) = u(x+r, t) - u(x, t)$, is furthermore the characteristic time scale τ_r of such r-eddies, is the typical energy dissipation rate distribution in the flow, $\varepsilon(x, t) = \nu u_{i|j}^2$, its mean $\varepsilon = \langle \varepsilon(x, t) \rangle$, and is spatial correlation over r-eddy distances $\langle \delta\varepsilon(x+r)\delta\varepsilon(x) \rangle$, with $\delta\varepsilon(x, t) = \varepsilon(x, t) - \langle \varepsilon \rangle$. Here, $u(x, t)$ is the Eulerian velocity field of the flow and $u_{i|j}$ the derivative of the i-th velocity component with respect to its variable component x_j.

As we know today, all these quantities seem to obey power law behaviors as a function of the eddy size r, e.g.

$$\langle v^2(r) \rangle \propto r^\zeta . \tag{1}$$

2 Historical Remarks

It is one of the great achievements of science in the first half of the last century to realize the importance of such power laws or *scaling behavior* in physical systems, if these are hierarchically ordered, and if all scales interact strongly. Such hierarchical structures are also called *scale free*, because except of externally given upper or lower length scales there is no internal characteristic scale in the system.

In turbulence the upper scale is the external, geometrical length scale L at which the energy is permanently injected, either by shearing of the boundaries or under pressure drop. The lower, so-called inner length scale is defined as the one at which viscosity becomes dominant; viscosity is described by the kinematic viscosity v, units mm^{2}s^{-1}, being about 1 for water or 15 for air.

Power laws in isotropic, homogeneous turbulent flow, which is an idealized notion of real flow, have been proposed by several authors independently in the forties of the last century, C.F. von Weizsäcker being one of them. In 1941 two papers by Kolmogorov and by Oboukhoff appeared in the USSR [1, 2], in which $\zeta = 2/3$ and the corresponding exponent $-5/3$ for the wave number spectrum $E(k) \propto k^{-5/3}$ were derived.

In 1945, C.F. von Weizsäcker and W. Heisenberg, during their unintended stay in Farmhall, UK after the war, developed a similar scaling theory of turbulent flow. They submitted in December 1946. Their manuscripts were published in the first volume of Z. Physik after the war, in 1948 [3,4]. During this long delay time between deriving and publishing their theory they, of course, became aware of the previously not available results by Kolmogorov and Oboukhoff. The physical idea was the same: In turbulent fluid flow it should be the stationary rate of flow of energy through this open system, which is the only relevant control parameter together with the lack of a preferred scale. Such scale freedom is implied by the nonlinear character of the equations of motion, the Navier–Stokes equations.

If the energy flow rate, expressed in terms of the energy input rate $\propto U^3 L^{-1}$ or of the energy dissipation rate ε, all taken per unit mass, thus $[\varepsilon] = [U^3 L^{-1}] = $ m^{2}s^{-3}, is the only relevant control parameter for the eddy energy $\langle v^2(r) \rangle$ besides the eddy size r itself, and if scaling, i.e., a power law like (1) is assumed to hold, dimension counting is sufficient to derive

$$\langle v^2(r) \rangle = b\varepsilon^{2/3} r^{2/3} . \tag{2}$$

Analogously, power counting of units leads to the spectral law

$$E(k) = C_K \varepsilon^{2/3} k^{-5/3} . \tag{3}$$

Here the Kolmogorov and Oboukhoff constants b and C_K are (via Fourier transform) related by $b = 4.822 \times C_K$. b is in the range around 8.4.

Von Weizsäcker and Heisenberg also became aware of another (short) note, in fact the abstract, of a contribution of Lars Onsager to the 1945 meeting of the New York Physical Society, in which he stated in a few lines that under control of the energy dissipation rate the turbulent spectrum is $\propto k^{-5/3}$ [5]. These historical remarks may shed light on the progress of turbulence research in that time.

3 Turbulent Diffusion

It was first Lewis Fry Richardson who ingeneously very early found and introduced scaling behavior when describing turbulent fluid flow [6, 7]. He studied particle pair dispersion in the earth's atmosphere and observed that the diffusivity K, as usual measured in cm²s⁻¹, evidently depends on the scale r of the particle pairs' separation r,

$$\frac{K(r)}{\text{cm}^2\text{s}^{-1}} = 0.6 \left(\frac{r}{\text{cm}}\right)^{4/3}. \qquad (4)$$

In [6] the constant was given to be 0.2, in [7] the three dimensional nature of the diffusion process was taken into account and the constant is reported to be 0.6. $K(r)$ varies over about 11 orders of magnitude, while the length scales include the flights of the seeds of dandelion with their puffy parachutes, the flights of the balloons in the competitions starting at the beach of Brighton, UK, as well as observing tornado trajectories, and several other examples. Altogether these range from centimeters to thousands of kilometers, 10^0 cm $< r < 10^8$ cm.

Richardson missed the Kolmogorov, Oboukhoff, von Weizsäcker, Heisenberg, Onsager scaling of $\langle v^2(r) \rangle$ and of $E(k)$ by a hair's breadth. Namely, from (4) one can conclude that $K(r)/r^{4/3}$ is a scale independent constant of dimension cm$^{2/3}$s^{-1}, saying that its third power cm²s⁻³ is a scale independent constant, which characterizes the mean turbulent atmosphere. It can be interpreted as an energy (per mass) in cm²s⁻² per time in s, i.e., as the input or output rate of energy, ε. Equation (4) then reads

$$K(r) = C\varepsilon^{1/3}r^{4/3}, \text{ with } C\varepsilon^{1/3} = 0.6\,\text{cm}^{2/3}\text{s}^{-1}. \qquad (5)$$

The Richardson constant C, according to very recent measurements or numerical simulations [8–11], is of order 0.5 ± 0.3. This corresponds to a mean atmospheric dissipation rate $\varepsilon \simeq 0.2\,\text{cm}^2\,\text{s}^{-3}$. A mean field like theory, as it shall be described in Sect. 5, which can be considered as a refined version of the basic ideas in [1,3], gives in relaxation approximation $C \approx 13$ (see [12, 13]). Recently it became clear that dynamical memory effects are strong and cannot be ignored in the time correlation decay [14, 15]. That can be taken into account within a mode coupling theory. In mode coupling approximation $C \approx 0.5$ through ≈ 0.8 is obtained [15].

4 Scaling Law of Turbulence

C.F. von Weizsäcker [3] analyzed the idealized state of isotropic, homogeneous turbulence primarily by position space considerations. He subdivided the flow volume into a hierarchically nested set of cubes having linear scales $L_0 > L_1 > \cdots > L_n > \cdots$. Surprisingly, although the range of validity of the scaling cascade of eddies of decreasing sizes L_n has been addressed by von Weizsäcker, he did not introduce the lower limit L_{vis} of the spectral range, which is provided by the viscosity. One can do this by comparing the nonlinear, inertial term in the Navier–Stokes equation, $u_{\text{vis}}L_{\text{vis}}^{-1}u_{\text{vis}}$, with the viscous term, $\nu u_{\text{vis}}L_{\text{vis}}^{-2}$. These terms balance if $u_{\text{vis}}L_{\text{vis}} = \nu$. Using that the velocity on scale L_{vis} either is given by the inertial range formula (see later, (6), or cf. (2)) $u_{\text{vis}} \sim (\varepsilon L_{\text{vis}})^{1/3}$ or by the viscous range expression $u_{\text{vis}} \sim \sqrt{\varepsilon/\nu}L_{\text{vis}}$, one obtains $L_{\text{vis}} \sim (\nu^3/\varepsilon)^{1/4}$ in both cases.

From measurements and numerical data we conclude that L_{vis}, the lower length scale of the scaling range, is about 10 times of what is called today the inner, viscous or Kolmogorov scale $\eta \equiv (\nu^3/\varepsilon)^{1/4}$, i.e., $L_{\text{vis}} \approx 10\eta$.

Each scale L_n is chosen to be by a factor of $\delta < 1$ smaller than the previous one L_{n-1}, i.e., $L_n = L_{n-1} \times \delta = L_0 \times \delta^n$. The velocity field is considered as a superposition of contributions of these boxes, $\boldsymbol{u} \simeq \boldsymbol{u}_0 + \boldsymbol{u}_1 + \cdots + \boldsymbol{u}_n + \cdots$. Each u_n is a successive average of $\boldsymbol{u}$ over the cubes of order n, having the meaning of L_n-size motions. In addition it is time averaged. The spatial derivatives of the u_n are described by u_n/L_n. The energy loss on scale L_n then is $S_n \sim \nu_n \sum_{m=0}^{n}(u_m/L_m)^2$, the sum depending on the large scale gradients with $m \leq n$ and on the effective kinematic viscosities $\nu_n = L_n \sum_{m=n+1}^{\infty} u_m$ depending on the small scale velocity contributions u_m with $m > n$. Numerical constants of order unity are assumed to be independent of the scale labels m or n.

The basic assumption in von Weizsäcker's argument then is, that the energy loss on any scale L_n has to be independent of the scale number n, i.e. $S_n = S$, some constant value. That comes from the stationarity of the turbulent flow: The energy, which is put in at the largest scale, must on average also be lost on the smallest one. Also, S cannot depend on the arbitrary boxing with the L_n and not on the decomposition into the u_n. The scaling Ansatz $u_{n+1} = \xi u_n$ etc. straightforwardly then leads to $S_n \sim u_n^3/L_n$. Under the n-independence of $S_n = S$ this immediately implies the famous power law

$$u_n \sim L_n^{1/3}. \tag{6}$$

The corresponding energy spectrum in wave number space, $k_n \sim L_n^{-1}$, is

$$E(k_n) \sim k_n^{-5/3}. \tag{7}$$

From this mixture of argument and modeling by simple formulae, which attest a deep insight into the physics of turbulent flow, together with the velocity scaling von Weizsäcker also derived the turbulent viscosity $\nu_n \sim L_n u_n \sim L_n^{4/3}$ and the characteristic time scales $\tau_n \sim \ell_n/u_n \sim \ell_n^{2/3}$ for the L_n-eddy turnovers.

The reader will realize that these arguments have been repeated since again and again. They are used in textbooks, see e.g. [16, 17], they have been generalized to cope with non-space filling intermittent clustering of turbulent activity, the so-called β-model, and even the modern multifractal models [17, 19] argue along this line, generalizing the von Weizsäcker modeling appropriately.

Thus the von Weizsäcker argument from 1945/8 can be found ever since to model developed turbulence. In [3] also a careful discussion is given on the limits of validity and why the Reynolds number has to be large enough. These qualitative estimates may be less convincing but serve to emphasize the essence: turbulent fluid flow is an open system, whose external control parameter is the rate of flow of energy per mass, which maintains the turbulence. And that quantitatively the spectral law (6) does not presume more than the counting of units, he expressed by writing "dass die mathematische Form des Gesetzes kaum mehr als eine Dimensionsbetrachtung voraussetzt" [3, p. 621]. An important notion is the *turbulent exchange* ("turbulenter

Austausch" [3]), what is called turbulent viscosity today, and its scale dependence $v_n \sim L_n u_n$. As usual, the kinematic viscosity is the product of the relevant scale and the relevant velocity. The molecular kinematic viscosity is given analogously by $\nu = \lambda_F v_{\text{th}}$, the product of the mean free path λ_F and the thermal velocity v_{th}, while the turbulent viscosity measures the product of the n-eddy scale L_n and the n-eddy velocity u_n.

Von Weizsäcker's paper finishes with a comparison of his results with some data, first of atmospheric turbulence, then of the motion of interstellar matter in the Milky Way. Let us consider the latter. $L_0 \approx 30\,000$ light years is the distance of the sun from the center of the Milky Way [20]. The typical diameter L_n of an average nebula is about 30 light years. With the 1/3 scaling of the velocity one finds $v_n = v_0 \times (L_n/L_0)^{1/3} = v_0/10 = 30\,\text{km s}^{-1}$ if for the rotational velocity of the Milky Way at the sun's position the value $v_0 = 300\,\text{km s}^{-1}$ is used. This result for v_n agreed satisfactorily with the observations.

5 Mean Field Theory of Turbulence

The energy distribution $\langle v^2(r) \rangle$ of r-eddies can be measured and compared with equations (2) and (3). While the scaling argument of C.F. von Weizsäcker did not allow to fix the prefactors b or C_K, this was achieved in Heisenberg's paper [4], published back-to-back with von Weizsäcker's [3]. It was formulated in Fourier instead of position space as done in the latter one. The section of [4] dealing with the spectrum is a standard reference today; the section on C_K presented a rather sophisticated argument that did not find widespread notice in later work, but Heisenberg succeeded here to derive a value for C_K.

A conclusive and coherent theory of turbulence has to be based on the Navier–Stokes equation

$$\partial_t u_i = -(\boldsymbol{u} \cdot \nabla) u_i - \text{grad}_i\, p + \nu \Delta u_i + f_i\,, \qquad (8)$$

together with incompressibility div $\boldsymbol{u} = 0$. A derivation of equation (2) from the Navier–Stokes equation (8) in a mean field type approach will be presented now; it has been derived in [21]. It leads to the 2/3 scaling of eq. (2) including the numerical factor in satisfactory agreement with the experimental one $b_{\text{exp}} = 8.4$. The reader will realize, how many aspects of this theory can be found in von Weizsäcker's and Heisenberg's seminal papers. The following sketch of the theory for the eddy energy distribution or, what is the same, the structure function $D(r)$, emphasizes these analogies; mathematical details can be found in [21], cf. also [22, 23].

The first step is to introduce proper objects which correspond to the r-eddies u_n on scale $L_n \hat{=} r$. This is done by defining the *super-r-scale* field

$$\boldsymbol{u}^{(r)}(\boldsymbol{x}, t) \equiv \int_{V_r} \frac{d^3 y}{V_r} \boldsymbol{u}(\boldsymbol{x} + \boldsymbol{y}, t) \equiv \langle \boldsymbol{u}(\boldsymbol{x} + \boldsymbol{y}, t) \rangle_{\boldsymbol{y}}^{(r)}\,, \qquad (9)$$

together with the *sub-r-scale* field

$$\tilde{u}^{(r)}(x, t) = u - u^{(r)} . \tag{10}$$

The y-integral extends over the sphere with radius r and volume V_r, located at position x. While $u^{(r)}(x, t)$ contains all scales larger than r, the $\tilde{u}^{(r)}$-field contains the smaller ones. It varies rapidly in space and usually has a smaller amplitude. The super-r- and sub-r-scales are approximately statistically independent. The scale boundary r is allowed to vary continuously through all scales, including the inertial subrange ISR (in which the nonlinear term of the Navier–Stokes equation dominates) as well as the viscous subrange VSR (in which the viscous term of the Navier–Stokes equation is dominant).

The energy in the super-r-scales is closely connected with the r-averaged structure function $D(r) = \langle\!\langle (u(x+r, t) - u(x, t))^2 \rangle\!\rangle = \langle\!\langle v^2(r) \rangle\!\rangle$. The double brackets $\langle\!\langle \cdots \rangle\!\rangle$ denote the statistical average, the single brackets $\langle \cdots \rangle_y^{(r)}$ indicate the V_r-average. It is

$$\langle\!\langle (u^{(r)})^2 \rangle\!\rangle = \langle\!\langle u^2 \rangle\!\rangle - \frac{1}{2} \langle\langle D(y_1 + y_2)\rangle_{y_1}^{(r)}\rangle_{y_2}^{(r)} . \tag{11}$$

While the subscale velocity field evidently is Galilean invariant, the superscale energy is – up to a constant that vanishes under differentiation – expressed by the also Galilean invariant structure function $D(r)$. Using the sub- and super-fields therefore leads to a Galilean invariant theory in position space. The r-dependence for $r \to 0$ will be analytical, smooth. No singularities will occur. Note the similarity of the decomposition $u(x, t) = u^{(r)}(x, t) + \tilde{u}^{(r)}(x, t)$ with the Reynolds decomposition. But note also the difference. The two terms are **not** obtained by time averaging the flow field but by position space averaging over an r-sphere with variable radius r, appropriate to describe r-eddies in isotropic flow. Both terms show the full time dependence.

To calculate the power balance of the superscales, $\langle\!\langle u^{(r)} \partial_t u^{(r)} \rangle\!\rangle = 0$, one r-averages the Navier–Stokes equation (8). The resulting $u^{(r)}$-equation contains, among others, a contribution $\langle (\tilde{u}^{(r)} \cdot \nabla) \tilde{u}^{(r)} \rangle_y^{(r)}$, describing the acceleration due to the contribution of the (small) subscales. The following terms enter the power balance:

$$0 = \langle\!\langle u^{(r)} \cdot f \rangle\!\rangle + \nu \langle\!\langle u^{(r)} \Delta_x u^{(r)} \rangle\!\rangle - \langle\!\langle u^{(r)} \cdot \langle (\tilde{u}^{(r)} \cdot \nabla) \tilde{u}^{(r)} \rangle_y^{(r)} \rangle\!\rangle . \tag{12}$$

The first term is the energy input into all scales above r. But since the external stirring force f, which produces the turbulence and which represents the boundary effects, depends on the external scale L only and contains no modes with $r < L$, varying r in $u^{(r)}$ has no influence on the first term, which therefore is independent of r. It represents the whole input of energy and must be equal to the total energy dissipation rate, $E_{in} = \varepsilon$.

The second term in (12) is the viscous dissipation $-E_d^{(r)}$. With (11) one finds

$$E_d^{(r)} = (\nu/2) \langle\langle \Delta_{y_1} D(y_1 + y_2)\rangle_{y_1}^{(r)}\rangle_{y_2}^{(r)} . \tag{13}$$

The third term, denoted as $-E_t^{(r)}$, describes the energy transfer balance from the super-r-scales to the sub-r-scales. To calculate it, one needs the equation of motion for $\tilde{u}^{(r)}$ and integrates this formally along the Lagrangian trajectory from $-\infty$ till time t. The resulting expression of type $\langle\!\langle u^{(r)} u^{(r)} \tilde{u}^{(r)} \tilde{u}^{(r)} \rangle\!\rangle$ is factorized under the mentioned assumption of statistical independence of the sub- and superscales into $\langle\!\langle u^{(r)} u^{(r)} \rangle\!\rangle \langle\!\langle \widetilde{u}^{(r)} \widetilde{u}^{(r)} \rangle\!\rangle$. $\widetilde{\boldsymbol{u}}^{(r)}$ fluctuates on a much smaller scale then $\boldsymbol{u}^{(r)}$. Also, since the time decay is dominated by the (fast) small eddies, the large eddies are taken care of only in equal time approximation, giving a contribution $N^{(r)} \times \langle \partial \partial D \rangle_y^{(r)}$. The prefactor $N^{(r)}$ is the Lagrangian time correlation function integral

$$N_{jk}^{(r)} = \int_{-\infty}^{0} d\tau \, \langle\!\langle \widetilde{u}_j^{(r)}(\boldsymbol{x},t) \widetilde{u}_k^{(r)}(\tau|\boldsymbol{x},t) \rangle\!\rangle \,. \tag{14}$$

This can be interpreted as the Kubo formula for a transport coefficient, here the *eddy viscosity* exerted by the sub-r-scales on the super-r-scales. $N^{(r)}$ is a Galilean invariant object, i.e., not spoiled by advection.

The time correlation decay has been studied in [24, 25] by continued fraction transform. In 1-pole approximation $N^{(r)}$ is

$$N^{(r)} \sim D^2(r)/(2\varepsilon)\,, \quad r \in ISR\,. \tag{15}$$

It might be useful to indicate the argument leading to this result. Obviously $N^{(r)} \sim D \times$ decay time. The latter is the inverse of the decay rate, $\langle \mathcal{L} \rangle = \langle\!\langle v d_t v \rangle\!\rangle / \langle\!\langle v v \rangle\!\rangle$. The denominator again is the structure function D, the numerator becomes $2\varepsilon - v \Delta_r D(r) \approx 2\varepsilon$, from which then expression (15) results. It remains to convince oneself of $\langle\!\langle v d_t v \rangle\!\rangle = v \langle\!\langle v \Delta_x v \rangle\!\rangle \approx 2\varepsilon$: start from $v \partial_{r_i} \partial_{r_i} \langle\!\langle (u_j(\boldsymbol{x}+\boldsymbol{r}) - u_j(\boldsymbol{x}))(u_j(\boldsymbol{x}+\boldsymbol{r}) - u_j(\boldsymbol{x})) \rangle\!\rangle$. It thus is this Kubo formula (14), which introduces ε as the relevant control parameter for the eddy viscosity!

The power balance then reads

$$E_t^{(r)}(\{D\}) + E_d^{(r)}(\{D\}) = \varepsilon\,. \tag{16}$$

All contributions in (16) are expressed in terms of averaged structure functions $D^{(r)}$, of v, and of ε. In power law ranges, the r-averages of the D's reproduce these power laws although with other prefactors. We obtain the final equation for $D(r)$:

$$\left[\frac{D^2(r)}{b^3 \varepsilon} + v \right] \frac{3}{2} \frac{1}{r} \frac{dD(r)}{dr} = \varepsilon\,. \tag{17}$$

The first term represents the eddy viscosity $v_{\text{turb}}(r)$ or the energy transport between the super- and the sub-scales. The second one describes, how the molecular viscosity contributes. Both add up to the total energy dissipation. It is $v_{\text{turb}}(r) \sim D^2(r)/\varepsilon \sim \varepsilon^{1/3} r^{4/3}$. This eddy viscosity thus has von Weizsäcker scaling, cf. (27) of [3].

In [21] we calculated b numerically from the full power balance equation (16), starting from $D = 0$ for $r = 0$ and integrating to $D(r)$ for arbitrary r. We found $b = 6.3$, near to the experimental value $b_{\exp} = 8.4$.

It is straightforward to integrate the simplified equation (17). Suffice it to consider the two limiting cases. In the viscous subrange the v-term dominates, immediately leading to

$$D(r) = \frac{1}{3} \frac{\varepsilon}{v} r^2\,, \quad 0 \le r \le 10\eta, \quad \text{VSR}\,. \tag{18}$$

In the inertial subrange the eddy viscosity $\nu_{\text{turb}}(r)$ dominates, implying

$$D(r) = b\varepsilon^{2/3}r^{2/3}, \quad 10\eta \leq r \leq L, \text{ ISR}. \tag{19}$$

The complete solution, starting from $D(0) = 0$, can be found after integrating (17). One obtains

$$D^3(r) + 3\nu\varepsilon b^3 D(r) = b^3\varepsilon^2 r^2. \tag{20}$$

See Fig. 1 for an impression of $D(r)$ versus r together with some measured data points. It seems obvious that the offered position space, Lagrangian type theory describes the structure function or eddy energy distribution rather well. The theory has no cutoff, no singularity, needs no scaling ansatz, and needs no adjustable parameter.

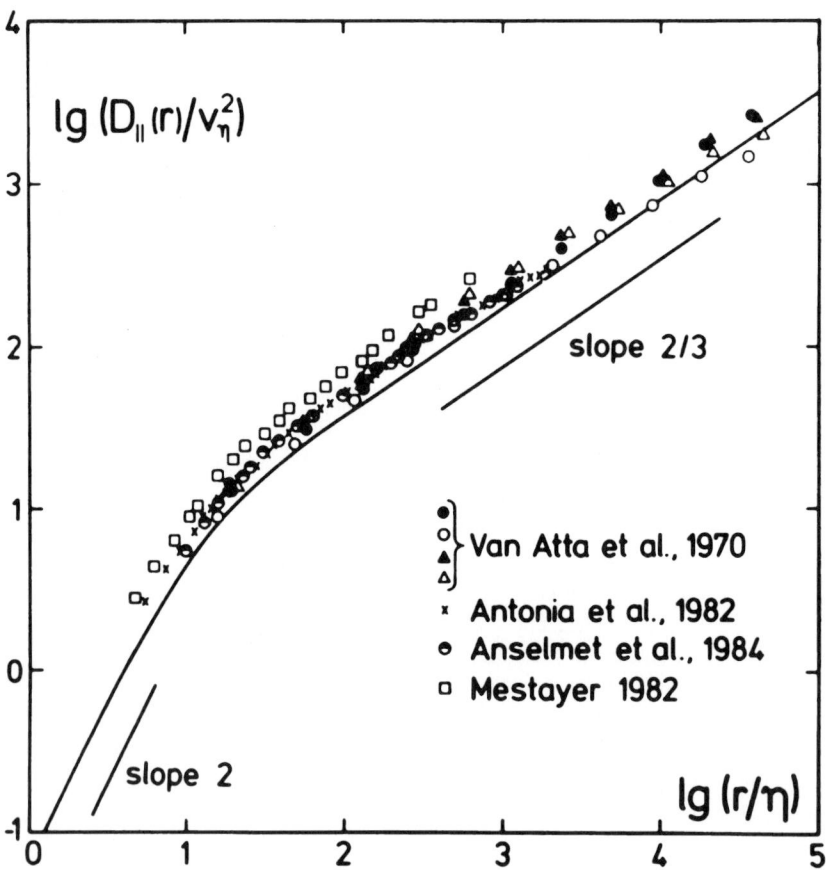

Fig. 1. Longitudinal part of the second order structure function $D_\|(r) = \langle\!\langle (u_\|(x+r,t) - u_\|(x,t))^2 \rangle\!\rangle$ versus r from theory together with data, from [21].

6 Temperature Structure Function

One might ask, if the scaling methods presented in the previous chapters can be applied to other turbulence problems. A successful example is thermal advection, if the turbulent flow field and the temperature field can be considered as isotropic and homogeneous.

Let us study the passive scalar case. $\theta(x, t)$ be the temperature field. Its equation of motion reads as

$$\partial_t \theta(x, t) = -(u \cdot \nabla)\theta + \kappa \Delta \theta + f_\theta . \tag{21}$$

Here $u(x, t)$ is the given flow field, κ the transport coefficient, in this case the thermal diffusivity, and f_θ the forcing of the θ-field, physically due to the boundary conditions. The temperature dissipation is described by the rate $\varepsilon_\theta = \kappa \langle\!\langle \mathbf{grad}\,\theta \cdot \mathbf{grad}\,\theta \rangle\!\rangle$, having dimension $[\varepsilon_\theta] = K^2\,s^{-1}$ (with K = degree Kelvin). From right to left the terms in (21) describe the external forcing, the diffusive spreading, and the advection by the flow.

The structure function of interest now is $D_\theta(r) = \langle\!\langle (\theta(x+r, t) - \theta(x, t))^2 \rangle\!\rangle$. The fluid's flow field is described as before by its own structure function $D(r)$; in the passive scalar case this is considered as given. We again can introduce the variable range decomposition, i.e., we divide the passive scalar field into the averages over r-spheres, $\theta^{(r)}(x, t) = \langle \theta(x+y, t) \rangle_y^{(r)}$, together with the sub-$r$-scale field $\widetilde{\theta}^{(r)} = \theta - \theta^{(r)}$. Applying the method presented in Sect. 5 we obtain, cf. [26],

$$\left(\beta_\theta \frac{D^2(r)}{\varepsilon} + \kappa\right) \frac{3}{2}\frac{1}{r}\frac{dD_\theta}{dr} = \varepsilon_\theta . \tag{22}$$

This equation for the temperature structure function $D_\theta(r)$ is the analogue of the corresponding equation (17) for the velocity structure function. The parameter β_θ can be obtained from the numerical integration of the full equation for D_θ, including all r-averages. We calculated the value $\beta_\theta = 9.93 \times 10^{-3} \approx 1/100$. The first term in (22) describes the advection of heat, the second its diffusive spreading, and the r.h.s. is the thermal dissipation rate. Equation (22) is an approximation; because it is linear in D_θ and the velocity structure function $D(r)$ is considered as given, it can be solved explicitly:

$$D_\theta(r) = \varepsilon_\theta \sqrt{\nu/\varepsilon}\,(2/3)\,Pr \int_0^{r/\eta} \frac{a\,da}{1 + \beta_\theta\,Pr D^2(a)/v_\eta^4} . \tag{23}$$

$Pr = \nu/\kappa$ denotes the Prandtl number of the fluid, $\eta = (\nu^3/\varepsilon)^{1/4}$ the Kolmogorov inner length scale, $v_\eta = (\varepsilon\nu)^{1/4}$ the Kolmogorov velocity. Equation (23) is parameter free and valid for all Pr and for all scales r, and therefore represents a complete description of the passive scalar structure function.

Depending on Pr and on scale r there are several ranges. First, if $\beta_\theta\,Pr\,D^2/v_\eta^4$ is much less than 1, we have the *diffusive subrange* without advective effects.

$$D_\theta(r) = \varepsilon_\theta \sqrt{\nu/\varepsilon}\,(Pr/3)\,(r/\eta)^2 = \varepsilon_\theta\,r^2\,/\,3\kappa , \tag{24}$$

valid for $r \leq (9/(\beta_\theta\, Pr))^{1/4}\, \eta \equiv r_1$, if the fluid is in the VSR. The thermal diffusion dominates, $D \sim r^2$ is in the viscous range and therefore small.

If Pr is sufficiently small, but looking on larger scales r, the second term in the denominator of (23), $\beta_\theta\, Pr\, D^2/v_\eta^4$, may become large. Then we enter a second range, denoted as the *advection-inertial subrange*. Here we observe the temperature spreading by advection through the r-eddies.

$$D_\theta(r) = b_\theta\, (\varepsilon_\theta/\varepsilon^{1/3})\, r^{2/3}\,,\ \ b_\theta = 2.53\,,\ \ (Pr/3b_\theta)^{-3/4}\, \eta \leq r\,. \tag{25}$$

Under these conditions we find $D_\theta \propto r^{2/3}$ and thus the same 2/3-exponent as in the velocity structure function. This exponent $\zeta_\theta = 2/3$ is directly imported from the eddy energy distribution $\langle\!\langle v^2(r) \rangle\!\rangle = D(r) \propto r^{2/3}$.

For sufficiently large Pr the temperature structure function may develop a plateau, i.e., an r-independent third range, denoted as *advective-viscous subrange*. Since Pr is supposed to be large, the term $\propto Pr\, D^2(a)$ in the denominator of (23) is already large even for $r \in$ VSR, i.e., $D(r) \propto r^2$. That leads to

$$D_\theta = \varepsilon_\theta\, \sqrt{v/\varepsilon}\, (\pi/2)\, \sqrt{Pr/\beta_\theta}\,. \tag{26}$$

In this range the θ-transport is mediated by advection, i.e., the second term in the denominator of (23) contributes, but the flow still is in the viscous subrange.

For further details and previous references the reader is referred to [26]. As an interesting information it is pointed out, that the turbulent Prandtl number, which is defined as $v_{\text{turb}}(r)/\kappa_{\text{turb}}(r)$, the ratio of eddy viscosity and eddy heat diffusivity, is independent of r and has the value $Pr_{\text{turb}} = b_\theta/b \approx 0.40$. The experimental value is about $3/8.4 = 0.36$.

7 Outlook

The scaling behavior of turbulent fluid motion including the nondimensional constants in the structure function and in the diffusion law may thus be considered as being clarified to some extent. The deep insight that the relevant control parameter of turbulent fluid flow is the energy dissipation rate per mass, ε, together with the counting of units, as found by Kolmogorov, Oboukhoff, von Weizsäcker, Heisenberg, and Onsager, was in important step. The next step has been to determine the dimensionless prefactors together and simultaneously with the 2/3 scaling exponent, with interesting success, as I have decribed.

Where are we today?

It is known since long – in fact since soon after von Weizsäcker's paper had been published – that there seems to be a deviation $\delta\zeta$ from the exponent 2/3 in (1), (2), see [27–29] and many other authors since then,

$$\zeta = 2/3 + \delta\zeta\,,\ \ \delta\zeta \approx 0.03\,. \tag{27}$$

This is attributed mainly to the strong fluctuations of the local energy dissipation rate $\varepsilon(x, t)$ together with the nonlinear dependence of most physical quantities on ε. It is not only the second order structure function, but also the pth order ones

$$D_p(r) = \langle\langle v_\parallel^p(r)\rangle\rangle \propto r^{\zeta_p}, \tag{28}$$

which seem to show deviations from the classical scaling $\zeta_{p,0} = p/3$,

$$\zeta_p = p/3 + \delta\zeta_p. \tag{29}$$

There are several models to describe the deviations $\delta\zeta_p \neq 0$, denoted as log-normal-, as β-, or as multifractal models [18, 19, 28, 29], see also [17]; and there are similar further models in the literature. It is only the most recent development, which raises hope and excitement that one can understand and possibly calculate the intermittency corrections $\delta\zeta_p$ eventually from the nonlinear Navier–Stokes equation of motion (8). Most progress has been obtained for passive scalar (or passive vector) advection. For a report on the present state see [30].

The intermittency theory of turbulence has to go considerably beyond the ideas of von Weizsäcker and the other founders of turbulence scaling laws. To determine $\delta\zeta_p$ it is no longer possible to count only units. Dimensionally the Kolmogorov scaling $p/3$ exhausts the power law equations. Any deviation $\delta\zeta_p$ enforces that one has to introduce a physical length scale explicitly into the expressions for the structure functions,

$$D_p(r) \propto \varepsilon^{p/3}\, r^{p/3}\, (r/\ell)^{\delta\zeta_p}. \tag{30}$$

There are arguments that ℓ should be the outer length scale L rather than the inner, Kolmogorov scale η. One of the most direct ones is based on the experimental observation that $\delta\zeta_p$ as a function of p is convex downwards, $\zeta_{p+q} \geq \frac{1}{2}(\zeta_{2p} + \zeta_{2q})$. From the Schwarz inequality $|\langle\langle v_\parallel^{p+q}\rangle\rangle| \leq \sqrt{\langle\langle v_\parallel^{2p}\rangle\rangle \langle\langle v_\parallel^{2q}\rangle\rangle}$ together with the power law ansatz one finds $(r/\ell)^{\zeta_{p+q}} \leq (r/\ell)^{(\zeta_{2p}+\zeta_{2q})/2}$. Thus convexity downwards is obtained, if $r/\ell \leq 1$ for all r, i.e. if ℓ the largest available length scale $\ell = L$ is taken.

Also beyond Kolmogorov, von Weizsäcker, etc. scaling theory is the surprising experimental and numerical evidence of anisotropy in the scaling behavior. One finds different scaling of the longitudinal and transversal structure functions, $\zeta_{p,\parallel} \neq \zeta_{p,\perp}$, cf [31–36].

An interesting idea to describe and understand this observation is to expand the tensorial structure function $D_{ij\ldots k} = \langle\langle v_i v_j \cdots v_k\rangle\rangle$ in terms of the $SO(3)$ eigenfunctions [36, 37] and to find clean scaling for the $SO(3)$-invariant amplitudes in the various ℓ, m sectors, $\propto r^{\zeta_{\ell,m}}$.

Neither from the theoretical nor from the experimental point of view is fully developed turbulent fluid flow really understood from the Navier–Stokes equations as far as intermittency and anisotropy beyond Kolmogorov, Oboukhoff, von Weizsäcker, Heisenberg, Onsager scaling is concerned. We know today of several sources for deviations from pure scaling in the accessible large but finite Reynolds number range up to order 10^6 to 10^7. The effects of the physical boundaries are still visible as we have learned from high Rayleigh number thermal convection [38, 39]. The deviations from classical scaling $\zeta_{p,0} = p/3$, anyhow small and measurable with only rather limited accuracy, may contain several physical effects. Thus there is new and even higher excitement and mystery in the turbulence age beyond Kolmogorov, von Weizsäcker, etc.

This contribution in honor of Carl Friedrich von Weizsäcker has been devoted entirely to fully developed turbulence, as it was von Weizsäcker's interest in [3]. The reader might appreciate another remark, concerning the mechanism for the onset of turbulence. More than a century after Osborn Reynolds first observed it in experiment, this question has seen considerable recent progress. Heisenberg, in his thesis [40], was rather disappointed that he could not solve the instability puzzle. Meanwhile quite another, new and surprising mechanism for the onset of turbulence has been found [41–43], the *nonnormal-nonlinear* transition scenario.

The physics of turbulence thus has seen considerable progress but remains challenging and highly exciting, waiting for an even deeper stage of insight.

References

1. A.N. Kolmogorov: *Compt. Rend. Acad. Sc. (C. R. Acad. Nauk) USSR* 30, 301 (1941)
2. A.M. Oboukhoff: *Compt. Rend. Acad. Sc. (C. R. Acad. Nauk) USSR* 32, 19 (1941)
3. C.F. von Weizsäcker: *Z. Phys.* 124, 614 (1948)
4. W. Heisenberg: *Z. Phys.* 124, 628 (1948); *Proc. R. Soc. (Lond)* 195, 402 (1948)
5. L. Onsager: *Phys. Rev.* 68, 286 (1945)
6. L.F. Richardson: *Proc. R. Soc. (London)* A110, 709 (1926)
7. L.F. Richardson: *Beitr. Phys. Atm.* 15, 24 (1929)
8. M.C. Jullien, J. Paret and P. Tabeling: *Phys. Rev. Lett.* 82, 2872 (1999)
9. S. Ott and J. Mann: *J. Fluid Mech.* 422, 207 (2000)
10. G. Boffetta and I.M. Sokolov: *Phys. Rev. Lett.* 88, 094 501 (2002)
11. T. Ishihara and Y. Kaneda: preprint (2002)
12. S. Grossmann: *Ann. d. Phys. (Leipzig)* 47, 577 (1990)
13. S. Grossmann and I. Procaccia: *Phys. Rev.* A29, 1358 (1984)
14. S. Grossmann and C. Wiele: *Z. Phys.* B103, 469 (1997)
15. S. Grossmann: preprint, Marburg 2002
16. L.D. Landau and E.M. Lifschitz: *Hydrodynamik*, 5th edition. Akademie Verlag, Berlin 1991
17. U. Frisch: *Turbulence – The Legacy of A.N. Kolmogorov*. Cambridge University Press, 1995
18. U. Frisch, P.-L. Sulem and M. Nelkin: *J. Fluid Mech.* 87, 719 (1978)
19. G. Parisi and U. Frisch. In: M. Ghil, R. Benzi and G. Parisi (eds.): *Turbulence and Predictability in Geophysical Fluid Dynamics*. Proc. Intern. School of Physics "Enrico Fermi" 1983, Varenna, Italy. North Holland, Amsterdam 1985, pp. 84–87
20. This corresponds to 3×10^{22} cm; in [3] there is a misprint here
21. H. Effinger and S. Grossmann: *Z. Phys.* B66, 289 (1987)
22. H. Effinger and S. Grossmann: Selfsimilarity of Developed Turbulence. In: W. Güttinger (ed.): *The Physics of Structure Formation: Theory and Experiment*. Proc. Conference 27.10.–1.11.1986, Tübingen, Germany. Springer, Berlin 1987, pp. 346–351
23. S. Grossmann: Diffusion in Fully Developed Turbulence: A Random Walk on a Fractal Structure. In: L. Garrido (ed.): *Fluctuations and Stochastic Phenomena in Condensed Matter*. Proc. Sitges Conference 25–31 May, 1986. Lecture Notes in Physics 268. Springer, 1987, pp. 287–314

24. S. Grossmann and S. Thomae: *Z. Phys.* B49, 253 (1982)
25. D. Daems, S. Grossmann, V.S. L'vov and I. Procaccia: *Phys. Rev.* E60, 6656 (1999) and chao-dyn/9811024
26. H. Effinger and S. Grossmann: *Phys. Fluids* A1, 1021 (1989)
27. G.K. Batchelor and A.A. Townsend: *Proc. R. Soc. (London)* A199, 238 (1949)
28. A.M. Oboukhoff: *J. Fluid Mech.* 13, 77 (1962)
29. A.N. Kolmogorov: *J. Fluid Mech.* 13, 82 (1962)
30. G. Falkovich, K. Gawędzki and M. Vergassola: *Rev. Mod. Phys.* 73, 919 (2001)
31. A. Noullez, G. Wallace, W. Lempert, R.B. Miles and U. Frisch: *J. Fluid Mech.* 339, 287 (1997)
32. S. Grossmann, D. Lohse and A. Reeh: *Phys. Fluids* 9, 3817 (1997)
33. B. Dhruva, Y. Tsuji and K.R. Sreenivasan: *Phys. Rev* E56, R4928 (1997)
34. I. Arad, B. Dhruva, S. Kurien, V.S. L'vov, I. Procaccia and K.R. Sreenivasan: *Phys. Rev. Lett.* 81, 5330 (1998)
35. I. Arad, L. Biferale, I. Mazzitelli and I. Procaccia: *Phys. Rev. Lett.* 82, 5040 (1999)
36. I. Arad, V.S. L'vov and I. Procaccia: *Phys. Rev.* E59, 6753 (1999)
37. S. Grossmann, A. von der Heydt and D. Lohse: *J. Fluid Mech* 440, 381 (2001)
38. S. Grossmann and D. Lohse: *J. Fluid Mech.* 407, 27 (2000)
39. S. Grossmann and D. Lohse: *Phys. Rev.* E66, 016305 (2002)
40. W. Heisenberg: *Ann. d. Phys. (Leipzig)* 74, 577 (1924)
41. L. Boberg and U. Brosa: *Z. Naturforsch.* A43, 697 (1988)
42. L.N. Trefethen, A.E. Trefethen, S.C. Reddy and T.A. Driscol: *Science* 261, 578 (1993)
43. S. Grossmann: *Rev. Mod. Phys.* 72, 603 (2000)

From Dust Disks to Planetary Systems

Thomas Henning

1 The Scheme – From Fiction to Reality

The origin and evolution of planets and planetary systems and their relation to the origin of life is one of the most fundamental and fascinating problems contemplated by mankind. The Kant-Laplace hypothesis, which implies that the formation of planets occurs in a flattened nebula around a young stellar object, still forms the basis of our modern understanding of planet formation. The nearly co-planar and almost circular orbits of the planets in our Solar System support this hypothesis. Numerical simulations of the collapse of dense molecular cloud cores imply that disks are a natural by-product of star formation. They also have demonstrated that the formation of a single star-disk system or a binary with a circumbinary disk depends on the initial conditions in the cloud cores. In contrast to the Kant-Laplace hypothesis, catastrophe theories involve interaction with other stars as the trigger mechanism for the formation of planetary systems. These theories are no longer seriously considered for the formation of bound planetary systems. However, they have again gained great importance as a possible mechanism for the production of "free-floating" substellar objects. These may form by the interaction of star-disk systems in dense stellar clusters or by the interaction of planets among each other.

In his 1943 paper "Über die Entstehung des Planetensystems", C.F. von Weizsäcker introduced the concept of turbulent friction as the driving force behind the evolution of protoplanetary disks [27]. In a protoplanetary disk, mass is transported inwards towards the central star and angular momentum outwards in a system where the viscosity is caused by turbulent motion. This process offers an explanation of why 0.13% of the mass of the solar system carries 99.5% of the angular momentum. Convective flows were thought to play a major role in shaping the solar system. Whether this is indeed true or whether magneto-rotational instabilities are the driver of turbulence will be discussed in the next section.

The growth process in protoplanetary disks starts with submicron-sized particles which contain about 1% of the total disk mass [3]. Little of the growth can take place in the dense molecular cloud cores from which stars form because the process is too slow under such physical conditions. In the first stage of dust evolution in disks, grains grow by collisional coagulation. The relative velocities of the small grains of the initial dust population are produced by Brownian motion. These velocities are very small and of the order of 10^{-3} ms^{-1}. Experiments and numerical simulations have demonstrated that particles stick by van der Waals forces when they collide with these velocities and form relatively open fractal-like structures (e.g. [4]). In later stages, larger particles start to sediment towards the disk midplane. Relative velocities occur through differential settling and radial drift motions. Larger (metre-sized) bodies get collisional velocities up to 100 m/s and it remains to be seen which processes can finally lead to growth at such large collisional velocities where fragmentation is to be expected. Nature has to cross a size scale of 10 orders of magnitude before km-scale planetesimals form (see Fig.1 for dimensions during the planet formation process). The envisioned Safronov–Goldreich–Ward instability in a dust sublayer, which immediately leads to the formation of planetesimals, does not hold up under detailed investigations [25]; the timescale for the formation of the dust layer is too long and the velocity dispersion required for the instability to occur is smaller than what can be expected.

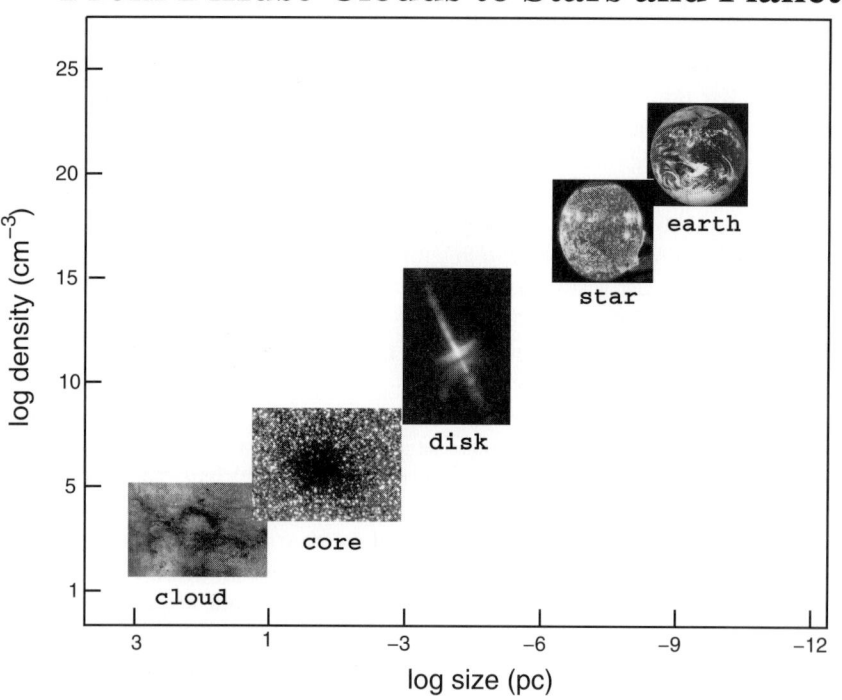

Fig. 1. Dimensions and densities during the development from a molecular cloud to a planet. The astronomical length scale of 1 parsec (pc) is equivalent to 3.1×10^{16} m or 3.3 light years. Based on an idea of J. Alves

Planetesimals decouple from the gas and grow further by gravitational attraction – we enter the accretional stage of the evolution. In the terrestrial planet zone, planet embryos of mass $\approx 10^{23}$ kg are formed. Runaway growth of these objects due to dynamical friction and three-body interactions results in the production of smaller planets. They grow further by pairwise accretion until a stable configuration with large enough spacing of the orbits is reached. The prevalent view of how giant planets form involves two stages. First a solid core of 10–30 Earth masses is accreted, and later it captures the gas from its environment and grows further until no material is left. Alternative theories for the formation of giant planets involve gravitational instabilities in a massive and cold protoplanetary disk [5].

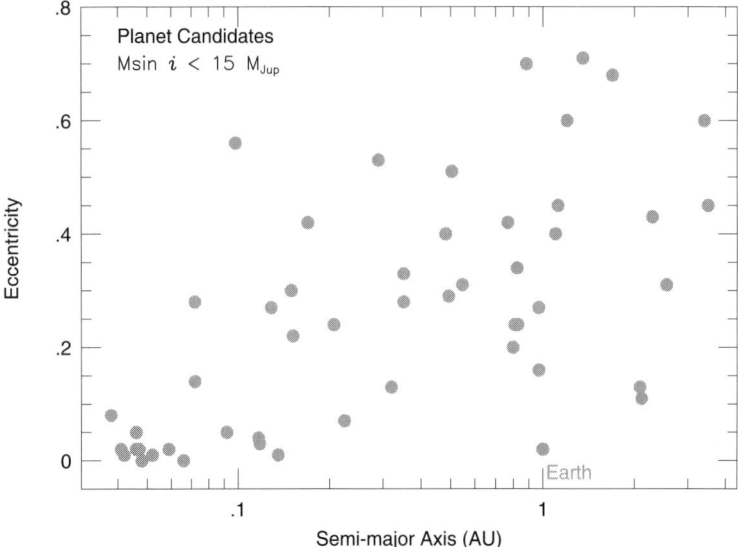

Fig. 2. Distribution of eccentricities vs. semi-major axis

In recent years, this picture of planet formation has been brought from imagination to reality. New infrared array detectors, space-based infrared missions, especially the IRAS mission, and sensitive millimetre-wave telescopes have allowed us to find the first evidence for protoplanetary disks around other young low-mass stars. More recently, imaging of such disks became possible, mostly with the *Hubble Space Telescope* and adaptive optics (AO)-assisted observations from the ground. In 1995, M. Mayor and D. Queloz announced the first detection of a planet around a solar-type star [22]. This breakthrough in searching for other worlds came through high-precision radial velocity searches, which reached the accuracy to discover planets through induced variations in stellar reflex motions. Since the first detection, about 1000 stars have been surveyed. This is nearly a complete sample of solar-type stars within a distance of 30 pc from the Sun. With the survey, more than 100 planets – including 11 multiple planetary systems – have been detected. The observed planets are very different compared with the planets of our Solar System. The observations showed the presence of massive planets in close orbits (fractions of the Sun-Mercury distance) and planets with high orbital eccentricities around other stars (see Fig.2). Although the radial velocity method favours the detection of massive objects close to the star, their very existence was unexpected and challenges theories of planet formation. The team of G. Marcy and P. Butler found the first Jupiter-like planet, orbiting the Sun-like star 55 Cancri at nearly the same distance as the Jovian system orbits the Sun [9]. 55 Cancri was already known to have one planet with a minimum mass slightly smaller than the mass of Jupiter, orbiting the star at a distance only one-tenth that from Earth to Sun. The second planet orbits 55 Cancri at 5.5 AU, comparable to Jupiter's distance from the Sun of 5.2 AU. The slightly elongated orbit has a period of about 13 years, comparable to Jupiter's orbital period of 11.86 years. Its mass should range between 3.5 and 5 times the mass of Jupiter.

2 Protoplanetary Disks – The First Steps

The first evidence for the presence of disks around low-mass stars came from the detection of excess thermal infrared and millimetre emission produced by small solid particles orbiting the objects. Millimetre surveys [2, 14] imply that at least 50% of the nearby T Tauri stars are associated with disks. Their masses are typically of the order of 0.01 solar masses and their outer radii are around a few hundred AU (1 AU = distance between Sun and Earth). Detailed studies of disk morphology and kinematics require measurements with high spatial resolution. The solar system has a diameter of about 100–150 AU, including the disk-shaped region of the Kuiper Belt. The Kuiper belt contains numerous icy bodies which are pristine remnants of the solar nebula. In the nearest star-forming regions such as the Taurus-Auriga complex, the Solar System, with a size of 100 AU, would have an angular size of only 0.7 arcseconds. Therefore, the resolution required to resolve these objects is clearly on the subarcsecond scale. It can only be provided by AO-assisted imaging on large telescopes, the *Hubble Space Telescope* (HST) with its stable point spread function (PSF), or infrared and millimetre interferometers. The HST delivered images of dust disks in Orion as silhouettes in front of the bright ionized hydrogen complex in this star-forming region. In addition, disks traced by scattered light can be found in HST images. Fig.3 shows images of the disk and environment of the Herbig Be star HD 100546, which is located at a distance of 103 pc and has an age of 10 Myr [13]. The optical coronographic data were obtained with the Space Telescope Imaging Spectrograph (STIS) on board the HST.

Millimetre interferometry has revealed a few examples of disks around young low-mass stars where Keplerian rotation of the gas masses around the star could be detected, although the kinematics in the circumstellar environment are usually more complicated. Fig.4 shows the structure and kinematics of a disk in the Bok globule CB 26 [19]. Bok globules are small molecular clouds which form low-mass stars. This disk is optically thick at millimetre wavelengths inside 120 AU, and has a symmetric warp beyond 120 AU, an outer radius of about 200 AU, and a mass of

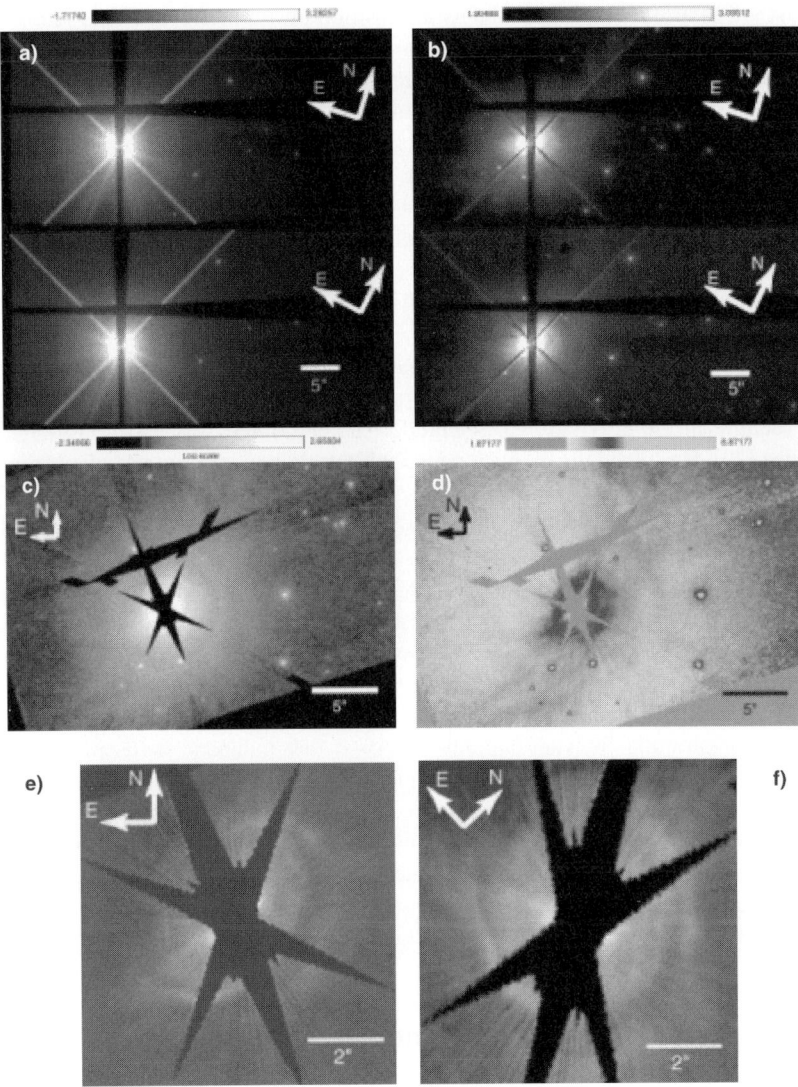

Fig. 3. Imagery of the disk and environment of the Herbig Be star HD 100546. (**a**) Reduced STIS data prior to PSF subtraction, displayed in false colour with a logarithmic scale and at two different position angles. The PSF dominates these raw data. (**b**) After PSF subtraction. Nebulosity can be traced from 0.5 arcseconds out to 10 arcseconds from the star at all position angles. (**c**) A field of 30 arcseconds ×25 arcseconds containing combined data from both spacecraft orientations. The irregular polygons in the image are due to voids in the data coverage caused by the combined effects of stellar diffraction spikes and the structure of the STIS coronographic wedge. (**d**) Clumpy structure on the 1–2 arcsecond scale revealed after filtering the image from (c). (**f**) After rotating the data from (e) to orient the region of maximum elongation horizontaly, the deprojected image, assuming a system inclination of 49 degrees. After Grady, Polomski, Henning et al. (2001)

a least 0.1 solar masses. It is probably a transition object between deeply embedded protostellar disks and the more evolved protoplanetary disks around T Tauri stars.

Although the detected disks share many of the properties expected for a "protoplanetary" nebula, direct evidence for grain growth and the presence of planetesimals or planets in the disks is still difficult to get with present-day instrumentation. The detection of warped disks, gaps in disks, gas accretion coming from comet collisions or evaporation, and a change of the scattering and absorption properties of dust particles during growth may finally reveal the evolution of disks towards planetary systems. There are already a number of observations which indicate grain growth and the presence of planets in more evolved disks, such as the warped disk around the star β Pictoris.

We have already stressed that protoplanetary accretion disks are "machines" for the transport of angular momentum. The purely hydrodynamical case for a turbulent gas rotating in the gravitational potential of a central star was first considered by R. Lüst [20]. He solved the relevant equations, taking into account turbulent friction. N.I. Shakura and R.A. Sunyaev introduced the concept of α viscosity where the viscosity was expressed in terms of the sound speed (subsonic turbulence), the disk scale height (largest eddies smaller than the disk thickness), and a dimensionless parameter, α, which is smaller than 1 [23]. This formulation leads to a radial velocity which is highly subsonic, and to a timescale for the radial evolution of the disk. Observations suggest that the timescale of protoplanetary disk evolution is on the order of 10^6 years and that the α value should be around 10^{-2}.

The driving instability behind the turbulent motion could not be identified, although Shakura and Sunyaev already proposed that magnetic fields may play an important role. The discovery of the importance of magneto-rotational instability (MRI) provides the basis for a self-consistent mechanism for the production of turbulence in disks [1]. Purely hydrodynamical convection is no longer considered to be the driving mechanism behind turbulence. Despite this progress, the development of a comprehensive theory of rotating, partly ionized gas under the action of central gravity remains one of the great goals of modern astrophysics.

3 Detection of Planets and Planetary Systems

The detection, imaging, and spectroscopy of extrasolar planets is a challenge even for astronomical observations with modern large telescopes, adaptive optics, and interferometry. The "cold" planets are located at distances larger than 1 AU and shine only thanks to the reflected light of the central star. The luminosity contrast at visual wavelengths between the Sun and Jupiter amounts to 8–9 orders of magnitude. This contrast problem can be solved by a combination of high-order adaptive optics with spectroscopy or polarimetry, a large space-based telescope such as the *Next Generation Space Telescope* – now called *James Webb Space Telescope* –, or nulling interferometry where the light of the central star is phase-shifted and "nulled". All these techniques are presently being developed, but are not yet in operation.

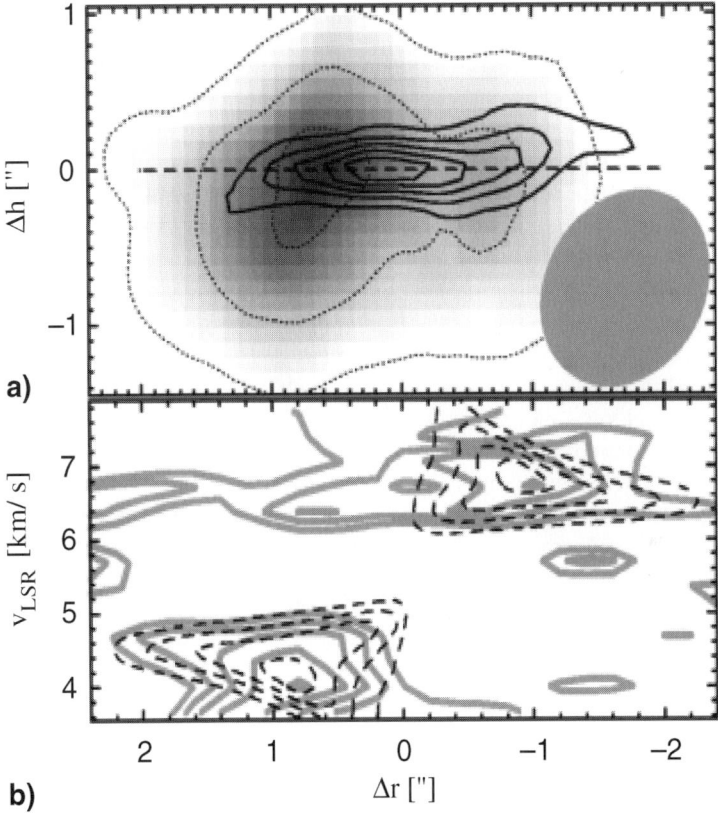

Fig. 4. (a) Integrated emission in the rotational transition $J = 1 - 0$ of the ^{13}CO molecule in grayscale (dotted contours give different intensity levels) with 1.3 mm dust continuum contours overlaid. A dashed line represents the plane of the disk. The gray ellipse indicates the beam size of the observations. (b) Position-velocity diagram along the disk major axis. The thick contours indicate different levels of the maximum intensity of the observed velocity field. The expected emission from a Keplerian disk around a star with a mass of 0.35 solar masses is represented by dashed contours. After Launhardt & Sargent (2001).

Therefore, indirect detection techniques are currently being used. The radial velocity method was particularly successful in detecting extrasolar planets and planetary systems. According to Kepler's laws, planets and a star orbit a common centre of gravity. The velocity of the Sun as the result of its gravitational interaction with Jupiter amounts to 12,5 m/s. The radial velocity of the central star can be measured by the periodic motion of spectral lines due to the Doppler effect. With an iodine absorption cell as reference, one can measure wavelength changes ($\Delta\lambda/\lambda$) with a precision of 3 m/s. This technique favours the detection of massive planets close to the central star and gives only a lower limit for the mass of the planet if the inclination of the orbit is not known.

The first detection of an extrasolar planet around the star 51 Peg in 1995 revealed the presence of a rather exotic object [22]. The orbital period of the 51 Peg planet, with a lower mass limit of 0.47 Jupiter masses, was only 4.23 days. The planet must be located at a distance of only 0.05 AU from the central star. We now know of the presence of about 80 planets and 7 multiple planetary systems. The eccentricities of the orbits range between 0 and 0.67 and the lower limit of their masses between 0.16 and 13 Jupiter masses. There are different types of objects: close-in-giants or "hot Jupiters" within 0.1 AU with nearly circular orbits (example: 51 Peg), planets with high eccentricities (example: 70 Vir), and objects with orbital parameters similar to Jupiter (example: 47 UMa). Most of the planets orbiting within 0.1 AU have nearly circular orbits caused by tidal circularization.

There is not yet a clear consensus on what the upper mass limit of planets is. The deuterium burning limit at 13 Jupiter masses is often used, although there may be a more physical formation criterion which would lead to a different mass limit. However, observations show that the mass function of substellar companions grows rapidly toward smaller masses. Companions with masses larger than 10 Jupiter masses are extremely rare.

In the case of a planet in an eclipsing orbit, the inclination angle of the orbit must be close to 90 degrees and the mass can be directly determined. This is the case for the planet orbiting HD 209458, where a transit was detected [6, 10, 15]. The mass of this object amounts to 0.69 Jupiter masses and there is no doubt that this is a planet. The radius of 1.35 Jupiter radii is in good agreement with models for giant planets which are strongly irradiated. The density estimate is about $0.35\,\text{g/cm}^3$, less than the density of Saturn. In the case of HD 209458b recent investigations indicate the presence of sodium absorption by the atmosphere of the planet [11].

In a few cases precise measurements and a detailed analysis of the radial velocity data led to the detection of more than one planet orbiting the central star. An example is the system around the star υ Andromedae [8]. It contains three planets with minimum masses of 0.71, 2.11, and 4.61 Jupiter masses. The 0.71 Jupiter mass planet is located on a nearly circular orbit at a distance of 0.06 AU and the two other planets are at 0.83 and 2.5 AU (semimajor axis) with larger eccentricities of 0.18 and 0.41, respectively.

4 Interactions – Numerical Simulations

The observational data clearly show that planets with masses larger than the mass of Jupiter exist. In addition, we find objects in close orbits with high masses and planets with large eccentricities. Before these detections, it was generally assumed that giant planets could not grow much beyond a mass of 1 Jupiter mass because of gap opening, and that they were only located in outer regions of planetary systems, because otherwise not enough mass would be available for their growth.

The observations point to the interaction of planets with their disks. The general idea is that giant planets indeed form further out in the protoplanetary disk (at several AU from the star) and migrate inwards. This radial motion of the planet through the disk is primarily caused by gravitational torques acting on the planet. The gravitational interaction of the planet with the disk leads to spiral density waves, which emanate from the planet. The disk is no longer axisymmetric, which results in a net torque on the planet. Numerical simulations draw a detailed picture of this process [7, 16]. The typical migration time scale for Jupiter-mass planets is on the order of 10^5 years, well in agreement with analytical estimates based on linear theories, although numerical simulations with high spatial resolution show first deviations from analytic estimates of this time scale [12]. The results of such a simulation are shown in Fig. 5.

Depending on the mass of the planet, it can open a gap due to angular momentum conservation. Material closer to the star is pushed inwards and material further out moves outwards. Despite opening of the gap, accretion of matter through the gap is still possible and the planet can grow up to 5 to 10 Jupiter masses. A number of mechanisms have been proposed to halt the migration of the planet, including disk dissipation and inner hole formation by the stellar magnetic field.

The explanation for large eccentricities may well come from planet-planet interactions [18, 26]. This would also imply that in systems where planets with large eccentricties are observed, additional planets should be detected. The stability of the systems strongly depends on the planetary masses and orbital inclinations. Instability should occur when the planets come closer than the tidal radius (Hill radius). This may also lead to the ejection of a planet from the system. Orbital resonances, common among planetary satellites and asteroids in the Solar system, may be an important factor for structuring certain planetary systems. The first 2:1 resonance among major planets was detected for a pair of planets, orbiting the star GJ 876 [21]. Numerical simulations suggest that different disk-driven migration speeds may bring the planets together and finally lead to locking in a 2:1 resonance [17, 24].

These considerations demonstrate that disk-planet and planet-planet interactions certainly play an important role in shaping planetary systems and deserve further investigation as a new aspect of theories of planet formation and the evolution of planetary systems.

5 Tools for the Future

The detection of protoplanetary disks and extrasolar planets is a major breakthrough in our understanding of the formation and evolution of planetary systems. These detections initiated a strong interest in this new and rapidly evolving field of modern astrophysics.

Apart from radial velocity searches and transit observations, other methods such as the measurement of positional displacements (high-precision astrometry), gravitational microlensing, pulsar timing, direct imaging and spectroscopy, and nulling interferometry - both from the ground and from space – will open other parameter re-

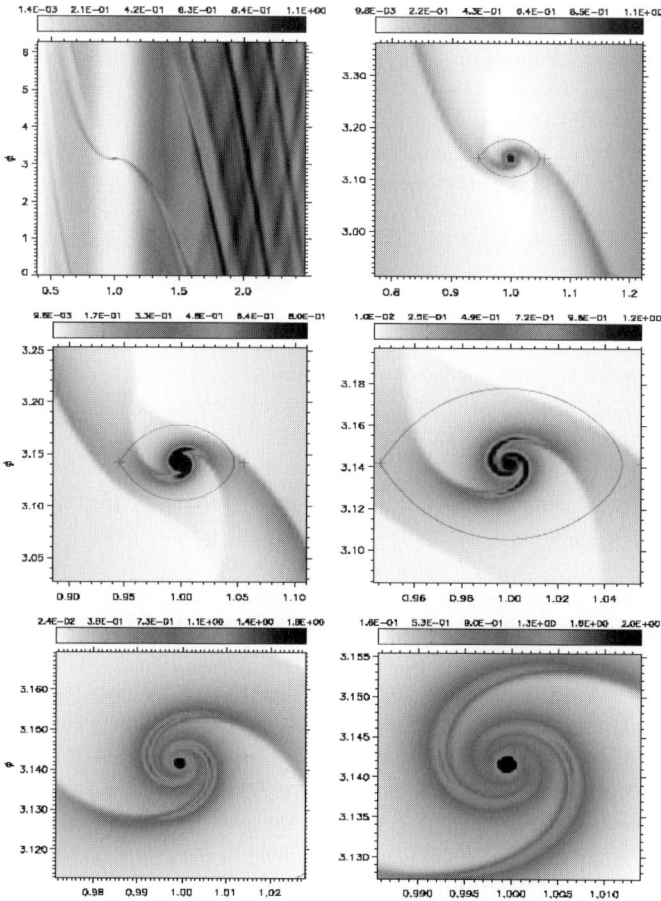

Fig. 5. Numerical simulation of the interaction of a planet with a circumstellar gas disk. Shown is the surface density after 210 orbits of the planet. The abscissa indicates the radial distance from the central star, the ordinate the angular distance in the disk. The central star has the mass of the Sun and the planet has a mass of 0.5 Jupiter masses. The top-left pannel shows the whole simulated disk region. From left to right and top to bottom, the other panels zoom in onto the planet. The over-plotted curve represents the Roche lobe of the restricted three-body problem. Plus signs indicate the positions of L_1 and L_2 Lagrangian points, respectively, on the left and the right of the planet. After D'Angelo, Henning, and Kley (2002).

gions of the planet discovery space. Here we should mention that the first objects with planet masses were detected by pulsar timing around the pulsar PSR $1257 + 12$ [28].

In the coming years we will see strong progress in numerical simulations of planet formation, including detailed radiative transfer and a treatment of planetary structure together with the mass accretion process. Higher resolution will be provided by multi-grid calculations, and global 3-dimensional (magneto) hydrodynamical simulations including more than one planet will be performed. The coupling of dust grains with the motion of the gas will be a challenge for numerical simulations. The treatment of disk physics has moved from purely hydrodynamical calculations to a magneto-rotational instability as the cause of turbulence in disks. The Weizsäcker concept that turbulent viscosity drives disk evolution still forms the basis of our present understanding of protoplanetary disks and their evolution.

References

1. S.A. Balbus, J.F. Hawley: *Astrophys. J.* 376, 214 (1991)
2. S.V.W. Beckwith, A.I. Sargent, R.S. Chini and R. Güsten: *Astron. J.* 99, 924 (1990)
3. S.V.W. Beckwith, Th. Henning and Y. Nakagawa: *Protostars and Planets IV.* Univ. of Arizona Press, Tucson 2000, p. 533
4. J. Blum, G. Wurm, S. Kempf et al.: *Phys. Rev. Lett.* 85, 2426 (2000)
5. A.P. Boss: *Astrophys. J.* 563, 367 (2001)
6. T.M. Brown, D. Charbonneau, R.L. Gilliland, R.W. Noyes and A. Burrows: *Astrophys. J.* 552, 699 (2001)
7. G. Bryden, X. Chen, D.N.C. Lin, R.P. Nelson and J.C.B. Papaloizou: *Astrophys. J.* 514, 344
8. P.R. Butler et al.: *Astrophys. J.* 526, 916 (1999)
9. G.W. Marcy, P.R. Butler, D.A. Fischer et al.: *Astrophys. J.* 581, 1375 (2002)
10. D. Charbonneau, T.M. Brown, D.W. Latham and M. Mayor: *Astrophys. J.* 529, L45 (2000)
11. D. Charbonneau, T.M. Brown, R.W. Noyes and R.I. Gilliland: *Astrophys. J.* 568, 377 (2002)
12. G. D'Angelo, Th. Henning and W. Kley: *Astron. Astrophys.* 385, 647 (2002)
13. C. Grady, E.F. Polomski, Th. Henning et al.: *Astron. J.* 122, 3396 (2001)
14. Th. Henning, W. Pfau, H. Zinnecker and T. Prusti: *Astron. Astrophys.* 276, 126 (1993)
15. G.W. Henry, G.W. Marcy, R.P. Butler and S.S. Vogt: *Astrophys. J.* 529, L41 (2000)
16. W. Kley: *Mon. Not. Roy. Ast. Soc.* 303, 696 (1999)
17. W. Kley: *Mon. Not. Roy. Ast. Soc.* 313, L47 (2000)
18. G. Laughlin, F.G. Adams: *Astrophys. J.* 526, 881 (1999)
19. R. Launhardt, A. Sargent: *Astrophys. J.* 562, L173 (2001)
20. R. Lüst: *Z. Naturforschg.* 7a, 87 (1952)
21. G.W. Marcy, R.P. Butler, D. Fischer, S.S. Vogt, J.J. Lissauer and E.J. Rivera: *Astrophys. J.* 556, 296 (2001)
22. M. Mayor, D. Queloz: *Nature* 378, 355 (1995)
23. N.I. Shakura, R.A. Sunyaev: *Astron. Astrophys.* 24, 337 (1973)
24. M.D. Snellgrove, J.C.B. Papaloizou and R.P. Nelson: *Astron. Astrophys.* 374, 1092 (2001)
25. S.J. Weidenschilling: *Icarus* 116, 433 (1995)
26. S.J. Weidenschilling, F. Marzari: *Nature*, 384, 619 (1996)
27. C.F. von Weizsäcker: *Zeitschrift für Physik* 22, 319 (1943)
28. A. Wolszczan, D.A. Frail: Nature 355, 145 (1992)

Part III

The Unity of Nature and the Nature of Time

Science and Its Relation to Nature in C.F. von Weizsäcker's Natural Philosophy

Klaus Michael Meyer-Abich

1 The Meaning of Knowledge, Time, and History in the Philosophy of Nature

It may have been around the middle of the 1930s when Niels Bohr once met Max Planck – then already in his late seventies – at the Copenhagen railway station. They took a taxi and soon resumed their dispute on Bohr's and Heisenberg's interpretation of quantum theory. After a while Planck rejoined: 'But you must admit, Bohr, that a god-like eye could know both the particle's position and momentum!' Bohr replied: 'I do not think that this is a question of what a god-like eye can know, but of what you mean by knowing.'

Planck's view was that of classical physics. In fact, what a god-like eye would observe the world to be is a fairly good description of the kind of knowledge sought by a classical physicist. What he is looking for is a description of the world as it 'really' is, not a human or anthropomorphic view but a description that cannot be identified as somebody's description. Even for an agnostic physicist – which Planck wasn't – the god-like view is an appealing picture for that paradigm of classical physics.

Not only is the god-like eye of classical physics extraterrestrial or out of space but it is out of time as well. As for the Laplacean demon – which is essentially the same idea in a more secularised form – any present state of the world by means of the laws of nature implies its complete future. While it is a matter of daily experience to go for an overview from a distance in space, we are in a different situation with respect to time. In ordinary life future events may be expected but they are not known in advance, and often enough we are surprised by what really happens. Moreover – as we learn from the philosophy of history – any past as well is known only as the past of some present situation. Historians, therefore, do not simply add new chapters to the book of history as time goes by but rewrite the book again and again as situations change.

To go for an overview in time, as we are used to doing in space, is the basic innovation of classical physics. The idea, of course, came from astronomy, originally referring to the regular pace of the heavenly gods; but when their motions were identified as essentially 'mechanical' their reversible time came down to earth as well. With special relativity time finally seemed to have become simply a fourth dimension of space. Curiously enough, however, as Weizsäcker once pointed out, when Hermann Minkowski announced just this his wording was that "from now on" time is known to be nothing but that fourth dimension. Even in denying the prevalence of time, therefore, he did not get around to confirming its priority, even in four dimensions.

Thus what Planck meant by knowing was what a god-like eye would know, an eye that

- presently knows the future as if it had passed already, so that it can never be surprised,
- knows the past not as the backward perspective from some particular presence but as the growing accumulation of facts.

According to Planck's paradigm of knowledge physics describes The order of The world, the capital Ts denoting that there is but one of both, one order of the one world, and that human beings can achieve knowledge of that order 'as it really is' or as it is in itself, unencumbered by any human smell. History has been deleted from that order.

Niels Bohr did not answer Planck's question about the god-like eye because he did not share this ideal or paradigm of physical knowledge. What he meant by knowledge was *human* knowledge in some particular *situation* that he used to refer to as the conditions of observation. As in his famous discussion with Einstein, Bohr generally insisted on the reference of physical concepts to such conditions. As he put it, any concept is 'defined' only with respect to certain arrangements so that, for instance, the word 'momentum' is used meaningfully only when it refers to a situation in which the momentum can be observed, and the same is true when some position is considered. From this view, Planck's statement about the god-like eye was contradictory in itself because it referred to two mutually exclusive arrangements for the same object at the same time.

Bohr's philosophy may be termed pragmatic but is far from empiristic. The basic idea is not that reality depends on measurement but that *human* beings deal with *human* realities. As Bohr put it in a first reply to the Einstein–Podolsky–Rosen argument, the conditions to which the definition of a physical concept refers are "an inherent element of any phenomenon to which the term 'physical reality' can be unambiguously applied" [3, p. 65]. Physics does not just deal with matters but with matters of fact, or action[1]. In one of his more philosophical papers – which happened to be his contribution to the celebration of Planck's 50th doctoral anniversary – the core of Bohr's argument similarly was "the relative meaning of every concept" [1, p. 62], [2, p. 96]. More colloquially, he used to tell the story of the little boy who in a shop asked for mixed candy for two øre. The shopkeeper replied: 'For two øre you are getting two pieces, I am giving you a red one and a green one; go and mix them yourself'. This

---

[1] This is most easily said in German: "Die Naturwissenschaft handelt von Tat-Sachen, nicht nur von den Sachen" [8, p. 187].

shows the relative meaning of the concept 'to mix', not being applicable to very low numbers.

The anthropomorphism or human shape of human knowledge was one of the earliest discoveries in Greek philosophy. It was Xenophanes who observed that the Thrakians conceived God to be blue-eyed and red-haired while the Ethiopians imagined him to be black with Negroid lips. And this was not meant as a sceptical argument with respect to the existence of God but simply meant that any conception of reality is somehow fashioned by its origin. In modern times this was rediscovered by Spinoza and came to be the starting point of Kant's criticism.

Bohr did not pay special attention to the concept of time and this particular gap was filled by Carl Friedrich von Weizsäcker. His first fundamental discovery was that the discrepancy between the reversibility of statistical mechanics and the irreversibility of the Second Law was a matter of human interpretation [10, p. 275]. In other words, statistical mechanics still refers to the god-like eye while human experience is conditioned by a fundamental difference between past and future which is the dissimilarity of facts and probabilities. Therefore, to understand the future as that of growing entropy is an example of the conditions of experience being part of physical reality, or of physics to be *human* knowledge. Michael Drieschner has outlined the relevance of Weizsäcker's paper in his contribution to the present book, indicating that up to now it has not been adequately recognised in the interpretation of thermodynamics.

As a new idea of time had been the basic innovation in classical physics, Weizsäcker's discovery of the Second Law's anthropomorphism was bound to identify quantum theory as being based on a different concept of time. Even relativity theory dealt with absolute time in the sense of not being dependent on the humanity, or human shape, of knowledge. As a first step beyond that, Weizsäcker discovered the distinction between past and future as the 'relative meaning' (Bohr) of time in terms of humanity, not only in terms of this or that particular human situation.

A few years later, the second step was in the same direction but even more fundamental. This was the discovery – or rediscovery – of the humanity of physics in terms of history. This sounds much more pretentious than the words he used which simply read: *Nature is older than man, and man is older than science* [11, p. 8]. The idea, however, is that (1) man originated from nature so that human history is part of the history of nature, whereas (2) physics emerged from human history within the history of nature. Nature, therefore, among millions of species gave birth to one that turned his eyes on her by way of physics. Niels Bohr conveyed the same idea even more simply by his Ceterum censeo, namely that 'we describe nature of which we are part ourselves'. Yet he didn't consider this reflective structure as part of the history of mind in nature.

Bohr's thinking was close to Spinoza's as well as to Plato's philosophy of nature. Plato was the first to corroborate his political philosophy with respect to nature by checking whether the political order of his political 'state' complied with the order of nature into which it was to be embedded. Neither Plato nor Spinoza, however, considered this embodiment as a historical process. The idea of the history of nature emerged only in the 18th century when extinguished species were discovered. Also

Isaac Newton had left the origin of the planetary system to biblical creation while Immanuel Kant put forward a physical explanation which later on was completed by Laplace. On these lines Weizsäcker as well came to study the history of nature.

Kant had shied away from conceiving cosmic evolution to be continued by the evolution of living beings and even more from thinking to become a process in nature as Georg Picht has recently expressed it [9, p. 159]. Kant's former student Johann Georg Herder was the first to make that step in his "Ideas on the philosophy of history of mankind" (1784–91). Following Herder's pattern, Alexander von Humboldt in his "Kosmos" (1845–62) first described the universe at large, then proceeded to the solar system and identified the earth in a truly Copernican way. He then explained how life emerged from matter on this little planet and that among so many species finally man entered the picture. While this is not explored until the end of the first volume of "Kosmos", the second volume is devoted to the specific achievements of this particular species, namely to the natural history of mind ("Naturkunde des Geistes", I. 383f.) in art as well as in science. It is as if Weizsäcker's two statements – nature is older than man, and man is older than science – exactly referred to the two stages set apart by Humboldt. He seems never to have mentioned Herder or Humboldt as his spiritual ancestors with respect to the "natural history of human knowledge" [15, p. 98], but true ideas happen to be rediscovered again and again.

At the same time the historical anthropomorphism of physics – human shape of physics in the history of nature – tends to be frequently forgotten. This happened to Herder as well as to Humboldt when classical physics was alleged to describe The order of The universe with the two capital Ts. It may happen to Weizsäcker's rediscovery as well together with its footing in the Copenhagen interpretation of quantum theory. Weizsäcker ventured to go beyond Herder as well as Humboldt on this foundation, however.

2 The Crisis of the Atomic Bomb in the Natural History of Mind

Weizsäcker's starting point was that physics is knowledge and that quantum theory is that level of physics on which it can no longer be ignored that physics is knowledge (cf. [14, p. 160]). This is not only closely reminiscent of Bohr's favourite book title "Atomic physics and human knowledge" but is essentially the Copenhagen philosophy itself. His next step then was that knowledge of the future refers to the realm of possibility while knowledge of the past is concerned with facts. Finally he spelled out this distinction regarding the history of nature as it has been conceived in Herder's philosophy of history or in Humboldt's "natural history of mind". Considering the current dichotomy of science and the humanities, which was unknown and inconceivable to Humboldt, Weizsäcker's book title "History of Nature" [11] was a deliberate challenge to the anthropocentric idea of history in modern humanities (cf. [16, p. 365]). Historians ought to realise that thinking has become a process in nature, not only about nature in the human mind.

This led Weizsäcker one step further, however, and brought him into his particular approach to the foundations of physics. He discovered that Herder's and Humboldt's idea of nature's being prior to man and man to science was circular in the sense of reiterating itself. In fact, if nature became aware of herself by means of the human mind, this mind as well ought to become aware of itself in nature as emerging from her. Thus human knowledge of nature implied a retrieval of itself becoming aware of nature. Understanding the evolution of knowledge in the history of nature, therefore, offered a chance to overcome one of the basic inconsistencies in classical physics, namely to claim an explanation of the universe without offering the slightest indication that this same universe accommodated someone to describe it. This incompatibility has been pointed out again and again by Weizsäcker's friend Georg Picht. The opposite idea that a theory should not rule out the conditions of its own existence was termed "semantic consistency" by Weizsäcker (cf. [14, pp. 231f.], [17, p. 135]).

Now we are far from understanding the evolution of life and particularly the descent of man or even the emergence of reason. However, we know that modern physics is a historical fact so that it must have become possible in the course of evolution. The human mind, therefore, has been gifted by natural history with the talent to let nature become aware of herself by means of physics. Perhaps we should be prepared for quite different kinds of experience beyond physics (cf. [14, pp. 208/219]). For the time being, however, modern science seems to be our age in the natural history of mind. Yet it is an open question as to whether we should go on with science and technology or try to overcome it.

Modern weapons technology plus a lack of conservative knowledge in relation to excessive destructive knowledge towards nature have fostered doubts as to whether science and technology are heading in the right direction. While these doubts are raised in the public, the average scientist would not be concerned, having made a personal commitment to science. Yet there was a time when in science many physicists were concerned about their responsibility for the political implications of their scientific work. This was after World War II when the atomic bombs frightened humanity as it had never before been frightened by any implication of human knowledge. Moreover, the bomb was a direct outcome of so-called 'basic research' in a particular political situation so that the customarily alleged division between basic and applied science could not be pretended here. Weizsäcker was sensitive enough to realise that this had not simply happened by a tragic accident but that there was "a straight course" in the history of physics "from Galileo to the atomic bomb", and he never lost that awareness [13, p. 116], [15, p. 462], [17, p. 231].[2]

In a way physics has lost its innocence by experiencing its involvement in the atomic bomb. But what are the implications for the life and work of a physicist who so to speak has become adult by that experience? As Weizsäcker put it, he will feel responsible for understanding the relevance of science to the world's change and to adapt his research accordingly (cf. [16, p. 429]). There are several ways to do so, however.

---

[2] When I started to study physics in the 1950s, I happened to look into the court of a jail through the window of the lecture hall. I never lost the feeling that society finally might put up us physicists there. Yet I studied it because I wanted to become a philosopher of nature and felt that as such I ought to know what physicists knew about it or her.

One gets aware only of what one loves, Goethe wrote to a friend who did not love nature (to Jacobi, May 10, 1812). This was also Weizsäcker's leading idea in the first years after 1945, and he came back to it as a bridge to Georg Picht more than forty years later when he wrote an introduction to Picht's lecture on nature after his death [9, p. xv]. Picht always had reservations about Weizsäcker's later course of justifying physics. He could certainly have agreed, however, when Weizsäcker observed in 1947, for instance, that killing for the sake of knowledge leads to knowledge for the sake of killing [12, p. 181] or that any experiment presupposes the transition from loving to impersonal awareness. "If love is omitted from loving cognition, the Christian god changes into the Cartesian one and Christian soul into the subject of modern science ... Can we experiment where we love?" [12, p. 176] "Experimentation means to exercise power over nature. The command of power then becomes the ultimate proof of the correctness of scientific thought" [12, p. 172] Certainly control is not compatible with love.

At the end of his famous lecture on "The history of nature" in 1946 Weizsäcker as well had left no doubt that "the scientific and technological world of modern times is the result of man's venturing knowledge without love" [11, p. 132]. Also in his later work he never renounced that judgement. While primarily his conclusion was to abandon knowledge if it precludes love [11, p. 90], however, he changed his mind afterwards. It seems that talking to Karl Barth, the theologian, was decisive in this respect. Weizsäcker asked him whether he might continue as a physicist after that straight course had led from Galileo to the atomic bomb, and reported Barth to have answered: 'If you believe what Christianity customarily affirms but what nobody really believes: that Christ will return, then you may continue, or even you must do so' [15, p. 462]![3]

Weizsäcker certainly has liked to hear that he might go on with his "beloved physics" (l.c.), but what was really the meaning of Barth's answer? To believe in Christ's return can certainly not justify everything. Barth, therefore, must have meant: If you believe in Christ's return and can reconcile your work as a physicist with that belief, then go ahead. That reconciliation thus was the real issue so that Barth actually left Weizsäcker with his original question on a higher level. To put it in his own words: Having understood the relevance of knowledge to world change, how does this change my research? [16, p. 429]. Belief, as he put it [15, p. 475], is the unbounded devotion to the grace of reconciliation in love. How does that devotion go together with the physicist's devotion to particles and fields?

Weizsäcker responded to Barth's challenge on three levels. Primarily he took a leading role in observing the physicist's responsibility for the implications of the political availability of atomic weapons. Until recently it seemed that the industrialised nations might learn to ban the use of such weapons as means to solve political

---

[3] Barth died in 1968, but the meeting must have happened before 1958 because in that year I joined Weizsäcker's working group and stayed with him until 1972. He would certainly have told us if the meeting had been in those years. He used to refer to Barth's encouragement as a key guidance in his life from time to time and I remember him doing so only in terms of a past event.

conflicts, or even to use them as a stabilising factor in international relations (by second strike capabilities, the ABM treaty, and so on). Much of Weizsäcker's political and scientific work was devoted to these developments. Details are given in the paper of Götz Neuneck in this volume.

Secondly, because of his extreme capability to understand others Weizsäcker – like Goethe – has always had an aversion to joining parties. A partial blindness for the other side is always required to do so and he could hardly be blind for something. For instance, he was too intelligent to join the shallow optimism of the Baconians who trusted in mankind to become the subject of history by means of science and technology. On the other hand, he was devoted to power and completely averse to any political radicalism. In other words, he did not conceive himself as a lonely prophet so that he never joined any marginal group like the environmentalists, or the pacifists etc., inasmuch as such groups again were not immune to shallow optimism. The solution he found brought him close to Heidegger's later philosophy – 'later' also meaning: after his failure in joining a party. This became the second stage of facing Barth's challenge.

The ruling technocrats and the opposing marginalists, Weizsäcker observed, share two basic convictions, namely that (1) we are in a position to achieve goals that we are free to choose, and (2) that true knowledge is benign [15, p. 417]. Both assumptions are not necessarily true. According to Heidegger, truth has its own history, and the dangers of technology are part of that history, i.e. truth itself is bound up with danger. While the Baconians and the marginalists dispute about opposite goals it may well be that human goals are much less decisive in history than both groups assume. As Heidegger once put it: Both seem to be reckoning without the host. Let's face that only a god can save us.

I belief that this is true; but how could that salvation come about without the slightest compliance on our part? In the New Testament a radical distinction is made between those who bring their own small light and those who offer not even that. The former are admitted to the heavenly light that man cannot achieve by himself while the latter are damned. In fact, Heidegger attempted to bring his own light when he supported national socialism for a while. Having burnt his fingers, he was in a position to step back but this did not hold equally for others.

Weizsäcker agreed with Heidegger on the historical relevance of science and technology. Those who believe in alternatives, he felt, do not recognise the overwhelming fatality of that course from Galileo to the atomic bomb – the fate we cannot escape. Science determines our destiny, Weizsäcker argued, because "for the modern world it is its proper form of truth" [15, p. 425]. In Nietzsche's terms this truth sounds like the particular error without which we cannot live. Even if this has been true up to the atomic bomb – couldn't we live without that error beyond it? Yet – instead of letting Goethe finally get his chance – Weizsäcker insisted that science invariably "reveals the inevitability of modern thought's progress" [15, p. 461]. Thus far he joined Heidegger against the naive apologists as well as the critics of that dominating science. Yet he was confident of being capable of going one step further than Heidegger because this philosopher "had not completed the course of science" [15, p. 427].

Whether that fate of science and technology is rooted as far back as the Neolithic revolution, as Weizsäcker suggested in the same context [15, p. 431], may be disputed. In any case even for a better future we can only gradually leave the past behind. In this sense Weizsäcker did not have to renounce love as a better precondition of knowledge but perhaps was justified in concluding that its time was still to come. He did not say so but he may have felt that way. "The Cartesian subject is lonely" [12, p. 175], he observed, and "perhaps it will take a billion years to be not lonely any more" [17, p. 215]. I feel that it must not take that long, but Weizsäcker's attitude leads me to his final answer with respect to Barth's encouragement.

3 Understanding Explanation – Weizsäcker's Titanic Project, and Some Suggestions About It

Devotion in love is aware of the other as a 'you' and leads to *understanding,* be it of a thing, a plant, an animal or a fellow human being. As opposed to that, devotion to some other as an 'it' – whatever it may be – in knowledge is headed for an *explanation* of something. What scientists have in mind in general has come to be an explanation. Ethologists like Konrad Lorenz make an exception to that, and some animal or human psychologists do as well, but this does not change the picture fundamentally.

On a third level, Barth's reply to Weizsäcker's question may now be taken to refer to the compatibility of understanding and explanation. This is the way Weizsäcker had put the question already in the early years after the atomic bomb [12, p. 180] and how he developed it in his Gifford lectures [12, p. 60], [13, p. 136/153]. The relatedness he looked for became the idea to *understand physics, or to understand explanation*, in such a way that the fundamental theories were shown to be implications of one basic assumption. Essentially this meant to complete physics in principle, leaving only a lot of details for further consideration. Obviously, this was a titanic project, but titanism – as Weizsäcker remarked elsewhere [16, p. 32] – usually indicates a crisis, and this was his way of facing it.

In the first volume of his Gifford lectures Weizsäcker announced that his solution would be given in the second volume; however, this never appeared [13, p. 153]. A few years later he still felt the solution to be close at hand [14, pp. 202/206/213] and even in his later years each time I met him he would tell me about the ongoing progress of this great project of his life. Yet he never did achieve his goal. This volume shows how close he and others came.

True enough, Weizsäcker considered the concepts of science as 'completely dark and devoid of explanation' [14, p. 287]. This is not the way physicists generally feel but the fact that they can handle these concepts does indeed not explain what happens when they do so. Any physicist approached by Weizsäcker with the Socratic question as to whether he understands what he is doing would soon have to admit his ignorance. So far this has been the distinction between a philosopher and a physicist. Weizsäcker's idea now was to understand – and at the same time to explain and deduce – the concepts and laws of physics from the implicit assumptions in any physicist's knowledge, taking philosophy – what it really is – as the art to make implicit assumptions explicit. If he had succeeded in doing so the distinction between physics and philosophy would have vanished. He thought that he might thus beat Heidegger [15, p. 429].

To understand physics as such, different answers for different parts like thermodynamics or quantum theory could not be accepted, but what was the unity of physics? In the first place there seemed to be a historical tendency of unification which Heisenberg addressed when he conceived the history of physics as advancing in ever more encompassing "closed theories" [11]. The basic model was classical mechanics as a limiting case of special relativity theory. Weizsäcker discussed the different parts of physics with respect to that idea and came to the conclusion that physics indeed is approaching a *historical* unity. So far the most general theories seemed to be quantum theory as the theory of all possible objects, elementary particle theory as the theory of all really existing objects and cosmology as the theory of the one existing world [14, p. 221].

Assuming the historical unity of physics to be a future unification of those three theories, how could the *systematic* unity of that theory be conceived? The corresponding definition of classical mechanics was that it is the general theory of bodies moved by forces in space and time, of thermodynamics that it is the general theory of temperature and pressure, etc., but what was common to all of them? For a general determination Weizsäcker resumed Immanuel Kant's idea that the general laws of physics – like the conservation principles, for instance – just reflect the human forms of perception, as the irregular movements of the planets reflect our own movement of the earth. He extended this idea, however, from the general laws to cover the specific laws as well, so that the whole corpus of physics was to be developed as an explication of one fundamental idea of knowledge.

This was not supposed to identify physics with any knowledge whatsoever so that the spiritual development of mankind – or rather of nature by means of mankind – would come to an end when physics was completed (cf. [14, pp. 208/219]). What Weizsäcker conceived to be the particular definition of physics as such was that it is *dealing with objects in time*. Here *objects* are to be understood as empirical alternatives (which can be decided empirically), and *time* as the general horizon of being, or of nature itself. According to this assumption physics deals with yes-or-no findings of a subject in the time horizon of being. To do so means to perform physics, but the human mind is not restricted to yes-or-no observations. Love, for instance, is different. The only limit to objectification then is its being objectification [14, p. 290] or, in other words, the definition of physics expresses its sole limitation.

To presume that physics is 'nothing but' the explication of what is implied when somebody sets out to deal with objects in time was Weizsäcker's titanic approach to *understanding* what happens in a physical *explanation*, and in this sense to reconcile both forms of knowledge. Is it conceivable to reduce space as well as the Lorentz group, particles and fields, Hilbert space and Schrödinger's equation, local interaction, the mathematics of the S-matrix etc. to such simple concepts as time and object (cf. [14, p. 201])?

Weizsäcker followed this idea with a persistence comparable to that which he certified to the historical course of physics. His spiritual radicalism in this respect much more than outweighed his complete lack of political radicalism. And his approach indeed allowed him – though on the credit for a treasure yet to be found – an inverted physicalism, or even materialism, which seemed to get beyond the debate on reductionism. Because to all those who complained that living beings were not to be considered as 'nothing but' physical systems, he now could rejoin: But do you really know what a physical system is? And then he proceeded as had Plato, who similarly had asked those who refuted to consider the planets – whom they presumed to be gods – as nothing but glowing matter: But do you properly know what matter is? In his approach to elementary particle theory Plato then presented a paradigm for a theory that explained matter as consisting of immaterial structures (Tim. 53c-55c, Nom. 886d), and in the same sense Weizsäcker concluded matter to be properly spirit. Matter then appears to be spirit inasmuch as it allows objectification [14, p. 289], [15, pp. 188ff.].

Doubts, however, remain as to why the same spirit deals with itself in such opposite discriminations as are understanding on the one hand and causal explanation on the other. Physicians feel that dilemma most acutely when they treat their patients like biochemical systems and at the same time try to understand what's wrong with them personally. Weizsäcker posed that question himself but did not give an answer [14, p. 290]. Yet he maintained that apart from causal explanations we can also – according to Tschuang Tse's story – share the joy of the fish, or understand them [15, p. 101], because there is joy in nature and thus in connatural beings (cf. [8]). Indeed there is truth and morality in nature [14, p. 320], [16, p. 370]. If not, how could they have emerged in man? The moral nature of man, therefore, is – as Plato suggested – the whole of nature as it came forth in human shape. I share this enhanced naturalism and find myself on Weizsäcker's side against all those who consider nature as 'nothing but' nature and man as something better than her. *Nature is not 'nothing but nature'*. In man nature became man and in fish nature became fish so that there is no 'nothing but'. I would like to point out, however, that Weizsäcker's assessment of physics with respect to nature bears some limitation which is not always made explicit. He sometimes seems to suggest that only physics when completed will finally teach us what nature really is. From the priority of nature to man and of man to physics it follows indeed that nature becomes aware of herself in the human mind by means of physics, but the same is true for art, mythology, and religion. Understanding includes understanding of explanation – Weizsäcker's goal with respect to physics – but physics when understood would not include art, or religion, and certainly not love. Physics, therefore, does not disclose The unity of nature but only that particular unity which manifests itself in physics. I think that Weizsäcker would agree: he once said so himself in his introduction to Picht's lecture on nature [9, p. xiii]. The question as to how objectification and care – or the 'it' and the 'you' – are related should, therefore, be left for further consideration. Weizsäcker may have felt a similarity to the relationship of law and gospel here as it is considered in Christian theology. But while we learn from the law how much we stand in need of love, can we expect a similar lesson from physics?

A further suggestion refers to Weizsäcker's conjecture that science is "the hard core of occidental culture" [15, pp. 96/460f., et passim]. Classical science – and modern technology is still based on it – is a canon of rules in different chapters on how to dispose of things and beings to reach certain goals in the sensual world [8]. Aren't these goals external to science the main object of culture? As Weizsäcker himself observed, a culture cannot be sound when its means are ten times more carefully developed than the consciousness of the goals to be reached by them [15, p. 104]. Science, therefore, as a part of culture is a challenge to the whole. For instance, the atomic bomb, was and remains a challenge to political culture not to discharge international conflicts by means of violence any more. And the industrial economy has become so destructive that it can almost be no longer recognised as the economy of nations which once considered themselves to be cultivated people. In both cases culture must be developed as a counterpart to science and technology, or rather science and technology should be reembedded into an embracing and more holistic culture. They are not the hard core of an already existing culture but they should live up to become an element of future culture.

My final suggestion turns back to Weizsäcker's first fundamental discovery, i.e. to the physical relevance of the structure of time. It is true that future events are possible while past events are facts. Together with Georg Picht, Weizsäcker used to insist on this as if past events remained facts for an indefinite time but this is an additional assumption which is not needed for his approach to the foundation of physics. Now I do not hesitate to admit that, for instance, (1) the solar eclipse of 28 May 585 BC presently remains to be a fact. Also it could probably still be considered a fact that (2) in the Middle East a battle between the Lydians and the Persians (Medans) took place on that day and moreover that (3) both parties attributed the eclipse to their stupid war – which at that time had been going on for several years already – and were so frightened about this cosmic wrath that they immediately decided to settle their conflict peacefully. I feel, however, that the factuality of these facts is somehow waning or decreasing from the eclipse to the battle and particularly to the sudden settlement, inasmuch as the latter facts are remembered not least because of the eclipse. But now let's assume a blue sky on the morning of that day at the place of battle. Can this remain to be a fact? And if it does, what about the morning sky at the same place on 28 May 10,000, 100,000, 1,000,000 ... years ago, days that are not renowned for anything particular? Might the colours of those skies still be considered facts? This again depends on what we mean by knowing, but I cannot reasonably think so.

According to the philosophy of emergent evolution in history there is not The past which, for instance, every year grows by one year, but any past is the past of some particular present, and the same is true for the future. This means that the facts of the past and the possibilities of the future are limited by horizons from a present point on the curved course of history so that our views in both directions would be extended not further than the neighborhood of that particular present. If this neighborhood were defined, for instance, by a time horizon about 10,000 years in diameter, a past of about 5,000 years and a future of again about 5,000 years could be meaningfully attached to some particular present and the factual past as well as the possible future were limited by that horizon of foresight as well as of hindsight. To some extent this idea is even found in relativity theory insofar as any light cone is attached to a particular present and the 'big bang' is in fact nothing like a point but a blurred space – at least as long as the singularity is not dissolved by a future quantum gravitational theory.

The main reason for my scepticism with respect to the factuality of The past is that the invariability of the laws of physics seems to be a condition of the possibility of physics but yet is only a Cartesian idea. It presupposes the Cartesian dichotomy of mind and matter and falls with it. But do we really believe that only matter is subject to historical change while mind is not? And if it were, the historical change of the laws of nature could not again follow an invariable pattern. Instead of the Cartesian view I could rather imagine that the universe came into being like a work of art. Of course, some of these are produced according to given rules, or prescriptions, but these tend to stay behind. As Giordano Bruno once put it, a work of excellence in art, reveals its particular order when it is completed, but that order was not just embodied into matter but emerged only in the course of production. And should the cosmos not be regarded as a first class production?

References

1. N. Bohr: Wirkungsquantum und Naturbeschreibung. In: *Atomtheorie und Naturbeschreibung [1929]. Vier Aufsätze mit einer einleitenden Übersicht.* Springer, Berlin 1931, pp. 60–66
2. N. Bohr: The Quantum of Action and the Description of Natur (1929). In: *Atomic Theory and the Description of Natur. Four Essays With an Introductory Survey.* University Press, Cambridge 1934/1961, pp. 92–101
3. N. Bohr: [Letters to the Editor] Quantum Mechanics and Physical Reality (Copenhagen, June 29, 1935). In: *Nature* 136, July 13 (1935) p. 65
4. J.W. von Goethe: Letter to Jacobi, May 10, 1812. In: *Goethes Briefe. Hamburger Ausgabe in vier Bänden.* Ed. by K.R. Mandelkow. 2nd ed, C.H. Beck, München 1976, Vol. III, pp. 190f
5. W. Heisenberg: Der Begriff "Abgeschlossene Theorie" in der modernen Naturwissenschaft. In: *Dialectica* 2, No. 3/4, 331–336 (1948)
6. J.G. Herder: *Ideen zur Philosophie der Geschichte der Menschheit* [1784–91]. Ed. by M. Bollacher. Werke in zehn Bänden, Bd. VI. Deutscher Klassiker Verlag, Frankfurt a. M. 1989, 1214 pp.
7. A. von Humboldt: *Kosmos. Entwurf einer physischen Weltbeschreibung.* 5 Vol., Cotta, Stuttgart/Tübingen 1845–1862
8. K.M. Meyer-Abich: *Praktische Naturphilosophie – Erinnerung an einen vergessenen Traum,* C.H. Beck, München 1997, 520 pp.
9. G. Picht: Der Begriff der Natur und seine Geschichte. Mit einer Einleitung von Carl Friedrich von Weizsäcker. In: *Vorlesungen und Schriften.* Studienausgabe ed. by C. Eisenbart and E. Rudolph, Klett-Cotta, Stuttgart 1989, xv, 502 pp.
10. C.F. von Weizsäcker: Der zweite Hauptsatz und der Unterschied von Vergangenheit und Zukunft. *Annalen der Physik,* 36, 275–283 (1939); reprinted 1971, pp. 172–182
11. C.F. von Weizsäcker: *Die Geschichte der Natur. Zwölf Vorlesungen.* Hirzel, Stuttgart 1948, 138 pp.; *The History of Nature.* London 1951, 179 pp.
12. C.F. von Weizsäcker: *Zum Weltbild der Physik* [1943]. 8. Ed, Hirzel, Stuttgart 1958/1960, 378 pp.; *The World View of Physics.* London 1952, 219 pp.

13. C.F. von Weizsäcker: *Die Tragweite der Wissenschaft. Erster Band: Schöpfung und Weltentstehung. Die Geschichte zweier Begriffe*. Hirzel, Stuttgart 1964, xi, 243 pp.
14. C.F. von Weizsäcker: *Die Einheit der Natur*. Hanser, München 1971, 481 pp.; *The Unity of Nature*. New York 1980, x, 406 pp.
15. C.F. von Weizsäcker: *Der Garten des Menschlichen. Beiträge zur geschichtlichen Anthropologie*. Hanser, München 1977, 612 pp.; *The Ambivalence of Progress. Essays on Historical Anthropology*. Paragon House, New York 1988, 304 pp.
16. C.F. von Weizsäcker: *Wahrnehmung der Neuzeit*. Hanser, München 1983, 440 pp.
17. C.F. von Weizsäcker: *Der Mensch in seiner Geschichte*. Hanser, München 1991, 244 pp.

Note: Quotations as well as paraphrases from Weizsäcker's books are given in my translation so that the page numbers refer to the German editions.

C.F. von Weizsäcker's Philosophy of Science and the Nature of Time

Michael Drieschner

Carl Friedrich von Weizsäcker has always regarded the structure of time as being the most important aspect of his philosophy.[1] Since this Festschrift deals mainly with physics I want to emphasize the influence reflections about the structure of time have on understanding physics. This is one way how (good!) philosophy can help physicists to better penetrate the meaning of their own theories.

So as a consequence we deal here with the connection between science and the theory of time. We have to take into account two aspects of time that have been treated quite differently in tradition, namely:

1. Time in its modes of past, present, and future: It is characteristic for an event to change its features by these three modes "in time". Alternative version: each event changes its features by these three modes ... An event that begins with being *possible* becomes *actual*, and then it is *factual*, past. Scientists often regarded this aspect as purely subjective, with no legal place within science.

2. The sequence of events, and measurement of times, represented e.g. as a sequence and count of years or days or nanoseconds. The focus is in this case the sequence of events and the time lapse between them. They do not change in time. Thus in modern science time is a real parameter (t). This represents time in a way that is regarded "objective" in philosophy of science and in logic.

The distinction between these two aspects of time is closely related to the distinction between the A-series and the B-series of events, as described in the well-known paper by the McTaggarts.[2] Nevertheless even the McTaggarts suppose that time is a sequence of events both in future and in past. Thus their view is very close to

---

[1] This is hinted at in the title of his monumental work: "Zeit und Wissen" (time and knowledge), München (Hanser) 1992. He describes it in the introduction, e.g. pp. 27–32, in more detail in I,6; I,7.

[2] John McTaggart, Ellis McTaggart, The Unreality of Time. Mind XVII (1908), pp. 457–474.

the "ontology of classical physics" or, as Meyer-Abich calls it, treating the future as a past to come.[3]

It is C.F. von Weizsäckers contribution to an understanding of science having pointed out the importance of what the true character of future is. This paper is supposed to give an overview of this contribution.

1 Philosophy of Time

Time is so fundamental that a special effort is needed to describe its structure. "So what is time?" Augustine asks. "As long as nobody asks me I know it. But when somebody asks me, and I want to explain it to him, I do not know it."[4]

This is the strange character of time: Everything goes on "in time". But when I say so I am already using a spatial metaphor, for time is not like a container with events "in" it. Temporality is so all-embracing that I barely notice it, and I am almost unable to describe it, if not by spatial images.

Physics treat time like a spatial coordinate. Time is a parameter t that takes on real values from $-\infty$ to $+\infty$. This comes in rather handy for the usual calculations. The parameter t symbolizes the position of the hand on a clock at the time of an event. The laws of physics give us the possibility to *predict* events, depending on the position of that hand.

But the parameter t represents only a small aspect of what time can be. I cannot understand anything from it about my present life, about future being quite different from past, about the fact that I cannot move arbitrarily in time (as I can in space), and much more like that. Physicists are inclined to look at the parameter t as the original description of time, and at everything else that makes up time as "subjective accessories". From this fact we can understand that physicists, when they speculate about reality, think quite seriously about 'reversal of time' or about 'closed loops of time' or about 'several times at once' (where, again, 'at once' is a *temporal* characterization).

In the writings of the founders of philosophy, Plato and Aristotle, a term like 'time' is used only in passing. *Plato*, however, gives the famous definition according to which the world-craftsman, the demiourgos, makes time "of eternity that abides in unity, an everlasting like-ness moving according to number – that to which we have given the name Time."[5] Plato's text itself explains that what is meant is the course of days, months, and years. So this is apparently something similar to what modern physics use the parameter t for. Only for Plato the cyclic character of that process is important.

---

[3] K.M. Meyer-Abich: *Die andere Ordnung des Lebendigen*. In: R.-M.E. Jacobi, P.C. Claussen and P. Wolf (Hg.): *Die Wahrheit der Begegnung*. Würzburg 2001, pp. 347–366; here p. 351; cf. his paper in this volume

[4] Aurelius Augustinus: *Confessiones* XI, 14: "Quid est ergo tempus? Si nemo ex me quaerit, scio; si quaerenti explicare velim, nescio."

[5] ... ποιεῖ μένοντος αἰῶνος ἐν ἑνί κατ' ἀριθμόν ἰοῦσαν αἰώνιον εἰ κόνα, τοῦτον ὅν δὴ χρόνον ὠνομάκαμεν. Plato, (Tim. 37d7f). Translation: F.M. Cornford: *Plato's Cosmology*, London 1937.

Aristotle, on the other hand, derives time from motion in general; motion does not have to be cyclic. Motion, in turn, he derives from the pair of concepts *potential* and *actual*, fundamental for his philosophy. He defines motion thus: "The actuality of that which potentially is, as such, is motion."[6] This formulation has often been misunderstood, still today some English translations (and most German ones!) give, instead of 'actuality', e.g.: "the progress of the realizing"[7] or "realization of their potentiality"[8]. This translation looks more plausible at first sight, but it is of no use as a definition since the concept of 'realization' presupposes the very *process* that is to be defined.[9] – The definition by Aristotle, read correctly, is especially interesting because it associates *time* with *potentiality*, as we will do below as well.

Augustine, whose treatise on time we have quoted above, is known in the history of philosophy as the first one to treat time in the same way as it is mainly treated in the philosophy of today: He treats it as that strange structure of past, present, and future, where always I am *present* myself and looking from my point of view, which is *now*, onto past and future (which is, again, a spatial metaphor!).

Augustine takes up Aristotle's question how time can *be*; because the past is no more, the future is not yet, and the present is only the division between past and present. Augustine even takes up a formulation by Aristotle where he says that time is within soul. But for Augustine this gets a different color. Whereas Aristotle puts *counting* of time into soul, Augustine adds that future is within soul as an expectation, and past is within soul as a memory. In that way Augustine solves the problem in a quite different way from Aristotle's. He, so to speak, brings the "modern", subjective view into the philosophical discussion.

Before *Kant* could set a new beginning in the theory of time, there had been the invention of classical mechanics by Isaac Newton. Kant takes up Newton's theory. His new idea was to define time as the *form of inner intuition*. This makes time "subjective", on the one hand, according to his transcendental philosophy. On the other hand he was able to give time all the features that are necessary for the construction of classical Newtonian mechanics. Kant's theory of time, which is central for his philosophical system, remains therefor entirely within the framework of classical physics. Time is viewed essentially as a sequence of events, its structure as present, past, and future does not occur in Kant's writings. Thus neither in the Kant dictionary by Schmid[10] nor in that by Eisler[11] the keywords past or future are found. To this it fits perfectly that Kant, in treating causality, considers quite naturally nothing but deterministic causality. A different possibility, i.e. statistical causality, does not

---

[6] ἡ τοῦ δυνάμει ὄντος ἐντελέχεια, ᾗ τοιοῦτον, κίνησίς ἐστιν. Aristotle (Phys. 201a10f.) Translation: E. Hussey: *Aristotle's Physics*, Book III and IV. Oxford 1983. Cf. W. Wieland: *Die Aristotelische Physik*, Göttingen. 1961,²1970, p. 298.

[7] Ph.H. Wicksteed, F.M. Cornford: *Aristotle, the Physics*. London, Cambridge 1957.

[8] W.D. Ross: *Aristotle, Physics*. Oxford 1936.

[9] cf. Wieland, l.c.; E. Hussey, l.c.

[10] C.Chr.E. Schmid: *Wörterbuch zum leichtern Gebrauch der Kantischen Schriften*. Jena (Cröker) 1798.

[11] R. Eisler: *Kant Lexikon*. Berlin (Mittler) 1930.

appear until the 20th century. It did occur to Kant no more than to other scholars of the 18th century – we will come back to this.

An especially beautiful formulation of that determinism can be found in *Laplace*, interestingly enough in his treatise on probability: In Laplace's writings probability occurs only in connection with ignorance. The course of events itself is strictly determined in his description. He writes[12]: "An intelligence that for a given instant knew all the forces that animate nature, and the respective situation of all beings that constitute it, if, besides, it was vast enough to submit those conditions to analysis, it covered in the same formula the movements of the largest bodies of the universe and those of the lightest atom: Nothing would be uncertain for it, and future as well as the past were present before its eyes." – In respect to time it is characteristic for determinism that Laplace considers past and future as *present* before the eyes of such super-human intelligence. Time, in its temporal structure, disappears entirely in view of a strict determinism. Everything that happens is equally present in some higher sense.

It seems that for *Einstein* the space-time continuum, that gives, according to his theory of relativity, a place to everything that is real, is *simultaneously* present in the same sense as for Laplace. Einstein gives the shortest and most brilliant formulation in a letter of condolence after his friend Besso died, shortly before Einstein died himself[13]: "Now he has preceded myself a bit even in the farewell to this strange world. This does not mean anything. For us faithful physicists the separation between past, present, and future means nothing but an, although obstinate, illusion." – Here Einstein takes the framework of space-time, that is best suited for the description of measurable events, as the true reality. The structure of time, on the other hand, which we know before any physics, to him looks like an illusion.

In logic, time appears, if at all, in a similar manner. Traditionally logic treats propositions that are valid for all times. They claim, in this sense, "eternal presence". Modern formal logic has its application mainly in propositions of mathematics or logic itself, in any case in propositions that are not assigned to specific times. Even formal temporal logic, as e.g. A. Prior's "tense logic",[14] presupposes as self evident that time consists of a series of events that just "are there". They suppose that the future is nothing but the past to come. It is C.F. von Weizsäcker who, on the contrary, proposes by his idea of a "logic of temporal propositions" a proper status for temporality even in logic, especially for the logic of future. Up to now, though, Weizsäcker only gives programmatic sketches. It would be worthwhile developing those sketches into a system.

---

[12] P.S. de Laplace: *Essai philosophique sur les probabilités*. Paris 1814, p. 2. "Une intelligence qui, pour un instant donné, connaîtrait toutes les forces dont la nature est animée, et la situation respective des êtres qui la composent, si d'ailleurs elle était assez vaste pour soumettre ces données à l'analyse, embrasserait dans la même formule, les mouvements des plus grand corps de l'univers et ceux du plus léger atome: rien ne serait incertain pour elle, et l'avenir comme le passé, serait présent à ses yeux." (translation MD)

[13] P. Speziali (ed.): *Albert Einstein – Michele Besso. Correspondance 1903–1955*. Paris 1972, p. 537 (translation MD).

[14] A. Prior: *Past, Present and Future*, Oxford 1967.

2 Statistical Thermodynamics

In the middle of 19th century the idea began to be accepted that thermodynamics – beginning with the theory of gases, later thermodynamics in general – could be derived from the statistics of its smallest parts. For a gas, e.g., this is the statistics of its molecules. The great advantage of this statistical theory is that it is derivable from very general principles. So it teaches understanding thermodynamics, the science of steam engines, which looked very special in the beginning, as a general theory of approximate description of any physical system. The generality of the objections against that theory corresponded to the generality of the theory itself. Until now among these objections the "reversal objection" is felt to be particularly serious. This objection, first formulated by Lord Kelvin and J. Loschmid in 1875, says: The second law of thermodynamics says that the entropy of a system that is closed energetically as well as materially, can increase or stay the same, but cannot decrease. This describes e.g. two systems at different temperatures. Their temperatures converge when they are brought into contact, the temperatures will never become different by themselves. This "irreversibility" of thermodynamic processes can be derived from the mechanics of a system consisting of very many partial systems, e.g. from the mechanics of a gas that consists of very many (10^{23}!) freely moving molecules.

Now the problem is that mechanics is a reversible theory. This means that with every process the reverse process is possible as well, according to mechanics, where velocities have reverse direction and the sequence of the positions is reversed. Statistics adds to mechanics nothing but a reduction of detail in the description of processes, in such a way that only average values are retained. One cannot see how a reduction like this could change anything about the basic reversibility of the theory.

This problem is stated more precisely in the "reversal objection": Regard any development of a thermodynamic system, where entropy increases. Now imagine that in the basic mechanical system all velocities are reversed. Then also the thermodynamic states the system has just passed will be passed in reversed order, in such a way that entropy will decrease. Mechanically the latter process is possible as well as the former, but thermodynamically it is impossible. Thus it is not possible, says the reversal objection, that thermodynamics is derived from mechanics just by making the description incomplete.

Here the structure of time enters the scene. Boltzmann gives several arguments in defense of statistical thermodynamics. At first he says that usually, in considering thermodynamic systems, we start with a state of low entropy; thus, in regarding *all* mechanical systems that belong to this thermodynamic state, we find an overwhelmingly large probability for an increasing entropy. Later he explains his point of view in regard of the whole universe as follows[15]:

---

[15] L. Boltzmann: *Vorlesungen über Gastheorie*, 2 Vols. Leipzig 1898, 1896/98; 21910, 90: "Für das Universum sind also beide Richtungen der Zeit ununterscheidbar, wie es im Raume kein oben oder unten giebt. Aber wie wir an einer bestimmten Stelle der Erdoberfläche die Richtung gegen den Erdmittelpunkt als die Richtung nach unten bezeichnen, so wird ein Lebewesen, das sich in einer bestimmten Zeitphase einer solchen Einzelwelt befindet, die

"For the universe, the two directions of time are indistinguishable, just as in space there is no up or down. However, just as at a particular place on the earth's surface we call "down" the direction toward the center of the earth, so will a living being in a particular time interval of such a single world distinguish the direction of time toward the less probable state from the opposite direction (the former toward the past, the beginning, the latter toward the future, the end). By virtue of this terminology, such small isolated regions of the universe will always find themselves "initially" in an improbable state. This method seems to me to be the only way in which one can understand the second law – the heat death of each single world – without a unidirectional change of the entire universe from a definite initial state to a final state."

This is a particularly obvious manifestation of the prejudice of a typical physicist. He thinks that the only true description of the world is the one by equations of mechanics or thermodynamics, where the real parameter t governs processes. And on the other hand he thinks that past, present and future are "subjective" accessories of the individual, that have to be unmasked as soon as they lead to difficulties about the true, objective description by physical equations. – That Boltzmann calls his abstruse proposal the only method to think the structure of time can be read today as his admission of failure.

In the beginning of 20th century J.W. Gibbs completed Statistical Thermodynamics theoretically. Naturally he encountered the same problem, but he dismissed it rather pragmatically. He writes in his standard treatise[16]:

"But while the distinction of prior and subsequent events may be immaterial with respect to mathematical fictions, it is quite otherwise with respect to the events of the real world. It should not be forgotten, when our ensembles are chosen to illustrate the probabilities of events in the real world, that while the probabilities of subsequent events may often be determined from the probabilities of prior events, it is rarely the case that probabilities of prior events can be determined from those of subsequent events, for we are rarely justified in excluding the consideration of the antecedent probability of the prior events."

Here Gibbs hints, in a rather hidden way, at an idea that should later bring the solution of the problem: Probability is generally applied only to *predictions*, not to propositions on past events. It is possible, admittedly, to give propositions about past events a good sense, like e.g.: "Probably Napoleon was born in 1769". But the uncertainty we indicate by the word 'probably' does not refer to the past fact itself. For

---

Zeitrichtung gegen die unwahrscheinlicheren Zustände anders als die entgegengesetzte (erstere als die Vergangenheit, den Anfang, letztere als die Zukunft, das Ende) bezeichnen und vermöge dieser Benennung werden sich für dasselbe kleine aus dem Universum isolierte Gebiete, "anfangs" immer in einem unwahrscheinlichen Zustande befinden. Diese Methode scheint mir die einzige, wonach man den zweiten Hauptsatz, den Wärmetod jeder Einzelwelt, ohne eine einseitige Änderung des ganzen Universums von einem bestimmten Anfangs- gegen einen schließlichen Endzustand denken kann." – English edition: Lectures on gas theory; translated by Stephen G. Brush. Berkeley 1964; New York 1995.

[16] J.W. Gibbs: *Elementary Principles in Statistical Mechanics*. New York 1902. Reprint: Woodbridge, CT, 1981, p. 150–151

Napoleon was born in 1769 or he was not born that year; the fact exists, independently of whether we know it or not. What is uncertain is what we will possibly know, in *future*. Thus even when we assign probabilities to past facts we mean a *possibility*, our knowledge that may become real in future.

C.F. von Weizsäcker picks up this thread when he gives a refutation of the "reversal" objection in his paper of 1939[17]: The difference between past and future, which is characteristic for thermodynamics, does not mysteriously come into theory by an approximate description. It is rather ourselves who introduce this difference from outside, just in applying probability only to future. This appears to be so self-evident that nobody made it explicit before 1939. In 1971, when his paper was printed again, Weizsäcker himself writes: "When I wrote it I felt that I had set forth something rather trivial". He calls his text nothing but an attempt at explaining Gibbs' words.

In his paper C.F. von Weizsäcker begins by stating that Boltzmann's H-theorem does not imply a difference between past and future. What Boltzmann proves is that with any state of non-maximal entropy all neighboring states have, with overwhelming probability, higher entropy, i.e. past states as well as future ones. Past and future are entirely symmetric. From Boltzmann's assumption of thermodynamic probability (i.e. thermodynamic equilibrium) there follows rather "that a non-maximal value of entropy of a system we know nothing else about is, with overwhelming probability, a relative minimum of entropy", as Weizsäcker puts it.[18] The phrase "we know nothing else about" again indicates the assumption of equilibrium, i.e. of equal probability of all microstates.[19]

For a *prediction*, the original application of probability, this entails growth of entropy. For the past, however, we need additional considerations.

Suppose you know that the system you consider is in thermodynamic equilibrium. Then Boltzmann's considerations are immediately valid, a state of non-maximal entropy is most probably an extreme of a fluctuation. Often, however, we consider a system about the past of which we have or can infer some information. When I see, e.g., a pot of lukewarm coffee on a table I can be rather sure that the coffee was hot before and has cooled down, increasing its entropy. This conclusion seems reasonable, considering European household customs, i.e. from implied *facts* of the past. The idea, on the other hand, that lukewarm coffee could be the result of a fluctuation is absurd, considering imaginable past facts.

Thus the problem how the difference between past and future comes into Statistical Thermodynamics is resolved rather convincingly: We ourselves introduce that difference into our considerations. Once we have drawn out attention to this structure it is not mysterious any more. – It is a pity, though, that this solution, that has been given as early as in 1939, has not yet entered the discussion within the

---

[17] C.F. von Weizsäcker: Der zweite Hauptsatz und der Unterschied von Vergangenheit und Zukunft. *Annalen der Physik* 36(1939), 275. Reprinted in: Die Einheit der Natur. München 1971, p. 172–182

[18] l.c., p. 174

[19] A detailed discussion is found in: M. Drieschner: *Voraussage – Wahrscheinlichkeit – Objekt*. Berlin etc. 1979, p. 48–57, 215–219.

scientific community. Recent presentations[20] still reproduce Boltzmann's discussion, which apparently is unsatisfactory. Not even Gibbs' idea (of 1902!) has brought a modification of those presentations.

3 Probability

We have already mentioned the connection between probability and the structure of time when we dealt with thermodynamics: Future is the tense of possibility, and quantified possibility is probability. To put it more exactly: future events are *possible*; some of those possible events will be *actual*, then they will be present; and immediately afterwards they will be *factual*, facts of the past.

I am able to try to predict possible events, e.g.: "Tomorrow there will be an eclipse." But there are predictions that are not as unambiguous – this is a discovery of the 18th century – , predictions with a probability. What do we mean if we give a probability? What is, after all, *probability*?

There is the so-called classical definition of probability by Laplace, saying: "Probability is the ratio of the number of favorable cases to the number of possible cases."[21] This applies mainly to the combinatorial considerations that were usual in the beginning of probability theory, e.g. for calculating the chances in games of cards or of dice: The probability to draw a king in a deck of 52 cards is 1/13, namely 4 (the number of kings, the *favorable* cases) divided by 52 (the number of all cards, the *possible* cases). One sees quickly that this is not a proper *definition*: Laplace himself puts his "definition" under the condition that "we see no reason why one of those cases would occur more easily than any other one."[22] He could have put it more clearly: This applies if we suppose equally probable cases. Thus Laplace reduces unequal probabilities to equal probabilities, but he does not define the *concept* of probability. A true definition of probability, on the other hand, ran into all but insurmountable difficulties. The reason is the inherent vagueness of probability that cannot be removed by a definition, sharp as that definition might be: Apparently probability has to do with relative frequency. But it cannot simply be identified with relative frequency since the latter equals probability only roughly.

The problems with the foundation of objective probability led to the introduction of a *subjective* concept of probability.[23] The latter refers to the subjective assessment

---

[20] E.g. A. Grünbaum: *The Anisotropy of Time*, 1967. In: T. Gold and D.L. Schuhmacher (eds.) *The Nature of Time*. U.P. Cornell; or: A. Grünbaum, *Philosophical Problems of space and time*. (Ed. by R.S. Cohen and M.H. Wartofsky) Dordrecht, Boston ²1973 (Boston Studies in the Philosophy of Science. Vol. XII; Synthese Library.); similarly in: L. Sklar, *Physics and chance: philosophical issues in the foundations of statistical mechanics*. Cambridge 1996.

[21] P.S. de Laplace: Recherches sur l'intégration des équations différentielles aux différences finies et sur leur usage dans la théorie des hasards. In : *Mémoires de l'Académie Royale des sciences de Paris (Savants étrangers)* 7 (1773) 1776; reprinted in : *Œuvres complètes de Laplace*, Vol. VIII, Paris 1891, p. 69–197; this quote p. 146

[22] Laplace, loc. cit.

[23] Bruno de Finetti since the 1920s; in English: Probability, Induction, and Statistics. London (Wiley) 1972

for the degree of truth of a proposition, made explicit e.g. in the willingness to bet. According to this view the rules of probability theory contain nothing but the conditions for the consistency of such assessments. We can make them explicit in the condition that a bet has to be *fair*.

C.F. von Weizsäcker's contribution to this debate consists, to begin with, in his hint that, according to probability theory itself, the probability is the *expectation value* of the relative frequency.[24]

This insight is not in itself a *definition* of probability, but it contributes to a consideration of consistency. From our joint work a definition has resulted that sounds almost ridiculously simple: "Probability is a predicted relative frequency." – Here the relation to the structure of time becomes apparent: A probability statement always refers to future events. Even if its propositional content refers to the past, as in our example of the Napoleon's birthdate, probability refers to the future possibility that the assertion about the past fact *will* prove true.

For the concept of probability, as for thermodynamics, the inclusion of the structure of time gives amazingly simple solutions:

At first it is clear that predictions do not have to come exactly true. Probability theory itself gives a prediction for the mean *deviation* of relative frequency from the predicted value in an actual series of experiments. In order to specify this prediction, in turn, one can calculate the deviation of those deviations from their predicted value, for series of series of experiments, etc.[25] Thus probability has a hierarchical structure that can be continued as far as one likes. In this structure we can specify the place of the consistency considerations by C.F. von Weizsäcker. The expectation value he refers to in: "probability is the expectation value of the relative frequency" is derived from the probability of the next higher step in the hierarchy, namely the probability of the results of *series* of experiments.

The definition of probability we have given, as a predicted relative frequency, allows us to see the systematic place of the difference between objective and subjective probability: *Prediction* always contains a *subjective* element, predictions may turn out wrong. But predictions made in science are supposed to prove true empirically, i.e. to indicate *objective* facts. We could describe the relation thus: The subjective interpretation of probability emphasizes the character of proposition, of knowledge: the subjective opinion about what the relative frequency will be. For the objective interpretation, on the other hand, the emphasis is on the content of prediction, on the real future relative frequency that would confirm a true prediction.

Let me add a remark on the concept of probability in general. Our definition refers to one of many possible meanings of probability. There could be (and are) other ways to use this word. The structure of the argument is thus: Science deals with relative frequencies and their prediction. I find that what traditionally is called probability usually agrees with my concept of predicted relative frequency. And I also find that for

---

[24] C.F. von Weizsäcker, *Zeit und Wissen*, München (Hanser) 1992, part II, 4; C.F. von Weizsäcker, *Aufbau der Physik*, München (Hanser) 1985, Chapt. 3.

[25] Cf. M. Drieschner: *Moderne Naturphilosophie*, Paderborn 2002, in more detail in: M. Drieschner: *Voraussage, Wahrscheinlichkeit, Objekt*, Berlin etc. 1979.

this concept of predicted relative frequency some problems that are usually discussed in relation to probability find a solution. But it is quite possible that there are other concepts of probability that are not affected at all by this argument of mine.

Now we can ask whether probability is objective in the sense of a measurable quantity or not. The fact that probability is found in a measurement only approximately is not a counter-argument. For this is true for all measurable quantities. But for probability this inaccuracy is of very fundamental nature: What we measure is a relative frequency; and above we have seen that probability is *not* the same as relative frequency. If we want to interpret probability as a property of a physical system we have to treat it, apparently, as a kind of disposition, a "propensity", as Popper[26] calls it, to produce certain relative frequencies. This propensity does not appear directly as a result of a measurement. What we measure, the phenomenon, depends on the propensity in a well-known way, but it is not identical with it. – In the discussion of quantum mechanics we will come across such structures again.

In spite of the systematically unavoidable inaccuracies of the prediction of relative frequencies probability theory allows exact calculations with real numbers. How is that?

Probability theory has become pure mathematics since its axiomatization by A. Kolmogorov in 1933.[27] The crucial point in his axiomatics is banishing the problematic relation between probability and relative frequency entirely from mathematics into the "application". In his axiomatics he included only the *relations* among probabilities, that could be stated exactly and rather simply. In fact probability theory from the beginning dealt with nothing but relations among probabilities and their "consistency", as mentioned above.

Another brilliant simplification in Kolmogorov's work is his treatment of the so called product rule saying that the probability for event A *and* B is the product of the probability of A alone and the probability of B alone, provided the two events are *independent*. Kolmogorov does not have to give a criterion for the independence of events but he introduces the product rule by a *definition*. This definition reads something like: "We call two events A and B independent if the product rule is true for them." – Seen this way, probability theory is pure mathematics. Mathematicians put aside the problems we mentioned above as "application problems".

We want to introduce the opposite view as well, which proves probability theory to be a *science*. This is again aided by regarding the structure of time. For we can see, from the structure of time, that the most general law of nature is a probability law. Since this is a very special assertion in the framework of our investigations, I will explain it a bit more in detail.

What is a law of nature? – Reduced to the most general scheme every law of nature is a prescription how to get empirically testable predictions from the present state of affairs. Thus our assertion reads: The most general empirically testable

---

[26] Karl Popper: *The propensity interpretation of the calculus of probability and the quantum theory*. In: S. Körner (ed.): *Observation and Interpretation*. London 1957

[27] A. Kolmogorov: *Grundbegriffe der Wahrscheinlichkeitsrechnung*. Berlin 1933. English: *Foundations of the Theory of Probability*, Chelsea Publish. Company, New York, 1950.

prediction is a probability statement. Above we introduced probability as a predicted relative frequency. So now we assert: The most general empirically testable prediction predicts a relative frequency.

Let me give a short sketch of the argument[28]: We can give, evidently, unambiguous "simple" predictions, yes or no. But one could also think of specifying predictions like: "Sometimes yes, sometimes no" in a way to make them empirically testable. This kind of predictions should be *general*, just as the simple ones. This means that they cannot apply, e.g., to a definite number or a definite sequence of yes and no. For this would again be a simple prediction, only for a more complex experiment. – The only general prediction that specifies "Sometimes yes, sometimes no", it turns out, is the prediction of a relative frequency. This means it is a probability.

We then can derive the well-known rules of probability calculus from the definition "predicted relative frequency". In doing so we cannot, as Kolmogorov, introduce *independence* by a definition. But we can specify the independence of events A and B by the condition that the predicted relative frequency for A is the same if either B is the case or if non-B is the case. With those premises we can derive Kolmogorov's theory. Using the structure of time the definition of probability given turns out to be the basis for the whole theory of probability.

We have seen above that we are able to consider at most, as a property of the system, the *propensity* to produce certain relative frequencies. A relative frequency, in turn, is a property of an actual series of measurements. This could be, e.g., a series of 14 throws of a dice, and the result could twice be "1"; whereas the corresponding probability, the propensity of the system to produce the result "1", could have been 1/6. This latter disposition, the propensity, is usually (as in our example) not confirmed exactly by the actual frequency. But the disposition is valid, by its definition, for *any* actual series of experiments. – What does that mean?

It has often been argued if one can apply probability to single events or only to series of events. This is contending about a goat's wool: One can assign probability to the class of all possible series of experiments; but with the same right one can assign probability to one experiment, as representative for that class; for it is constitutive for that class that its series consist all of one and the same type of experiment. – In this description it is a problem, which experiments are "of the same type", with that same probability; it is a question of the skill of the one who devises the experiment to ensure that "same type" for all experiments.

4 Quantum Mechanics

Quantum mechanics can be interpreted as a generalized probability theory. We can understand it much better, again, in considering the structure of time, as introduced by C.F. von Weizsäcker into the interpretation of quantum mechanics.

Kolmogorov's axioms of (classical) probability calculus allow a generalization to a quantum mechanical probability theory. Kolmogorov bases his axioms on the set $\mathcal{F}$

---

[28] The argument is presented in more detail in: M. Drieschner: *Voraussage, Wahrscheinlichkeit, Objekt*, Heidelberg 1979 (in German).

of random events, where every random event is represented by a set of *elementary* random events. His first axiom reads:

"I. $\mathcal{F}$ is a field of sets."

A field of sets is what is today called a Boolean lattice (of sets). For quantum mechanics we instead use as a first axiom:

"I'. $\mathcal{F}$ is a lattice of closed subspaces of Hilbert space."

The difference between these two axioms contains all differences between classical physics and quantum mechanics; Kolmogorov's other axioms remain the same. Those differences become clearer, again, from considering the structure of time. In fact, basing the theory on a lattice of subspaces instead of a field of sets entails a fundamental *indeterminism*.[29]

Indeterminism mirrors future's peculiarity as contrasted with the past: Future events are *possible*; in general with every event also alternative events are possible: future is open. The quantum mechanical lattice of propositions can be understood most easily as an expression of open future, as a lattice of *predictions*. We find that in this lattice it is never possible to make *all* propositions with certainty, i.e. with probability 0 or 1. There will always be propositions with probability *between* those two extreme values. This is the fundamental indeterminism of quantum mechanics.

One could, in principle, treat the classical lattice of propositions (Kolmogorov's field of sets) as a lattice of predictions as well. (Although in classical physics we can suppose that all predictions can be made "in principle" with certainty, i.e. with probability 0 or 1. Thus probabilities other than 0 or 1 must be due to our ignorance – as Laplace says in his classical formulation of determinism.) In this view the classical lattice of propositions is a degenerate case of the quantum mechanical lattice of propositions which, "accidentally", contains only probabilities 0 and 1.[30] So in Weizsäcker's view of the structure of time, quantum mechanics and classical physics admit a uniform classification, namely as theories of predictions.

Difficulties arise if one comes from the other side, the side of classical physics that presume that all predictions are certain. Such predictions can as well be understood as descriptions of properties that are there *within themselves*. If I can predict with certainty that I will find, e.g., planet X in position y, then I can as well say: "Planet X *is* really in position y". So predicting the result of a measurement has turned into stating a fact. For classical physics these are, as we can easily see, equivalent. But in quantum mechanics, with its fundamental indeterminism, this does not work any more.

This is apparently the source of many problems for someone who is used to the "ontology of classical physics", and this is where the dissatisfaction of "classical" physicists with quantum mechanics comes from.

In the same spirit "realism" in the interpretation of quantum mechanics asks what the *reality* described by quantum mechanics really is, or what lies *behind* the quantum

---

[29] M. Drieschner: *Voraussage, Wahrscheinlichkeit, Objekt*. Berlin etc. 1979.

[30] Technically speaking the quantum mechanical lattice of propositions becomes a classical one when there is a complete superselection rule, i.e. when no superposition of states is possible, and therefore all observables are compatible among each other.

mechanical description. – We are trying to answer this difficulty with recourse, again, to the structure of time:

The primary purpose of a physical theory is generating empirically testable predictions rather than a description of existing reality. In case the predictions can be made with certainty they can be reformulated, as we have seen above, as a description of reality. But this very possibility is excluded in quantum mechanics – if we exclude, for the moment, rather far-fetched variants like Bohmean mechanics.

In a second step we can specify the question of reality in a deeper way: Certainly every prediction presupposes a fundamental reality that allows describing the facts that form the basis for the prediction, and finally those facts that confirm or disprove the prediction. Niels Bohr calls this essential requirement for the interpretation of quantum mechanics the "necessity of classical concepts".

Fundamental difficulties result from this necessity of the classical concepts, which I can only sketch here: Quantum mechanics gives nothing but probabilities for the results of possible experiments. If we want to describe unambiguously any arrangements and results of measurements we need a language, concepts to describe reality, facts. Niels Bohr says we must be able to describe what we have done and what we have learnt. This is impossible in quantum mechanics alone, for this purpose we need the concepts and theories of classical physics.

But here a problem rises: Quantum mechanics was introduced, it was finally felt, as a relief, because it describes phenomena classical physics was unable to describe. Whenever the results of the two theories differ, quantum mechanics is right, classical physics is wrong. Then how is it possible that the (true) quantum mechanics presupposes, in the end, the (wrong) classical physics?

The practical physicist has an easy answer to this question: Where classical physics is needed for quantum mechanics, namely for describing arrangements and results of measurements, the two theories agree in a very good approximation. Thus we can assume the validity of quantum mechanics and still, in a good approximation, use the concepts of classical physics. For all practical proposes (FAPP, as an acronym) this is quite all right.

The philosopher, though, particularly if he is mathematically and logically minded, wants to know it more precisely. This "FAPP" may suffice for the practically working physicist, but the logician must conclude: Approximately correct means, if you take it seriously, wrong. So the whole theory is apparently inconsistent!

The discrepancy that shows up here also formally appears in the description of the process of measurement. Here we cannot present the theory of measurement in every detail, but let me at least sketch a rough outline.

The process of measurement in quantum mechanics is interesting mainly because the theory is indeterministic. This means that before measurement several results are *possible*, but after measurement only one of the results has become *actual*. It is true, this occurs in classical physics as well. But there we can console ourselves with the thought that "in itself" already before the measurement there existed but one possibility, and that it was only our ignorance that forced us to take more than one possibility into consideration. But in quantum mechanics even with the most exact description there remain, in general, more than one possibility for the result of a measurement. The theory is fundamentally indeterministic.

The *state* of the system leaves open, before measurement, several possibilities with their corresponding probabilities. After measurement this diversity is reduced to one single possible case. This case then has got (for an immediately following second measurement) probability 1. That change of state is called the "reduction of the wave packet".

There are "realists" among physicists (or, still more, among philosophers of science) who look for a physical mechanism that brings about this change of physical state. But if we take the structure of time seriously, as explained above, we can see that we do not need a physical mechanism. In fact, a prediction with more than one possibility *means* nothing but that in the end one of those possibilities will be realized, the others not. That is what is meant by the predictions of quantum mechanics. The change in description after measurement is the physicist's own decision. He could just as well continue with the old description and keep all prior possibilities for further predictions, including corresponding probabilities. Then he would waive the chance of using the information won by the experiment, but that latter description would be as valid as the first "reduced" one. It is obvious that there *can* be no physical mechanism within the described system for a decision of the one who describes that system.

There is a possibility to waive information from measurement in the formal description of the process of measurement as well. We would presuppose for that description that a measurement has actually taken place, and that, consequently, one of the possibilities has become actual; but the information *which* one has become actual is waived. Then our description would contain all possible outcomes with the respective probabilities, it would be a "mixture" of states.

The most interesting point in this description is that this *mixture* is different from the *state* that results from the initial state of system + measuring instrument by the measuring interaction according to the Schrödinger equation. A long discussion has shown that here is a fundamental problem we cannot get rid of by simple tricks. The generation of a mixture described above has also been called – misleadingly – "reduction of the wave packet". I rather recommend using the more precise "disappearing of the interference terms". For the difference between the two descriptions is that the correlations between system and measuring apparatus, which are present after the measuring interaction in the *state* description, have disappeared in the *mixture*. This change is usually called the *cut* between system and measuring apparatus.

We can look upon this fundamental problem of the theory of measurement as the formal expression of the problem of *classical concepts* mentioned above: If we want to describe unambiguously what we have measured we have to waive the remaining correlations between system and measuring apparatus. With a good measuring apparatus this can be done quite easily. For in that case the interference terms are so small that they play no role for any practical purpose ("FAPP"); so there again is no problem for the practical physicist. But if we look closely we see that those interference terms, however small they may be, will always exist, they will never be zero exactly. Thus, strictly speaking, it is a mistake to neglect them. Eugene Wigner, who discussed this problem very carefully,[31] finally could offer no other way out than adding small non-linear parts to quantum mechanics that make the interference terms disappear within a short time after interaction.

In my opinion Wigner's solution is wrong. Again Weizsäckers analysis of the structure of time helps to solve the problem:

We are dealing with predictions within *physics*. Physics, however, contain *approximations* in its very foundations. This is seen rather easily: We can do physics only if we can deal with objects independently of their environment. But "in reality" there are no separate objects; everything is related with everything. One can see this already in celestial mechanics: Conceptually isolating e.g. a planet in the solar system from the totality of celestial bodies means an approximation. Strictly speaking, according to physics itself, every celestial body, however far it may be, has an influence on our planet from gravitation alone, not mentioning other types of interaction. These influences are so small that they can be practically neglected, but strictly speaking they are there. Treating the planets only under the influence of the sun and neighboring planets is an approximation and so, strictly speaking, it is wrong. If we did not use that approximation, however, we could not do physics at all. And, above all, we cannot describe a strictly independent object at all: At least an interaction with the measuring apparatus must exist in order that the object can be an object *for me*.

The approximation introduced by neglecting the interference terms is of exactly the same sort: We neglect the very small interaction that relates system and measuring apparatus still after measurement. Thus we introduce an approximation of the same sort as we have introduced in the very foundations of physics. – Translating this into the language of the structure of time means: We can give, fundamentally, only approximate predictions. This is true from their probability character alone, since probability propositions cannot be verified exactly. But it is true because of the fundamental approximation character of physics as well, which says: We only can give predictions about isolated objects which, strictly speaking, do not exist.

During the last decades many proposals have been published to solve this problem, e.g. under the name of "consistent histories" or of "decoherence".[32] Those proposals amount to the same solution under a different name, namely to the old suggestion to neglect the interference terms ("FAPP"). Unfortunately, the authors of such recent proposals give the impression that they could now offer, differently from the old authors, an *exact* solution of the problem. I am full of understanding, since making big noise is part of the business. But this claim would mean more than one could make good.

A common argument against the view put forward here reads like this: "One who puts so much emphasis on *predictions* has only eyes for the possibilities of manipulation, he has an 'instrumentalist´s' view of nature. Genuine philosophy of

---

[31] E. Wigner: *Remarks on the Mind-Body Question*. In: I.J. Good (ed.). The Scientist Speculates, London 1961. New York 1962, p. 302; reprinted in: E.P. Wigner: *Symmetries and Reflections*. Bloomington and London 1967, p. 171–184.

[32] Cf. e.g. the works of Detlev Dürr or Roland Omnes and their collaborators.

nature should inquire more deeply, namely about what the *real* basis is, maybe hidden, of the outer appearances." A suggestion like this is, not easily recognized, founded again on the ontology of classical physics. For it presupposes that "in itself" and *behind* the appearances, there is something else that perhaps does not show itself easily but whose description is the genuine goal of philosophically oriented science.

A program like this may be understandable from the point of view of classical physics. But there is nothing to justify it in this generality. For if we ask, according to empirical science, about the general structures of reality, we ask about an *objective* description in the spirit of this science. This means we ask about a description that *everyone* could in principle verify *at any time*. But if we will be able to verify a proposition empirically this proposition must be a prediction: We must be able to look if it is true *after* it has been made. This is what we brought out by our analysis of the structure of time. It is a specialty of classical physics that such predictions can also be formulated as descriptions of reality *in itself*. What makes this speciality possible is the fact that in classical physics with maximal knowledge all predictions can be made with probability 1 or 0. Where we cannot presuppose that any more, as e.g. in quantum mechanics, there is no such *reality* in itself any more. But objective description is still possible. – Anyone who calls this view, from the perspective of the ontology of classical physics, "instrumental", spoils every chance of understanding a more generally objective description of reality.

We see that C.F. von Weizsäcker's analysis of the structure of time is not only helpful for the interpretation of science but that it is indispensable for that task. Apparently the results of science become entirely incomprehensible for anybody who tries to keep the structure of time out of his interpretation.

What Is Missing? –
The Fundamental Role of Time
in C.F. von Weizsäcker's Conception of Physics
and Some Insights from Modern Neuroscience

Eva Ruhnau

1 Time and Experience

C.F. von Weizsäcker's conception of physics is based on the idea that the laws of physics should express nothing but the preconditions of the possibility of (scientific) experience. He follows the Kantian idea that the predictions of the physical laws, which are predictions of future measurements, are well founded at present and will prove valid *in* experience because they formulate the necessary preconditions *for* experience. Quantum theory is one (or even the) basic theory in physics. This leads him to the following hypotheses:

Hypothesis I

Quantum Theory is valid for all scientific experience because it poses preconditions for the possibility of scientific experience itself.

"Physics rests on experience. This term is used here for scientific experience, i.e., for the finding and testing of general rules or laws, valid in the field of perception, mainly of sense perception. The rules and laws are hypothetically assumed to express knowledge. Knowledge means that somebody knows something, in more abstract terms that a subject knows an object. There is no knowledge without a knower, there is no knowledge without a known". [1]

Scientific knowledge presupposes distinctions. The building blocks of scientific knowledge are n-fold alternatives, i.e., sets of n statements one of which will turn out to be true if empirically tested. The most basic alternative is the 2-fold one, a simple yes-no question or binary alternative.

Hypothesis II

The factorization of knowledge into empirically decidable alternatives – into observables – is one precondition of the possibility of scientific experience.

The question then arises: what is experience? In C.F. von Weizsäcker's view, the meaning of experience is to learn from the past for the future. In experiencing, in learning from the past for the future, we gain knowledge. Furthermore, scientific experience is defined as extracting rules from past observations and applying these rules to predict future observations. Whenever we do physics, we do it now, and we apply our past experience for predicting what is going to happen. In other words: experience already presupposes the triad structure of facts, events, and possibilities.

Hypothesis III

Time with its three modes of past, present and future is another precondition of the possibility of experience.

Past events are factual, whereas future events cannot be predicted with certainty. Future events can only be predicted as probabilities. The second law of thermodynamics, for example, cannot be derived from the H-theorem without the assumption that only future entropy values of presently known systems are derived from the H-theorem, and never past entropy values [2, 3]. Empirically testable probabilities are predictions of relative frequencies for the results of measurements. The concept of probability as quantification of possibilities turns out to be a central concept in the reconstruction of physics. The acts of measurement create discrete facts.

These three hypotheses of C.F. von Weizsäcker (which are mutually dependent) are strongly tied to the basic philosophical problems underlying the unresolved debate about the interpretation of quantum theory. Therefore, a careful analysis should consider in detail the separation into subject and object of knowledge and the related topic of observation, and the role of the now as transition point between past and future.

Let me take as starting point of this analysis the following citation of C.F. von Weizsäcker:

"The difference between past facts and future possibilities corresponds to the possibility that subjects can acquire knowledge ... We cannot avoid the dual question, what is the role of physics as subjective knowledge of nature, and what is the role of subjects as parts of nature ... Quantum theory, for the first time in the history of physics, has made it inevitable to discuss the role of the conscious subject of knowledge". [4]

These three terms – subjective knowledge, subjects as part of nature and consciousness – have certainly caused a lot of confusion and inaccuracy in the debate about quantum theory. Therefore, some words of caution are necessary. Subjective knowledge does not mean solipsism. Scientific knowledge is always intended to be intersubjective. However, scientific knowledge has to be distinguished from pure registration of a phenomenon; scientific knowledge relates to a subject and an object of knowledge. As the upholder of scientific knowledge, the word "subject" already includes some time-bridging persistence.

To consider subjects as part of nature demands the question whether these subjects can be described by the same laws of nature, which are known by the subjects. On the one hand, this leads to investigations of whether a theory is universal or not. On the other hand, it leads to the question whether quantum theory is a necessary prerequisite to understand human consciousness or not. To spell it out very strictly, I think that at present there is no realistic hint that the brain sciences need any explanatory input from quantum theory. This may change in the future, but at the moment, all discussions in these directions are merely speculation. It seems, for example, that the explanation of the experience of consciousness with a still unknown theory of quantum gravity is based on a categorial mistake. What could be interesting are ideas about *structural analogies* between questions in quantum theory and those in neuroscience. These analogies may point to fundamentally similar processes in the creation of objects and meaning. However, this is true for an abstract theory and interpretation of observation and meaning only.

Concerning the term "consciousness", I would like to discuss some insights from modern neuroscience in this paper. As pointed out, the reader should be strictly aware that these insights will *not* be presented, for example, to suggest an explanation of the collapse of the wave function via consciousness, and vice versa.

2 Subject/Object Separation and Observation in Physics

The decision how to define an "observer" and something "observed", the separation into subject and object of knowledge, provides the customary basis of the scientific method. Usually, this act of decision falls into oblivion. Classical science may even be defined as the collection of theories where oblivion of this partition is fundamental for its success. As we know, quantum theory corrects this point of view. The description of the observed depends on the determination of a frame of reference or a universe of discourse. Scientific knowledge rests on the fact that – within such a context – decidable alternatives exist. As long as the definition of the object of scientific inquiry via its properties is an unproblematic matter, the act of measurement as a decision between alternatives presents no problem, in principle. But quantum theory indicates that the definition of an object may only be an approximation.

What about the observer itself? Does the theory apply to the observer, as it should, if it is to be regarded as a universal theory? To obtain a complete description, the observer has to know the truth-values of all propositions, including propositions concerning his own state. However, if the theory is rich enough to admit an internal observer, the requirement of logical consistency implies incompleteness [5–7]. This leads to an infinite regress reflecting the fact that the act of separation into observer and observed is not a genuine part of the theory; it is a metatheoretical choice. The decisive act of division of a wholeness (something without parts) into a part, which observes and a part being observed can also be approached by investigation of internal and external perspectives (for details see [8]). In the following, the term "system" denotes the entity being referenced within a theoretical discourse.

Let us call an attribute of a system, defined by the description of an apparatus to measure it an *observable*. An observable, the measurement of which can only give one value of the set $(0, 1)$, is called a *proposition*. Let S_1 denote a system. Let S_2 be a system together with a measuring apparatus A. If a measurement has taken place, then S_2 has registered the truth-value of a proposition of S_1 defined by the measuring device A. What has taken place is a registration, and (under specific conditions) a record has been produced. Registration does not necessarily need the presence of human beings; S_2 may be identical with A. What is registered is the final outcome of a yes/no inquiry and not a superposition of possible results. The question is whether we should already call S_2 an *observer* or not.

Many confusions in the interpretation of the measurement process have their roots in the failure to distinguish between registration and observation. Has the collapse of the wave function taken place if it is registered, but a human being does not yet observe this registration? Do we need the consciousness of Wigner's friend for the collapse to happen? In this context, C.F. von Weizsäcker has made clear statements in favour of the Copenhagen interpretation (CI): "CI is a minimal semantics for quantum theory in as far as it renounces any additional metaphysical assumption, limiting ourselves on what we can actually know. Precisely in order to avoid any speculation on consciousness, Bohr decided, not to describe the measuring process itself by quantum theory. In the moment, in which the observer is informed on the result of a measurement he will have to apply this new knowledge and to use a new wave function. Since these measuring results which can be classically described can be simply and straightforward observed within the frame of classical physics, no analysis of mental processes is needed." [4]

Here we see a clear distinction between registration and observation. The wave function is a catalogue of conditional probabilities for possible future events, conditioned by the information a subject possesses through the preparation of the system. This catalogue, and hence the wave function, must be changed by new information about the system, i.e., by new measurements implying new conditions for the probabilities. The one who changes the catalogue is the observer/subject. For him, the results of measurements denote information in the context of his scientific theory and knowledge.

To be more precise, an observer (with his catalogue of knowledge) can be connected to the distinction of internal and external perspectives. Let us call a system without an external observer an *endosystem*. This notation refers to the ontic aspect of a system. An endosystem is an entity existing independently of an observer's knowledge about it. Furthermore, an endosystem cannot be observed from outside. In other words, it cannot be observed as a whole. However, an internal observer is not excluded. Let us call an observer external to a system an *exosystem*. This refers to the epistemic aspect. These definitions have the consequences that an exosystem cannot observe an endosystem defined within the same conceptual frame.

Let us call an endosystem containing one or more exosystems a *universe of discourse*. Let us call the observed systems *objects*. This act of separation into observer and object of observation may be carried out in different ways. One and the same endosystem may be represented as a variety of universes of discourse. However, using a Gödel-type argument, it can be shown that the act of partition cannot be verified within the endosystem. Its status is that of a hypothesis whose truth-value can be fixed only on a meta-level. In other words, representation of an endosystem as a universe of discourse opens an infinite regress.

The measurement problem in quantum theory can now be described in the following way: Let S denote a quantum system and A the apparatus to measure an observable of S. At the beginning, both systems are in their initial states; they are uncorrelated, being two independent parts of a composite system. Which steps are necessary to transform the system S into an object? First, S and A have to interact. This interaction can be described quantum theoretically. It evolves the composite system into a correlated state. Consequently, the probability distribution for the alternative outcomes of the measurement is not a classical probability distribution. The quantum correlated state contains alternatives, which are never observed to coexist. These quantum correlation terms are an expression of the fact that the coherent state has to be considered as an endosystem.

To get observable results one has to get rid of the correlation terms, i.e. the excess information about the wholeness of the quantum world. For this, one needs a superselection rule to determine a preferred frame of reference for the measuring apparatus such that the probability distribution of the alternative outcomes becomes a classical one.

One possibility of causing such an effective superselection rule is by interaction of the endosystem with the environment [9]. The endosystem ceases to be an endosystem; it changes to an open system, which – together with its environment – constitutes a universe of discourse. The quantum system has become an object. This is the description in the conceptual frame of the observer. Furthermore, the observer has to admit that something irreversible has happened; possibilities were transformed to facts. However, quantum system, measuring apparatus and environment, together, can again be considered as an endosystem. The selection of a preferred frame is only effective, i.e., infinite regress is circumvented, if the environment is large enough to contain enough degrees of freedom to dissipate the quantum correlations. In this sense, object formation and irreversibility are only approximations.

3 The Now as Transition Point Between Past and Future

In C.F. von Weizsäcker's view, experience already presupposes the triad structure of facts, events, and possibilities, i.e., the structure of time as past, present and future. Measurements produce facts. Predictions concern possible future events. Facts are constituted in the present and are, retrospectively, described as past events with respect to instants, which have already passed.

C.F. von Weizsäcker connects facts to discreteness and possibilities to continuity in the following way: "We can know the past as far as we know facts from it. Describable facts can be counted. Past facts known to us are a discrete manifold. For the future, we can only estimate possibilities. In classical physics and in traditional quantum theory we describe them by probabilities or even complex functions. These concepts are known as continuous. However, the present only consists of discrete events out of such a continuum, events that later remain known as facts. How does our prognostic vision of the future connect these possible discrete events with the continuum, with which we conventionally describe possibilities?" [4]

A possible answer to this question is what C.F. von Weizsäcker calls "multiple quantization". It is based on a consistent application of the concept of probability as the expectation value of the relative frequency of events in an ensemble (see [10]).

Concerning this point of view, I do not want to go into details. However, I would like to mention that the conjunction of facts with discreteness might presuppose hidden assumptions about the process of observation itself. This is briefly illustrated with the following argument.

How is time usually treated in physics? Is it something given? Is it an object? Can time be measured? In the quantum theoretical formalism, time cannot be measured like other observables, time is not an operator. Do clocks measure time? Clocks measure durations as long as a cyclic process and an irreversible "memory" structure to count the number of cycles exist.

Mathematically, time is modeled as a (strictly) ordered set. Referring to the elements of this set as instants and interpreting their ordering as earlier-later relation leads to (physical) time as a set of points in a linear-successive order. Furthermore, this set is considered to form a continuum – the continuum of the real numbers. This is based on the idea that between two time points there is always a third time point. However, in measuring temporal data we do not observe points of time; what we observe are, for example, the factual positions of the hands of clocks. In other words, we measure time by observing facts. Therefore, taking the linear continuum as the mathematical model of time can also lead to the following definition of time: *continuous time is the abstract structure of unlimited observability.*

With respect to C.F. von Weizsäcker's view on continuity and discreteness, this point of view is just the opposite one. However, it may reveal some hidden hypotheses and features guiding our ideas about time. Unlimited observability leads to the following problem: The superposition of possible quantum states is transformed through observation to a factual existing state. However, what happens if a system is constantly observed, i.e., if the time intervals between successive observations are steadily decreased? Quantum theory predicts that such a system under permanent observation cannot change its state even though such a state might be extremely unstable. This is called quantum Zeno effect (see for example [11, 12]). Continuous observation results in a fixation of the state of the system.

Physics is a science based on observations. The above argument seems to indicate that there might be limits to observations. These limits are not limits of measuring instruments. The individual processes themselves seem to generate such limits. To overstate the argument one could probably say: (F)actualization and observation of events are complementary.

Actualization is necessary to establish facts. An event or fact, which is actual, is embedded in a Now. Or, vice versa, the Now is determined by events. This circularity is not only a problem of physics; it can be traced throughout the history of philosophy. The usual step to break the circle consists in the construction of an appropriate metalevel.

The time of physics is lacking any concept of the Now. Embedded in the paradigm of the Cartesian distinction between *res extensa* and *res cogitans*, the extended and the knowing substance, the concepts of time and Now usually reflect the Cartesian division between matter and mind. In this paradigm, the Now occurs in physical time only as a transition point between earlier-later, or past and future. Whereas in mental time, the phenomenon of an experienced Now occurs, which is itself transient. In the Cartesian view, the Now belongs to subjective consciousness and has no place in objective physics.

This peculiarity of the Now was also the topic of a discussion between Einstein and Carnap. "Einstein said that the problem of the Now worried him seriously. He explained that the experience of the Now means something special for man, something essentially different from the past and the future, but that this important difference does not and cannot occur within physics. That this experience cannot be grasped by science seemed to him a matter of painful but inevitable resignation." [13]

In the last decades, several experimental and technical paradigms have been developed in modern neuroscience to deal with the experience of the Now. In the following, I want to discuss some of the ideas and experimental results. These findings give new insights with respect to C.F. von Weizsäcker's hypothesis that time with its three modes of past, present and future is a precondition of the possibility of experience.

4 Temporal Logistics of the Brain I

Dealing with subjects' conscious reports on temporal experiences, subject and object of observation coincide. These reports reveal that there is no simple and unique mapping between external physical time and internal experience of time (see for example [14]). The objective and subjective temporal order of events can be different. I suggest that this is not only a feature of brain activity and self-awareness, but points to a fundamental conceptual peculiarity of any theory of observation and knowledge, where subject and object are not mutually independent.

In the following, I want to demonstrate which temporal constraints shape our conscious experience. Let us choose as an *external* initial condition the presentation of two stimuli (flashes of light or tones) in a well-defined temporal order with (external) temporal distance dt. The *internal* experience of an observer of these stimuli reveals a whole hierarchy of elementary temporal percepts.

If the external temporal interval, dt, is below a certain value, the coincidence threshold, the two stimuli coincide and only one event is being observed. This *coincidence threshold* is different for different sensory modalities and is connected to the different transduction times, i.e., the time it takes to transform the sensory signals into neuronal activities. If we increase dt, exceeding the coincidence threshold for all modalities, but staying below a temporal distance of approximately 30 milliseconds, two events are being observed. However, it is not possible to decide which event is the first, and which is the second one. Below this *order threshold*, no before-after relation can be experienced. If we increase the temporal distance, dt, above the order threshold, the (outer) temporal direction between the two events is perceived as first and second event.

It is important to note that the order threshold is the *same* for all modalities. Therefore, the perception of temporal order seems to be connected to a central processing mechanism. The hypothesis is that the brain creates and is structured by zones of co-temporality [15–17]. The external temporal duration of such zones is approximately 30–40 milliseconds. It is assumed that supra-threshold stimuli that reach the sensory surface initiate neuronal oscillations with a period that corresponds to the duration of these zones of co-temporality. The occurrence of such oscillations is based on re-entrant connections between different modules [18, 19]. Zones of co-temporality are characterized by the fact that temporal relationships between events cannot be defined; this corresponds with respect to the direction of time to *zones of atemporality*, where all available information is treated as co-temporaneous.

As the exact temporal central availability within such a time window does not play any role, information can be collected from different regions of the brain and from different sense organs within such an interval. Why are such binding processes necessary?

5 Separation and Integration in the Brain

To obtain a subjective representation of mental phenomena, specific logistical requirements have to be met. In order to function, a brain has to reach a certain level of *activation*. Another logistical requirement is the *organization* of the spatially distributed mental functions. Specific mental functions are represented *locally* in the brain. Localization of functions means that a defined neuronal process (inter alia characterized by specific neurochemical mechanisms) relates a mental function to a spatially defined area. Mental functions are available in modules. The whole repertoire of the mental is represented in this way, i.e., percepts, memories, feelings and volitional acts. Experimental verifications of this thesis are mainly based on observations with brain-injured patients. Pieces of the entire repertoire of mental functions can be lost if specific neuronal algorithms are no longer available. In other words, specific lesions of the brain result in defined – *inter*individually similar – functional losses. The loss of mental functions is their proof of existence.

The spatial segregation of mental functions throughout the brain leads to the question how integrated subjective experiences are possible. Each mental act is characterized by ("simultaneous") activities in different areas of the brain. In particular, PET (positron emission tomography) studies show that there are always *several* spatially distinct areas of the brain that exhibit higher activity during defined psychological tasks (for example in reading a text). In general, each mental act can be characterized by a specific pattern of spatially distributed activities within neuronal assemblies. This observation indicates that neuronal mechanisms, which provide the integration between the distributed activities, are required in order to result in the experience that each subjective phenomenon is only *one* – not many.

It is useful to distinguish different operational levels of integration. The problem of integration is already met at the *intra*sensory level. Perceived objects usually cover different areas of brightness or tones. But reaction times depend on the intensity of the given stimuli. For example, a difference of light intensity of approximately two logarithmic units corresponds to a difference in transduction time of about 20 milliseconds – with decreasing transduction time corresponding to increasing intensity. In the auditory domain, a difference in transduction times of about 20 milliseconds occurs with respect to intensities of 45 and 90 decibel [20]. In addition, intrasensory object integration of different qualities like colour or form has to be ensured.

However, *inter*sensory integration is also necessary. It is known from reaction time experiments that simple auditory reaction time is considerably shorter than simple visual reaction time (with a difference of about 30 milliseconds). This can be traced back to the biophysical difference of transduction for both sensory modalities. Furthermore, the distance of the sound source from the subject is important in the auditory case and affects auditory reaction time. Auditory reaction time is prolonged if the stimulus is removed from the subject. This prolongation is due to the time the sound wave takes to reach the subject. In the visual case, however, the time the light wave needs from source to target can be neglected because of the much higher velocity of light. Because of the difference in sound and light velocities and the different transduction times it is calculated that at approximately 10 meters distance auditory reaction time corresponds to visual reaction time. This distance can be called the "horizon of simultaneity" [16]. The temporal central availability in the two sensory domains is co-temporaneous for objects at this distance from the perceiving subject. Therefore, perception of an object by means of auditory and visual stimuli as *one* requires intersensory integration.

In summary, these results demonstrate that the central temporal availability of stimuli within one sensory modality and between different modalities is constantly changing. Why then do we perceive identity of objects, why do we usually not perceive the disintegration of an object identity into many separate identities corresponding to different contexts of perception? Here, time is the glue and the clue in solving this problem.

As a solution of the integration or binding problem many experiments using a whole range of totally different experimental paradigms suggest the following hypothesis:
The adirectional temporal zones or zones of co-temporality (of 30–40 millisecond duration in external time) define *elementary integration units* (EIUs). Within these zones, activation from different functional units of the brain is correlated.

Which kind of processes could yield such specific temporal structures of information processing? It is assumed that the occurrence of adirectional temporal zones is based on relaxation oscillations of neuronal assemblies (connected via mutual re-entrant maps between different modules). Supra-threshold stimuli reaching the sensory surface initiate such neuronal oscillations. The elementary integration units are defined as successive periods of these oscillations. Thus each period corresponds to a temporal window of the brain. Within such a window, all available information is treated as co-temporaneous, thus obtaining intra- and intersensory integration. Within such elementary integration units, no direction of time can be perceived. A before-after relation is available only if at least two temporal windows are provided.

6 Object Formation in the Brain

Electrophysiological findings of coherent oscillations in the gamma frequency range of 30–60 Hz force the idea that spatially segregated activities of the brain might be integrated via coherent cortical oscillations [21,22]. Let us take, for example, a moving object. Those parts of the visual scene belonging to the object are distinguished through spatiotemporal coherence of motion, contours and colour. How does the brain "know" that spatiotemporal coherence is a property of objects, which can be used to differentiate (figure from ground) and to integrate (different features)?

In the development of cognitive structures, connections between neurons where patterns of activity are correlated are stabilized. Such selective coupling is generated because neurons which are activated simultaneously (via spatiotemporal coherent features of objects) tend to strengthen their synaptic connections. Such groups of nerve cells form *feature detectors*; they are, so to speak, measuring devices of the relevant features. The associated neurons oscillate with a frequency of about 40 Hz if the appropriate features of an object are available. Spatially segregated feature detectors can, in the case of coherent properties of the perceived object, synchronize their rhythmic activities.

It is important to note that if synchronized oscillations are supposed to provide a solution of the binding problem, the question of their functional effectivity arises. If we do not want to postulate the existence of a homunculus in the brain observing synchronisation, the mass activity generated by the synchronized activities of neurons has to be functionally effective. What could be an appropriate experimental paradigm to study the *functional relevance* of coherent oscillatory phenomena? An answer to this question will be given in the last section of this paper.

A structural analogy between the measurement problem in quantum theory or the transition from possibilities to facts and the problem of an observer (homunculus) within the brain can be shown now. Both topics can be embodied in the question "how do objects come into existence?" Like in classical physics, regarding an object and a brain as obviously given and separated entities escapes the crucial question. Instead, an adequate structure to begin with is to think of the brain of a newborn organism as a network of unspecified neurons receiving stimuli from the environment. Yet the environment is not known as separated. In the context of the previous endo/exo discussion, the basic structure of a network together with its chaos of changes forms an endosystem. Can it specify itself to a universe of discourse?

At this early stage the brain does not have any measuring apparatus (feature detector) at its disposal. However, the brain does not operate in an empty space. To a certain extent, the environment provides spatiotemporal stabilities and symmetries to which the organism has to adapt to behave adequately. The chaos of neuronal changes caused by stimuli from external sources has to be categorized. Systems external to the brain have to be "measured" by the brain. How could this be achieved?

As mentioned above there is evidence for coherent states of the brain which facilitate differentiation and integration. These coherent integrative states of the brain can be described as temporal correlations of neurons established by re-entrant processes between different brain modules. Different levels of integration correspond to symmetries and unities of external objects. The brain considered as endosystem constitutes a "measuring apparatus" adapted to (and partially reflecting) spatiotemporal continuities and symmetries of its environment.

But a measuring apparatus is not yet an observer. Registration of coherence in a yes/no manner is necessary. This could be modeled by effectors being realized as threshold units (combined with motor neurons) connecting occurrence of synchronization to behavior. This would imply that the endosystem brain has become a universe of discourse containing exosystems observing objects. Note that these objects are *inherent in* the endosystem brain!

7 Temporal Logistics of the Brain II

In Section 4, it was shown how the experience of the succession of perception could be understood. However, what about the perception of succession? It is characteristic for human perception (and action) that successive events can be bound together to generate a new content – like hearing a musical motif or processing visual motion. Other integration mechanisms than the one described above are necessary to comprise mental events into perceptual units or Gestalts. Actually, two other binding operations in the time domain have to be distinguished, the conscious Now and the duration.

At a basic level a temporal integration process is observed that links several successive time windows (of 30 milliseconds) together up to intervals of approximately 3 seconds, providing the foundation to form perceptual units. This binding process is characterized by the fact that it automatically links events together for up to several seconds, but it can do so also for shorter intervals if it is required. This operational principle implies that the temporal binding process is syntactically closed up to approximately 3 seconds and semantically open for shorter intervals. Such processes constitute the formal basis of the experienced subjective Now. Experimental support for such an automatic integration process comes from reproduction studies of temporal intervals [15]. If subjects have to reproduce the duration of intervals, intervals of approximately 3 seconds are reproduced accurately. Shorter intervals are slightly overestimated, whereas in the reproduction of longer intervals underestimation is observed. Temporal segmentation in the 3 second range is also discovered in language processing [16, 23] or in the temporal organization of intentional acts [24]. All these results based on different experimental paradigms provide clear evidence that the *conscious Now* – language and culture independent – has a duration of approximately 3 seconds. The experienced Now is not a point but is extended [25].

On the highest level of binding contents of consciousness are linked together. The integration intervals of 3 seconds, the Now-moments, serve as a formal basis for the representation of these contents, however, as the logistical prerequisite they do not select "what" is represented and "how" the represented information is linked together. The subjective *continuity of experience* is presumably the result of a *semantic* connection of what is represented within each 3 second time window. Continuity of experience seems to require already semantic aspects, i.e. the concept of meaning. The observation that continuity can break down as in the case of schizophrenia implies that under normal circumstances a specific neuronal process is responsible for the semantic nexus.

8 Experience and Time

In summary, concerning time what can we learn from the brain? On a formal level, time appears to be of discrete nature. Syntactically constitutive for time are two temporal discrete mechanisms:

- a high frequency mechanism providing elementary integration units; the external duration of these adirectional zones is of the order of 30 milliseconds;
- a low frequency mechanism defining intervals of integration of 3 seconds, which serve as the logistical basis of the conscious Now; within these intervals sequential information is bound together independent of content.

In this way, we have a temporal syntax at our disposal, i.e. windows of co-temporality without internal temporal succession (adirectional temporal zones) and a limited capacity of integration of successive units of co-temporality (Nows). However, with regard to content, time seems to be continuous as a consequence of semantic binding operations.

As was mentioned earlier, synchronized oscillations generating elementary integration units have to be functionally effective. Otherwise, one cannot avoid postulating the existence of a homunculus in the brain observing synchronisation. The former question was: what could be an appropriate experimental paradigm to study the *functional relevance* of coherent oscillatory phenomena?

All data supporting the proposed model of elementary integration units are mainly results of reaction time studies and are strictly time locked to stimulus onset. This suggests that we have to take into account *complete sensory-motor cycles*. If coherent oscillations are functionally effective, this effectiveness should be relevant and prove itself on the *output* side. Brains are open systems embedded into a specific environment. Permanent observation or measurement would correspond to a continuous transposition of incoming (afferent) into outgoing (efferent) pattern of activity. Such a flow of motion would yield the goal-oriented behavior of the organism, as a totality, impossible. Therefore, the coordinated action of the whole organism requires zones, which are free of observation, or zones of co-temporality.

In the last decades, significant experimental and theoretical prejudices in favour of cognition and the sensory part of awareness are manifest. This, I claim, makes the homunculus unavoidable. In the context of the cognitive paradigm, neuronal pattern of activity cannot reflect, per se, a holistic aspect. Necessarily, the question "holism for whom or what?" is raised. Correction of the cognitive prejudice and appropriate consideration of the motor aspect could presumably solve the problem. Functional effectivity should be relevant at the efferent side. The homunculus dissolves in sensory-motor integration. In brief: *perception and motion constitute an inseparable unity.*

Our brain, like the brain of any other organism, does not develop while sitting around and watching. The whole organism and also its parts are in permanent motion. For example, within a few seconds after artificially fixating the motion of our eyeballs, we are no longer able to see anything. As described above, feature detectors – the measuring instruments of the brain – would not develop if the organism (the baby) would not move and grasp.

At the phenomenological level we usually observe (!) observation and action as separated. It seems that in the flow of action, no observer can constitute himself. Observation must interrupt and break this flow. The 3 second windows, the logistical basis of the conscious Now, exhibit this very clearly. Within the subjective Now of 3 seconds, percepts, memories, feelings and volitional acts form a unity; only *retrospectively* are they conceptually disconnected. In the lived presence, in experiencing, there is no separation of mental contents. Only in the act of reflection, i.e., after transforming oneself into the observer of the own mental acts, does the separation of mental contents take place. From the position of the observer, from a position outside the lived experience, the unity of mental contents becomes evaluated and split into different categories. The dividing line between present and past is observation.

The knowledge we are usually referring to is *explicit knowledge*. Modern brain science has revealed that this is only one type of knowledge. There is also *implicit (tacit) knowledge*, the ability of doing something, the knowledge of action. And there is *personal knowledge* expressing itself in pictorial scenes imprinted into our memory via strong mental impressions. Explicit knowledge is the knowledge of the observer only.

In speaking, describing, defining, and in forming concepts, we are always separating the world into two, into observer and observed, subject and object of knowledge, internal and external. Such structurings into conceptual pairs, or dichotomies, are not rational results of universes of discourse, but their generative precondition. Both aspects of a conceptual pair are complementary. A basic dichotomy in modeling reality is the conceptual pair: constancy and change. To perceive change, something has to be invariant. Vice versa, constancy needs change as its background to be observed. However, within the conceptual realm of western philosophy, the complementarity of constancy and change is turned into a hierarchy. The staticity of being is chosen as dominant, and the dynamics of change becomes a derived and secondary concept. This transformation of complementarity into hierarchy is the essential (though generally forgotten) precondition of scientific realism and objectivism. As a result, the purpose of knowledge is the construction of a perceptual entirety of objects and properties – not the comprehension of a world created by the dynamical unity of opposites.

Caused by the dominance of constancy, we have no adequate framework to talk about functions and processes. We try to get a grasp of processes with the same formal equipment that we use to classify things. Actually, we do not treat processes like objects – which are considered to be identical with themselves in time – yet, we do deal with processes as successions of states in time. Usually processes are described as temporal succession of objective states. They are defined as objects + time.

This subordination of the dynamics of reality to objective categories of being forms the basis of the success of modern science. However, what is missing? What could we learn from the most dynamic and most complex system in the universe – the human brain? May be, in our scientific theories, we should consider the concept of "change" as essential as the concept of "being". [26] It is fascinating how the brain encompasses both aspects, the dynamical change and the constancy of neuronal structures. The creation of time Gestalts in the brain from dynamics is rooted in the action of the whole (organism). The temporality of the brain reveals the complementarity of experience and reflection, the complementarity of living and describing.

In lived experience, there is no subject-object separation. Such separation occurs retrospectively only when experience has turned into observation (the explicit foundation of science). Temporality has to be regarded as necessarily experiential and not – like in a Kantian sense – as a precondition of the possibility of experience. On the contrary, experience is seen as a precondition of the possibility of Temporality.

References

1. C.F. von Weizsäcker: Reconstruction of Quantum Theory. In: L. Castell, C.F. von Weizsäcker (eds.): *Quantum Theory and the Structures of Time and Space*. Vol. 6, Carl Hanser Verlag, München 1986, pp. 7–40
2. C.F. von Weizsäcker: Der zweite Hauptsatz und der Unterschied von Vergangenheit und Zukunft. *Ann. d. Phys.* (5) 36, 275–283 (1939)
3. T. Görnitz, E. Ruhnau and C.F. von Weizsäcker: Temporal Asymmetry as Precondition of Experience – The Foundation of the Arrow of Time. *Int. J. of Theoret. Physics* 31, 37–46 (1991)
4. C.F. von Weizsäcker: Time – Empirical Mathematics – Quantum Theory. In: H. Atmanspacher, E. Ruhnau (eds.): *Time, Temporality, Now*. Springer, Berlin 1997, pp. 91–104
5. K. Gödel: Über formal unentscheidbare Sätze der Principia Mathematica und verwandter Systeme I. *Monatshefte für Mathematik und Physik* 38, 173–198 (1931)
6. H. Primas: Mathematical and philosophical questions in the theory of open and macroscopic quantum systems. In: A.I. Miller (ed.): *Sixty-two Years of Uncertainty*. Plenum Press, New York 1990, pp. 233–257
7. E. Ruhnau: Logica, tempo e libertà – un'utopia? In: E. Donini (ed.): *Tempi con/fusi*. Milella, Lecce 1984, pp. 181–201
8. E. Ruhnau: The Now – a hidden window to dynamics. In: H. Atmanspacher, G. Dalenoort (eds.): *Inside versus Outside. Endo- and Exo-Concepts in the Sciences*. Springer, Berlin, pp. 293–308
9. W.H. Zurek: Decoherence and the transition from quantum to classical. *Physics Today* Oct., 36–44 (1991)
10. C.F. von Weizsäcker: *Aufbau der Physik*. Hanser, München 1985, p. 306
11. R. Omnès: Consistent interpretations of quantum mechanics. *Rev. of Mod. Physics* 64, 339–382 (1992)
12. M.C. Fischer, B.M. Gutiévrez and M.G. Raizen: Observation of the quantum Zeno and anti-Zeno effects in an unstable system. *Phys. Rev. Lett.* 87, 040 402 (2001)
13. R. Carnap: Autobiography. In: P.A. Schilpp (ed.): *The philosophy of Rudolf Carnap*. Open Court, La Salle 1963, p. 37

14. D. Dennett, M. Kinsbourne: Time and the observer: the where and when of consciousness in the brain. *Behavioral and Brain Sciences* 15, 183–247 (1992)
15. E. Pöppel: Time perception. In: R. Held, H.W. Leibowitz and H.L. Teuber (eds.): *Handbook of sensory physiology* Vol. VIII. Perception. Springer, New York 1978, pp. 713–729
16. E. Pöppel: *Grenzen des Bewußtseins. Über Wirklichkeit und Welterfahrung.* Deutsche Verlagsanstalt, Stuttgart 1985; Engl. Ed.: *Mindworks. Time and Conscious Experience.* Hartcourt Brace Jovanovich, Boston 1988
17. E. Ruhnau, E. Pöppel: Adirectional temporal zones in quantum physics and brain physiology. *Int. J. of Theoret. Physics* 30, 1083–1090 (1991)
18. G.M. Edelman: *The Remembered Present. A biological theory of consciousness.* Basic Books, New York 1989
19. O. Sporns, J.A. Gally, G.N. Reeke Jr. and G.M. Edelman: Reentrant signaling among simulated neuronal groups leads to coherency in their oscillatory activity. *Proceedings of the National Academy of Sciences USA* 86, pp. 7265–7269 (1989)
20. E. Pöppel, E. Ruhnau, K. Schill and N. von Steinbüchel: A hypothesis concerning timing in the brain. In: H. Haken, M. Stadler (eds.): *Synergetics of cognition.* Springer Series in Synergetics, Vol. 45. Springer, Berlin, 1990 pp. 144–149
21. R. Eckhorn, R. Bauer, W. Jordan, M. Brosch, W. Kruse, M. Munk and H.J. Reitboeck: Coherent oscillations: a mechanism of feature linking in the visual cortex? *Biological Cybernetics* 60, 121–130 (1988)
22. C.M. Gray, W. Singer: Stimulus-specific neuronal oscillations in orientation columns of cat visual cortex. *Proceedings of the National Academy of Sciences USA* 86, 1698–1702 (1989)
23. F. Turner, E. Pöppel: The neuronal lyre: poetic meter, the brain and time. *Poetry (USA)* August, 277–309 (1983)
24. M. Schleidt, I. Eibl-Eibesfeldt and E. Pöppel: A universal constant in temporal segmentation of human short term behavior. *Naturwissenschaften* 74, 289 (1987)
25. E. Pöppel: Temporal mechanisms in perception. *International Review of Neurobiology* 37, 185–202 (1994)
26. E. Ruhnau: The Deconstruction of Time and the Emergence of Temporality. In: H. Atmanspacher, E. Ruhnau (eds.): *Time, Temporality, Now.* Springer, Berlin 1997, pp. 53–69

Irreversibility via Semichaos

Georg Suessmann

1 Stochastic Entropy

The processes described by Statistical Mechanics, with the exceptions of stationary processes, are those that approach equilibrium and are thus irreversible, even though the underlying dynamics of the invisible molecular motions is invariant under time reversal. Here we want to restrict ourselves to a discussion of non-relativistic classical particle dynamics in Hamiltonian form without friction. The following considerations should also apply to non-relativistic quantum mechanics, mutatis mutandis, despite the additional problems relating to measurement processes that are inherent to quantum mechanics. Within the framework of reversible mechanics, we therefore ask how one may understand the breaking of temporal symmetry, at the micro level, when passing from pure dynamics to its statistical variant.

We first recall that, in general, a sub-sector of symmetrical dynamics can be of lower symmetry than that of the complete dynamics. One can see this most simply using a limiting case of a particular solution of the relevant equations of motion. The dynamical symmetry only states that the complete manifold of solutions is transformed as a whole by an operation of congruence. Thus an individual solution is not necessarily transported into itself but only into some element of the set. In short, particular solutions will generally not be symmetrical even in cases where the laws are symmetrical. A well-known example is the strong anisotropy of the planetary system in spite of the perfect isotropy of Kepler's laws. Accordingly, we can assume that Statistical Mechanics admits a non-symmetrical sub-set of the complete set of possible initial conditions for micro-dynamical motion. With this remark the paradox is resolved, at least in principle. However, there remains the problem of showing the manifest occurrence of such asymmetry for the initial or final conditions.

This concerns, above all, the entropy theorem, according to which the entropy S of a thermally isolated system, not yet in internal thermostatic equilibrium, increases with a probability close to certainty. A prominent example of this behavior is Boltzmann's *Eta-theorem* concerning the stochastic state variable $H = -(1/k)S$ for highly dilute gases. Boltzmann's argument, however, did not make clear at what point or how he had introduced his justification for breaking the temporal symmetry, so that doubts were raised soon after its proposal.

2 Objections

The two most important and famous objections were, on the one hand, the so called "global-time" *objection of recurrence* by Zermelo, and, on the other the "local-time" *reversibility paradox* by Loschmidt. The objection of recurrence appeals to *quasi-ergodicity*, or almost periodic trajectories for each spatially bounded system of particles, which had been proved by Poincaré. (Boltzmann himself had even postulated the strict hypothesis of *ergodicity* – which later proved to be untenable for topological reasons, and according to which even each trajectory should be *strictly* periodic). However, in that case H (or *Eta*) would in the long run increase just as frequently as it would decrease. Consequently, one could *not* claim that an increase of *Eta* could be categorically excluded. Opposing this, Boltzmann insisted that these recurrence times had super-astronomical length, so the mathematical recurrence would be irrelevant in any physical sense. If, however, insistence is put on mathematical exactness, one could add the following. In the by far longest time periods, the mathematical gas is in a state of maximal probability (or entropy), whereas *Eta* is at its global minimum. The eons of increasing *Eta* are as rare in comparison to the incessantly repeated hibernation and are as short as those of decreasing *Eta*, and this is, as we know, especially true in our finite *cosmos* (as distinct to an infinite model universe). For the temporal antipodes, local-cosmic notions of "past" and "future" appear with exchanged roles. The H-theorem is to be read accordingly so that the time variable t always indicates the actually *present* (that means cosmological-local) measure of time. Such an understanding of the H-theorem, one that Boltzmann indicated at the end of his "Lectures on the Theory of Gases", makes the objection of recurrence pointless.

The objection of reversibility weighs all the heavier, as it attacks our problem so to speak head on, as became apparent at the end of Sect. 1. This causes an inversion, that is to say a transformation which inverts each particle momentum according to $p \Rightarrow \mathbf{T}p := -p$. However, according to the Hamiltonian equations of motion this immediately implies that the trajectory of the system of particles are run through in opposite temporal order: $\rho(t) \Rightarrow \mathbf{T}\rho(t) := \rho^T(-t)$ with $\rho^T(t; q, p) := \rho(t; q, -p)$. If this is applied to Boltzmann's entropy $S = -kH$, the result is that S has to increase with time at exactly the time when it should decrease. It therefore follows that S must remain constant in time (in contradiction to the H-theorem). Boltzmann thus appears to be obliged to prohibit the operation of time reversal $\mathbf{T}$; but how is he able to justify this interdiction? The story goes that Boltzmann answered the objection of his colleague Loschmidt, that one could just reverse the momenta and thus the sign of dH/dt, with the response: "It is you, dear colleague, who reverses the momenta!" By this joke, Boltzmann wanted to express that the reversal of momenta converts a natural and easily attainable state into an artificial, highly improbable one. The unnatural character of this $\mathbf{T}$ is apparently expressed in a strange tenacity of the state

Tρ (which is created from an ordinary initial state ρ) in following a concentrating teleology that is not motivated by any dynamics and which, therefore, appears to put out of action normal causality.

However, this argument had to miss its objective by a long way, as thermodynamics had in the mean time left the engineering character of the time of Carnot far behind. As the issue was now the validity of a *theoretical* proof, Loschmidt was well able to invert all those many particles, (he had been the first to give an estimate of their huge number), in his mind at one stroke. When Boltzmann admitted that the reversed state was not utterly impossible albeit by an overwhelming measure improbable, he used an intuitive, qualitative notion of "probability", which neither satisfied the definition given by Laplace nor its later generalization by Kolmogorov; here, at most, we might speak of something such as "plausibility". In particular, we must keep in mind that the time reversal transformation **T** leaves the Liouville-Lebesgue measure of probability *invariant*: $\rho(q, p) = \rho^{\mathbf{T}}(q, p)$ with $d(p) = d(\mathbf{T}p)$. A state of the Boltzmann gas, which is allowed by the laws of physics and markedly different from a state of thermostatic equilibrium, has objectively an extremely low probability, be its possibility as plausible to us as we would like; its temporal mirror image is, however, no less probable but *equally* improbable, in spite of our subjective estimates which rely on habit. Suppose we toss a coin a thousand times and obtain "heads" at each throw today, but tomorrow throw all "tails", then the optimist will declare today's event as "logical" but tomorrow's event as "paradoxical", whereas the pessimist will judge it in just the opposite way. According to Hume, one should not declare the facts that we are familiar with as necessary by nature. Equally, we should realize that plausible initial values, just because they appear to be "simple" to our understanding, are not a bit more probable than their corresponding final values.

Thus Loschmidt's objection against Boltzmann's *H*-theorem turns out to miss the point, as Boltzmann assumed, (unfortunately not explicitly, although *in fact*), for his invisible molecular motion is in a state which is highly improbable, as it is a very particular initial state. The improbability is marvelously disguised by its simplicity; (just as though we had considered a long series of throws of six as *probable*, for the reason that this is a happy result and is so *easily* communicated). In any case, this improbability is not increased in any way by time reversal **T** and this "time reflection" is only excluded by us because, so to speak, behind the veil of invisible thermal motion, it leads away from the initial (though supposedly highly improbable) state, which violates the initial conditions.

3 Molecular Semichaos

Unfortunately, by his concept of a *"molecular chaos"* Boltzmann nourished misunderstandings, which have gone around in this area for more than a hundred years. With this (rather natural) concept he had intended to underpin his *Stoßzahlansatz*, which leads to the notion of a cross section that is valid to this day and which (especially in quantum mechanics) has been applied on innumerable occasions. Boltzmann's idea of chaos, whereby in an "ideal", i.e. highly diluted gas, there should be no essential dynamical correlation between point-like molecules, has astonishingly, always been taken as plausible, not least to the fact that it appears to express a desirably simple and therefore probable initial state. How such concepts can be erroneous has already been expounded in the previous two paragraphs. Here, we must add that they have no validity in respect to their statistical claims.

In this context, the following historical remark might be of interest. *Robert Boyle*, a highly remarkable *Sceptical Chymist* and principal founder of the theory of gases (as well as our modern theoretical chemistry), had introduced for air the notion of a "gas" in place of the notion of "elements", a concept that had been used since antiquity (with its newer variants) with verbal recourse to the Greek word *chaos*. By this he wanted to express that there is no order amongst the molecules of the gas. We shall see that this expresses (only) a partial truth.

The (alleged) absence of correlations, according to the current picture of molecular chaos, is expressed in the statistics of molecules by simply equating the two particle distribution F at the points of phase space $s_m = (q_m, p_m)$ with $m = 1, 2$ with the product of the distribution f of two single particles:

$$F(s_1, s_2) = f(s_1) \cdot f(s_2) . \tag{1}$$

As this *statistical independence* amongst molecules is supposed to be exact and generally valid, it leads, as we shall see presently, to an outright contradiction to the dynamical laws for the statistics in elastic collisions. According to the laws of collision, if two partners in a collision are not correlated shortly before a collision, they must be after collision, if the ideal gas is still *inhomogeneous*, (i.e. not already at constant particle density). This fact expresses an element of *order* which is superimposed onto the chaos and which represents a structure, a "descent" from past to the future which must be properly analyzed.

However, before we turn to the relevant calculation, we shall briefly refer to what *C.F. von Weizsäcker* proposed on the temporal asymmetry of thermodynamics. According to his perception, a structural difference between past and future is *a priori* inherent in time, which is thus independent of all experience. The past is a fact while future events must be regarded as open possibilities. With this structure, the concept of probability is always *prognostic* and should never be used in a *retrospective* sense. However, couldn't we ask, for example, how probable it is, according to present state knowledge, that Cesar had indeed crossed to Britain? Shouldn't we be allowed to ask for the probabilities of events of the Ice Age or even antediluvian times? Be this as it may, the validity (or invalidity) of Boltzmann's Stosszahlansatz exists totally independently from what is meant semantically from stochastic notions. The issue here is not about stochastics but simply statistics, which means only about (large) numbers. Corresponding arguments for Boltzmann's hypothesis on chaos are all the more valid, simply because they concern neither prognosis nor retrospect but only the currently existing *correlations*, whereas time comes into play only through the (*deterministic*) and *reversible* equations of motion.

The unavoidable weakening of the chaos postulate, which we propose to call the hypothesis of *limited chaos* or, in short, *semi-chaos*, amounts to the occurrence of the inequality

$$F(s_1, s_2) \neq f(s_1) \cdot f(s_2) \tag{2}$$

that is essential in the scattering region. We shall study this matter within a very simple $(1+2)$-dimensional model of an ideal gas, that goes back to Lorentz and the Ehrenfests, and which for the sake of brevity we wish here to call the *game-board model*.

4 The Model

As is familiar from movie film strips, our *model time* T consists of entire multiples of an *elementary* or *cyclic time* θ_0. Let us briefly note, that this discrete model of time, together with its spatial counterpart, removes the basis of the paradox of motion put forward by Xenon. The conventional kinematics of e.g. astronomy, road traffic or the Schrödinger's wave function is also free of inner contradictions when a continuous spacetime is assumed.

Accordingly, our *model space* S is built up on a lattice, as is familiar from board games, having a lattice *grid length* h_0. In its simplest version we take a (much extended) square grating of the kind used in the game board "Go".

In order to simplify the calculations we shall, in general, replace the time variable t and the space coordinates (x, y) by their dimensionless equivalents $t^* := t/\theta_0$ and $x^*, y^* := (x, y)/h_0$ respectively. To simplify the formula we provisionally suppose in addition that $\theta_0 := 1$ as well as $h_0 := 1$, so that $t := t^*$ as well as $(x, y) = (x^*, y^*)$. To re-establish the conventional expressions with dimensions, one only needs to substitute t by t^* and (x, y) by (x^*, y^*). According to our conventions $\mathsf{T} = \mathbb{Z}$ and $S = \{0, \ldots, J\} \times \{0, \ldots, J\}$ having a very large integer number J. This grid of integer numbers S is then to be complemented by the (corresponding) half integer *intermediate grid* $C := \{\frac{1}{2}, \ldots, J - \frac{1}{2}\} \times \{\frac{1}{2}, \ldots, J - \frac{1}{2}\}$, which is required for the simplest formulation of the laws of motion.

The entire game board $S \cup C$ is populated by two different kinds of tokens, the first kind are mobile *discs* M, which move in S, and the others are N static *rhombi*, which are placed in C. We, of course, assume, that M and N are very large numbers.

Basically, the *laws of motion* affirm that the discs move independently from one another, whilst being reflected by the rhombi and walls. We now give formal expression to this.

5 The Equations of Motion

Due to the lack of interaction amongst the discs we can mainly restrict the discussion, at least for the moment, to the simplest case $M = 1$, that of a single moving particle. Its state of motion is, as is customary in canonical mechanics, a *point in phase space* $s := (z, w)$ where $z := (x, y)$ is the position and $w \in (N, E, S, W) =: W$ with $N = (0, +1), E = (+1, 0), S = (0, -1), W = (-1, 0)$ represents the *momentum* of our disc. (Descriptively, w represents the *direction of the wind* and W the *wind rose*.)

Let us designate the *step* in the time laps from t to $t+1$ by an index $+1$, something we can represent by the defining equation $s_{+1}(t) := s(t+1)$. It is now only a question of expressing s_{+1} algebraically through s.

Two cases occur in regard to the position of the disc in respect to whether it is in the *direct neighborhood* of a rhombus or not. Accordingly, we divide the phase space Γ of our gas into the two parts $\Gamma_\pm = \Gamma_+ \cup \Gamma_-$ and Γ_0. Here Γ_- stands for the sub-case immediately before the collision of the disc on the rhombus, whilest Γ_+ stands for the sub-case immediately after the collision. With this notation we can establish that the particle will move free of forces in Γ_0 and in Γ_+, whilest in Γ_- it confronts an immediate reflection by the adjacent rhombus. This we have now to cast into calculational form.

To this end we first concentrate on the case of a single rhombus, which we can assume to be located at the position $(a, b) =: c \in C$ (in an immobile position). We can characterize our three cases as follows:

$$
\begin{aligned}
|z - c| > 1 & & & & \text{in case} & & s \in \Gamma_0 \\
|z - c| = 1 & & \text{and} & w \cdot (z - c) > 0 & \text{in case} & & s \in \Gamma_+ \\
|z - c| = 1 & & \text{and} & w \cdot (z - c) < 0 & \text{in case} & & s \in \Gamma_-
\end{aligned}
\qquad (3)
$$

The notion of distance which is used here $|(x, y) - (a, b)| := |x - a| + |y - b|$ is that of a taxi driver's geometry. So we obtain as the *law for a one time step* the formulas

$$
(z, w)_{+1} \begin{cases} = (z + w, w) & \text{for} \quad (z, w) \in \Gamma_0 \cup \Gamma_+ \\ = (z, w + 2z - 2c) & \text{for} \quad (z, w) \in \Gamma_- \end{cases}
\qquad (4)
$$

From here one can easily derive the corresponding statements for the backwards step from $(z, w)_{+1}$ to (z, w) or, equivalently, from (z, w) to $(z, w)_{-1}$,

$$
(z, w)_{-1} \begin{cases} = (z - w, w) & \text{for} \quad (z, w) \in \Gamma_0 \cup \Gamma_- \\ = (z, w - 2z + 2c) & \text{for} \quad (z, w) \in \Gamma_+ \end{cases}
\qquad (5)
$$

and in the general case, the iterated formula for $(z, w)_\tau$ having arbitrary displacement in time $\tau \in \mathbb{Z}$, in which the complete *trajectory* $s_t^{\mathbb{Z}}$ has an *initial value* $s_0 = s$, that can be freely chosen.

For later use we still note the *explicit parametrization* of the two critical regions of phase space Γ_- and Γ_+:

$$
\Gamma_\pm(c) := \{(z, w) : 2z - 2c = \pm w + w' \quad \text{with} \quad w' \perp w\},
\qquad (6)
$$

where we can also write for the *collision parameter* $(1/2)w'$

$$
w' \in (w \perp) := \{w\mathbf{R}, w\mathbf{L}\}
\qquad (7)
$$

in which the matrices $\mathbf{R}$ and $\mathbf{L}$ represent a rotation of 90° to the right or left respectively.

The case of more than one disc can easily be taken care of by letting the index m run from 1 to M, which we attach to the phase space variables z and w as well as to their components x, y and u, v respectively. Similarly we deal with the numerous rhombi: Their spatial coordinates are given by $(a_n, b_n) =: c_n$ with $n \in \{1, ..., N\}$ and accordingly we write

$$\Gamma_0 = \Gamma_{01} \cup ... \cup \Gamma_{0N} \quad \text{and} \quad \Gamma_\pm = \Gamma_{\pm 1} \cup ... \cup \Gamma_{\pm N} \qquad (8)$$

for the three phase space domains in which equations of motions occur. The rhombi are supposed not to touch one another: $|c_n - c_{n'}| > 2$ in case $n \neq n'$.

6 Distribution of Discs in the Case of a Single Rhombus

Let $f(t; z, w)$ be the number of discs, suitably averaged, which are found at time t with value $w \in W$ of the wind rose at the lattice point z. The total number of discs is then gained by the normalization

$$\sum_{z \in S} \sum_{w \in W} f(t; z, w) = M \qquad (9)$$

where, of course, everywhere $f(t; z, w) \geq 0$. The time dependence of this density is determined by the fact that the number of particles is not altered by the phase space current. For a single time step this amounts to an equation of the Liouville type, namely $f(t + 1, s_{+1}) = f(t, s)$ or $f(t, s) = f(t - 1, s_{-1})$. Inserting the mechanism given above for s_{+1} and s_{-1} we find for the *statistical equations of motion* respectively:

$$f(t; z, w) \begin{cases} = f(t - 1; z - w, w) & \text{for} \quad (z, w) \in \Gamma_- \cup \Gamma_0 \\ = f(t - 1; z, w - 2z + 2c) & \text{for} \quad (z, w) \in \Gamma_+ \end{cases} \qquad (10)$$

Here, c stands for a typical position of a rhombus c_n, with the position of a disc $z(:= z_m)$ remaining limited to the close environment of c, so that confusion with neighboring rhombi c' is not possible.

For the following evaluation it is advisable to add these two equations and use the functions χ which characterizes the domain,

$$\begin{aligned} f(t; z, w) =& f(t - 1; z - w, w)\chi(z, w; \Gamma_0 \cup \Gamma_-) \\ & + f(t - 1; z, w - 2z + 2c)\chi(z, w; \Gamma_+) . \end{aligned} \qquad (11)$$

Here $\chi(s, \Gamma) := 1$ for $s \in \Gamma$ and $:= 0$ for $s \notin \Gamma$. In addition, we have to pay attention to the fact that the domains Γ_-, Γ_- and Γ_0 implicitly depend on the parameter c, which is not modified by collisions.

To enhance the transparency, on the right of (11) we add and subtract a term which is proportional to $\chi(z, w; \Gamma_+)$,

$$\begin{aligned} f(t; z, w) =& f(t - 1; z - w, w) - f(t - 1; z - w, w)\chi(z, w; \Gamma_+) \\ & + f(t - 1; z, w - 2z + 2c)\chi(z, w; \Gamma_+) . \end{aligned} \qquad (12)$$

According to (6) the χ-factor implies the expression $2z - 2c - w = w'$ for the associated scattering parameter. To ease the handling of the arithmetic, we now express the χ values by a Kronecker δ:

$$f(t; z, w) = f(t-1; z-w, w)[1 - \sum_{w' \in (w\perp)} \delta(2z - w - w', 2c)] + \sum_{w' \in (w\perp)} f(t-1; z, -w')\delta(2z - w - w', 2c) \ . \tag{13}$$

We recall that the position of a disc z must be restricted to the closest position of rhombus $c \in \{c_1, \ldots, c_N\}$.

7 Distribution of Rhomboid-Disc Pairs

To clarify the transition from a typical rhombus to the whole system of rhombi we add the expression for the density of discs f of the relevant position of a rhombus (c) or of all positions of the rhombi $\{c_1, \ldots, c_N\}$ as parameters of a potential. Up to now, we have only discussed the *local* situation by means of $f(t; z, w; \ldots, c, \ldots)$. Now we shall investigate the *global* problem by means of $f(t; z, w; c_1, \ldots, c_N)$. For this we must replace the value of the parameter c by c_N in (13) and then sum over the collision terms. We then obtain the *statistical equations of motion*

$$\begin{aligned} f(t; z, w; c_1, \ldots, C_N) &= f(t-1; z-w, w; c_1, \ldots, c_N) \\ &- \sum_{n-1}^{N} \sum_{w' \in W} [\sigma(w', w) f(t-1; z-w; w; c_1, \ldots, c_N) \\ &- \sigma(w', w) f(t-1; z; -w'; c_1, \ldots, c_N)] \delta(2z - w - w', 2c_n) \end{aligned} \tag{14}$$

in which we have introduced the collision *cross section*

$$\begin{aligned} \sigma(w', w) &= 1 - (w \cdot w')^2 = \sigma(w, w') \\ &= \sigma(-w', w) \ . \end{aligned} \tag{15}$$

Its symmetry with regard to the two momenta is often called *micro-reversibility* (or *detailed balance*). In concrete terms, for parallel momentum vectors we have $\sigma(w', w) = 1$, and for corresponding orthogonal vectors $\sigma(w', w) = 0$.

Concerning the collision centers $c_1, \ldots, c_N$, it is obvious to regard our system of rhombi as a "frozen gas", the result of an infinitely heavy molecule ($M = \infty$) limit. According to this model, we introduce, in an analogous way to $f(t; z, w)$ the mean rhomboid-density $g(c) \geq 0$ with normalization

$$\sum_{c \in C} g(c) = N \ . \tag{16}$$

With this definition, summations over the N variables c_N can easily be replaced by "discrete integration" over the single variable c. This perception expresses a reduced knowledge of the positions of the rhombi, according to a decrease of the associated *negentropy* $kH := k\sum_{c \in C} g(c) \log g(c) \leq \log N$.

The assumption of an unperturbed coexistence of the two one particle distributions $f(s)$ and $g(c)$, which we provisionally made, however, follows Boltzmann's perfect "molecular chaos", which is untenable, as we saw in Sect. 3. We are therefore forced to replace the product $f(s)g(c)$, which formulates the statistical independence of both particles, by a more general two particle distribution $F(s, c)$. For this distribution we may establish the relative normalization as

$$\sum_{c \in C} F(t; s, c) = N f(t; s) \qquad (17)$$

and positiveness as $F(t; s, c) \geq 0$. Formally, $f(s; c_1, ..., c_N)$ can be regarded as a microscopic, non-entropic special case or, better, limiting case of $F(s, c)$ with an additional factor $g(c; c_1, ..., c_N) = \sum_{n=1}^{N} \delta(c - c_n)$. However, we want to think of the discrete function F as a rather slowly varying mean value, as a macroscopic bird's eye view of the micro-state, as is known from atomic theories of materials. As such a mean value of our $F(s, c)$ can be represented through a linear combination of many $f(s; c_1, ..., c_N)$ cases.

In this manner we may generalize (14) in such a way that everywhere $f(z, w; c_1, ..., c_N)$ is replaced by $F(z, w; c)$,

$$F(t; z, w; c) = F(t-1; z-w, w; c)$$
$$-\sum_{w' \in W} [\sigma(w', w) F(t-1; z-w; w; c) \qquad (18)$$
$$-\sigma(w', w) F(t-1; z; -w'; c)] \delta(2z - w - w', 2c_n) .$$

On averaging using the operators $N^{-1} \sum_{c \in C}$ in accordance with (17), $\delta(2z - w - w', 2c)$ immediately goes over into 1, and we obtain

$$f(t; z, w) = f(t-1; z-w, w)$$
$$-\sum_{w' \in W} \left[\sigma(w', w) F(t-1; z-w; w; z - \frac{1}{2}w - \frac{1}{2}w') \qquad (19) \right.$$
$$\left. -\sigma(w', w) F(t-1; z; -w'; z - \frac{1}{2}w - \frac{1}{2}w') \right] .$$

This is, of course, not a closed equation of motion for f alone. We shall now obtain such an equation, as an approximation, in the form of a restricted phase space, by approximating F by the product form $f \cdot g$.

8 The Time Asymmetric Diffusion Equation

On the basis of the initial considerations, in connection with (1) and (2), we take as an approximation that, *before* the collision, the two collision partners (the colliding disc and the rhombus which is hit) are *stochastic pairs* that are strangers to one another, so that their mutual scattering parameters w' are, statistically, uniformly distributed. We express this (approximatation to) uncorrelatedness by the localized factorization:

$$F(t; s; c) = f(t; s)g(c) \quad \text{for} \quad (s; c) \in \Gamma_-, \tag{20}$$

in an approximation which we can expect to become increasingly better the more dilute the (almost) "ideal" model gas becomes,

$$M + N \ll J^2. \tag{21}$$

In this approximation we have concentrated ourselves on the small part Γ_- of our phase space $\Gamma = S \times C$, for which $(z, w; c) \in \Gamma_-$ is equivalent to $(z, w) \in \Gamma_-(c)$. According to (6), this is equivalent to the orthogonality $2c - 2z + w \perp w$.

Substituting (20) into (19) and respecting the micro reversibility (15), we obtain through the help of the difference equation

$$f(t; z, w) = f(t-1; z-w, w) - \sum_{w' \in W} \sigma(w, w')$$
$$\times g(z - \frac{1}{2}w - \frac{1}{2}w')[f(t-1; z-w, w) - f(t-1; z, -w')] \tag{22}$$

the desired "transport equation" for the mobile gas of discs in the frozen gas of rhombi, although only in a preliminary discretized form. The "transport" which is meant here is a kind of diffusion, which always runs in *one* direction of time: from past to future.

We shall see that, in general, the transition from (19) to (22), which was mediated by (20), has now moved away from invariance under inversion, something which had been valid up to this point. This break of symmetry in time is apparently due to the fact that equation (20) is taken as valid only for the constellations (Γ_-) shortly before the collision but not for analogous constellations shortly after collision. Our Ansatz (20) mathematizes (for each instant t) the hypothesis, which we call in short "*semi-chaos*". Its consequence is the *Stoßzahl-theorem* which is due to Boltzmann and according to (19) implies that the frequency of transitions of momentum from w' to w is given through the collision cross section $\sigma(w, w')$.

For the sake of simplicity, we have assumed the rhombi to be stationary and furthermore have assumed their spatial distribution to be constant:

$$g(c) = g_0 = N/J^2. \tag{23}$$

In respect to temporal tendencies this simplification is neutral.

9 The Transition to the Continuum

We had envisaged the transition to the continuum previously in (17), but had not yet carried it through. The transition can be achieved through extreme cases in which

$$1 = \theta_0 \ll \omega^{-1} \quad \text{and} \quad 1 = h_0 \ll k^{-1}, \tag{24}$$

where ω designates the predominant frequencies and k the predominant sequences. In this transition the microscopically small eccentricities w and w' disappear with respect to the macroscopic coordinates of position z and c, so that in the limiting case the values z and $z - w$ coincide with the value of $c = z - \frac{1}{2}w - \frac{1}{2}w'$, $(z, z - w \to c,$ and analogously $t - \theta_0 \to t)$.

In addition, we shall assume that the function $f(t; z, w)$, which is continuous in t and z, is even smooth in these two variables (having a continuous derivative), so that we can apply Taylor's theorem. However, beforehand, for the sake of enhanced clarity, it is advisable to subtract the term $f(t - 1; z, w)$ on both sides of the difference equation (22). In addition, we shift the spatial gradient term to the left side of the equation in order to isolate the flow terms from the collision terms, so that (22) now takes the clearly arranged form

$$[f(t; z, w) - f(t - 1; z, w)] + [f(t - 1; z, w) - f(t - 1; z - w, w)]$$
$$= -g_0 \sum_{w' \in W} \sigma(w, w')[f(t - 1; z - w, w) - f(t - 1; z, w')] \quad (25)$$

which invites a differential quotient. The limit, according to (24), provides the *transport equation*

$$(\partial_t + w \cdot \partial_z) f(t; z, w)$$
$$= -g_0 \sum_{w' \in W} \sigma(w, w')[f(t; z, w) - f(t; z, w')] \quad (26)$$

in the style of Boltzmann. It is still invariant against time translations $f(t; z, w) \Rightarrow f(t - \tau; z, w)$, but no longer against time reflection $f(t; z, w) \Rightarrow f(t - \tau; z, -w)$. The collision term on the right hand side changes sign relative to the flow term on the left hand side of (26).

10 The H-Theorem

The breaking of reflection symmetry is most clearly revealed by the monotonic increase of the entropy $S = -kH$ with time, where Boltzmann's *Eta* is, in our case, given by the integral

$$H(t) := H_0 + \int d(z)^2 \sum_{w' \in W} f(t; z, w) \log f(t; z, w) \,. \quad (27)$$

Its change in time is obtained from (26) by a partial integration over z with the assumption that the boundary terms vanish: $dH(t)/dt = -\int d(z)^2 \sum_{w \in W} f(t; z, w) \log f(t; z, w) g_0 \sum_{w' \in W} \sigma(w, w')[f(t; z, w) - f(t; z, w')]$. Due to micro reversibility, from (15), the factor $\log f(t; z, w)$ can be replaced by its antisymmetrized variant with respect to w and w'. As the logarithm is a strictly monotonic function one then finds that the integrand, and with it the integral, cannot be positive:

$$dH(t)/dt \leq 0 \quad (28)$$

and is equal to zero exactly when the distribution is strongly isotropic everywhere: $f(t; z, w) = (1/4) f_0(t; z)$ for all t, z and w.

From an integration over the square of the gradient ∂_z^2 one realizes that a momentarily isotropic distribution can only exist if it is at the same time homogeneous. Therefore, a *thermostatic equilibrium*, $\partial_t f(t; z, w) = 0$, can exist exactly when the distribution is independent of both w and z at all times: $F(t; z, w) = f_{+\infty}$. This state is, of course, only reached asymptotically (however, exponentially decreasing), $f(+\infty; z, w) = f_{+\infty}$. In a closed system the final state, which is the future limiting state, is particularly simple, and for just this reason can be regarded as "carrying particularly little information" that is to say is utmost "entropic" or "chaotic" and in this is a quantitative sense of *disorder* or *lack of structure*. With increasing "simplicity" of the state, the measure of disorder is increased.

11 The Correlations

The last statement may be surprising as the molecular statistics, as we have found, seems to have to a certain extent the opposite tendency, whereby the molecular order is at first not reduced by collisions, but increased in an essential respect. This drive for structure building apparently becomes overcompensated by the collisions' strong drive to level out. (We are reminded here of the shuffling of cards.) Here we do not want to follow any further the increase of the effectively assumed phase space by it being split up, but shall restrict the analysis to the correlation between the collision partners. Prior to their collision, they had been strangers to one another. Afterwards they have a common history.

We ask for the counterpart of (20) for its mirror image in time $(s; c) \in \Gamma_+$, that is $s \in \Gamma_+(c)$. Through this restriction, on the right hand side of (18) only the third term, the gain term remains, whilst the flow and loss terms drop out. In this case:

$$F(t; z, w; c) = \sum_{w' \in W} \sigma(w, w') F(t-1; z, -w'; c) \delta(2z - w - w', 2c) \quad (29)$$

where the right hand side, differently to the left hand side, belongs to an *incoming* configuration:

$$(z, w; c) \in \Gamma_+ \quad \text{and} \quad (z, -w'; c) \in \Gamma_- \quad (30)$$

according to (6), as $2z - 2c + (-w') = w \perp (-w')$ due to the δ-factor. Therefore using (20), we can factorize each term on the right as

$$F(t-1; z, -w'; c) = g(c) f(t-1; z, -w') \quad (31)$$

and in the case $(z, w; c) \in \Gamma_+$, therefore

$$2z - 2c - w \perp w \,. \quad (32)$$

Under these conditions and due to (15), we find the result

$$F(t; z, w; c) = g(c) f(t-1; z, w + 2c - 2z) \,. \tag{33}$$

This is different from (20) or (31) as it cannot be factorized in the original sense, that is having a factor $f(s)$ that is independent of the parameter c. We thus obtain a *correlation coefficient*

$$k_+(t; s; c) = F(t; s; c)/g(c) f(t; s) \quad \text{for} \quad (s; c) \in \Gamma_+ \tag{34}$$

which is, generally, not only different from 1, but depends on s and c as well.

We show this using the example of a particularly simple distribution, namely a distribution that is independent of the spatial coordinates: $f(t-1; (x, y), (u, v)) := f_0 + u f_1$, in which the west wind ($u = +1$) has a higher air density than that of the east wind ($u = -1$), $0 < f_1 < f_0$. Due to (20) this choice proves to be a stationary state,

$$f(t; z; w) = f(t-1; z; w) = f_0 + u f_1 \tag{35}$$

with $w = (u, v)$. From (33) we conclude

$$F\left[t; \left(a - \frac{1}{2}u, b + \frac{1}{2}v\right), (0, v); (a, b)\right]$$
$$= f\left[t-1; \left(a - \frac{1}{2}u, b + \frac{1}{2}v\right), (u, 0)\right] g(c) = f_0 + u f_1 \,. \tag{36}$$

which leads to

$$k_+\left[t; (x, y), (0, v); \left(x + \frac{1}{2}u, y - \frac{1}{2}v\right)\right] = (f_0 + u f_1)/f_0 \neq 1 \,. \tag{37}$$

In this way, the correlation after collision $k_+ - 1 = \pm f_1/f_0$ not only differ remarkably from the analogously defined but vanishing correlation prior to collision $k_- - 1 = 0$, but depends very sensitively on the tiny collision parameter $(1/2)u = (1/2)u\theta_0 = \pm(1/2)h_0$.

12 Heredity

The search for the origin of these one-sided correlations leads to the spatial variations of the distribution f in the intermediate environment and farther out from the collision point z. The length scales which determine these variations are, in the first place, the *mean distance between rhombi* $d_0 = h_0 g_0^{(1/2)}$, given by $d_0^2 N = h_0^2 J^2$ and thereafter the *mean free path* $l_0 = h_0 g_0^{-1}$ given by $h_0 l_0 N = h_0^2 J^2$, in which

$$g_0 \ll 1 \tag{38}$$

as a result of (23) and (21). Thereby, the hierarchy of (24) is supplemented and sharpened to

$$h_0 \ll d_0 \ll l_0 \ll k_{-1} \tag{39}$$

whereby d_0 is the proportional mean (or the geometric mean) of h_0 and l_0. As is known, for typical cases of three dimensional gases, $h_0:d_0 \simeq d_0:l_0 \simeq l_0:k^{-1}:10^1$. We can visualize d_0 and l_0 as mean radii of geometric, respectively dynamic neighborhoods, whereas h_0 can be interpreted in our model (amongst others) as mean radii of repulsive scattering forces.

The fact, that even in a not perfectly ideal gas the free paths l_0 are much longer than the distances d_0 and still longer than the scattering radii h_0, signifies that most collisions lead to a migration into distant regions. There, the remnants of correlations from the past, originating at the beginning of the long journey from that expulsion, hardly play a role any more in respect to the encounters to be expected with any foreign partner. The free running particles, which are on the move or are still in the far field Γ_0 immerse rather rapidly into the neutral chaos of the force free molecules of the gas once again. We therefore do not anticipate any notable pre-correlations for such collisions. That is to say: our molecular population reproduces, with sufficient approximation, its semi-chaotic characteristic from one scattering generation to the next. This very remarkable *heredity* of semi-chaos, resulting from collisions, is a hitherto unrecognized property of molecular dynamics in connection with assumed statistics.

When these statistics are made more precise using stochastic concepts, one obtains a Liouville–Gibbs probability distribution ρ, which, apart from the thermodynamic limit, is inhomogeneous in regard to space and non-stationary with respect to time, but in such a way that all the local proportions, in the critical phase space regions, are to a large extent constant and remain so. The balancing of pressure by diffusion in a layer of cotton wool can perhaps serve to illustrate this matter.

If this is extrapolated by heredity into the distant past, one arrives of necessity at the cosmological boundary conditions that, as was explained in the initial paragraphs of our analysis, suppose an initial state of extraordinarily low entropy, in stark contrast to the final state of high entropy.

13 Appendix: Boundary Conditions

In the description of our model we had mentioned the associated boundary conditions and used them, but had not specified them. The simplest modeling postulates are, of course, elastic reflections from each of the four walls. We thus require, for example, at the western wall ($x = 0$) the following equations of motion:

$$\begin{aligned} f[t; (0, y), (+1, 0)] &= f[t-1; (0, y), (-1, 0)], \\ f[t; (x, y), (u, 0)] &= f[t-1; (x-u, y), (u, 0)] \\ \text{for} \quad x > 0 \quad \text{as well as for} \quad x &= 0 \quad \text{with} \quad u = -1 \,. \end{aligned} \tag{40}$$

Accordingly, the corresponding boundary conditions for the remaining three walls are at hand; one only needs to replace $(0, y)$, $(\pm 1, 0)$ by (L, y), $(\pm 1, 0)$ or $(x, 0)$, $(0, \pm 1)$ by (x, L), $(0, \pm 1)$.

A different, still simpler model consists of imposing periodic boundary conditions in place of boundary conditions. This implies closing the square through a congruent identification of opposing pairs of sides of a torus, which is a figure having no boundary.

Remarks on Fractional Time

Rudolf Hilfer

*In dankbarer Erinnerung an die Jahre 1983–1986
Herrn Prof. Dr. Carl Friedrich von Weizsäcker
anläßlich seines 90. Geburtstages gewidmet*

1 Introduction

It is not possible to repeat an experiment in the past. The underlying philosophical truth in this observation is the difference between certainty of the past and potentiality of the future. This difference is discussed, for example, in C.F. von Weizsäcker's papers [1, 2], and it was often pointed out by him in our discussions in the years 1983–1986 in the Starnberg institute. The perennial philosophical problem related to this difference between past and future is the question whether time is real or not. Compared to the difference between past and future, the dogmatic time reversibility of mechanical processes in physics appears as a secondary and derived property. Let us therefore assume for the purposes of this paper that the asymmetry of time is more fundamental than the reversibility of time implied by the limited validity of mechanical equations.

If the asymmetry of time is placed above all other laws of physics (e.g. also above the law of energy conservation) then the so called problem of irreversibility becomes reversed (see [3]): Instead of explaining how irreversible behaviour (such as diffusion) arises from reversible mechanical laws, one must now explain why time reversible equations appear so frequently in physics. In this paper I intend to show that this "reversed irreversibility problem" can be solved, and that, surprisingly, the answer has nothing to do with statistical mechanics or the second law, thereby avoiding the ongoing debate [4] about Boltzmann's views.

Let me therefore postulate the following statement as an empirical and fundamental law of nature:

Every time evolution of a physical system is irreversible.

The mathematical structures corresponding to irreversible time evolutions are abstract Cauchy problems and operator semigroups [5]. In fact postulating irreversibility as the starting point of physics stands in stark contrast to most fundamental theories of physics including Weizsäcker's "Urtheorie" [6, 7]. In these theories one starts from a symmetry group containing the time translations as a subgroup. In this way one postulates, by Noether's theorem, energy conservation and reversibility as the fundamental starting point.

In the following I want to show that there is no contradiction between time translation invariance of physical observations and the fundamental nature of time irreversibility. On the contrary, both facts can be reconciled. Let me first discuss basic requirements for time evolutions in physics. Next some mathematical consequences will be drawn that have been discussed earlier in [3, 8]. Finally I shall attempt to interpret the consequences in the light of the deliberations of C.F. von Weizsäcker exposed e.g. in [1, 2].

2 Requirements for time evolution operators

2.1 Semi-group property

A physical time evolution $\{T(\Delta t) : 0 \leq \Delta t < \infty\}$ is defined as a one-parameter family (with time parameter Δt) of bounded linear time evolution operators $T(\Delta t)$ on a Banach space B. This family of operators is a representation of the semi-group of nonnegative real numbers $(\mathbb{R}_+, +)$ (time durations), i.e. it fulfills the conditions

$$T(\Delta t_1)T(\Delta t_2)f(t_0) = T(\Delta t_1 + \Delta t_2)f(t_0) \qquad (1)$$

$$T(0)f(t_0) = f(t_0) \qquad (2)$$

for all $\Delta t_1, \Delta t_2 \geq 0$, $t_0 \in \mathbb{R}$ and $f \in B$. The elements $f \in B$ represent time dependent physical observables, i.e. functions on the time axis $\mathbb{R}$. In the following B is chosen as the set $B = L^1(\mathbb{R})$ of Lebesgue-integrable functions on the time axis in a vector space (e.g. $\mathbb{R}^n$) with norm $\|f\| = \int |f(t)| dt$. Note that the argument $\Delta t \geq 0$ of $T(\Delta t)$ has the meaning of a time duration, while $t \in \mathbb{R}$ in $f(t)$ means a time instant.

Equations (1) and (2) define a physical time evolution as a representation of the semigroup $(\mathbb{R}, +)$ of real numbers, not as a group. The inverse elements $T(-\Delta t)$ are absent. This reflects the fundamental difference between past and future, in the sense that it is not possible to evolve into the past.

The linear operator A defined by

$$D(A) = \left\{ f \in B : \underset{\Delta t \to 0+}{\text{s-lim}} \frac{T(\Delta t)f - f}{\Delta t} \text{ exists} \right\} \qquad (3)$$

and

$$Af = \underset{\Delta t \to 0+}{\text{s-lim}} \frac{T(\Delta t)f - f}{\Delta t} \qquad (4)$$

is called the infinitesimal generator of the semigroup. Here s-lim $f = g$ is the strong limit and means $\lim \|f - g\| = 0$ as usual.

2.2 Continuity

A physical time evolution should be continuous. This requirement is realized mathematically by the assumption that

$$\text{s-lim}_{\Delta t \to 0} T(\Delta t) f = f \qquad (5)$$

holds for all $f \in B$, where s-lim is again the strong limit. Semigroups of operators satisfying this condition are called strongly continuous or C_0-semigroups.

2.3 Homogeneity

Homogeneity of time is used here to mean two requirements: First it requires that observations are independent of the initial instant or position in time, and secondly it requires arbitrary divisibility of time durations and self-consistency for the transition between time scales. This second condition will be discussed in detail below in Sect. 2.5.

Independence of physical processes from the position of initial instant on the time axis means that physical experiments and processes are not changed if they are *ceteris paribus* shifted in time. Sometimes one rephrases this as: All time instants are equivalent. Or one says that there are no preferred or distinguished time instants. This formulation can easily be misunderstood. The requirement that the difference between past and future is fundamental implies that the present is special, and in this sense the start of an experiment is always special, even though it can be shifted to another position. Mathematically this distinction of the starting instant t_0 is a consequence of the semigroup properties in (1) and (2). The special character of the initial instant is familiar in biological systems where it corresponds to the time instant of birth.

The requirement that the initial instant can be shifted to any other instant is expressed mathematically as the requirement of invariance under time translations. As a consequence one demands commutativity of the time evolution with time translations. One demands

$$\mathcal{T}(\tau)T(\Delta t)f(t_0) = T(\Delta t)\mathcal{T}(\tau)f(t_0) = T(\Delta t)f(t_0 - \tau) \qquad (6)$$

for all $\Delta t \geq 0$ und $t_0, \tau \in \mathbb{R}$. Here the translation operator $\mathcal{T}(t)$ is defined by

$$\mathcal{T}(\tau)f(t_0) = f(t_0 - \tau). \qquad (7)$$

Note that τ can also be negative. This means that physical experiments in the past have the same outcome as in the present. This can be checked in the present with the help of documents (e.g. a video recording), irrespective of the fact that the experiment cannot be repeated in the past.

2.4 Causality

Causality of the physical time evolution requires that the values of the image function $g(t) = (T(\Delta t)f)(t)$ depend only upon values $f(s)$ of the original function with time instants $s < t$.

2.5 Homogenous Divisibilty

The semigroup property (1) implies that for $\Delta t > 0$

$$T(\Delta t) \ldots T(\Delta t) = [T(\Delta t)]^n = T(n\Delta t) \tag{8}$$

holds. Homogeneous divisibility of a physical time evolution is the requirement that there exist rescaling factors D_n for Δt such that with $\Delta t = \overline{\Delta t}/D_n$ the limit

$$\lim_{n \to \infty} T(n\overline{\Delta t}/D_n) = \overline{T}(\overline{\Delta t}) \tag{9}$$

exists und defines a time evolution $\overline{T}(\overline{\Delta t})$. The limit $n \to \infty$ corresponds to two simultaneous limits $n \to \infty$, $\Delta t \to 0$, and it corresponds to the passage from a microscopic time scale Δt to a macroscopic time scale $\overline{\Delta t}$.

3 Consequences from the requirements

The requirement (6) of homogeneity implies that the operators $T(\Delta t)$ are convolution operators. Let T be a bounded linear operator on $L^1(\mathbb{R})$ that commutes with time translations, i.e. that fulfills (6). Then there exists a finite Borel measure μ such that

$$(Tf)(s) = (\mu * f)(s) = \int f(s-x)\mu(\mathrm{d}x) \tag{10}$$

holds [9, p. 26]. Applying this theorem to physical time evolution operators $T(\Delta t)$ yields a convolution semigroup $\mu_{\Delta t}$ of measures $T(\Delta t)f(t) = (\mu_{\Delta t} * f)(t)$

$$\mu_{\Delta t_1} * \mu_{\Delta t_2} = \mu_{\Delta t_1 + \Delta t_2} \tag{11}$$

with $\Delta t_1, \Delta t_2 \geq 0$. For $\Delta t = 0$ the measure μ_0 is the Dirac-measure concentrated at 0.

The requirement of causality implies that the support $\operatorname{supp}\mu_{\Delta t} \subset \mathbb{R}_+ = [0, \infty)$ of the semigroup is contained in the positive half axis.

The convolution semigroups with support in the positive half axis $[0, \infty)$ can be characterized completely by Bernstein functions [10]. An arbitrarily often differentiable function $b : (0, \infty) \to \mathbb{R}$ with continuous extension to $[0, \infty)$ is called Bernstein function if for all $x \in (0, \infty)$

$$b(x) \geq 0 \tag{12}$$

$$(-1)^n \frac{\mathrm{d}^n b(x)}{\mathrm{d}x^n} \leq 0 \tag{13}$$

holds for all $n \in \mathbb{N}$. Bernstein functions are positive, monotonously increasing and concave. The set of all Bernstein functions forms a convex cone containing the functions that are positive and constant.

The characterization is given by the following theorem [10, p. 68]. There exists a one-to-one mapping between the convolution semigroups $\{\mu_t : t \geq 0\}$ with support on $[0, \infty)$ and the set of Bernstein functions $b : (0, \infty) \to \mathbb{R}$ [10]. This mapping is given by

$$\int_0^\infty e^{-ux} \mu_{\Delta t}(dx) = e^{-\Delta t b(u)} \tag{14}$$

with $\Delta t > 0$ and $u > 0$.

The requirement of homogeneous divisibility further restricts the set of admissible Bernstein functions. It leaves only those measures μ that can appear as limits

$$\lim_{n \to \infty, \Delta t \to 0} \underbrace{\mu_{\Delta t} * \ldots * \mu_{\Delta t}}_{n \text{ factors}} = \lim_{n \to \infty} \mu_{n\overline{\Delta t}/D_n} = \overline{\mu}_{\overline{\Delta t}} \tag{15}$$

Such limit measures $\overline{\mu}$ exist if and only if $b(x) = x^\alpha$ with $0 < \alpha \leq 1$ and $D_n \sim n^{1/\alpha}$ holds [3, 11, 12].

The remaining measures define the class of fractional time evolutions $T_\alpha(\Delta t)$ that depend only on one parameter, the fractional order α. These remaining fractional measures have a density and they can be written as [3, 8, 13–15]

$$T_\alpha(\Delta t) f(t_0) = \int_0^\infty f(t_0 - s) h_\alpha\left(\frac{s}{\Delta t}\right) \frac{ds}{\Delta t} \tag{16}$$

where $\Delta t \geq 0$ and $0 < \alpha \leq 1$. The density functions $h_\alpha(x)$ are the one-sided stable probability densities [3, 8, 13–15]. They have a Mellin transform [16–18]

$$\mathcal{M}\{h_\alpha(x)\}(s) = \frac{1}{\alpha} \frac{\Gamma((1-s)/\alpha)}{\Gamma(1-s)} \tag{17}$$

allowing to identify

$$h_\alpha(x) = \frac{1}{\alpha x} H_{11}^{10}\left(\frac{1}{x} \left| \begin{array}{c} (0, 1) \\ (0, 1/\alpha) \end{array}\right.\right) \tag{18}$$

in terms of H-functions [17–20]. Their Laplace transform is

$$\mathcal{L}\{h_\alpha(x)\}(u) = e^{-u^\alpha}. \tag{19}$$

The infinitesimal generators of the fractional semigroups $T_\alpha(\Delta t)$

$$A_\alpha f(t) = -(D^\alpha f)(t) = -\frac{1}{\Gamma(-\alpha)} \int_0^\infty \frac{f(t-s) - f(t)}{s^{\alpha+1}} ds \tag{20}$$

are fractional time derivatives of Marchaud–Hadamard type [21, 22]. This fundamental and general result suggests to generalize physical equations of motion by replacing the integer order time derivative with a fractional time derivative as the generator of time evolution [3, 13].

For $\alpha = 1$ one gets $h_1(x) = \delta(x-1)$, the Dirac-measure at $x = 1$. In this case the fractional semigroup $T_1(\Delta t)$ reduces to the conventional translation semigroup $T_1(\Delta t) f(t_0) = f(t_0 - \Delta t)$. This observation provides an answer to the question from the introduction why ordinary first order time derivatives appear so frequently in the equations of physics. The theory sketched here can be made more precise and detailed. From the detailed theory one finds that the special case $\alpha = 1$ occurs more frequently in the limit (15) than the cases $\alpha < 1$, in the sense that it has a larger domain of attraction. The fact that the semigroup $T_1(\Delta t)$ can often be extended to a group on all of $\mathbb{R}$ provides an explanation for the seemingly fundamental reversibility of mechanical laws and equations.

4 Philosophical remarks

The homogenous divisibility of time could perhaps be viewed as a mathematical expression of the philosophical idea that C.F. von Weizsäcker calls "umfassende Gegenwart" [2, p. 612ff]. Homogeneous divisibility formalizes the fact that a verbal statement in the present tense presupposes always a certain time scale. For this reason the present should not be thought of as being a time instant as a certain time period or duration, no matter how short it is [3, 8].

Fractional time evolutions seem to be related also to the subjective human experience of time. In physics the time duration is measured with periodic clocks by counting the number of periods. Contrary to this the subjective human experience of time corresponds to the comparison with an hour glass, i.e. with a nonperiodic reference. Correspondingly most humans experience a fixed physical time interval (e.g. an hour, a day, a year) during senescence as being much shorter than in their infancy or adolescence. The reason is that the reference process to which the fixed interval is compared is one's own life time. It seems that a time duration is considered "long" if it is comparable to the time interval that has passed since birth. In other words the rapidity of perception decreases with age. It seems that this phenomenon is reflected in fractional stationary states defined as solutions of the stationarity condition $T_\alpha(\Delta t) f(t) = f(t)$. Hence the existence of fractional time evolutions requires a generalization of concepts such as "stationarity" or "equilibrium". This outlook could be of interest for nonequilibrium and biological systems [3, 8, 13–15].

Finally, also the special case $\alpha \to 0$ challenges philosophical remarks. In the limit $\alpha \to 0$ the time evolution operator degenerates into the identity. This could be expressed verbally by saying that for $\alpha = 0$ "becoming" and "being" coincide. For $\alpha = 0$ physical states evolve in time by being mapped to themselves, hence without change. This paradoxical situation reminds me of C.F. von Weizsäcker's thoughts in [2, Kap. 13] concerning certainty of the future and potentiality of the past.

References

1. C.F. von Weizsäcker: Der zweite Hauptsatz und der Unterschied von Vergangenheit und Zukunft. *Ann. Physik* 36, 275 (1939)
2. C.F. von Weizsäcker: *Aufbau der Physik*. Hanser, München 1985
3. R. Hilfer: Fractional time evolution. In: R. Hilfer (ed.): *Applications of Fractional Calculus in Physics*. World Scientific, Singapore 2000, p. 87
4. J. Lebowitz: Macroscopic laws, microscopic dynamics, time's arrow and Boltzmann's entropy. *Physica A* 194, 1 (1993)
5. R. Butzer and H. Berens: *Semigroups of Operators and Approximation*. Vol. 145 of *Die Grundleheren der mathematischen Wissenschaften in Einzeldarstellungen*. Springer, Berlin 1967
6. L. Castell and W. Heidenreich: Dynamical groups of binary alternatives. In: L. Castell, M. Drieschner and C.F. von Weizsäcker (eds.): *Quantum Theory and the Structures of Time and Space*, Vol. 4. Carl Hanser Verlag, München 1981, p. 64
7. C.F. von Weizsäcker: The present status of the theory of urs. In: L. Castell, M. Drieschner and C.F. von Weizsäcker (eds.): *Quantum Theory and the Structures of Time and Space*, Vol. 4. Carl Hanser Verlag, München 1981, p. 7
8. R. Hilfer: Foundations of fractional dynamics. *Fractals* 3, 549 (1995)
9. E. Stein and G. Weiss: *Introduction to Fourier Analysis on Euclidean Spaces*. Princeton University Press, Princeton 1971
10. C. Berg and G. Forst: *Potential Theory on Locally Compact Abelian Groups*. Springer, Berlin 1975
11. B. Gnedenko and A. Kolmogorov: *Limit Distributions for Sums of Independent Random Variables*. Addison-Wesley, Cambridge 1954
12. H. Bergstöm: *Limit Theorems for Convolutions*. Wiley, New York 1963
13. R. Hilfer: Classification theory for anequilibrium phase transitions. *Phys. Rev. E* 48, 2466 (1993)
14. R. Hilfer: Fractional dynamics, irreversibility and ergodicity breaking. *Chaos, Solitons & Fractals* 5, 1475 (1995)
15. R. Hilfer: An extension of the dynamical foundation for the statistical equilibrium concept. *Physica A* 221, 89 (1995)
16. V. Zolotarev: Mellin-Stieltjes transforms in probability theory. *Theor. Prob. Appl.* II, 433 (1957)
17. W. Schneider: Stable distributions: Fox function representation and generalization. In: S. Albeverio, G. Casati and D. Merlini (eds.) *Stochastic Processes in Classical and Quantum Systems*. Springer Verlag, Berlin 1986, p. 497
18. R. Hilfer: Absence of hyperscaling violations for phase transitions with positive specific heat exponent. *Z. Physik B* 96, 63 (1994)
19. C. Fox: The G and H functions as symmetrical Fourier kernels. *Trans. Am. Math. Soc.* 98, 395 (1961)
20. A. Prudnikov, Y. Brychkov and O. Marichev: *Integrals and Series*, Vol. 3. Gordon and Breach, New York 1990
21. B. Rubin: *Fractional Integrals and Potentials*. Longman, Harlow 1996
22. R. Hilfer: *Applications of Fractional Calculus in Physics*. World Scientific Publ. Co., Singapore 2000

The Dynamics of Modelling

Gérard G. Emch

1 Introduction

It is with some awe that I accepted the invitation to contribute this homage to Professor von Weizsäcker. Although he had graciously told me of his personal association with Switzerland, I do not remember whether I had mentioned, during our conversations at Tutzing, that I had first known *of* him in the late 1950s and early 1960s when the Swiss people were to vote on an initiative to put in their Constitution a clause that Switzerland would never obtain nor maintain a nuclear arsenal. A graduate student at the time, I attended the meetings of a group of physicists and theologians in Geneva who were studying the writings of a "Physicist, now Ordinarius in Philosophy at Hamburg." It is only much later that I had the privilege to participate to the Tutzing meetings and to know the physicist he always was, and the influential philosopher of science he had become. I keep his *Aufbau der Physik* [28] as a memento of these encounters, next to the official proceedings [3].

The major philosophical preoccupations in von Weizsäcker's writings on the foundations of physics are mainly with quantum mechanics; my two contributions to the Tutzing meetings were indeed along this line [4].

Nevertheless, for methodological purposes which I will describe below, I wish here to focus instead on the emergence of thermodynamical behaviour from statistical mechanics. Consequently, even where von Weizsäcker addresses issues in probability theory and formal logic, my incursions in these territories present different aspects of these landscapes. One of the traits I appreciate in von Weizsäcker is his ability to see the shared grounds in approaches that others may perceive to be hopelessly distant from his own. I feel confident therefore that he will accept the exceptions just mentioned, and thus the more formal aspects of this paper, although I am aware of his own methodological choice as stated in the preface to the *Aufbau* [28]: *"Jedenfalls habe ich niemals die hermetish-abweisende Knappheit gesucht, die in der Mathematik verbreitet ist."* Characteristically of von Weizsäcker's approach, the first section of the first chapter of the *Aufbau* begins with a question: *"Was ist die Wahrheit der Physik?"* followed one page later by another question *"Wie ist Theorie möglich?"* to which he adds: *"Sie folgt niemals mit logischer Notwendigkeit aus der Erfahrung."*

I address here an allied question, namely the nature of models in their relations with developing theories. My approach borrows occasionally from edifices built by two philosophers once associated with von Weizsäcker, namely Erhard Scheibe who, in the beginning of his career, was von Weizsäcker's Assistant in Hamburg, and Imre Lakatos who, in the last years of his life, was a member of the Scientific Board of von Weizsäcker's Max-Planck-Institute at Starnberg. From Scheibe [20], I learned much of what I understand about the philosophers' views on the reduction process in the physical sciences; several of these concerns are addressed in the book I wrote with Chuang Liu *The Logic of Thermostatistical Physics* [5]. Here, I will transpose, adapt and modify some of the elements in the "rational reconstruction" process Lakatos documented in his *Proofs and Refutations: The logic of Mathematical Discovery* [14].

The current investigation began as I was preparing the introductory commentary I was asked to write for Volume 6 of the *Collected Work of Eugene Wigner* [31]. I noticed then that Wigner was using the word "model" in two different connotations. I could generally associate the first with the praxis of theoretical physicists, while I found the second to be characteristic of the endeavours of mathematical physicists.

While the above refers to "what people do" – the kinematics – the dynamics comes in when we ask "why people do what they do". The models I call *H-models* are mainly directed at deriving specific predictions or applications of an accepted theory. In contrast, the purpose of my *L-Models* is to verify the internal consistency of an axiomatized theory. H-models explore the limits of this world, L-models the limits of possible worlds. Both H-models and L-models are expected to suggest modifications to the theory.

The classification is admittedly soft. Nevertheless, I suggest that it has three merits. The first is normative: it recognizes a difference of purpose between theoretical physics and mathematical physics; to the difference in the purpose of modelling is necessarily associated a difference in methodology: what is acceptable or required for one purpose might not be so for the other; for instance the demonstration of predictive powers *vs.* the need for mathematical rigour. I argue that the classification has a second merit: it is descriptive and its malleability allows models [and scientists] to migrate from one class to the other, thus allowing to trace some of the driving forces that participate to the dynamics of the re-construction of scientific theories. This classification still has a third merit: it bears on Reichenbach's distinction between the "context of discovery" and the "context of justification". In my discussion of models in their context of justification, I appeal to the distinction made in formal logic between *syntax* and *semantics*.

2 Some Models from Thermophysics

To delineate the above distinctions, I follow a phenomenological approach along a path cut through what the practitioners do. My first steps on this path were through [31]; I explored it further in collaboration with Chuang Liu as we chose for [5], the laboratory facilities provided by thermostatistical physics. In the present section, I bring up five cases which are representative for my purpose. The technical details and specific bibliography are readily available in [5], where many other examples drawn from both equilibrium and non-equilibrium thermophysics are also discussed.

2.1 Thermodynamics

The historical background, *i.e.* the context of discovery, unfolds on a backdrop of pragmatic evidence, or at least of what the founders consider to be such evidence: the "caloric", "cycles between thermal reservoirs", "Victorian prohibitions" ... and all. This baroque assemblage of tacit assumptions, shifting concepts and profound insights is hardly the "mature science" ready for Hilbert's axiomatization programme [11, 6th problem], [12]. Yet a concise syntax for this nascent science is provided by Clausius in a sequence of publications between 1850 and 1879 [5, p. 64] where he recognizes explicitly that besides the usual differential forms that are now called exact, there exist differentials of another type. Let me describe, in the spirit of Reichenbach's context of justification, the syntax Clausius delineates.

The rules of the differential calculus are assumed on the domain

$$D = \{(V, T) \in \mathsf{R}^+ \times \mathsf{R}^+\}. \tag{1}$$

Two differential forms, η and τ, are distinguished:

$$\eta = \Lambda_V dV + C_V dT \quad \text{and} \quad \tau = p dV \tag{2}$$

where Λ_V, C_V and p are smooth, yet unspecified, real-valued functions of (V, T); and where η and τ satisfy the property that for all smooth simple, closed countour $\Gamma \subset D$:

$$\int_\Gamma (\eta - \tau) = 0 \quad \text{and} \quad \int_\Gamma \frac{1}{T}\eta = 0. \tag{3}$$

This syntax allows to build a theory without reference to the interpretation of its terms: the following general theorems follow immediately from the above axioms.

Theorem 1. $\exists$ *smooth functions U and S from D to R such that $\eta = dU + \tau$ and $\eta = TdS$.*

Theorem 2. $\partial_T (\Lambda_V - p) = \partial_V C_V$.

Theorem 3. $\Lambda_V = T (\partial_T \Lambda_V - \partial_V C_V)$.

Theorem 4. $\Lambda_V = T \partial_T p$.

The consistency of the axioms is established by a L-model, the ideal gas. It is defined by two constitutive equations

$$p(V, T) = R\frac{T}{V} \quad \text{and} \quad \partial_T C_V = 0. \tag{4}$$

where R is a positive constant. The first of these equations already entails:

Scholium 1. $\Lambda_V = p$, $\partial_V U = 0$, $\partial_V C_V = 0$.

The two constitutive equations together then entail the following constraints.

Scholium 2. *The function C_V is constant.*

Scholium 3. $U(T) = C_V (T - T_o) + U(T_o)$.

Scholium 4. $S(V, T) = C_V \ln(T/T_o) + R \ln(V/V_o) + S(V_o, T_o)$.

Other L-models have been constructed to show the mutual independence of the axioms. Short of another name, I will refer generically to the framework just obtained, namely axioms, theorems and L-model, as an *open theory*.

We have reached the stage of Hilbert's *"Chairs, Tables and Beer-mugs"*. To move further than these, the semantics is to be extended beyond mathematics into physics. The first step in this direction is to recognize that the ideal gas, seen so far as a L-model, is also the H-model associated with the names of Boyle, Mariotte and Gay–Lussac. Compare then: (a) the empirical features described by these experimentalists within the motivating backdrop loosely mentioned in the beginning of this subsection; and (b) the terms appearing in the axioms and the H-model, together with their mathematical consequences. This confrontation suggests the following identifications: V with "Volume", T with "Temperature", η with "Heating", τ with "Working", p with "Pressure", C_V with "Specific Heat", U with "Internal Energy", S with "Entropy", R with the "Universal Gas Constant", the empirical value of which is 8.314... $J \deg^{-1} \text{mole}^{-1}$.

The ideal gas was actually the only H-model for many years. Thomson even argues that Clausius' considerations are limited to this model. To make things worse, this model is incapable of serving as a H-model for the vapour-liquid process occurring in the steam engine, the prime motivation of Carnot who nevertheless thinks routinely in terms of the ideal gas model. Despite Thomson's resistance, the open theory of Clausius now supersedes this particular model. Note again that the theorems are consequences of the axioms themselves, while the scholia belong to the particular instanciation, the ideal gas. I wish to postpone to Sect. 2.4 below the mild modifications, in both syntax and semantics, necessary to accommodate typical non-ideal gases.

2.2 The Speed of Sound

In the *Principia,* Newton gives the basic theory of harmonic motion, recognizes that sound is a compression wave in a material medium, air in his case; and expresses the speed v of the sound wave in terms of the compressibility β of the gas through which it propagates, the density ρ of the gas, and its pressure p:

$$v = \sqrt{\frac{1}{\rho \beta}} \quad \text{with} \quad \beta = -\frac{1}{\rho}\frac{\partial \rho}{\partial p}. \tag{5}$$

Granted the larger context of the *Principia* no L-model is called for here. But Newton needs now a H-model for the functional relation between the density ρ and the pressure p. Newton's H-model consists of one assumption and two hypotheses. The assumption happens to be incorrect, and the hypotheses plainly falacious; Westfall writes, even more plainly, about fudging [29].

The one assumption is that the phenomena is isothermic, *i.e.* occurs at constant temperature. Newton does not warn his reader that he has no direct evidence for this, and that he makes the assumption only because it is simple enough to allow him to use Boyle's law (1662), namely:

$$pV = \text{const} \quad i.e. \quad \rho \propto p. \tag{6}$$

From (5) and (6), Newton obtains

$$v = \sqrt{\frac{p}{\rho}}. \tag{7}$$

Newton tests the validity of his formula by getting separately the experimental values of v, the speed of sound, and the other two directly measurable quantities that appear in (7), the pressure p and density ρ of air; the numerical values he measures for v and $\sqrt{p/\rho}$ vary somewhat from one edition to another (1687, 1694, 1726). He finds for $\sqrt{p/\rho}$ a value between 968 and 979 English feet per second. He measures directly the speed of sound from the return time of the echo in the ambulatory of Trinity College. In the first two editions he reports for v values between 920 and 1109 English feet per second, a margin of error compatible with the value he had found for $\sqrt{p/\rho}$, thus vindicating (5) and (6), *i.e.* theory and H-model.

No fudging, so far. In the third edition, however, Newton uses for v the value measured by Derham in 1705–08 from the time elapsed between the flash of cannons fired at a distance of 12 miles, and the later arrival of the sound. This experimental set up allows a much better precision, and Newton feels compelled to accept $v = 1142$ English feet per second, which is *not* compatible anymore with $\sqrt{p/\rho} = 968$ nor 979 English feet per second. Newton also calls on the value found in 1700–02 by Sauveur – the founder of musical acoustics in France – namely 1070 French feet per second, which fits with the new figure, $v = 1142$ English feet per second. The empirical data therefore seemed to *falsify* the theory as implemented by the proposed H-model, albeit with a discrepency of less than 20%:

$$\frac{1142}{979} = 1.166 \leq \gamma_N \leq 1.180 = \frac{1142}{968}. \tag{8}$$

To escape this falsification, Newton indulges in overt fudging and sets up to fabricate two extravagant hypotheses. Firstly, he invokes what he calls the "crassitude" of air, namely the fact that the particles of air are of finite size; to this he attributes an increment of 109 English feet per sec. Secondly, he estimates that the water vapour suspended in the air contributes a multiplicative factor of 21/20. Now

$$(979 + 109) \times \frac{21}{20} = 1142. \tag{9}$$

The agreement with the experimental values of Derham and Sauveur is right on the dot: the formula is verified with a precision better than one part in a thousand! This is too wonderful to be true: there could be no way for Newton to justify with such a precision the estimates involved in his two corrections.

In a 1759 memoir, Lagrange noticed that something was amiss with Newton's theory and/or H-model, without however bringing out the key of the mystery, namely that Newton's reliance on Boyle's law was misplaced: the sound phenomenon is too rapid to be isothermic as the heat generated by compression has no time to go anywhere; the processus is thus adiabatic. But this requires more than what was available in Newton's time, namely a decent theory of heat. Recognizing this, Laplace (1816) and Poisson (1823) replace in Newton's argument Boyle's law (6) on the isotherms by

$$pV^\gamma = \text{const} \quad \text{with} \quad \gamma = \frac{C_p}{C_V}. \tag{10}$$

This is known today as the Laplace–Poisson law for the adiabatic changes; it is obtained by setting instead $dS = 0$ in Scholium 4: C_V, resp. $C_p = C_V + R$, are the specific heats at constant volume, resp. constant pressure. Their values are $7/2$ and $5/2$ for air. Keeping Newton's theory (5), this new H-model corrects Newton's formula (7) to read:

$$v = \sqrt{\frac{p}{\rho}} \sqrt{\gamma} = v_{\text{Newton}} \cdot \sqrt{\gamma}. \tag{11}$$

Compare now

$$\sqrt{\gamma} = \sqrt{1.4} = 1.183 \tag{12}$$

with (8): Newton's experimental data are remarkably good, especially in view of the paucity of the experimental techniques available in his times. He is deceived into fudging by his H-model, while his theory is adequate to the task when used in conjunction with the Laplace–Poisson H-model.

Finally, let me end this account with two light notes. First, a technological application: nowadays, a routine experimental set-up consists in measuring the speed of sound through the gas to determine β from Newton's formula (5), with the Laplace–Poisson caveat that β is the *adiabatic* compressibility. Second, a piece of literary criticism: in a communication before the Tenth International Congress on the Enlightenment (Dublin, July 1999), Antoinette Emch-Dériaz and I propose to use Newton's blatant fudging as a touchstone to assay the independence of the French translator – and annotator – of Newton's *Principia*, Madame du Châtelet, who was working in the late 1740s with some help from the mathematician Clairaut. None of her comments addresses explicitly the difference of tone involved in Newton's fudging; had she detected it, astutely she hides it in an unconspicuous typographical artifice.

2.3 Probability

Descriptions of games of chance can be traced back to the earliest civilizations. This phenomenology is rationalized as a theory in the period extending roughly from Pascal

(1623–1662) to Laplace (1749–1827). It took another 100 years for its syntax to be distilled by Kolmogorov in the earliy 1930s. Measure-theory provides a mathematical L-model, although the disposal of sets of measure zero has a contentious history.

For the semantics, Kolmogorov is satisfied by mentioning briefly von Mises' work on the interpretation of probabilities as limiting frequencies. However, the basic concept introduced by von Mises in the 1920s, namely the *Kollectiv,* is not as mathematically straightforward to implement as it sounds. The recursive function theory in mathematics is motivated in part by the difficulties encountered in specifying how to *select* representative subensembles. By the late 1930s, a L-model is finally obtained for the Kollectiv; but soon afterwards Ville, a student of Borel, shows that the theory does not exclude a H-model with martingales, namely a Kollectiv where sequences of relative frequencies approach a limit in a one-sided manner, thus allowing a bettor to win consistently. Note in passing that von Neumann, apparently not aware of these difficulties, invokes the analogy with the classical Kollectiv when, in his *Foundations of Quantum Mechanics,* he appeals to ensembles of similarly prepared systems from which representative subensembles are to be selected for measurement purposes.

In the late 1930s, independently of the developments in the von Mises approach, de Finetti proposes a completely different semantics, qualifying as a L-model, where probabilities are interpreted in terms of fair bettings, thus bringing probability back to its origins in the den of gamblers and bookies. The new twist however is the appeal to concepts that were later to be refined, developed and formalised by engineers and mathematicians working in the related areas of information theory and complexity theory.

Except for a brief allusion to von Neumann, my discussion has been limited so far to classical probability theory. At the syntactic level, this theory calls *events* the measurable subsets of a measure space. Intersection, union and complementation, together with the distributive rule, equip the space of events with the structure of Boolean algebra. The semantics is provided by interpreting events as *yes-no propositions,* intersection as *and*, union as *or* and complementation as *negation*. In the quantum realm, the Heisenberg uncertainty relations prohibit the Boolean distributivity from being an empirically justifiable axiom. The axiom must be weakened to orthomodularity. The events are now the projectors of a von Neumann algebra.

Particular L-models are the classical Boolean lattices $\mathcal{F}(\Omega)$ of the measurable subsets of measure spaces Ω; they are atomic when the event space is discrete, the points of such a space being interpreted as pure states.

The old quantum theory considers the particular L-models provided by the irreducible von Neumann algebras $\mathcal{B}(\mathcal{H})$; their lattice consists of all closed subspaces of a Hilbert space $\mathcal{H}$; they are atomic, with the one-dimensional subspaces identified with the pure states. Von Neumann algebras that are direct sums of such algebras are called for in the description of quantum systems exhibiting superselection rules [30]. $\mathcal{B}(\mathcal{H})$ is said to be a factor of type I_n or I_∞ depending on whether $\mathcal{H}$ is of dimension $n < \infty$ or infinite. When and only when $n < \infty$ the lattice satisfies moreover an additional condition, namely modularity. This condition is also satisfied by the lattice of projectors of type II_1 von Neumann factors; this L-model is not atomic. There is only two more types of von Neumann factors, the factors of type II_∞ and III; the latter appear routinely in the description of systems with infinitely many degrees of freedom.

The alleged opposition perceived between quantum and classical probability theories pertains to the context of discovery: traditional textbooks consider only one of two extremal models: $\mathcal{B}(\mathcal{H})$ for quantum mechanics and $\mathcal{F}(\Omega)$ for classical mechanics. Modern research programmes recognize that many interesting systems are not representable by either of these models, but can be handled within the general framework of von Neumann algebras. I shall mention results along those lines in Sect. 2.5. Beyond these L- and H-models, most of these programmes however, leave aside an essential *semantic* problem, namely a description of the measuring process that would improve on the von Neumann basic description which many of us find unsatisfactory. I intend to discuss elsewhere the extend to which the von Neumann algebra models – with continuous superselection rules – accomodate contingency in a consistent quantum temporal logic such as that pioneered by [26,27]. Much remains to be done. Yet, the centrality of such a project – for the interpretation of quantum mechanics as well as for the implementation of "correcting" learning algorithms in computer science – is supported by the sweeping historical survey [18] of what is classically understood by "temporal logic".

2.4 The Vapour-Liquid Phase Transition

The oldest phenomenological reference I found for phase transitions is Job 38:30. Less poetic, but certainly more scientific empirical descriptions were reported by Andrews in 1869; specifically, the existence of a critical point characterized by the appearance of an opalescent "glow", the flattening of isotherms in the vincinity of the critical point; and the horizontal portion of the low-temperature isotherms, corresponding to the coexistence of distinct phases: liquid and gas. In 1873 van der Waals offered a H-model:

$$\left[p + a \left(\frac{N}{V} \right)^2 \right] [V - Nb] = NkT \qquad (13)$$

where a and b are substance-specific constants which model two effects: a the attractive interaction between the molecules, and b the hard-core that prevent the molecules from penetrating each others. Note that when a and b are neglected, the van der Waals equation contracts to the Gay–Lussac law.

Equation (13) exhibits a critical temperature T_c above which the isotherms are stable, in the sense that $\partial_V p \leq 0$; in fact $\partial_V p < 0$. Below T_c, the isotherms present an instable interval where $\partial_V p > 0$. On the isotherm $T = T_c$, $\partial_V p$ is strictly negative everywhere except for exactly one critical value of $V = V_c$ at which $\partial_V p = 0$. Let $p_c = p(V_c, T_c)$, $v = V/N$ and $v_c = V_c/N$. Note that there are only two theoretically adjustable parameters in (13), namely a and b; whereas there are three experimentally accessible quantities, namely v_c, T_c and p_c among which there is therefore some relation. Indeed, one verifies that $v_c = 3b$, $kT_c = 8a/27b$ and $p_c = a/27b^2$, so that

$$\frac{p_c v_c}{kT_c} = \frac{3}{8}. \tag{14}$$

This *prediction* of the van der Waals H-model is fairly well *verified* in real substances as diverse as Water (.230), Carbon dioxide (.275), Hydrogen (.304) and Helium (.308). In constrast, the substance-specific constant a varies 200-fold across the substances just listed; the other constant, b, is more tame and varies only within a range 1:1.5. The relation (14) is known as the law of corresponding states (1880). The acceptance of its universality played an essential motivating role in the quest for the liquefaction of Helium (1908). In terms of the reduced variables $\tilde{v} = V/V_c$, $\tilde{T} = T/T_c$ and $\tilde{p} = p/p_c$, the van der Waals equation takes the universal form

$$[\tilde{p} + 3\tilde{v}^{-2}][\tilde{v} - \frac{1}{3}] = \frac{8}{3}\tilde{T}. \tag{15}$$

The van der Waals H-model suffers however from two physical shortcomings. First, the unstability $\partial_V p > 0$ has to be corrected by an *ad hoc* thermodynamical argument – the Maxwell construction – to obtain the coexistence plateau, and thus isotherms that are piecewise analytic only; the latter idealization can be accomodated by a mild generalization of Clausius axioms. Second, the account of the microscopic nature of the fluid is very crude; it improves with Ornstein's mean-field approximation in 1908; some of the basic ideas are also advanced by Weiss in 1907 in his H-model for ferromagnetism. However, both of these models require interaction ranges that are unrealistic. To Lenz' initiative belongs the first L-model for phase-transitions; but, in 1925, his student Ising treats only the 1-dimensional version of the model and finds correctly that it does not exhibit a phase-transition; however he unluckily infers that the 2- or 3-dimensional versions of the model would lead to the same result.

Several steps are still necessary to pass from this background to reliable understandings and predictions based on microscopic theories and models.

First, one has to recognize that the emergence of thermodynamics in statistical mechanics requires the so-called thermodynamical limit in which volume and number of particles tend to infinity while the density is kept fixed. As late as the 1937 van der Waals Congress, the physics community is still so sharply divided on the issue that an informal, if somewhat whimsical and certainly inconclusive, vote is called by Kramers.

Second, one has to understand the role of dimensionality in the manifestation of cooperative behaviour in spite of the short-range potential in the inter-atomic interactions. General theoretical results establish that, even in the thermodynamical limit, the 1-dimensional Ising model cannot exhibit a phase transition. While most of these no-go theorems were originally proven with classical statistical mechanics in mind, Araki's treatment of the 1-dimensional version of the Heisenberg model shows that quantum theory is not the cure. The nearest-neighbor Ising model on a 2-dimensional lattice is the first L-model to prove that a description of phase transitions emerges from a statistical mechanics that (a) includes the thermodynamical limit, and (b) considers systems which are at least two-dimensional. These developments extend essentially from Onsager's solution (1944) and Yang's proposal (1952) to the mathematically careful treatment by Schultz, Matthis and Lieb (1964). As a H-model however, the Ising model still fails: its description of the behaviour in the immediate vincinity of the critical point does not agree with empirical evidence.

The last step is to introduce the ideas of scaling and the renormalisation semigroup, a theory that finds its L-model in the Dyson hierarchical model. The later also makes predictions that are in much better agreement with empirical data, thus endowing it with some credibility as a H-model. As such however, it is peculiar in that the unrealistically long-range and artificially hierarchical structure of the interactions favour the emergence of co-operative behaviour so drastically that the one-dimensionality constraint is effectively overcome.

Nevertheless, at this point, the statistical mechanics of a large class of phase transitions has reached the stage of a *"closed theory"* (see Sect. 3). The essential ingredients are in place, namely the thermodynamical and scaling limits. Their consistency is proven by L-models. The concepts allow to picture how the desired macroscopic phenomena emerge from known microscopic interactions. H-models exist for a variety of physical circumstances in which phase transitions are exhibited, including those that are accompanied by spontaneous symmetry breaking. Theoretical predictions on universal relations between critical coefficients are verified by empirical data, and the agreement even extends to the individual values of many of these coefficients. The modelling of the geometric segregation of coexisting pure phases is in progress; for a brief introduction, see [5, pp. 418–419], and for details, see [17]. The residual complexity of computations and experiments is unlikely to be the veil that hides the necessity for improvements requiring revolutionary new approaches rather than "small" changes.

2.5 Superconductivity

Discovered in 1911, superconductivity is accompanied by a great variety of different manifestations which obtain their first coherent microscopic explanation only in 1957 with the BCS model. The theoretical context was provided by the quantum theory of solids. The most compelling evidence that the explanation resides in electron-electron interactions mediated by phonons had come in 1950 with the prediction and laboratory confirmation of the isotope effect

$$T_c \sqrt{M} = C \qquad (16)$$

where T_c is the critical temperature of the onset of superconductivity, M is the atomic mass and C is nearly a constant across the four isotopes of Mercury.

With that background, the BCS-theory starts as a H-model, namely a quantum system enclosed in a finite cubic box Λ of edge $2L = V^{1/3}$ with Hamiltonian

$$H_\Lambda = \sum_{p,s} \epsilon(p) a_s(p)^* a_s(p) + \sum_{p,q} b(p)^* \tilde{v}(p,q) b(q) \qquad (17)$$

$a_s(p)^*$ is the creation operator for an electron of momentum p and spin s up ($\uparrow$) or down ($\downarrow$); and $a_s(p)$ the corresponding annihilation operator. Similarly,

$b(p)^* \equiv a_\uparrow(p)^* a_\downarrow(-p)^*$ is the creation operator for a pair of electrons of opposite momentum and spin, a so-called *Cooper pair*. And $b(p)^* \tilde{v}(p,q) b(q)$ describes the energy transfer during a collision between Cooper pairs, with

$$\tilde{v}(p,q) = V^{-1} \int d\xi d\eta \, e^{ip\xi} \, v(\xi, \eta) \, e^{-iq\eta} \tag{18}$$

which is assumed to satisfy

$$\left. \begin{array}{c} \sum_q |\tilde{v}(p,q)| < \infty \quad ; \quad v \equiv \int d\xi d\eta |v(\xi, \eta)| < \infty \quad ; \quad v(\xi, \eta)^* = v(\eta, \xi) \\ \text{and thus:} \quad |\tilde{v}(p,q)| < \dfrac{v}{V} \to 0 \quad \text{as} \quad V \to \infty \, . \end{array} \right\} \tag{19}$$

In the mean-field approximation, the Hamiltonian can be diagonalized to give

$$H_\Lambda = \sum_{p,s} E(p) \, \gamma_s(p)^* \gamma_s(p) \, . \tag{20}$$

The *quasi-particles* created and annihilated by $\gamma_s(p)^*$ and $\gamma_s(p)$ are obtained through a Bogoliubov–Valatin transformation of the form:

$$\left. \begin{array}{l} \gamma_\uparrow(p)^* = u(p)^* a_\uparrow(p)^* - v(p)^* a_\downarrow(-p) \\ \gamma_\downarrow(-p) = v(p) a_\uparrow(p)^* + u(p) a_\downarrow(-p) \end{array} \right\} . \tag{21}$$

The one-particle energy of these quasi-particles is given by

$$E(p) = \left[\epsilon(p)^2 + \Delta(p)^* \Delta(p)\right]^{1/2} \tag{22}$$

where the energy-gap $\Delta(p)$ satisfies the self-consistency equation

$$\Delta(p) = -\sum_q \tilde{v}(p,q) \frac{\Delta(q)}{2E(q)} \tanh\left(\frac{1}{2}\beta E(q)\right) \, . \tag{23}$$

Finally, the coefficients $u(p)$ and $v(p)$ in (21) are solutions of the equations

$$\left. \begin{array}{ll} u(p)^* u(p) = \dfrac{1}{2}\left[1 + \dfrac{\epsilon(p)}{E(p)}\right] I & u(p)^* u(p) + v(p)^* v(p) = I \\ \text{and} & \\ v(p)^* v(p) = \dfrac{1}{2}\left[1 - \dfrac{\epsilon(p)}{E(p)}\right] I & u(p)^* v(p) = -\dfrac{1}{2}\dfrac{\Delta(p)}{E(p)} I \end{array} \right\} . \tag{24}$$

The functional dependence of the energy gap $\Delta(p)$ on the inverse temperature $\beta = 1/kT$, obtained from (23), compares well with the empirical data. This result, and the general success of the model in describing the several phenomena associated with superconductivity, leads to a suspicion that the model might be *exactly soluble*, and thus be a L-model. However, there would seem to be two conceptual obstacles to this. First, the Hamiltonian (17) is invariant under the gauge transformations

$$a_s(p) \to e^{i\theta} a_s(p) \tag{25}$$

whereas the Hamiltonian (20) is not. Second, there is the incongruity of having the energy spectrum of (20) depending on temperature whereas no temperature dependence had been put in the definition of the original Hamiltonian (17).

The way out of these dilema was found by [8], and polished in a series of papers during the following years. It comes from the treasure chest of the mathematical theory of C^*-algebras and their representations.

First, the C^*-algebra here is the C^*- inductive limit $\mathcal{A}$ of the local C^*-algebras $\mathcal{A}_\Lambda$ generated by the Fermi operators $a^*(p)$ and $a(p)$ relative to the finite boxes $\Lambda \subset \mathsf{R}^3$.

Second, to every state ρ on a given C^*-algebra $\mathcal{A}$ corresponds canonically a representation $\pi_\rho(\mathcal{A})$ of $\mathcal{A}$ – through the GNS construction which is akin to the Wightman reconstruction theorem in quantum field theory. Here the state ρ is obtained as the thermodynamical limit of the local canonical equilibrium state ρ_Λ corresponding, for the inverse temperature β, to the Hamiltonian H_Λ. This limit exists, even though the limit of the Hamiltonian itself does not make sense since the energy becomes infinite. A limiting time-evolution can also be defined.

Third, there is a critical temperature T_c above which the representation so constructed has trivial center, the energy gap vanishes and the gauge invariance of the theory is preserved. Below T_c however, the center $\mathcal{Z}_\rho$ of the representation $\pi_\rho(\mathcal{A})$ is non-trivial; the energy gap is a non-vanishing element of $\mathcal{Z}_\rho$, and it is not a multiple of the identity. The gauge group S^1 acts transitively over $\mathcal{Z}_\rho$, with $\Delta(p) \to e^{i2\theta}\Delta(p)$, $u^*(p) \to e^{i2\theta} u^*(p)$, $v(p) \to v(p)$ and $\gamma_s(p) \to e^{-i\theta}\gamma_s(p)$. The gauge symmetry of the theory is thus preserved. The representation $\pi_\rho(\mathcal{A})$ decomposes uniquely in a direct integral, over S^1, of factor-representations, each of which is interpreted as a pure thermodynamical phase; in each of these the energy gap is a non-zero c-number satisfying the self-consistency condition (23). The pure thermodynamical phases, obtained through this decomposition, are not individually invariant under the gauge group S^1; but the manifold $\mathcal{M}_T$ of all the pure phases corresponding to the same temperature $T < T_c$ is stable under the action of the gauge group; in fact S^1 acts transitively and freely on $\mathcal{M}_T$. The thermodynamical limit has been controlled. BCS thus provides a L-model for a phase transition accompanied by the spontaneous breaking of a continuous symmetry group.

A major break-away from this closed theory was suggested in 1987 by the discovery of high T_c ceramic superconductors.

3 A Longer View

Philosophers and physicists often fail to understand each other because they are speaking different languages which happen to have the same words. There is a long tradition for this, even within mathematics, see indeed [15, p. 15] from which the previous sentence is a transposition. To the dismay of philosophers, "theory" and "model" are often used interchangeably by reputable, hard-nose physicists; *e.g.* the *TOE (Theory of Everything)* and the *String model* in the high-energy physics community.

Hertz [10] uses the words *"Bild"* and *"Darstellung"*, as well as *"bildliche Darstellung,"* for something that suggests quite different categories, *e.g.* "cognitive schemes or models", "theory of models", "Hertz's models", and "Hertz's theory" – all in the space of less than three pages in [13, pp. 140–142]; Janik and Toulmin sort these out with much care to show the influence of Hertz's presentation [10] on Wittgenstein's suggestions for what language can do and cannot do. Recall in this context [32, 2.1]: *"Wir machen uns Bilder der Tatsachen."*

Hertz specifies what he requires from the *Bild* which he aspires to circumscribe: it must be "logically permissible", "correct" and "appropriate". Namely: (1) "We should at once denote as inadmissible all images which implicitly contradict the laws of our thought ... What enters into the images in order that they may be permissible, is given by the nature of our mind." (2) "We shall denote as incorrect any permissible images, if their essential relations contradict the relations of external things ... What enters the images for the sake of correctness is contained in the results of experiments, from which the images are built up." (3) "What is ascribed to the images for the sake of appropriateness is contained in the notations, definitions, abreviations, and, in short, all that we can arbitrarily add or take away." [10, pp. 2–3]

Clearly Hertz's guidelines for his "images" refer to something larger than either "theory" or "model" in the acceptations used in the present paper. "Admissibility" is close to what I call the internal consistency of the syntax: axioms with their L-models. "Correctness" involves what I refer to as H-models with their physical semantics; see [26, pp. 297, 330–1], [27, pp. 132–133]. As for the notion of "appropriateness", it pertains to the criteria of elegance and/or economy in the formulation of the whole enterprise. Thus from the two vernacular translations of the German *Darstellung,* namely "representation" and "depiction", I believe that "representation" is too restrictively associated to the logical notion of "model" or to the mathematical use of the word, as in "representations of groups"; "depiction" is closer to the encompassing sense in which Hertz uses the term; I propose to assimilate it into the Heisenberg-von Weizsäcker notion of *abgeschlossene Theorie* or "closed theory", *i.e.* a theory that has reached such a stage of completeness that "it cannot be improved upon by means of small changes"; see for instance [9], [24, Sect. II.3], [28, Sects. 6.1, 6.7, 6.11.b, 14.2], [27, pp. 131–132], [21, pp. 224–5], [22, 23].

Hertz achieves a closed theory in [10]. Its limitations are instructive. Hertz, whose name is universally associated with laboratory evidence for electromagnetic waves, mentions his dissatisfaction with electromagnetism as one of the motivations for his *Mechanics.* Moreover, "Time, Space and Mass" is the title of the first chapter in each of the two parts of his *Mechanics,* completed in 1894 as he was dying. Yet, the quantum appears in 1900, empirically motivated by a discrepency between theory and experiment on electromagnetic black-body radiation. Moreover, time, space and mass were rejected as separate absolutes in 1905, in a daring generalization to the universe from the invariance Lorentz (and Poincaré) had noticed in the Maxwell equations for electromagnetism. Heuristic escape and logical criticism notwithstanding, these breaks are neither H-models nor L-models of pre-existing theories: they function as previews of yet-to-be formulated theories. It is in the nature of closed theories that only drastically can one break away from them: *"Heisenberg hat gesagt, nur ein wahrer Konservativer könnte ein wahrer Revolutionär sein."* [28, p. 287]. Planck and Poincaré come to mind; the revolutions which they help bring forth came when they were well advanced in their career: in 1885 both had become Professors, Planck at the University of Kiel, Poincaré at the University of Paris ... as for the iconoclast Einstein, in 1900 he had just graduated from the ETH-Zurich; life never seemed to give him the opportunity to develop into much of a conservative.

In contrast, my discussion of the dynamics of modelling applies primarily to the refinements of ideas leading, from the inside, towards a closed theory. Hence my referring to *open theories* for the theories I deal with: they are evolving parts of a grand design yet-to-come, and they are expected to undergo mathematical and physical adjustments, both at the level of the syntax and of the semantics. The driving forces are less drastic; the changes they cause are not revolutionary, but more subtle. I am trying to catch things mid-stream, and there are *no* clean beginnings. For one thing, every experimentalist has heard the age-old warning, see *e.g.* [1, 2, 28], that no empirical evidence is independent from some theoretical a priori. Consequently, in my enterprise, Reichenbach's context of discovery [19] must include a variety of disparate elements lurking from a poorly organized backdrop of "facts", "con-" and "-cepts", against which the play is played, namely the story of a dynamics by which tentative formulations of syntax and semantics undergo successive re-formulations. Clausius (see Sect. 2.1 above), like Newton (as discussed in [33]), did not just deduct hypotheses to be checked; they absorbed and distilled a great deal of background information. Only when these tasks have been performed, do the models take on roles paralleling those served in mathematics by proofs and refutations as parts of the process of conjecture refinement [14].

Having proposed to distinguish the H- and L-models appearing in the praxis of physics, I must elaborate briefly on their different functions and on what they achieve.

H-models explore the connections between the syntactic framework and the physical semantics. Their dynamic resources are more forceful in their demands for adjustments, so much so that one could say, about them also: *"Heuristic is concerned with language dynamics, while logic is concerned with language statics."* See [14, p. 93], who cites [6].

L-models anchor the language through the logical semantics of the syntax. They also call for re-formulations tending to an economy of principles (independence of the axioms) and ensuring internal consistency. *"The virtue of logical proof is not that it compels belief, but that it suggests doubts."* See [14, p. 48], who refers to [7]. In that sense, I believe that *not all* interesting models are heuristic; compare again to the original question asked within the mathematical context in [14, p. 91]

Closer scrutiny into the history of mathematics requires some adjustments/revisions to Lakatos' views on mathematical research programmes: many theorems start as ambitious conjectures that later are cut down, bent, or amplified to match what the mathematician is able to prove. This contingent dynamics is discussed in [15] and documented with a contemporary example in [16]. To emphasize similarities and contrasts with physics, let me first note that in mathematics, if the dynamics of conjecture reformulation lies in the ingredients that the mathematician chooses or finds necessary for his proofs, then the driving forces observed in the discussion of open theories in mathematics, as well as the beauty of closed theories, are to be found in the proof of their theorems. Even axiomatization is driven – in part – by the search for the most general conditions under which the central theorems can be proven.

Now, the examples discussed in Sect. 2 indicate that these may not be the rules governing physics research programmes: there the dynamics of modelling is driven by a combination of sources, the nature of some of which are quite different from those prevailing in mathematics, while others are only similar, but with some twists. The H-models of physics are contingent in a way that the heuristic conjectures of mathematics are not. Yet, internal contradictions do appear in both, and in both they have to be cleared by logical arguments: proofs in mathematics, L-models in physics. While falsification has a place in both, its reliability in mathematics is more immediate; it is much harder to reach a concensus on whether and when the physical semantics has been purged from all the ambiguities that may falsify the test for falsification itself. It is in part this complexity – more than their maturity – that makes physical disciplines fertile ground for axiomatisations, Hilbert's or any other.

Sometimes the harmony of a string quartet, where no member claims precedence over another, is enhanced by the adjunction of a fifth instrumentalist. Similarly, in the give-and-take between physics, theoretical physics, mathematical physics and mathematics, the voice of philosophy can contribute clarity to the ensemble. Certainly, the oft-repeated logo "Mathematics is the language of science" recesses into perspective when we recall that Galileo's contemporaries included the likes of Gabrielli, Monteverdi and Schutz. Yet, I would rather paraphrase a modern phenomenologist in the philosophy of mathematics [16], writing *physics* when she writes *mathematics*: "Clearly, the phenomenological philosophy of physics requires both philosophical and physical sophistication." Coming from the other side of the bridge, I can only concur and hope that this, my homage to Professor von Weizsäcker, will be construed as a contribution to the critique of the methodology of complementary scientific research programmes. Their confluence are the passing manifestations by which one grabs glimpses of the elusive unity of physics [22, 24, 25].

References

1. C. Bernard: *Introduction à la médecine expérimentale*. J.-B. Baillère, Paris 1865
2. P.W. Bridgman: *The Logic of Modern Physics*. MacMillan, New York 1927; Arno Press, New York 1980; *Reflections of a Physicist*. Philosophical Library, New York 1955
3. L. Castell, M. Drieschner and C.F. von Weizsäcker: *Quantum Theory and the Structures of Time and Space*, Tutzig, 1980 and 1982, Vols. 4 and 5. Hanser, Munich 1981 and 1983
4. G.G. Emch: *Quantum Measurement Processes: A Gedanken Experiment*. In [3], 1981, pp. 137–151; *A Derivation of the Wigner-Moyal Correspondence and its Extension beyond Flat Spaces*. In [3], 1983, pp. 19–33
5. G. Emch and C. Liu: *The Logic of Thermostatistical Physics*. Springer, Heidelberg 2002

6. L. Felix: *L'aspect moderne des mathématiques*. 1957; Engl. transl.: *The Modern Aspect of Mathematics*. Basic Books, New York 1960
7. H.G. Forder: *The Foundations of Euclidean Geometry*. Cambridge Univ. Press, 1927; Dover, New York 1958
8. R. Haag: The mathematical structure of the Bardeen–Cooper–Schrieffer model. *Nuovo Cim.* 25, 287–298 (1962)
9. W. Heisenberg: Der Begriff "Abschlossene Theorie" in der modernen Naturwissenschaft. *Dialectica* 2, 331–336 (1948). Engl. transl.: The Notion of a "Closed Theory" in Modern Science. In: W. Heisenberg: *Across the Frontiers*. Harper & Row, New York 1974, pp. 39–46
10. H. Hertz, *The Principles of Mechanics Presented in a New Form*, with a Preface by H. von Helmoltz, and an Introduction by R.S. Cohen. MacMillan, London and New York 1899
11. D. Hilbert: *Sur les problèmes futurs des mathématiques*. In: *Comptes rendus du Deuxième congrès international des mathématiciens*, Paris 1900. Gauthier-Villars, 1902, pp. 58–114
12. D. Hilbert: Axiomatischen Denken. *Mathematischen Annalen* 78, 405–415 (1918)
13. A. Janik and S. Toulmin: *Wittgenstein's Vienna*. Simon & Schuster, New York 1973
14. I. Lakatos: *Proofs and Refutations: The logic of Mathematical Discovery*. Cambridge Univ. Press, Cambridge 1976
15. B.P. Larvor: Lakatos as Historian of Mathematics. *Philosophia Mathematica* 7, 42–64 (1997)
16. M. Leng: Phenomenology and Mathematical Practice. *Philosophia Mathematica* 10, 3–25 (2002)
17. B. Nachtergaele: Interfaces and droplets in quantum lattice models. In: A. Fokas, A. Grigorian, T. Kibble and B. Zegarlinski (eds.): *International Congress on Mathematical Physics*, London 2000. International Press, Boston 2001, pp. 243–249
18. P. Ohrstrom and P.F.V. Hasle: *Temporal Logic: From Ancient Ideas to Artificial Intelligence*. Kluwer, Dordrecht 1995
19. H. Reichenbach: *Experience and Prediction: an Analysis of the Foundations and Structure of Knowledge*. Univ. of Chicago Press, Chicago 1938
20. E. Scheibe: *Between Rationalisn and Empiricism: Selected Papers in the Philosophy of Physics*. Springer, Heidelberg 2001
21. E. Scheibe: Paul Feyerabend und die rationalen Rekonstruktionen. In: P. Hoyningen-Huene and G. Hirsch (eds.): *Wozu der Wissenschaftphilosophie? Positionen und Fragen zur gegenwärtigen Wissenschaftphilosophie*. De Gruyter, Berlin 1988. The pages cited in our text are from the Engl. transl. in [20, pp. 212–227]
22. E. Scheibe: C.F. von Weizsäcker und die Einheit der Physik. *Philos. Natur.* 30, 126–145 (1993); Engl. transl. in [20, pp. 54–68]
23. E. Scheibe: Heisenbergs Begriff der abgeschlossenen Theorie. In: B. Geyer et al. (eds.) *Werner Heisenberg. Physiker und Philosoph*. Heidelberg 1993; Engl. transl. in [20, pp. 136–141]
24. C.F. von Weizsäcker: *Die Einheit der Natur*. Hanser, Munich 1971; Engl. transl.: *The Unity of Nature*. Farrar Strauss Giroux, New York 1980
25. C.F. von Weizsäcker: The Unity of Physics. In: T. Bastin (ed.): *Quantum Theory and Beyond*. Cambridge Univ. Press, 1971, pp. 229–262
26. C.F. von Weizsäcker: Probability and Quantum Mechanics. *Brit. J. Phil. Sci.* 24, 321–337 (1973)
27. C.F. von Weizsäcker: Preconditions of Experience and the Unity of Physics. In: P. Bieri, R.P. Horstmann and L. Krüger (eds.): *Transcendental Arguments in Science: Essays in Epistemology*. Reidel, Dordrecht 1979

28. C.F. von Weizsäcker: *Aufbau der Physik*. Hanser, Munich 1985
29. R.S. Westfall: Isaac Newton and the Fudge Factor. *Science* 179, 731–738 (1973)
30. G.C. Wick, A.S. Wightman and E.P. Wigner: Intrinsic Parity of Elementary Particles. *Phys. Rev.* 88 101–103 (1957)
31. E.P. Wigner: Philosophical Reflections and Syntheses (Annotated by G.G. Emch). Vol. 6 of: A.S. Wightman and J. Mehra (eds.): *The Collected Work of Eugene Paul Wigner*. Springer, Heidelberg 1995; 2nd printing: 1997
32. L. Wittgenstein: Logisch-Philosophische Abhandlung. *Ann. d. Naturphilos.* 14, 185–262 (1921); with Engl. transl.: K. Paul: *Tractatus Logico-Philosophicus*. Trench, Trubner & Co., London; Harcourt, Bruce & Co., New York 1922
33. J. Worrall: The Scope, Limits, and Distinctiveness of the Method of 'Deduction from the Phenomena: Some Lessons from Newton's 'Demonstrations'. In: *Optics. Brit. J. Phil. Sci.* 51, 45–80 (2000)

Part IV

The Structure of Quantum Theory and Its Interpretation

An Introduction to Carl Friedrich von Weizsäcker's Program for a Reconstruction of Quantum Theory*

Thomas Görnitz and Otfried Ischebeck

1 The Motivation for a Reconstruction of Quantum Theory and the Elements of the Program

In spite of his familiarity with quantum theory, which he successfully applied to the properties and processes of nuclei and stars, Carl Friedrich von Weizsäcker remained dissatisfied with the theory as it is commonly presented and understood. For him neither the closed mathematical framework nor the power of numerical calculation could become a substitute for inquiring into the understanding of quantum theory. Weizsäcker's dissatisfaction is even more remarkable when we recall his close professional relation and friendship with one of the founders of quantum theory, Werner Heisenberg.

Carl Friedrich von Weizsäcker learned from Bohr not to submit quantum theory to a mathematical formalism and to recipes of procedure without a clear underlying physical interpretation. Instead, physical and theoretic reasoning should be made on the basis of phenomena and their representation by the general notions of quantum theory. The mathematical apparatus should be applied only parsimoniously and with constant caution. Through Heisenberg and Bohr, he also had first hand reports on the debates of the Copenhagen school with its great opponents, Erwin Schrödinger and Albert Einstein.

In his dissatisfaction, C.F. von Weizsäcker could refer to notable companions as R.P. Feynman, who confessed [8]: *"There was a time when the newspapers said that only twelve men understood the theory of relativity. I do not believe that there ever was such a time. There might have been a time when only one man did, because he was the only guy who caught on, before he wrote his paper. But after people read the paper, many understood the theory of relativity in some way or other, certainly more than twelve. On the other hand, I think I can safely say that nobody understands quantum mechanics."*

Weizsäcker's program for a reconstruction of quantum theory was inspired by Heisenberg's proposal to introduce a fundamental length of about 10^{-13} cm into

---

* We thank Michael Drieschner for comments.

quantum physics [10]. Weizsäcker tried to incorporate this fundamental length into a modification of geometry at this scale, so that quantum theory should determine geometry in the microscopic world as General Relativity does in the macroscopic world.

In winter 1953–54 in a seminar at the Max Planck Institute in Göttingen, the attempts to modify quantum theory were discussed. C.F. von Weizsäcker took part in this seminar together with Heisenberg, Peter Mittelstaedt, Erhard Scheibe und Georg Suessmann. The apparent failure of all attempts for a modification of quantum theory led to the task of a reconsideration of the success of the theory, which Weizsäcker then took up in his program for a reconstruction of quantum theory. He published the essential questions and the approach to his reconstruction program in *Komplementarität und Logik* [16], which is dedicated to Bohr's 70th birthday. *Komplementarität und Logik II* [17] followed in 1958 together with *Komplementarität und Logik III* [18], which was published with E. Scheibe and G. Süssmann.

In *Komplementarität und Logik*, a "logic of complementarity" is abstracted from quantum theory whose core is Birckhoff and v. Neumann's quantum logic [1]. It is understood as the physical quantum theory, expressed as a logical formalism. The contribution of P. Mittelstaedt in this volume shows the role which quantum logic can play for an analysis of physics. Weizsäcker links the process of quantization to the passage from classical propositional calculus to the "logic of complementarity" and second quantization to the passage to metalogic. Weizsäcker went on to design a "temporal logic" which would be the common basis of logic, as the theory of time-bridging propositions and schematic proofs, and of physics, as the theory of predictions and empirical decisions. In Kantian terms, this temporal logic would have to formulate the preconditions for the possibility of experience [6]. Weizsäcker reports on his work in logic in *Aufbau der Physik* [19] and *Zeit und Wissen* [20].

The papers on *Komplementarität und Logic* introduce the *ur*, the quantized binary alternative and quantum bit of information, as the basic concept in the reconstruction. The *ur-hypothesis* was formulated that particles and fields, as well as space can be built from urs. This means a construction of matter and fields out of quantum information. Binary alternatives were already introduced by v. Neumann into the mathematics of quantum theory in the form of projection operators, from where they obtain a role for quantum logic, but only an indirect role for physics. Weizsäcker, however, gives binary alternatives a basic physical significance. Physical objects should be built from urs. The ur-hypothesis follows the rule that mathematical objects of physical theory should be representative of physical objects as much as possible.

In Hamburg and later in Starnberg, Weizsäcker collaborated with Michael Drieschner [3, 4] for the reconstruction of quantum theory by axiomatic methods. Lutz Castell and his collaborators in Starnberg used group theoretical methods, mainly relying on the conformal group, to explore the consequences of the ur-hypothesis for the structure of particles and fields. Paul Roman analyzed the cosmological evolution out of an original ur-soup. Reports on the work of the physics group at the Max Planck Institute in Starnberg are contained in the six conference reports *Quantum Theory and the Structures of Time and Space*, referred to as QTSTS 1–6 [2]. Weizsäcker continued to work on the link of ur-theory with quantum field theory and cosmology together with Thomas Görnitz and Dirk Graudenz (see the bibliography of Weizsäcker's publications given in the appendix). The interpretation of quantum theory as presented by Weizsäcker's program in terms of information theory was further developed in Holger Lyre's dissertation [12].

The program of reconstruction in its present form has five elements which are successively presented in Sects. 2 to 6 below:

- Discussion of the role of philosophy for quantum theory and its place in philosophy;
- The implications of the fundamental role of time for the structure of quantum theory;
- A systems of axioms which link the philosophical basis and the implications of the fundamental role of time to the mathematical structure and the physical content of the theory;
- The link of the theory as formulated by the axioms with physical objects as space, particles, fields and cosmology;
- An interpretation which links quantum theory to other parts of physics, other sciences and to philosophy.

Weizsäcker's two books, *Aufbau der Physik* [19], which will be translated into English in 2003, and *Zeit und Wissen* [20], give comprehensive expositions of his scientific program. Parts IV and V of this volume provide new contributions.

2 Quantum Theory and Philosophy

2.1 A Philosophical Analysis of the Universal Role of Quantum Theory in Physics

Carl Friedrich von Weizsäcker saw that the fundamental questions of modern physics, particularly of quantum theory, are a modern form of the basic questions which were posed by philosophers in many different ways since the time of the Greek schools. Physics, with quantum theory in its center, is thus seen as embedded in the philosophical tradition of the search for knowledge and the conditions of knowledge. In essence, the work is reactive in the tradition of philosophy: "Do I really understand what I have just done and said?"

By its nature philosophy inquires about the totality and about the unity of being and knowledge. The unity of nature is an underlying hypothesis of physics, and man is part of nature. With quantum theory, scientific theory attains this level of universality, which incorporates the conditions for attaining scientific knowledge by experiments. The experience with quantum theory is the prime motive of the physicist's belief in the unity of physics, which, in turn, expresses a belief in the unity of nature (see K.M. Meyer-Abich's contribution in part III). How close physics has come to the goal of a unifying theory for the totality of physics is not clear; some consider the current efforts as the end game of physical theory. Steven Weinberg [15] speaks of the "dream of a final theory of physics".

Science is a reflection on experience. Philosophy is a reflection on a reflection. Kant continued Plato's reflection on the inseparability of object and conscious subject in the act of perception. In Kant's analysis the limitation in the knowledge of objects determined by acts of perception is a basic condition of human knowledge. Correspondingly, it is in quantum theory that physics is confronted with the fact that the result of a measurement may depend on the measuring process itself.

Plato and Aristotle knew already that universal laws cannot be logically deduced from an always incomplete selection of empirical findings. Hume emphasized that universal laws are supposed to hold in future experience which is not yet available when we formulate the laws. Kant held that universal laws will hold in experience only if they express no more than the necessary conditions without which experience would not be possible at all. He considered logic and mathematics, Euclidian space and Newtonian time (although Kant does not speak of space explicitly as *Euclidian* and of time as *Newtonian*), and concepts like substance and causality as preconditions of scientific experience. His system of the description of nature is unable to resist the impact of modern physics; yet, his search for definitions and postulates which may be considered as necessary conditions of empirical knowledge remains valid in the attempt to formulate the basis of quantum theory.

The all-encompassing validity of quantum theory is not to be regarded as a simple empirical fact, but a problem in itself. Therefore, quantum theory, according to C.F. von Weizsäcker, is primarily a theory of the possibility of physical statements. The high degree of generality of quantum theory gives its basic postulates a position which resembles Kant's notion of *a priori* knowledge: We cannot know *a priori* that experience is possible, we can only know what must be the case for experience to be possible. Statements by quantum theory are on *information* on systems. Information replaces *substance* and *space* of Kant's philosophy.

It is not clear that an ultimate theory of physics, if such a theory can be found at all, would be fully defined by preconditions for experience. It might contain still other elements, and it may even change all our concepts in such a manner that the very concept of preconditions to experience might have to be replaced by a better one. But it appears as a meaningful heuristic principle that we should seek for plausible preconditions of experience as building blocks for the axiomatic treatment of a comprehensive theory. Birkhoff and v. Neumann [1] had intended to explain abstract quantum theory by simpler physical and epistemological principles. Their idea of a quantum logic belongs to this line of research, as does G.W. Mackey's theory of imprimitivity [13].

With Carl Friedrich von Weizsäcker's conception of quantum theory, physics again becomes an integral part of the philosophical analysis of the world, a place which it had held in the philosophical system of Aristotle. Natural science and philosophy become united again after their separation in modern times. This integration of philosophy with physics as a "natural philosophy" arises as a necessity from quantum theory. Philosophy will benefit from the integration of quantum theory. It is not quantum theory which has to justify itself to a tribunal in which philosophy takes the role of judge. It is to the philosophers to submit their systems to quantum theory in a philosophical process in which quantum theory acts as a witness.

Quantum theory represents in the modern sciences the inquiry of philosophy into the totality of knowledge, including the conditions for acquiring knowledge. In order to formulate the philosophical implications of the theory as exactly as possible, its systematic reconstruction is needed. Carl Friedrich von Weizsäcker felt that quantum theory, as commonly understood among physicists, is still seen through the eye glasses of classical physics. As quantum theory holds the place of the basic theory, it must be developed without recourse to classical physics. Classical physics can then be deduced as a limiting case.

2.2 The Method of the Circular Walk

Carl Friedrich von Weizsäcker's methodology for dealing with questions of the totality of experience and knowledge is the *"circular walk"*. The circular walk has to be traveled several times. Each turn can lead to new knowledge, possibly also knowledge of the conditions of knowledge. The journey can come to an end when the knowledge acquired includes that knowledge cannot increase any further, but most probably the journey will never come to an end. It is, however, not essential where the journey starts, as every place will be reached. The circular walk expresses the experience from the history of philosophy that the "same" questions will be asked again. However, through increase in knowledge, they never remain truly the same questions. The method of the circular walk should be distinguished from the notion of the philosophical system, which sets out from an unquestionable basic knowledge and proceeds to the deduction of all further knowledge.

When the circular walk returns us to a place already visited before, the knowledge acquired since the last visit ought to be consistent with the knowledge at the time of the previous visit; otherwise, the circular walk would not lead to consistent knowledge. Weizsäcker speaks of the requirement for *"semantic consistency."* The notion of semantic consistency of physical theory refers both to the mathematical structure of a theory and its experimental verification. Physical theories can be graded according to their degree of semantic consistency [20, p. 138–139]:

- A theory is called *semantically consistent in the narrow sense* when no semantic inconsistencies can be found within the theory;
- A theory is called *closed* when no semantic inconsistencies are possible within a stated domain of validity. The domain of validity can become enlarged or restricted by experience. Classical mechanics and non-relativistic quantum mechanics belong to this class;
- A theory is *complete* when no semantic inconsistencies at all are possible. This can only happen in an ultimate theory of physics.

With regard to natural science the circular walk relies on the facts that nature is older than man and man is older than natural science. Man can thus be understood from the history of nature and the way man understands nature can be understood from human history. Quantum theory then closes the circle by reinstating man into nature by a scientific theory. Classical physics, as developed in European culture on the basis of Descartes' distinction between *res extensa* und *res cogitans*, cannot close this circle. Its strong separation of objects and observers presents a mathematical framework for Descartes' view of nature which has opened a deep abyss between the material and the spiritual aspects of reality.

The overwhelming successes of science in the material world left no place for spirit in science. This process could be reversed by a quantum theory based on the foundations which Weizsäcker has given, where matter and energy are united with information. On this basis it is possible that consciousness becomes a genuine part of natural science [9]. Under the view point of measurability, consciousness can be assigned to information. If there is no fundamental difference between these three concepts, then there is no need to exclude one of them from science. We will return to this idea in Sect. 6.

3 Time, Probability and Quantization

3.1 Time and Probabilities

Quantum theory incorporates the *modi* of time in its probabilistic structure, as discussed by M. Drieschner and E. Ruhnau in Part III of this volume. However, in the formal representation of quantum theory, time appears only as a one-dimensional continuous parameter, and in most dynamical laws, future and past can be inverted without altering the theory's semantic consistency. Therefore, the integration of the *modi* of time into quantum theory is generally not discussed in textbooks. Physicists are traditionally concentrated on the *objects* of knowledge, on reality, while philosophers are trained to study the structure of *phenomena*, i.e. the way reality is given to *us*. Hence there is no difficulty in pointing out to philosophers that the modes of time reign as a precondition of experience. Aristotle, Saint Augustinus, Bergson, and Heidegger have studied the phenomenon of time far more carefully than is needed for physics [21].

Quantum theory deals with probabilities of outcomes of future measurements ("future modalities"). The indeterminacy of quantum theory is thus an expression of the central aspect of time that the future is not fully determined, that past and future are not identical under a time reversal. In Weizsäcker's view quantum theory incorporates the full nature of time, while classical theory does not. Classical physics incorporates the modi of time only in a rudimentary way, for example by suppressing advanced solutions for electromagnetic radiation coupled to material sources.

Probabilities as used in quantum physics are thus intimately linked to the notion of time. The definition of probabilities which is appropriate for quantum theory therefore turns out to be: probability is the expected value of relative frequencies (see the contribution by M. Drieschner). This definition of probabilities implies that an ensemble is formed over an ensemble. Thus, it is a regressive rather than a circular definition.

Probabilities are thus a kind of presence of the future; the future is present in the form of probabilities. Weizsäcker expresses the conviction that the basic sense of the notion of probabilities refers to future modalities and that all uses of the term probability implicitly refer to future modalities. Even a sentence like, "there has probably been rain yesterday," implies the condition, "if tomorrow there would be a verification of yesterday's weather."

3.2 Probabilities and Quantization

The physical statements about an alternative are statements on complex probability amplitudes of future outcomes of measurements. Probabilities are defined as expected values of relative frequencies. Measurement therefore implies a formation of an ensemble, which again constitutes a physical system subject to the postulates of quantum theory. The recursive definition of probabilities for quantum theory indicates that quantization is a general process of formation of ensembles according to the particular rules of probability theory. Successive iteration leads to a multi-layered structure of quantum theory by multiple quantization [18].

The first step of quantization leads from a finite set of binary alternatives to a finite dimensional state space. Quantized binary alternatives are called *urs*. This process was baptized as *naive quantization*. The next step leads from the vectors in a complex two-dimensional space to functions over these vectors. After a Fourier transform these functions fulfil the usual wave equations in a four-dimensional real space and can be interpreted as an equation for a quantum particle. A further step of quantization leads to the theory of an interaction-free quantum field.

4 Postulates for the Basic Structure of Quantum Theory

4.1 The Strategy for the Reconstruction of Quantum Theory

The general nature of quantum theory suggests to find its foundation by a system of axioms. In physics the axiomatic method must aim at a double goal. On the one hand, a set of axioms ought to fulfil the conditions known from mathematical axiomatics: to be consistent, non-redundant, and sufficient to deduce the theory from the axioms. On the other hand, the axioms themselves ought to permit a physical interpretation in which they would seem as plausible as possible. The axiomatic treatment of quantum theory in Weizsäcker's reconstruction program is therefore not primarily understood as a distillation of the general mathematical apparatus which is applied in the theory. It is understood as the construction of a bridge from the underlying concepts from philosophy, mathematics and experiments to the theory in its common presentation.

Care must, however, be taken in what is understood here as quantum theory. Weizsäcker divides it into *abstract*, *concrete* and *full* quantum theory, which amounts to a division of the reconstruction program into three steps.

Step 1: Abstract quantum theory

Historically, abstract quantum theory has been found by subtracting the common ground of the two theories of the states of the hydrogen atom, which were discovered in 1926 by Schrödinger and Heisenberg. Abstract quantum theory, as codified by John v. Neumann, can be characterized by the following four theses AQTh 1–4 (AQTh stands for "Abstract Quantum Theory"), whose explanation is the aim of step 1 in the program of reconstruction. The term "theses" distinguishes these from the "postulates" given in Sect. 4.3.

AQTh 1 *Hilbert space*: The states of each object are represented by rays in a Hilbert space.

AQTh 2 *Probability metric*: The square of the absolute value of the inner product of two normed vectors φ and ψ of the Hilbert space is the conditional probability for finding the state corresponding to ψ, if the system is in the state corresponding to φ.

AQTh 3 *Rule of composition*: Two coexisting objects A and B can be understood as a composed object, whose Hilbert space is the tensor product of the spaces of A and B.

AQTh 4 *Dynamics*: Time is described by a real coordinate t. The states of an object are functions of t, given by a unitary transformation $U(t)$ of the Hilbert space into itself.

At this level, the theory does not suppose the existence of a three dimensional space, space-time symmetries, particles and fields.

Step 2: Concrete quantum theory

It is Weizsäcker's aim to deduce, as far as possible, *concrete* quantum theory from the abstract theory, possibly under some additional hypotheses. This project, known as the *ur-hypothesis* (see Sect. 5 of this introduction and Part V of this volume), sets out to deduce space, space-time symmetries, elementary particles and their interaction, as well as cosmology from the structures of time and abstract quantum theory together with a minimal number of additional hypotheses. Only when this program has been essentially completed can step 3 be tackled:

Step 3: Full quantum theory

This step should unite abstract and concrete quantum theory in one set of postulates.

Presently, the axiomatic construction of abstract quantum theory as given by Weizsäcker and his collaborators is not in a fully satisfactory form, as it still contains a number of conditions which cannot be understood as conditions of the possibility of experience; or, perhaps equivalently, the semantic consistency of the theory is not yet visible.

Four different paths were taken or appear possible (see Chapt. 8 of *Aufbau der Physik*):

1. via probabilities and the lattice of propositions;
2. via probabilities applied directly to the vector space of states;
3. via probability amplitudes applied to the vector space of states;
4. via binary alternatives (*ur alternatives*) applied to the vector space of states.

The first path refers to the traditional axiomatic treatment of quantum theory by Birkhoff and v. Neumann [1], followed by Jauch [11] and others, also by Drieschner [3, 4]. We shall present in more detail the last of these paths following the exposition given by Drieschner, Görnitz and Weizsäcker [5] and shall proceed in three steps (Sects. 4.2–4.4):

- Verbal *definitions* of some terms.
- *Postulates for abstract quantum theory* which express the essence of quantum theory in the language prepared by the definitions.
- *Conclusions from the postulates* which aim at deducing the structure described by the four "theses" AQTh 1–4. We shall see that the "deduction" is not yet logically strict and needs a few additional assumptions of simplicity.

4.2 Definitions

Experience: Experience means to learn from the past for the future.

Facticity of the past, possibility of the future: Past events are objective facts, independent of an actual knowledge of them. Future events exist only as possibilities.

Temporal statements: A temporal statement (briefly "statement") is a verbal proposition (or a mathematical proposition with a physical meaning) referring to a moment in time.

States: States are recognizable events. A state is what is the case when some temporal statement is true. States at different times can be identical; it is meaningful to ask whether we observe now the same state as at a certain time before.

"State" is a central notion in traditional quantum theory. The definition uses the undefined term "event" which explicitly refers to time and the fact that events can be subsumed under descriptive concepts, logically speaking under universals, e.g. "The event of the rising sun."

Conditional probability: Let x and y to be two states. Then $p(x,y)$ is the probability that, if x is a present state, y will be found as the future state if searched for.

Alternatives: An n-fold alternative is a set of n mutually exclusive states, exactly one of which will turn out to be present if and when an empirical test of this alternative is made.

The two last definitions contain the dependence of quantum-theoretical predictions on observation in the phrases "if searched for" and "test of this alternative." In classical physics these two specifications would be unnecessary. The specifications are more modest than their classical omission. The given formulation would still permit the classical assumption that the specifications are superfluous.

Connection: Two states x and y are called connected if there is a law of nature determining their conditional probabilities $p(x, y)$ and $p(y, x)$.

Separability: Two states are called separable if they are not connected.

If the connection is transitive, i.e. if the existence (by law of nature) of probabilities $p(x, y)$ and $p(y, z)$ implies the existence of a $p(x, z)$, then connection is an equivalence relation, defining a partition of the class of all states into subclasses of mutually connected states. The logical position of the assumption that connection is transitive seems to be as follows: it is only relevant if there are separable states. It then implies the existence of classes of mutual connection as used in postulates PQTh 1 and PQTh 2. Vice versa, these three assumptions imply the transitivity of connection. It is presupposed that $p(x, y) = 0$ implies $p(y, x) = 0$.

4.3 Postulates for Abstract Quantum Theory

Three postulates for abstract quantum theory are introduced:

PQTh 1 *Separable alternatives*: There are alternatives whose states are separable from nearly all other states. "Nearly" will be defined as meaning all states not connected with the states of the alternative by postulate 2.

PQTh 2 *Indeterminism*: If x and y are two connected, mutually exclusive states $\{p(x, y) = p(y, x) = 0\}$, there are states z which are not logically constructed from x and y by mere logical operations and which possess conditional probabilities $p(z, x)$ and $p(z, y)$ none of which is equal to zero or to one.

PQTh 3 *Kinematics*: The conditional probabilities between connected states are not altered when the states change in time: $p[(x, t), (z, t)] = p[(x, 0), (z, 0)]$.

The postulates can be divided into two kinds. Postulates of the first kind express conditions of possible experience, that means conditions of human knowledge. They are called "epistemic." Postulates of the other kind formulate principles, which are hypothetically considered of general validity in the domain of the relevant reality. These postulates are called *realistic*. The method of the circular walk implies that a definite distinction between epistemological and realistic postulates is not possible. Human knowledge is seen as an event in reality and "reality" becomes a human concept. Yet, the phenomenon of scientific history, that there are closed theories, permits a distinction between epistemological and realistic postulates, which is relative to a given theory.

The "alternative" of PQTh 1 represents the elementary notion of the quantum theoretical concept of *observable*. The postulate is epistemological in a double sense: (1) it formulates a necessary condition for the applicability of the notions of quantum theory, (2) it is hardly imaginable how scientific experience should be possible without distinguishable, empirically decidable alternatives.

A testable law ought, at least, to predict probabilities, and a testable probability is a conditional probability: "if x, then in the average y in the fraction $p(x, y)$ of performed tests." Yet quantum theory itself implies that postulate PQTh 1 is not strictly true. In the composition of alternatives (and hence of objects) all states which are not products of states of the components make the probabilities of one state space dependent on the choice of the alternative decided in the other one. If we refer the Einstein–Podolsky–Rosen argument to *non-locality*, then this not only applies to location in position space but to any decidable alternative. We can call this statement the *essential holism* of quantum theory. Thus postulate PQTh 1 defines only the limited human approach towards the analysis of the wholeness of reality.

Quantum theory is not actually introduced before postulate PQTh 2. – The analysis is commonly restricted to finite alternatives. Countable infinite alternatives are however used in "concrete quantum theory" for the definition of a particle. This means that an infinite number of finite alternatives are connected as "possible properties of one object." An example is the decomposition of the Hilbert space of a free particle into the sum of the finite spaces of the possible values of its angular momentum. A similar procedure is repeated in quantum field theory, combining an infinite set of Hilbert spaces of possible single particles.

In classical physics one would say that connected states are states of the same object. We begin with the logical concept of an alternative and aim at reconstructing the concept of an object. This will not be fully achieved by abstract quantum theory. The concept of approximate separability and, hence, of separable objects presupposes a concept of position space for its semantic consistency; thus, it presupposes "concrete quantum theory".

In fully developed quantum theory the postulate PQTh 2 turns out to be equivalent to the principle of superposition. In *Aufbau der Physik* [19] it was called the postulate of extension. It is a phenomenological description of the open future.

In the formulation of the postulate states (or statements) were explicitly rejected which can be constructed from the given x and y by logical operations, i.e. by "and" and "or". Negation is in fact used in the definition of mutual exclusivity, but it is not either employed for the definition of states like z. This is only a methodological decision.

4.4 Conclusions from the Postulates

1. *State space*: The set of states connected with a separable alternative is called its state space. With innocuous generality we assume: The state spaces of all separable n-fold alternatives A_n is assumed to be isomorphic: $S(n)$. A state $z \in S(n)$ defines n conditional probabilities $p(z, x_i)$ where x_i ($i = 1, \ldots, n$) are the states defining the n-fold alternative; $\Sigma_{i=1}^{n} p(z, x_i) = 1$.
2. *Completeness*: For any mathematically possible set of values $p(z, x_i)$ there is a state z in $S(n)$.
3. *Equivalence of states*: All states in $S(n)$ are equivalent. Else their distinction would be an additional alternative connected with A_n. A_n would hence not have been separable.
4. *Symmetry group*: The equivalence of the elements of $S(n)$ is expressed by a symmetry group $G(n)$ which preserves the conditional probabilities between them. Due to conclusion 3, $G(n)$ must be a continuous group.
5. *Alternatives in $S(n)$*: Due to conclusion 3, there exists a $p(x, y)$ between any two states x and y of $S(n)$. The equivalence of all states in $S(n)$ further implies that any $z \in S(n)$ is a member of a precisely n-fold alternative of mutually exclusive states of $S(n)$.
6. *Metric in $S(n)$*: As an assumption of simplicity we suppose $G(n)$ to be a simple Lie group. There are two simple Lie groups preserving a relation of mutual exclusion between precisely n normalized vectors by preserving a metric: $O(n)$ and $U(n)$. Thus we assume $S(n)$ to permit a faithful irreducible representation in an n-dimensional vector space $V(n)$, $G(n)$ being either orthogonal or unitary.

The states of $S(n)$ will then correspond to normalized vectors in $V(n)$, i.e. to one-dimensional subspaces.

Here only linear representations are considered. The consequences of non-linear representations are not studied, which might lead beyond traditional quantum theory. Furthermore, $O(n)$ preserves a linear metric and $U(n)$ a sesquilinear one. Linear metrics can be constructed on vector spaces over the real and complex numbers and over the quaternions; sesquilinear metrics exist in spaces over the complex numbers and the quaternions. As the quaternions can be represented by 2×2 matrices of complex numbers and because there are no groups for the quaternions which are different from the groups over real or complex vector spaces, we suppose that there is no need to construct a quantum mechanics over quaternions.

7. *Dynamics*: According to postulate 3, the change of state in time must be a one-parameter subgroup $D(t)$ of $G(n)$. A particular choice of such a subgroup is called a law of dynamics.
8. *Preservation of state*: If a state is to be recognizable in time, there must exist a possible law of dynamics which keeps this state constant.
9. *Complexity*: The generator of $D(t)$, as defined in conclusion 7, must, according to conclusion 8, permit diagonalization. This is universally possible only if V is complex, and, due to the metric, a Hilbert space. Hence $G(n) = U(n)$.
10. *Composition*: Two alternatives A_m and A_n are simultaneously decided by deciding their Cartesian product $A_{m \cdot n} = A_m \times A_n$. $A_{m \cdot n}$ defines the Hilbert space $V(m \cdot n) = V(m) \otimes V(n)$.

We make the following remarks on these conclusions:

Conclusions 1–4: Symmetry of the state space

The traditional problem of quantum axiomatics is how to arrive at linearity, i.e. at a vector space. On path 1, a lattice of events is introduced, which leads to a projective geometry and hence to the embedding vector space. The approach on path 4 confines itself to pure states and introduces the vector space as a representation space of their symmetry group. The symmetry is not just postulated but implied by separability. Since separability is only an approximation, it cannot be maintained that abstract quantum theory, founded on it, should be a final theory; it could turn out to be an approximation to an even more holistic theory.

Conclusion 3: Equivalence of states

This conclusion can be reached in two steps. First the states x_i which define the alternative are mutually equivalent since their distinction would be a connected alternative. Second, all states of $S(n)$ are equivalent, since the definition of the x_i as members of an alternative refers to observation from outside, hence to interaction with other alternatives. As long as $S(n)$ is strictly separated, it will not be determined whether a given state $z \in S(n)$ is an element of an alternative which would be decided by introducing outside interaction. From this second step it follows that there exists a $p(x, y)$ between any two states x and y of $S(n)$.

Conclusions 5–6: Metric

The possibility of an essentially different, more complicated mathematical structure must be admitted that fulfils the postulates. The empirical success of quantum mechanics favors the choice presented in conlusion 6.

Conclusions 7–9: Dynamics and complex numbers

It is a traditional problem why quantum theory, if represented in a vector space, should use a complex space. Pauli [14] in his answer to Ehrenfest [7] starts out from a second order differential equation in time (reversibility) and from the postulate that there should be a probability density for position. Reversibility is introduced by postulate 3 and conclusion 6, i.e. by postulating dynamics to be described by a group. Pauli's probability postulate is replaced by the definition of conditional probabilities and by postulating that it should be mathematically possible to define a law of dynamics that would preserve the state.

Conclusion 8: State preservation

The assumption of state preservation might be qualified as another "assumption of simplicity." Why should it be impossible to measure time by the steadily moving hands of a clock? What only is assumed is that there should be a "possible" law of dynamics making the hands stand still. And it is to be admitted that in a real vector space of even dimension a one-dimensional orthogonal group will not keep any state constant at all. Thus the choice between $O(n, R)$ and $U(n)$ seems to be in favor of the latter.

Conclusion 10: Composition

The generalized EPR-non-locality (see the comment on postulate 1) turns out to be a consequence of postulate 2.

5 The Ur-Hypothesis

The subject of quantum theory is information in time on physical objects. Information is composed of quantized binary alternatives, the *urs* or *Uralternativen* (*ultimate alternatives*). Today quantum bits are no longer seen as a strange construct, but their ontological character is generally not yet accepted.

From classical physics we learn that particles, fields, but also space and the cosmos are physical objects. Quantum theory must be interpreted in the language of these objects. The ur-hypothesis sets out to construct all physical objects and structures from binary quantized alternatives. Its first aim is to understand the three-dimensional structure of space. If any object can be decomposed into urs, then the symmetry group of urs is essential also for these objects. This group is $SU(2)$ which is locally isomorphic to $O(3)$ the rotation group in $\mathbb{R}^3$. Therefore any object has to be invariant under $O(3)$ and can be described in a three-dimensional space. This reasoning provides a further manifestation of Weizsäcker's principle that mathematical objects of the theory should be become identified with physical objects as far as possible. In Part V of this volume, H. Saller takes a critical position towards Weizsäcker's deduction of space from the symmetry of urs.

Fock space quantization of fourfold urs, the doubled quantum bits, produces all relativistic particles (see Castell's papers in QTSTS 1 and 2). Representations of the Poincaré group for massive and massless particles can be constructed and a link to the cosmological evolution has been established (see part 5).

6 On the Interpretation of Quantum Theory

The discovery of quantum theory reminds us of Columbus who discovered the Americas by following old maps which did not contain these lands. Columbus first reached islands and later he found the continent. He could only interpret his discovery in terms of the old maps and thought he had arrived close to Japan. Only gradually the newly discovered world became integrated into the general picture of the world, and became the New World. Similarly, the interpretation of quantum theory has to integrate its discoveries into the picture of the world of physics and of science in general. This task will never come to an end.

6.1 The Relation of Quantum Theory to Classical Physics

For historical reasons the debate on the interpretation of quantum theory has centered more on the *concrete* than on the *abstract* parts of the theory. The debate has mourned over the sacrifice which had to be made when abandoning classical theory. Many physicists consider the novel aspects of quantum theory as a loss, expressed, for example, by Heisenberg's uncertainty relations and by the fact that results of measurement are only given by probabilities. The dispute between Bohr and Einstein revolved around the question whether this loss was worth the gain in correspondence between theory and experiment. But the contrary is true: classical physics presents a limiting case of quantum physics and, thus, carries less information than quantum physics. Quantum physics presents an increase in knowledge.

The ontology of classical physics deals with four realities: time, space, bodies and forces. Bodies and forces were later formalized as particles and fields. Abstract quantum theory, in its reconstructed form, retains only one of these four realities: time. It abstracts from space, bodies and forces. But quantum theory must prove to be semantically consistent with the classical realities. Concepts of classical physics must be used for being able to say what quantum theory speaks about. These concepts attain a precise sense only in a theoretical context which quantization alters radically.

6.2 The Role of the Continuum

In the program of reconstruction, the mathematical continuum is used as measure of time and as the set of possible values for probability. Quantum theory was historically made necessary by the "ultraviolet catastrophe", i.e. by the impossibility of thermodynamic equilibrium in the infinity of degrees of freedom of a classical mechanical continuum. In the mathematics of quantum physics, possibilities are represented by continua, facts are discrete. Thus quantum theory seems to replace the idea of an actually existing physical continuum by the continuum of probabilities, i.e. of possibilities. This recalls the traditional view from Aristotle to the times of Gauss, that infinity means no more than indefinite possibility (of counting, of dividing, etc.). Cantor defended his idea that the natural numbers form an actually infinite set (from which he proceeded to higher cardinalities) against the objections of the Aristotelians by the remark: "If you can actually count on indefinitely, the set of *possible* numbers is actually infinite". Cantor also thought of infinity (and, consequently, continuity) as an infinity of possibilities. We might, therefore, think of quantum theory precisely as the adequate theory of physical continuity.

6.3 The Measurement of Physical Events and the Copenhagen Interpretation

We consider the Copenhagen Interpretation not as one of several possible interpretations of a self consistent theory called "quantum theory", but as the attempt at giving minimal semantics to the formalism of quantum theory without which one would not know how to apply the formalism to reality at all, i.e. without which it would not yet be a theory in the sense of physics. Precisely for this reason it has never been possible to codify the Copenhagen Interpretation, since semantics shares the ambiguities of everyday language. Semantics develops with the application of the theory.

We compare Weizsäcker's approach with the early Copenhagen Interpretation in three respects: the correspondence with classical physics, the statistical interpretation of the wave function and the role of the observer.

Heisenberg's quantum mechanics was a mathematical model of the theory at which Bohr's correspondence principle had aimed. Observable quantities such as position, momentum and energy, were known from classical physics but now obeyed different mathematical laws which, for large quantum numbers, implied classical mechanics as a limiting case. Bohr insisted later that observations in quantum theory ought to be described in classical terms. The approach of the reconstruction program is not one of correspondence. It begins with abstract quantum theory without any empirical or classical specification of observables. This specification is reserved for the ensuing step of concrete quantum theory. However, it still keeps the essence of Bohr's insistence on classical terms. The corresponding, though far more restrained statement is postulate PQTh 1.

As with the Copenhagen Interpretation, the statistical nature of quantum theory is accepted as given. The amplification lies in the analysis of time as a precondition of experience. Probability is the only available scientific description of the open future which we know today.

The criticisms of the Copenhagen Interpretation were mainly connected with the role of the observer. The "collapse of the wave function by observation" is the crucial expression: quantum theory seems to describe knowledge instead of reality. In Weizsäcker's view, most of the dissatisfaction derives from an unwieldy phraseology. The wave function is the catalogue of those probabilities which are mathematically implied by the knowledge gained in an experiment. Probabilities in physics are conditional and change by the awareness of new conditions. "Knowledge" means to know a reality. The difficulty arises from a neglect of the difference between past and future. Past facts can be known; possibilities for the future are guesswork, guided by probabilities which can be empirically tested as expected values of relative frequencies in an ensemble. In this regard, it is not the primary concern wether the wave function is a real entity or only an accumulation of bits. A "collapse of the wave function" does not occur in this interpretation. If the models, which attribute such "collapse" to decoherence by interaction of a system with its environment, can ultimately corroborate Schrödinger's opposition against "quantum jumps", an armistice with the Copenhagen Interpretation should be possible.

6.4 The Separability of Alternatives

The theory has been built on the postulate that separable alternatives are possible. By its structure, quantum theory is a theory of the inseparable totality of the physical world. The important question then becomes: How can separable alternatives arise in physics? Why is postulate PQTh 1 so good an approximation in most known cases? It is a question relating to the semantic consistency of the complete quantum theory. In the frame of our physical and cosmological knowledge, the answer is: the cosmos is almost empty. It is therefore possible to observe particles as almost free objects. The concepts of disentanglement and decoherence can enrich this suggestion and help in finding the way towards the full quantum theory which has been envisaged in Sect. 4.

6.5 The Individuality of Processes

One of the astonishing facts in quantum theory is the non-locality of states. Even localized interaction on a part of it has instantaneous consequences for the whole state. A system can manifest such effects only if it is isolated from its environment. Bohr speaks in this case of an individual process. Individuality means also the inseparability of processes described by quantum theory. Bohr speaks of the limited possibility to separate physical processes, which is characterized by the quantum of action.

The individuality of processes marks the limits of classical-quantum correspondence and of the complementary usage of classical notions. With quantum theory, as reconstructed by Carl Friedrich von Weizsäcker, the conception that reality is a non-local individual process is totally compatible. Its last and true ingredients are the *urs*, quantum bits. In this connection, matter appears to dissolve into information. Weizsäcker's ideas keep the door open to a wider philosophical concept of matter. The book by Görnitz and Görnitz [9] pursues these lines of considerations.

References

1. G.A. Birkhoff and J. v. Neumann: The Logics of Quantum Mechanics. *Annals of Math.* 37, 823–843 (1936)
2. L. Castell, M. Drieschner and C.F. von Weizsäcker: *Quantum Theory and the Structures of Time and Space*, Vols. 1–6, Hanser, Munich, 1975, 1977, 1979, 1981, 1983, 1986
3. M. Drieschner: *Quantum Mechanics as a General Theory of Objective Prediction*. Thesis, University of Hamburg, 1970
4. M. Drieschner: *Voraussage-Wahrscheinlichkeit-Objekt. Über die begrifflichen Grundlagen der Quantenmechanik*. Lecture Notes in Physics 99. Springer, Berlin, 1979
5. M. Drieschner, Th. Görnitz and C.F. von Weizsäcker: Reconstruction of abstract quantum theory. *Intern. Journal of Theor. Physics* 27, 289–306 (1988)
6. M. Drieschner: Is (Quantum) Logic Empirical? *Journal of Philosophical Logic* 6, 415–423 (1977)
7. P. Ehrenfest: *Zeitschrift f. Physik* 78, 555 (1932)
8. R.P. Feynman: *The Character of Physical Law*. MIT Press, Cambridge, MA 1965, Chapt. 6: Probability and Uncertainty
9. Th. Görnitz and B. Görnitz: *Der kreative Kosmos*. Spektrum, Heidelberg, 2002
10. W. Heisenberg: Über die in der Theorie der Elementarteilchen auftretende universelle Länge. *Ann. Physique* 32, 20 (1938)
11. J.M. Jauch: *Foundations of Quantum Mechanics*. Addison-Wesley, Reading MA, 1968
12. H. Lyre: *Quantentheorie der Information*. Thesis, Ruhr-Universität Bochum, 1996; also published as: Quantentheorie der Information. Springer, Wien, 1998
13. G.W. Mackey: *Mathematical Foundation of Quantum Mechanics*. Benjamin, New York, 1963
14. W. Pauli: Einige die Quantenmechanik betreffende Erkundigungsfragen. *Zeitschrift für Physik* 80, 573 (1932)
15. S. Weinberg: *Dreams of a Final Theory*. Pantheon Books, 1993
16. C.F. von Weizsäcker: Komplementarität und Logik. *Naturwissenschaften* 42, 521–529, 545–555 (1955)
17. C.F. von Weizsäcker: Die Quantentheorie der einfachen Alternative (Komplementarität und Logik II). *Zeitschrift für Naturforschung* 13a, 245–253 (1958)
18. C.F. von Weizsäcker, E. Scheibe and G. Suessmann: Mehrfache Quantelung (Komplementarität und Logik III). *Zeitschrift für Naturforschung* 13a, 705–721 (1958)
19. C.F. von Weizsäcker: *Aufbau der Physik*. Hanser, Munich, 1985
20. C.F. von Weizsäcker: *Zeit und Wissen*. Hanser, Munich, 1992
21. G.J. Withrow: *The Natural Philisophy of Time*. Clarendon Press, Oxford, 2nd ed. 1980

Interpreting Quantum Mechanics –
in the Light of Quantum Logic

Peter Mittelstaedt

Summary. In contrast to the Copenhagen interpretation we consider quantum mechanics as universally valid and query whether classical physics is really intuitive and plausible. – We discuss these problems within the quantum logic approach to quantum mechanics where the classical ontology is relaxed by reducing metaphysical hypotheses. On the basis of this weak ontology a formal logic of quantum physics can be established which is given by an orthomodular lattice. By means of the Solèr condition and Piron's result one obtains the classical Hilbert spaces. – However, this approach is not fully convincing. There is no plausible justification of Solèr's law and the quantum ontology is partly too weak and partly too strong. We propose to replace this ontology by an ontology of unsharp properties and conclude that quantum mechanics is more intuitive than classical mechanics and that classical mechanics is not the macroscopic limit of quantum mechanics.

1 Classical and Quantum Physics – Their Respective Roles

Even today, 75 years after the discovery of quantum mechanics many quantum physicists are convinced that the Copenhagen interpretation of Niels Bohr is still the right way for understanding quantum physics. According to this interpretation we have to distinguish two distinct worlds, the quantum world of microscopic entities and the classical world of our everyday experience which is subject to classical physics. In the quantum world we are confronted with many strange features, complementarity, nonindividuality, nonlocality, and the loss of determinism. However, the apparatuses which measure and register the properties of the quantum system as well as the human observer, who reads the observed data are parts of the classical world that is free from the quantum physical absurdities mentioned. Hence the experimental physicist is always on the save side of classical physics. For describing and interpreting quantum physics we can use common language and classical logic.

We will query this doctrine here for several reasons. Firstly, during the last decades it became obvious that quantum mechanics is not restricted to the microscopic world of nuclei, atoms, and molecules but can be applied also to macroscopic systems. The discovery of macroscopic quantum effects like superconductivity, superfluidity, macroscopic tunnelling, etc. are strong indications that quantum physics holds also in the macroscopic world. Moreover, the successful attempts to a quantum cosmology show that quantum mechanics can even be applied on the cosmological level, to the problem of the creation of the universe and to the universe as a quantum mechanical object. Hence it seems, that there are no serious doubts today that quantum mechanics is universally valid and can be applied to all objects from elementary particles to the entire universe.

The second reason is presumably even more important. In the Copenhagen interpretation quantum mechanics is considered to be less intuitive than classical mechanics and sometimes paradoxical, whereas classical mechanics is assumed to correspond to plausible reasoning and to intuitive results. This is, however, not entirely correct. What we call "intuitive" and in accordance with "plausible reasoning" corresponds to our everyday experience, to our pre-scientific experience in the macroscopic world. However, classical physics and in particular classical mechanics is not exactly the theory of this pre-scientific experience. Classical mechanics is loaded with many hypotheses which can be traced back to the metaphysics of the 17th and 18th century, i.e. to the period from *Bacon, Descartes, Leibniz,* to *Kant*. These metaphysical hypotheses are without any empirical counterpart, they exceed clearly our everyday experience. As examples we mention here the existence of an absolute time, the complete determination of objects, the strict causality law, and the law of conservation of substance. It is obvious that the consequences of these metaphysical hypotheses are not *per se* intuitive in the above mentioned sense.

In quantum mechanics we are confronted with a quite different situation. Quantum mechanics may be understood as a theory of the physical reality which is free from some of the metaphysical hypotheses mentioned, i.e. quantum mechanics dispenses with some metaphysical exaggerations of the classical theory. It is important to note that quantum mechanics can be obtained from classical mechanics merely by reducing the ontological premises without incorporating new empirical components. This will be demonstrated in detail within the framework of the quantum logic approach to quantum mechanics. Consequently, in quantum mechanics just those parts of classical mechanics are missing which are not intuitive and which do not correspond to plausible reasoning. This means that quantum mechanics is more intuitive than classical mechanics – a result which is paradoxical at first glance. It is obvious that this result together with the universal validity of quantum mechanics strongly invalidates the dualistic approach of the Copenhagen interpretation.

2 Reduction of Ontological Hypotheses

The ontology of a certain domain of physics contains the most general features of the external reality which is treated in the physical domain in question. In particular the ontology should contain the material preconditions for a pragmatics which allows for the constitution of a scientific language and thus for the formulation of physical experience. The ontology which is underlying classical mechanics will be called classical ontology and denoted by $O(C)$. We will briefly characterise this classical ontology.

According to $O(C)$ there are individual and distinguishable objects S_i and these objects possess elementary properties P_λ in the following sense. An elementary property P_λ refers to a classical object system such that either P_λ or the counter property $\bar{P}_\lambda$ pertains to the system. An elementary property P_λ can always be tested by measurement with the result that either P_λ or the counter property $\bar{P}_\lambda$ pertains to the object. Furthermore, objects are subject to the law of "complete determination" according to which "if all possible predicates are taken together with their contradictory opposites then one of each pair of contradictory opposites must belong to it" [7]. Hence an object S possesses each elementary property P either positive (P) or negative ($\bar{P}$). It follows from these strong requirements that objects can be individualised by elementary properties if impenetrability is assumed as an additional condition.

For objects of the external objective reality the causality law and the law of conservation of substance hold without any restriction. Since there exist an absolute and universal time which refers to all objects of the external reality, the temporal development of these objects and their time dependent properties are strictly determined by a causal law of nature which fulfils also the conservation of substance. The additional assumption, which was formulated by Newton, that there exists also an absolute space, is irrelevant for our problem since the absolute space is not used explicitly in the theories of classical physics. Hence, the absolute space hypothesis will not be incorporated into the classical ontology $O(C)$.

There are important objections against this classical ontology. Since the metaphysical and theological reasons of Newton are no longer relevant for a justification of the ontology we have to search for alternative reasons. Are the ontological assumptions intuitive and plausible in the sense mentioned above? This is obviously not the case. The strict postulates of the classical ontology are almost in accordance with our everyday experience, but the rigorousness of the assumptions mentioned exceeds obviously the more qualitative and less rigorous prescientific everyday experience. The strict causality law, the unrestricted conservation of substance and the existence of one universal time are beyond our daily experience. These and other hypotheses of the classical ontology must not be considered as intuitive and plausible.

The second argument refers to the experimental evidence of the mentioned hypothesis. There is no experimental indication that objects can always be individualised and reidentified at later times, simply since experiments which would confirm this assumption have never been performed in classical physics. In addition, the principle of complete determination mentioned above has never been tested with an accuracy which would allow to call the result a principle. Consequently, there is no justification for a strict causality law such that the present state of an object allows for predictions about all elementary properties. Hence we find that there is no empirical justification for the classical ontology $O(C)$. Instead, the classical ontology is based on hypotheses whose origin can be traced back to the metaphysics of the 16th and 17th century.

The classical ontology is neither intuitive and plausible nor is it justified by experimental evidence. Moreover, – what is more important – the classical ontology $O(C)$ is not in accordance with quantum physics. A quantum mechanical object system does not possess all possible elementary properties P_λ either positive (P_λ) or negative ($\bar{P}_\lambda$). It is not carrier of all possible properties. Instead, only a subset of all properties pertains to the system and can simultaneously be determined. These properties are often called "objective" properties and they pertain to the object like in classical ontology. From these restrictions it follows that in quantum mechanics no strict causality law can be established and that object systems cannot be individualised and reidentified by means of their objective properties.

We will not use here these empirical results for a reconstruction of an ontology for quantum phenomena. However, we learn from these considerations, that the classical ontology is not only based on classical metaphysics and partially hypothetical, but that classical ontology contains too much structure and too strong requirements compared with quantum physics. This observation offers the interesting possibility to formulate the ontology $O(Q)$ of quantum physics by relaxing and weakening some hypothetical requirements of the classical ontology $O(C)$. It is important to note that no new requirements must be added to the assumptions of the classical ontology. Quantum ontology can thus be formulated as a reduced version of the classical ontology $O(C)$:

$O(Q)$-1: If an elementary property P pertains to a system as an objective property, then a test of this property by measurement will lead with certainty to the result P. In addition, any arbitrary elementary property P can be tested at a given object with the result that either P or the counter property $\bar{P}$ pertains to the object system. (These requirements are in complete accordance with $O(C)$).

$O(Q)$-2: Quantum objects are not completely determined. They possess only a few elementary properties either positive or negative. Properties which pertain simultaneously to an object are called "objective" and "mutually commensurable".

$O(Q)$-3: For quantum objects there is no strict causality law, simply since the present state of an object system is never completely determined.

$O(Q)$-4: The lack of complete determination and of strict causality implies that quantum objects cannot be individualised and reidentified at later times.

The mutual relations between classical and quantum ontology are the key for the very intertheoretical relations between classical and quantum physics. This will become obvious in the quantum logic approach to quantum mechanics when quantum ontology is used as starting point for establishing a formal language and logic of quantum physics. This way of reasoning will be made explicit in the following section.

3 The Quantum Logic Approach – An a priori Justification?

3.1 Language, Semantics, and Pragmatics

On the basis of the quantum ontology $O(Q)$ described above we will establish a formal language of quantum physics. Let S be a proper quantum system and $A, B, \ldots$ elementary propositions which attribute predicates (properties) $P(A), P(B), \ldots$ to system S at times $t_1, t_2, \ldots$ Hence, we write for elementary propositions $A(S, t_1), B(S, t_2), \ldots$ According to $O(Q)$-1 we assume that for every elementary proposition A there exists a finite testing procedure which shows whether

$P(A)$ pertains to S or not. If $P(A)$ pertains to S at time t_1, then the proposition $A(S, t_1)$ is called to be "true", otherwise $A(S, t_1)$ is said to be "false". The assumption, that for every elementary proposition there is a testing procedure which decides between "true" and "false" means, that these propositions are "value definite". Hence, an elementary proposition can either be proved (with result A) or disproved (with result $\bar{A}$), where $\bar{A}$ is the counter proposition of A. Furthermore, we assume that after a successful proof of A new proof attempt leads with certainty to the same result, provided the time interval between the two proof attempts is sufficiently small. This requirement is again in accordance with $O(Q)$-1. since after the first test the property $P(A)$ pertains objectively to the system and can thus be tested with the certain result $P(A)$. This assumption means that there are repeatable measurement processes, which can be applied to the testing procedures. However, – and this is an important restriction of $O(C)$ – if after a successful proof of A, say, another proposition B is proved, then a new proof attempt of A will in general not lead to the previous result. Hence, we will not assume that two propositions A and B are in general simultaneously decidable. If accidentally two propositions A and B are always jointly decidable, we will call A and B "commensurable". In this case, after the proof attempt of B the result of the previous A-test is still available. However, in the general case the result of a previous test is only restrictedly available. The commensurability of two propositions A and B can be expressed by a "commensurability proposition" $k(A, B)$ which can be tested by a conveniant sequence of A-tests and B-tests. Correspondingly, the proposition $k(A, B)$ will be called "true" if A and B are commensurable and "false", if A and B are not commensurable.

On the basis of the set $\mathcal{S}_Q^e$ of elementary propositions and commensurability propositions we introduce the logical connectives by the possibilities to attack or to defend them, i.e. by the possibilities to prove or to disprove the connective. Here, we consider the sequential conjunction $A \sqcap B$ (A and then B) which refers to two subsequent instants of time t_1 and t_2 with $t_1 < t_2$ and the logical connectives $\neg A$ (not A), $A \wedge B$ (A and B), $A \vee B$ (A or B), and $A \to B$ (if A then B) – which refer to one simultaneous instant of time. The definitions of the sequential and logical connectives by attack-and defence schemes can be illustrated most conveniently by chronologically ordered proof trees. Correspondingly, in the proof tree of the sequential conjunction $A \sqcap B$, the first branching point corresponds to a A-test at t_1, the second one to a B-test at t_2. Note, that for the truth of $A \sqcap B$ the commensurability of A and B does not matter. However, for the proof trees of the logical connectives, which refer to one simultaneous instant of time, the commensurabilities of the elementary propositions play an important role. The concepts of truth and falsity of a compound proposition which is composed by the connectives can then be defined by success and failure in a proof tree, respectively. – For the details of this well established operational approach we refer to the literature [8, 9, 16].

Furthermore, we will define here binary relations between propositions. First, the proof equivalence $A \equiv B$ means that A can be replaced in any proof tree of a compound proposition by B without thereby changing the result of the proof tree. Second, the value equivalence $A = B$ means that A is true (in the sense of a proof tree) if and only if B is true. Third, the relation of implication $A \leq B$ can be defined by $A \equiv A \wedge B$. Hence, the two implications $A \leq B$ and $B \leq A$ imply the proof equivalence $A \equiv B$. Finally, we mention that $A \to B$ is true if and only if $A \leq B$ holds.

The full quantum language $\mathscr{S}_Q$ can then inductively be defined by the set $\mathscr{S}_Q^e$ of elementary propositions and the connectives mentioned. Together with the always true elementary proposition V, the always false elementary proposition Λ, and the three relations the language $\mathscr{S}_Q$ reads

$$\mathscr{S}_Q = \{\mathscr{S}_Q^e; \sqcap, \wedge, \vee, \to, \neg; V, \Lambda; \equiv, =, \leq\}.$$

The connectives are defined by attack-and defence schemes which can be illustrated by proof trees. In addition, it is important to note that for the sequential and the logical connectives there are value equivalent elementary propositions, the measurement devices of which can explicitly be constructed [15]. Hence, by induction with respect to the connectives one arrives at the result that for any finitely connected proposition A there is an elementary proposition A^e which is value equivalent to A, i.e. which satisfies $A = A^e$. For the algebraic structure of the language $\mathscr{S}_Q$ these value equivalent elementary propositions play an important role.

3.2 Quantum Logic

The semantics described here is a combination of a realistic semantics (for elementary propositions) and a proof semantics (for connectives). Hence, the truth of a compound proposition depends on the connectives contained in it as well as on the elementary propositions and their truth values. However, there are finitely connected propositions which are true in the sense of the semantics mentioned, irrespective of the truth values of the elementary propositions contained in it. These propositions are called *formally true*. – The precondition that measurements are repeatable implies that $A \to A$, the *law of identity*, is formally true. The value definiteness of elementary propositions implies that also finitely connected propositions are value definite and thus $A \vee \neg A$, the *tertium non datur law*, is formally true. In a similar way, it follows that $\neg(A \wedge \neg A)$, the *law of contradiction*, and $(A \wedge (A \to B)) \to B$, the modus ponens law, are formally true. – Formally true propositions can also be expressed by "formally true implications". E.g. the modus ponens law reads $A \wedge (A \to B) \leq B$. In addition, if we make use of the special propositions V (verum) and Λ(falsum), then the relations $A \leq V$ and $\Lambda \leq A$ hold for all propositions $A \in \mathscr{S}_Q$. The formal truth of a proposition A can then be expressed by $V \leq A$. E.g. the tertium non datur law reads $V \leq A \vee \neg A$ and the law of contradiction $A \wedge \neg A \leq \Lambda$.

There are two kinds of propositions $A \in \mathscr{S}_Q$. If a compound proposition contains in addition to elementary and commensurability propositions only the logical connectives $\wedge, \vee, \neg$ and $\to$, then it is called a "logical proposition". In the more general case, when the proposition contains also *sequential* connectives, in particular the sequential conjunction $\sqcap$, then it is called a "sequential proposition". In addition to the formally true logical propositions mentioned above, there are also formally true sequential propositions. The totality of formally true implications can be summarised in a calculus which contains "beginnings" $\Rightarrow A \leq B$ and rules $A \leq B \Rightarrow C \leq D$. Here, we distinguish the calculus $\mathbf{L}_Q$ of formally true *logical* propositions and the calculus $\mathbf{S}_Q$ of formally true sequential propositions. For the explicit form of these formal systems we refer to the literature [8, 16].

For an algebraic characterisation of the calculi $\mathbf{L}_Q$ and $\mathbf{S}_Q$ we consider the corresponding Lindenbaum–Tarski algebras, i.e. the algebra of equivalence classes which is given here by the algebra of value equivalent elementary propositions. The Lindenbaum-Tarski algebra of the calculus $\mathbf{L}_Q$ is given by a complete, orthomodular lattice L_Q. Subsets of mutually commensurable propositions constitute a Boolean sublattice $L_B \subseteq L_Q$ of the lattice L_Q [9]. Moreover, if the entire quantum language $\mathcal{S}_Q$ refers to one individual quantum system, then the lattice L_Q is atomic (where the atoms correspond to pure states) and fulfils the covering law [17]. In this case we denote the lattice by L_Q^*. The Hilbert lattice L_H of projection operators in Hilbert space [1] can be obtained from the lattice L_Q^* by adding the Solér law, the meaning of which is, however, still open [14]. It is well known that by means of a result by *Piron* [11] from the lattice L_H the three classical Hilbert spaces can be obtained and that for the complex numbers $\mathbb{C}$ quantum mechanics in Hilbert space is achieved.

3.3 Is Quantum Mechanics a priori Valid?

The described quantum logic approach to quantum mechanics which starts from the relaxed quantum ontology $O(Q)$ and leads finally to the quantum mechanical Hilbert space, is sometimes considered as an a priori justification of quantum theory [8]. The term "a priori" seems to be legitimated here, since the starting point of this approach are the most general preconditions of a scientific language of physics, i.e. the assumptions of the weak ontology $O(Q)$. However, this way of reasoning is not fully convincing. Firstly, up to now there is no plausible and intuitiv justification of Solér's law, which appears in the present approach as an additional *ad hoc* assumption. Hence, one could ask whether the quantum ontology $O(Q)$ which is partly too weak and partly too strong is really the right starting point. $O(Q)$ is too weak, since the main restriction of quantum ontology with respect to classical ontology, the complementarity requirement, is a very strong postulate. Two properties $P(A)$ and $P(B)$ which are not commensurable for accidental reasons are complementary in the sense that they cannot be tested simultaneously. Complementarity in this strong form must be required in quantum mechanics for sharp observables which are given by PV-measures.

However, even in quantum mechanics the strong complementarity requirement can be relaxed by the uncertainty principle making use of unsharp observables in the sense of POV-measures. POV-measures are the most general observables in accordance with the probability reproducibility condition which allow for a probability interpretation of quantum mechanics [4]. Two unsharp properties of a quantum system can be attributed jointly to the object, if the conveniently defined degrees of unsharpness of the two properties fulfil the Heisenberg uncertainty inequality [3]. Obviously, a quantum ontology $O(Q^u)$ which replaces the complementarity requirement by the uncertainty principle, is somewhat stronger than the original ontology $O(Q)$.

The ontology $O(Q)$ is not only too weak but – with respect another feature – also too strong. In accordance with the classical ontology $O(C)$ we assumed in $O(Q)$-1 that any elementary property P can be tested experimentally with the result that either P or the counter property $\bar{P}$ pertains to the system. This ontological precondition implies that elementary propositions of the quantum language are value definite and that the tertium non datur holds in quantum logic $\mathbf{L}_Q$ as a general theorem. However, the ontological precondition that any elementary property can be tested by experiment (with the result P or $\bar{P}$) exceeds the possibilities of Hilbert space quantum mechanics. Within the framework of the quantum theory of unitary premeasurements it follows that pointer objectification cannot be achieved for closed systems [10]. Hence, value definiteness of elementary propositions is incompatible with quantum mechanics in Hilbert space and must be relaxed in some sense. In this situation it suggests itself to begin with elementary propositions that are not value definite which follows from the ontological assumption that properties are in general unsharp, i.e. given by POV-measures[1] Hence it seems that also the second objection against the quantum ontology $O(Q)$ can be taken account of by the quantum ontology $O(Q^u)$ based on unsharp properties.

Hence, on the basis of the slightly modified quantum ontology $O(Q^u)$ a fresh start by means of unsharp properties seems to be a quite promising attempt. In a first step of this approach a formal language and logic of not necessarily value definite quantum mechanical propositions must be developed. A formal logic of this kind can be constructed in various ways, either in analogy to intuitionistic logic [8] or by modifying the algebraic structure. In recent years many interesting systems for unsharp propositions were proposed , intuitionistic quantum logic, unsharp quantum logic, many valued logic, unsharp quantum logic, etc. We mention here in paricular the work of Giuntini and Dalla Chiara [5,6].

However, it is still an open question whether in this way a consistent operational approach to quantum mechanics can be obtained. Up to now the logical systems mentioned were not yet reconstructed in an operational way starting from a formal language of unsharp propositions. Furthermore, the Lindenbaum-Tarski algebra of a logical calculus of unsharp propositions is not *per se* equivalent to the algebra E_H of effects in Hilbert space. We do not know which kind of law must be added to the algebra of unsharp propositions in order to obtain the effect algebra mentioned (it could be as complicated as the Solér law). Even the last step of an operational foundation of quantum physics, the way from the effect algebra to the Hilbert space requires more detailed investgations.

3.4 The Ontological Priority of Quantum Physics

Although the task of reconstructing quantum mechanics on the basis of a logic of unsharp propositions is not yet finally performed, we can draw already some

---

[1] It must be mentioned that up to now it is not yet quite clear whether the problem of pointer objectification can completely be solved by POV-measures [2].

interesting conclusions which refer to the interpretation of quantum mechanics. The basis of this approach, the (uncertainty) ontology $O(Q^u)$ is somewhat richer than the (complementarity) ontology $O(Q)$ but weaker than the classical ontology $O(C)$ of complete determination. This classical ontology is not only based on experience but also on several metaphysical hypotheses – which are weakened or cancelled in $O(Q^u)$.

Since these metaphysical hypotheses (complete determination, individuality, and full determinism) clearly exceed our everyday experience, and since we call phenomena intuitive and understandable if they are in accordance with this everyday experience, classical mechanics is not thoroughly intuitive. However, since the hypotheses contained in the classical ontology $O(C)$ are strongly reduced in the (uncertainty) ontology $O(Q^u)$ as well as in the (complementarity) ontology $O(Q)$, we expect that the implications of the quantum ontology $O(Q^u)$ are more intuitive and more plausible than the implications from the classical ontology.

In particular the quantum logic approach can further illustrate this result. The logical systems which follow from the ontologies $O(Q)$ and $O(Q^u)$ are based on weaker and less hypothetical pragmatic preconditions than the Boolean lattice L_B of classical logic. Hence the resulting quantum mechanics is more intuitive and more plausible than classical mechanics. In addition, since quantum mechanics is based on weaker premises than classical mechanics, it is nearer to the "truth" than classical mechanics. According to Popper [12, 13] a theory T is nearer to the truth than a theory T^* if T has more true consequences and less false consequences than T^*. Of course, quantum mechanics is only nearer to the truth than classical mechanics but not absolutely true since it depends still on contingent premises[2].

On the basis of these results we can formulate the role of classical mechanics. *Firstly*, classical mechanics is loaded with metaphysical hypotheses which clearly exceed our everyday experience. Since quantum mechanics is based on strongly relaxed hypotheses of this kind, classical mechanics is less intuitive and less plausible than quantum mechanics. Hence classical mechanics, its language and its logic cannot be the basis of an adequate interpretation of quantum mechanics. *Secondly*, classical mechanics is not the limit of quantum mechanics for macroscopic phenomena. Since quantum mechanics of closed systems does not explain the objectification, i.e. the classical behaviour of pointer values, classical mechanics cannot be the macroscopic limit of quantum mechanics. However, this argument which is still subject of controversial debates, is not the main reason. The essential argument which shows that classical mechanics is not the limiting case of quantum mechanics is based on the observation that classical mechanics is loaded with metaphysical hypotheses without any empirical counterpart. Since some of these hypotheses are explicitly eliminated in quantum theory, it is obvious that there is no approximation procedure which leads from quantum mechanics to classical mechanics. Classical mechanics

---

[2] Poppers terminology provides some technical problems if it is applied to two theories T and T^* which are partly false as quantum mechanics and classical mechanics. These problems can, however, be solved [18].

describes a fictitious world which does not exist in reality. The classical world is an illusion.

References

1. G. Birkhoff and J. von Neumann: The Logic of Quantum Mechanics. *Ann. of Math.* 37, 823–843 (1936)
2. P. Busch: Can unsharp objectification solve the measurement problem. *Int. J. Theor. Phys.* 37, 241–47 (1998)
3. P. Busch: Indeterminacy Relations and Simultaneous Measurements in Quantum Theory. *Int. J. Theor. Phys.* 24, 63–92 (1985)
4. P. Busch, P. Lahti and P. Mittelstaedt: *The Quantum Theory of Measurement*, 2nd. ed. Springer-Verlag, Heidelberg 1996
5. M.L. Dalla Chiara: Unsharp Quantum Logics. *Int. J. Theor. Phys.* 34, 1331–1336 (1995)
6. R. Giuntini: Brower-Zadeh logic and the operational approach to quantum mechanics. *Found. of Physics* 20, 701–714 (1990)
7. I. Kant: *Critique of Pure Reason*. Translated by N.K. Smith, Macmillan 1929, p. B600
8. P. Mittelstaedt: *Quantum Logic*. D. Reidel Publ. Co., Dordrecht 1978
9. P. Mittelstaedt: Language and Reality in Quantum Physics. In: *Symposium on the Foundations of modern Physics*, 1987. World Scientific, Singapore 1987, pp. 229–250
10. P. Mittelstaedt: *The Interpretation of Quantum Mechanics and the Measurement Process*. Cambridge University Press, Cambridge 1998
11. C. Piron: *Foundations of Quantum Physics*. W.A. Benjamin, Reading Mass. 1976
12. K. Popper: *Conjectures and Refutations*. London 1963
13. K. Popper: *Objective Knowledge: An Evolutionary Approach*. Clarendon Press, Oxford 1972
14. M.P. Solèr: Characterisation of Hilbert Spaces by Orthomodular Lattices. *Communications in Algebra* 23, no. 1, 219–243 (1995)
15. E.W. Stachow: Experimental approach to quantum logical connectives. To appear in: P. Weingartner (ed.): *Alternative Logics – Do Sciences need them?* 2002
16. E.W. Stachow: Logical Foundation of Quantum Mechanics. *Inter. Journ. of Theor. Phys.* 19, 251–304 (1980)
17. E.W. Stachow: Structures of quantum language for individual systems. In: P. Mittelstaedt and E.W. Stachow (eds.): *Recent Developments in Quantum Logic*. BI-Wissenschaftsverlag, Mannheim 1984, pp. 129–145
18. P. Weingartner: *Basic Questions on Truth*. Kluwer Academic Publishers, Dordrecht 2000

On the Interpretation of Quantum Theory – from Copenhagen to the Present Day

Claus Kiefer

Summary. A central feature in the Copenhagen interpretation is the use of classical concepts from the outset. Modern developments show, however, that the emergence of classical properties can be understood within the framework of quantum theory itself, through the process of decoherence. This fact becomes most crucial for the interpretability of quantum cosmology – the application of quantum theory to the Universe as a whole. I briefly review these developments and emphasize the importance of an unbiased attitude on the interpretational side for future progress in physics.

Ich bin nicht damit zufrieden, wenn man eine Maschinerie hat, die zwar zu prophezeien gestattet, der wir aber keinen klaren Sinn zu geben vermögen.

Albert Einstein in a letter to Max Born (3.12.1953)

1 Copenhagen Interpretations and Alternatives

A physical theory contains both a mathematical formalism and an interpretational scheme. Although their relation may be subtle already for classical physics, it becomes highly non-trivial in quantum theory. In fact, although the mathematical framework had been basically fixed by 1932, the debate about its meaning is going on. As we shall see, this is due to the possibility of having conflicting concepts of reality, without contradicting the formalism. This is why all physicists agree on the practical application of the formalism to concrete problems such as the calculation of transition probabilities.

It is often asserted that the orthodox view is given by the Copenhagen interpretation of quantum theory. This is the standpoint taken by most textbooks. What is this interpretation? It is generally assumed that it originated from intense discussions between Bohr, Heisenberg, and others in Copenhagen in the years 1925–27. However, there has never been complete agreement about the actual meaning, or even definition, of this interpretation even among its main contributors. In fact, the Copenhagen interpretation has remained until today an amalgamation of different views.

As has been convincingly argued in [1], it is the incompatibility between Bohr's and Heisenberg's views that sometimes gives the impression of inconsistencies in the Copenhagen interpretation.

Historically, Heisenberg wanted to base quantum theory solely on observable quantities such as the intensity of spectral lines, getting rid of all intuitive (*anschauliche*) concepts such as particle trajectories in space-time [2]. This attitude changed drastically with his paper [3] in which he introduced the uncertainty relations – there he put forward the point of view that it is the theory which decides what can be observed. His move from positivism to operationalism can be clearly understood as a reaction on the advent of Schrödinger's wave mechanics [1] which, in particular due to its intuitiveness, became soon very popular among physicists. In fact, the word *anschaulich* (intuitive) is contained in the title of Heisenberg's paper [3].

Bohr, on the other hand, gave the first summary of his interpretation in his famous lecture delivered in Como in September 1927, cf. [4]. There he introduced the notion of complementarity between particles and waves – according to von Weizsäcker the core of the Copenhagen interpretation [5]. As is well known, he later extended complementarity to non-physical themes and advanced it to a central concept of his own philosophy. Complementarity means that quantum objects are neither particles nor waves; for our intuition we have to use both pictures, in which the range of applicability of one picture necessarily constrains the range of applicability of the other. Heisenberg, as a mathematical physicist, was not at ease with such an interpretation. He preferred to use one coherent set of concepts, rather than two imcompatible ones, cf. [1]. In fact, it was known by then that particle and wave language can be converted into each other and are transcended into the consistent formalism of quantum theory. As Heisenberg wrote in his book [6]: "Licht und Materie sind einheitliche physikalische Phänomene, ihre scheinbare Doppelnatur liegt an der wesentlichen Unzulänglichkeit unserer Sprache ... Will man trotzdem von der Mathematik zur anschaulichen Beschreibung der Vorgänge übergehen, so muß man sich mit unvollständigen Analogien begnügen, wie sie uns Wellen- und Partikelbild bieten."[1]

Even Bohr's own interpretation did change in the course of time. This happened in particular due to the influence of the important paper by Einstein, Podolsky and Rosen in 1935 (EPR) [7]. Before EPR, an essential ingredient of his interpretation was the uncontrollable disturbance of the quantum system by the apparatus during a measurement. The analysis of EPR demonstrated, however, that the issue is not the disturbance, but the non-separability (the entanglement) of a quantum system over in principle unlimited spatial distances. Therefore, in his response to EPR [8], Bohr adopted a strong operationalistic attitude, concealing the crucial concept of entanglement. Taking the indispensability of classical concepts for granted, he argued that even without any mechanical disturbance there is an "influence of the very conditions which define

---

[1] "Light and matter are unique physical phenomena, their apparent double nature is due to the essential inadequacy of our language ...If one nevertheless wants to procede from the mathematics to the intuitive description of the phenomena, we have to restrict ourselves to incomplete analogies as they are offered by the wave and particle pictures."

the possible types of predictions regarding the future behavior of the system" [8], i.e. no simultaneous reality for measurements of noncommuting variables such as position and momentum should exist. This attitude leads eventually to the consequence that quantum systems would not possess a real existence before interacting with a suitable measurement device ("only an observed phenomenon is a phenomenon").

A somewhat ambigous role in the Copenhagen interpretation(s) is played by the "collapse" or "reduction" of the wave function. This was introduced by Heisenberg in his uncertainty paper [3] and later postulated by von Neumann as a dynamical process independent of the Schrödinger equation, see Sect. 2. Most proponents of the Copenhagen interpretation have considered this reduction as a mere increase of knowledge (a transition from the potential to the actual), therefore denying that the wave function is a kinematical concept and thus affected by dynamics. The assumption of a dynamical collapse would definitely be in conflict with Bohr's ideas of complementarity which forbid a physical analysis of the measurement process. To summarise, one can identify the following ingredients as being characteristic for the Copenhagen interpretation(s):

- Indispensability of classical concepts for the description of the measurement process
- Complementarity between particles and waves
- Reduction of the wave packet as a formal rule without dynamical significance

This set of rules has been sufficient to apply the quantum formalism pragmatically to concrete problems. But is it still sufficient? And is it satisfactory?

Modern developments heavily rely on the concept of entanglement, in order to describe satisfactorily the precision experiments that are now being performed in quantum optics and other fields. From the experimental violation of Bell's inequalities it has become evident that quantum theory cannot be substituted by a theory referring to a local reality. This is of course a consequence of the superposition principle – the heart of quantum theory. Among recent developments employing entanglement are quantum computation and quantum cryptography [9] and the reversible transition from the (coherent) superfluid phase to an (incoherent) Mott insulator phase in a Bose-Einstein condensate [10], during which interference patterns appear and disappear. The superposition principle is indispensable for describing $K - \bar{K}$ and neutrino oscillations. It would be hard to imagine how all this can be understood by denying any dynamical nature of the wave function and to interpret it as describing mere knowledge. Moreover, it is now clear, both theoretically and experimentally, that the classical appearance of our world can be understood as a dynamical process *within* quantum theory itself, without any need to postulate it. Therein, entanglement plays the crucial role. This will be discussed in the next section.

There have been, in the course of history, various attempts to come up with an alternative to the Copenhagen interpretation [11, 12]. Here I want to mention two of them, the de Broglie-Bohm interpretation and the Everett interpretation. Many other interpretations are some variant or mixture of these two and the Copenhagen interpretation. The consistent-histories interpretation, for example, contains elements from both Copenhagen and Everett, see e.g. Chapt. 5 in [13]. Different from these interpretations are attempts which aim at an explicit change of the Schrödinger equation in order to get a dynamical collapse of the wave function, see e.g. Chapt. 8 in [13]. Up to now, however, there is no experimental hint that the Schrödinger equation has to be modified.

In the de Broglie–Bohm interpretation (or "theory"), the wave function Ψ is supplemented with classical variables (particles and fields) possessing definitive values of position and momentum. Whereas the wave function obeys an autonomous dynamics (obeying the Schrödinger equation without additional collapse), the particle dynamics depends on Ψ (often called a guiding field). Assuming that the particles are distributed according to $|\Psi|^2$, the predictions of this theory are indistinguishable from the ordinary framework, at least within non-relativistic quantum mechanics. After a measurement, the particle is trapped within one particular wave packet with the usual quantum-mechanical probability. Because the other wave packets are spatially separated from it after the measurement, they can no longer influence the particle. This represents an apparent collapse of the wave function and the occurrence of a definite measurement result. In principle, however, the remaining packets, although empty, can interfere again with the packet containing the particle, but the probability for this is tiny in macroscopic situations.

John Bell called the Everett interpretation a Bohm interpretation without trajectories. In fact, Everett assumes just as Bohm that the wave function is part of reality and that there is never any collapse. Therefore, after a measurement, all components corresponding to the different outcomes are equally present. It is claimed that the probability interpretation of quantum theory can be derived from the formalism (which is, however, a contentious issue). Von Weizsäcker calls this interpretation [5] " ... die einzige, die nicht hinter das schon von der Quantentheorie erreichte Verständnis zurück-, sondern vorwärts über es hinausstrebt."[2] The open question is of course when and how these different components ("branches") become independent of each other. This leads me to the central topic – the emergence of classical behaviour in quantum theory.

2 The Emergence of Classical Properties in Quantum Theory

If classical concepts are not imposed from the outset, they have to be derived from the formalism, at least in an approximate sense. John von Neumann was the first who analysed in 1932 the measurement process within quantum mechanics. He considers the coupling of a system (S) to an apparatus (A), see Fig. 1.

Let the states of the measured system which are discriminated by the apparatus be denoted by $|n\rangle$ (for example, spin up and spin down), then an appropriate interaction Hamiltonian has the form (see Chapt. 3 in [13], or [14])

$$H_{\text{int}} = \sum_n |n\rangle\langle n| \otimes \hat{A}_n . \tag{1}$$

---

[2] " ... the only one that does not fall back behind the understanding already achieved by quantum theory but which strives forwards and even beyond."

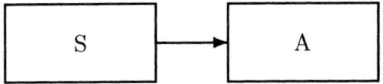

Fig. 1. Original form of the von Neumann measurement model

The operators $\hat{A}_n$, acting on the states of the apparatus, are rather arbitrary, but must of course depend on the "quantum number" n. Equation (1) describes an "ideal" interaction during which the apparatus becomes correlated with the system state, without changing the latter. There is thus no disturbance of the system by the apparatus – on the contrary, the apparatus is disturbed by the system (in order to yield a measurement result).

If the measured system is initially in the state $|n\rangle$ and the device in some initial state $|\Phi_0\rangle$, the evolution according to the Schrödinger equation with Hamiltonian (1) reads

$$|n\rangle|\Phi_0\rangle \xrightarrow{t} \exp(-iH_{\text{int}}t)|n\rangle|\Phi_0\rangle = |n\rangle \exp\left(-i\hat{A}_n t\right)|\Phi_0\rangle$$
$$=: |n\rangle|\Phi_n(t)\rangle. \quad (2)$$

The resulting apparatus states $|\Phi_n(t)\rangle$ are often called "pointer states". A process analogous to (2) can also be formulated in classical physics. The essential new quantum features now come into play when we consider a *superposition* of different eigenstates (of the measured "observable") as initial state. The linearity of time evolution immediately leads to

$$\left(\sum_n c_n |n\rangle\right) |\Phi_0\rangle \xrightarrow{t} \sum_n c_n |n\rangle |\Phi_n(t)\rangle. \quad (3)$$

But this state is a superposition of macroscopic measurement results (of which Schrödinger's cat is just one drastic example)! To avoid such a bizarre state, and to avoid the apparent conflict with experience, von Neumann introduced the dynamical collapse of the wave function as a new law. The collapse should then select one component with the probability $|c_n|^2$. He even envisaged that the collapse is eventually caused by the consciousness of a human observer, an interpretation that was later also adopted by Wigner. In the Everett interpretation, all the branches (each component in (3)) are assumed to co-exist simultaneously.

Can von Neumann's conclusion and the introduction of the collapse be avoided? The crucial observation [15], which enforces an extension of von Neumann's measurement theory, is the fact that macroscopic objects (such as measurement devices) are so strongly coupled to their natural environment, that a unitary treatment as in (2) is by no means sufficient and has to be modified to include the environment.

Fortunately, this can easily be done to a good approximation, since the interaction with the environment has in many situations the same form as given by the Hamiltonian (1): The measurement device is itself "measured" (passively recognised) by the environment, according to

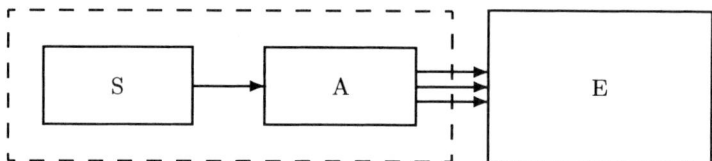

Fig. 2. Realistic extension of the von Neumann measurement model including the environment. Classical properties emerge through the unavoidable, irreversible interaction of the apparatus with the environment

$$\left(\sum_n c_n |n\rangle |\Phi_n\rangle\right) |E_0\rangle \xrightarrow{t} \sum_n c_n |n\rangle |\Phi_n\rangle |E_n\rangle. \quad (4)$$

This is again a macroscopic superposition, now including the myriads of degrees of freedom pertaining to the environment (gas molecules, photons, etc.). However, most of these environmental degrees of freedom are inaccessible. Therefore, they have to be integrated out from the full state (4). This leads to the reduced density matrix for system plus apparatus, which contains all the information that is available there. It reads

$$\rho_{SA} \approx \sum_n |c_n|^2 |n\rangle\langle n| \otimes |\Phi_n\rangle\langle \Phi_n| \quad \text{if} \quad \langle E_n | E_m \rangle \approx \delta_{nm}, \quad (5)$$

since under realistic conditions, different environmental states are orthogonal to each other. Eq. (5) is identical to the density matrix of an ensemble of measurement results $|n\rangle|\Phi_n\rangle$. System and apparatus thus seem to be in one of the states $|n\rangle$ and $|\Phi_n\rangle$, given by the probability $|c_n|^2$.

Both system and apparatus thus assume classical properties through the unavoidable, irreversible interaction with the environment. This dynamical process, which is fully described by quantum theory, is called *decoherence* [13]. It is based on the quantum entanglement between apparatus and environment. Under ordinary macroscopic situations, decoherence occurs on an extremely short timescale, giving the impression of an instantaneous collapse or a "quantum jump". Recent experiments were able to demonstrate the continuous emergence of classical properties in mesoscopic systems [16, 17]. Therefore, one would never ever be able to observe a weird superposition such as Schrödinger's cat, because the information about this superposition would almost instantaneously be delocalised into unobservable correlations with the environment, resulting in an apparent collapse for the cat state. The concept of decoherence motivated Wigner to give up his explanation of the collapse as being caused by consciousness [18]. In fact, decoherence makes it evident that living creatures play no particular role in the interpretation of quantum theory.

The interaction with the environment distinguishes the local basis with respect to which classical properties (unobservability of interferences) hold. This "pointer basis" must obey the condition of robustness, i.e. it must keep its classical appearance over the relevant timescales [19, 20]. Classical properties are thus not intrinsic to any object, but only defined by their interaction with other degrees of freedom. In simple

(Markovian) situations the pointer states are given by localised Gaussian states [21]. They are, in particular, relevant for the localisation of macroscopic objects.

To summarise, these developments have shown that classical concepts are not an indispensable input to the theory, as required by the Copenhagen interpretation, but a natural consequence of the theory itself when applied to realistic conditions.

3 Quantum Gravity

The modern developments discussed in the last section show that the main assumptions of the Copenhagen interpretation, such as complementarity and the demand for a priori classical concepts, are not obligatory. Still, one might argue, it could be possible to adopt this interpretation as a convenient background for pragmatic use.

While this may be true for ordinary laboratory situations, it may become impossible if quantum effects of the gravitational field are involved. It must be admitted that no effects of quantum gravity have yet been seen, or identified as such, and that no final consensus about such a theory has emerged. However, there exist many models with important applications in cosmology; without an appropriate and consistent interpretation, such models would be void of interest.

In ordinary quantum theory, time is given as an external parameter and not subject to quantisation. On the other hand, the gravitational field is described by Einstein's theory of general relativity, in which space and time are dynamical and not absolute. I have discussed elsewhere the reasons why one generally believes that gravity must be described by a quantum theory at the most fundamental level [22]. If one quantises a theory that classically possesses no absolute time – be it general relativity or some alternative theory – the ensuing quantum theory does not contain any time parameter at all. Since the role of time in general relativity is merely to parametrise spacetimes (the "trajectories" in the corresponding configuration space), the absence of time in quantum gravity is the consequence of the absence of trajectories in quantum theory. The central equation of quantum gravity is of the general form

$$\hat{H}\Psi = 0 , \tag{6}$$

where $\hat{H}$ is the total Hamilton operator of both gravitational and non-gravitational fields. The total quantum state is thus of a stationary form. Since quantum degrees of freedom are very sensitive to their environment (see Sect. 2), and since the dominant interaction in the Universe on large scales is gravity, one is immediately led to consider a quantum theory of the whole Universe – quantum cosmology [22]. Since the Universe naturally contains all of its observers, the problem arises to come up with an interpretation of quantum theory that contains no classical realms on the fundamental level.

How can a temporal dynamics be understood from the stationary equation (6)? It has been demonstrated that a concept of time re-emerges in a semiclassical limit as an approximate concept [20, 22]. In this limit, an effective time-dependent Schrödinger equation holds along formal WKB trajectories. Since different semiclassical branches usually decohere from one another, an observer cannot experience the other branches which only together form the one wave function Ψ in (6).

Clearly, the Copenhagen interpretation cannot cope with quantum cosmology. This is the reason why most people working in this field, at least implicitly, adopt the Everett interpretation, since it is hard to imagine from where a conceivable collapse could emerge. The problems that are addressed in quantum cosmology include the quantum origin of the Universe, the quantum probability for the occurrence of an inflationary phase, and the quantum-to-classical transition for primordial fluctuations which serve as the seeds for structure formation in the Universe, giving rise to galaxies [22]. With the advent of precision measurements for the spectrum of the cosmic microwave background radiation, some of these questions gain observational significance. It would thus be unsatisfactory to avoid a theoretical description of such processes just because some pre-conceived interpretation of quantum theory does not fit this purpose.

4 Conclusion

The Copenhagen interpetation needs classical concepts as prerequisites. On the other hand, quantum theory itself predicts the occurrence of decoherence through which systems such as measurement devices can appear classically to local observers. This can be, and has been, interpreted as a quantum justification for at least part of the original Copenhagen programme [19]. This explains, in retrospect, why the Copenhagen interpretations can serve as a background for pragmatic calculations, at least in non-gravitational situations.

The process of decoherence is based on the validity of the Schrödinger equation and can thus not describe any real collapse. For local experiments, this is definitely not needed, because decoherence can explain why we seem to observe a collapse or a "quantum jump". It would seem unnatural and ad hoc to introduce a real collapse at this stage, where an apparent collapse is predicted anyway by the Schrödinger equation. The concept of quantum jumps thus plays the role of epicycles in astronomy [23] – it describes naively what is observed, but it becomes redundant at the fundamental, theoretical, level.

Still, the measurement problem is not resolved for the total system including the environment. The only alternatives so far are to either assume an additional real collapse in violation of the Schrödinger equation (for which there is not yet any experimental hint), or to adopt an Everett interpretation with its simultaneous reality of all branches. It seems hard to imagine that an experimental decision between these alternatives can be made in the foreseeable future. However, it is important to keep an open mind and to avoid the burden of a pre-imposed interpretation – "sonst wäre ernstlich zu befürchten, daß es dort, wo wir das Weiterfragen verbieten, wohl doch noch einiges Wissenswerte zu fragen gibt" [24].[3]

---

[3] "Otherwise it would be seriously feared that just there, where we forbid further questions, there might still be something worth knowing that we might ask about."

References

1. M. Beller: *Quantum dialogue*. The University of Chicago Press, Chicago 1999
2. W. Heisenberg: Quantentheoretische Umdeutung kinematischer und mechanischer Beziehungen. *Z. Phys.* 33, 879–93 (1925)
3. W. Heisenberg: Über den anschaulichen Inhalt der quantentheoretischen Kinematik und Mechanik. *Z. Phys.* 43, 172–98 (1927); English translation in [12]
4. N. Bohr: The quantum postulate and the recent development of atomic theory. *Nature* 121 (suppl.), 580–90 (1928); Reprinted in [12]
5. C.F. von Weizsäcker: *Aufbau der Physik*. Hanser, München 1985
6. W. Heisenberg: *Physikalische Prinzipien der Quantentheorie*. Hirzel, Stuttgart 1958
7. A. Einstein, B. Podolsky and N. Rosen: Can quantum-mechanical description of physical reality be considered complete? *Phys. Rev.* 47, 777–80 (1935) Reprinted in [12].
8. N. Bohr: Can quantum-mechanical description of physical reality be considered complete? *Phys. Rev.* 48, 696–702 (1935); Reprinted in [12]
9. N. Gisin, G. Ribordy, W. Tittel and H. Zbinden: Quantum cryptography. *Rev. Mod. Phys.* 74, 145–95 (2002)
10. M. Greiner, O. Mandel, T. Esslinger, T.W. Hänsch and I. Bloch: Quantum phase transition from a superfluid to a Mott insulator in a gas of ultracold atoms. *Nature* 415, 39–44 (2002)
11. M. Jammer: *The philosophy of quantum mechanics*. Wiley, New York 1974
12. J.A. Wheeler and W.H. Zurek (eds.): *Quantum theory and measurement*. Princeton University Press, Princeton 1983
13. D. Giulini, E. Joos, C. Kiefer, J. Kupsch, I.O. Stamatescu and H.D. Zeh: *Decoherence and the appearance of a classical world in quantum theory*. Springer, Berlin 1996
14. C. Kiefer and E. Joos: Decoherence: Concepts and examples. In: Ph. Blanchard, A. Jadczyk (eds.): *Quantum future*. Springer, Berlin 1999, pp. 105–28; E. Joos: Elements of environmental decoherence. In: Ph. Blanchard, D. Giulini, E. Joos, C. Kiefer, and I.O. Stamatescu (eds.): *Decoherence: Theoretical, experimental, and conceptual problems*. Springer, Berlin 2000, pp. 1–17
15. H.D. Zeh: On the interpretation of measurement in quantum theory. *Found. Phys.* 1, 69–76 (1970); Reprinted in [12]
16. M. Brune et al.: Observing the progressive decoherence of the "meter" in a quantum measurement. *Phys. Rev. Lett.* 77, 4887–90 (1996)
17. C.J. Myatt et al.: Decoherence of quantum superpositions through coupling to engineered reservoirs. *Nature* 403, 269–73 (2000)
18. E.P. Wigner: *Philosophical reflections and syntheses*. Springer, Berlin 1995
19. W.H. Zurek: Decoherence and einselection. In: Ph. Blanchard, D. Giulini, E. Joos, C. Kiefer and I.O. Stamatescu (eds.): *Decoherence: Theoretical, experimental, and conceptual problems*. Springer, Berlin 2000, pp. 309–41
20. H.D. Zeh: *The wave function: It or bit?*, Contribution to the festschrift in honour of J.A. Wheeler. to appear, Online version: [quant-ph/0204088]
21. L. Diósi and C. Kiefer: Robustness and diffusion of pointer states. *Phys. Rev. Lett.* 85, 3552–55 (2000)
22. C. Kiefer: Conceptual issues in quantum cosmology. In: J. Kowalski-Glikman (ed.): *Towards quantum gravity*. Springer, Berlin 2000, pp. 158–87
23. E. Schrödinger: Are there quantum jumps? (Part I). *British Journal for the Philosophy of Science* 3, 109–23 (1952)
24. E. Schrödinger: Die gegenwärtige Situation in der Quantenmechanik. *Die Naturwissenschaften* 23, 823–28 (1935); English translation in [12]

Epistemic and Ontic Quantum Realities

Harald Atmanspacher and Hans Primas

Quantum theory has provoked intense discussions about its interpretation since its pioneer days. One of the few scientists who have been continuously engaged in this development from both physical and philosophical perspectives is Carl Friedrich von Weizsäcker [44]. The questions he posed were and are inspiring for many, including the authors of this contribution. Weizsäcker developed Bohr's view of quantum theory as a theory of knowledge. We show that such an epistemic perspective can be consistently complemented by Einstein's ontically oriented position.

1 Introduction

1.1 Einstein and Bohr on Realism

Most working scientists believe that there is an external world, which has the status of a reality to be explored by science. The goal of science is to achieve knowledge about how this external world is constituted and develops. Although scientific methodology requires observations and measurements for this purpose, the reality to be described is believed to "exist" independent of its possible empirical accessibility. This view is succinctly formulated by Einstein [9, p. 81]:

> "Physics is an attempt conceptually to grasp reality
> as it is thought independently of its being observed."

A realist stance of this kind does not exclude that acts of observation can have effects on an observed entity, and that a measurement of one property of an entity can lead to changes of another property. It simply claims that the "existence" of this entity is in some fundamental sense guaranteed, maybe even necessary as a precondition to observe or measure its properties.

On the other hand, there is an opposing, popular stance to the effect that quantum theory does not admit such an observation-independent realism. This view, which has been perpetuated in many modern monographs and textbooks, goes back to Bohr's claim that in quantum theory a realism with respect to measuring instruments is the only possible realism. According to Bohr:[1]

> "It is wrong to think that the task of physics is to find out how nature is. Physics concerns what we can say about nature."

The two quotations by Einstein and Bohr indicate a basic point of disagreement between the two in their ongoing conversations concerning the interpretation of quantum mechanics in the 1920s and 1930s.[2] Bohr focused on what we could know about and infer from observed quantum phenomena. By contrast, Einstein's position led him to consider Bohr's characterization of quantum theory as incomplete.

Both Einstein and Bohr did not clearly realize that they addressed different concepts of reality. Since they never made their basic viewpoints explicit, it is not surprising that they talked past each other in a number of respects. Both Bohr's operationalistic and Einstein's ontological concept of reality have their proper places in the study of matter. Both are legitimate and even necessary, but they must not be confused with each other. An observation-independent reality must be described in a conceptual and formal framework different from that used for observed facts and measurement.

1.2 Against Classical Prejudices

Quantum features such as non-commutativity, nonlocality, nonseparability and the possibility of entanglement have been forcing us to revise our classical ideas about the nature of matter. For a viable realistic interpretation of quantum theory, the concept of realism must not be associated with ideas taken over from classical physics, such as atomism, localization, separability, or similar preconceptions. Here are three examples.

It is most often assumed that a measurement of a property of a system results in one of the eigenvalues of this property. The underlying idea that every property has a sharp numerical value has been adopted from classical point mechanics without any sensible argument. Hence, an uncritical pointwise valuation of properties should be avoided.

A significant topic in the Bohr–Einstein discussions concerned Einstein's worries about the failure of separability in quantum theory. Yet, according to quantum theory the material world is basically a whole which does not consist of parts. The fact that modern quantum theory can successfully *describe* many aspects of the behavior of matter in terms of elementary systems and its interactions does not imply that matter *is composed* of such elementary systems.

The so-called "wave-particle dualism" from the early days of quantum mechanics led to the seemingly ineradicable misunderstanding that an electron is either a particle or a wave. By contrast, the formalism of quantum theory implies that there are infinitely many pairs of states of an electron which are neither wave-like nor particle-like, although they are dual in the same sense as wave-like and particle-like states.

---

[1] Quoted in [28].
[2] Compare [5] and [10].

1.3 Mathematical Codification of Quantum Theory

The first attempts to formulate quantum mechanics in a mathematically rigorous way were based on the fundamental uniqueness theorem for the irreducible representations of the canonical commutations relations over locally compact phase spaces [26, 39]. This codification, introduced by von Neumann [27], is restricted to quantum systems with only finitely many degrees of freedom and without classical observables. As a consequence, von Neumann's formalism does not comprise a proper description of the interaction of a charged particle with its electromagnetic field, the possibility of symmetry breakings, genuinely irreversible processes, or the existence of molecular (or other) classical observables.

Fortunately, there is no reason to identify quantum theory with the historical Hilbert-space or Fock-space representations. A more comprehensive description has been developed in the language of topological *-algebras. The so-called *algebraic quantum theory* is nothing else than a straightforward, mathematically precise and complete codification of the heuristic ideas of quantum mechanics of the pioneer days. It uses the same mathematical language for both classical and quantum systems. It is appropriate for microscopic, mesoscopic and macroscopic systems with finitely or infinitely many degrees of freedom. It covers not only classical point mechanics, classical statistical mechanics, and traditional quantum mechanics as special cases, but also systems conceived as fields, such as the electromagnetic field.

Algebraic quantum theory does not make use of ad hoc assumptions such as hidden variables or quantization procedures as they appear in other approaches.[3] In the Galilei-relativistic case, the rotation subgroup and the Weyl subgroup, describing space translations and velocity boosts, are all that is needed to generate the canonical commutation relations for position and momentum, for the orbital angular momentum, and for the spin angular momentum. Rather than interpreting quantum theory in terms of concepts of classical physics, the behavior of classical systems has to be explained in terms of quantum theory.

1.4 Individual and Statistical Descriptions

A basic issue for interpretations of quantum theory is the difference between *individual* and *statistical* descriptions. For instance, in Bohr's view quantum theory refers to an *individual system* or an *individual experiment*, whereas Einstein insisted that quantum theory refers to a statistical ensemble of experiments rather than individual systems. Is it possible to resolve this issue within algebraic quantum theory?

In classical mechanics, the mathematical formalism required for an individual description is different from the formalism required for a statistical description. While the individual description of point mechanics is usually formulated in terms of a (generally *nonlinear*) dynamics on an even-dimensional smooth manifold as

---

[3] We do not consider Bohmian mechanics, the consistent history proposal, the many-worlds view, the approach by Ghirardi, Rimini, Weber and Pearle, and related approaches, since we regard them as unsatisfactory.

a phase space, the statistical formulation uses Koopman's Hilbert-space formalism with a *linear* dynamics [19].

This example strongly suggests that the traditional Hilbert-space codification of quantum mechanics corresponds to Koopman's Hilbert-space formalism of classical mechanics. Their common statistical nature is reflected by the fact that they are both formulated in terms of the W*-algebraic probability theory. On the other hand, the algebraic description of *individual* physical systems does not use the topological structure of a W*-algebra,[4] but only the structure of C*-algebras with its algebraically defined norm topology.

2 Distinguishing Epistemic and Ontic Perspectives

2.1 Epistemology and Ontology

A crucial issue of any interpretational approach with respect to a scientific theory is the relation between elements of the theory on the one hand and elements of the domain of reality for which the theory is designed on the other. This so-called relation of reference cannot be exhaustively addressed within the scope of the scientific discipline concerned. For its proper discussion, genuinely philosophical issues must be taken into account explicitly.

One of the most general demands on a sound philosophical discussion is the distinction between epistemological and ontological statements. While ontology is a branch of metaphysics, epistemology comprises all kinds of issues related to the knowledge (or ignorance) of information gathering and using systems. The ways in which human beings process information (perceptually, cognitively, and otherwise), thus represent a set of central epistemological questions.

Ontology refers to the nature and behavior of systems as they are, independent of any empirical access. The different stances of Einstein and Bohr can thus be related to an ontological versus an epistemological emphasis. Their conflicting viewpoints demonstrate that particular problems in physics (as well as other natural sciences) make it important to distinguish the two perspectives.

For a proper discussion of interpretations of quantum theory, Scheibe introduced the notions of epistemic and ontic states of a system rather than addressing epistemological and ontological issues in the more general sense mentioned above [36, 37]. Discussing epistemic and ontic states of a system can be an epistemological or an ontological matter. For instance, referring to epistemic and ontic states of a system just as they are might be understood ontologically. Referring to descriptions in terms of epistemic and ontic states, briefly epistemic and ontic descriptions, might be understood epistemologically.

---

[4] A W*-algebra $\mathfrak{M}$ is a C*-algebra which is the dual Banach space of a Banach space $\mathfrak{M}_*$. The topology relevant for the W*-algebraic probability theory is the $\sigma(\mathfrak{M}, \mathfrak{M}_*)$-topology on $\mathfrak{M}$.

2.2 Epistemic and Ontic Descriptions

States of a system to which *epistemic descriptions* refer are called *epistemic states*. The mathematical representation of such states encodes empirically obtainable knowledge about them. If the knowledge about states and their associated properties is expressed by probabilities in the sense of relative frequencies for a statistical ensemble of independently repeated experiments, we speak of a *statistical description* and of *statistical states*.

Insofar as epistemic descriptions refer to a state concept encoding knowledge, they depend on observation and measurement. Such kinds of empirical access necessarily introduce a context under which a system is investigated. Therefore, properties associated with an epistemic state are *contextual*. General fundamental principles cannot be expected within epistemic descriptions.

States of a system to which *ontic descriptions* refer are called *ontic states*. They represent the system exhaustively, i.e. an ontic state is "just the way it is", without any reference to epistemic knowledge or ignorance. In this sense, ontic states are empirically inaccessible. The properties of the system are understood as *intrinsic* properties. As *individual states*, ontic states are the referents of *individual descriptions*.

The rational for an *ontic description* of quantum systems is not a nostalgic desire for a classical picture of reality behind quantum phenomena, but to understand quantum theory as a fundamental theory of the material world, based on first principles such as symmetries. Such principles are assumed to be quite universally valid, and they are used to provide a description of the material world which is as context-independent as possible or reasonable. We do not propose an ontic description as an absolutely context-free, ultimately fundamental theory of matter.

To the same degree to which ontic descriptions are context-independent, they hide the richness and variety of empirical reality. This manifests itself in the fact that pattern detection and recognition devices determine what is considered as relevant or irrelevant. Patterns are detected and recognized by rejecting information which is selected as irrelevant in particular contexts. Based on such contexts, *an epistemic state refers to the knowledge that can be obtained about an ontic state*.

2.3 Interplay of Epistemic and Ontic Approaches

The very characterization of *epistemic descriptions* in terms of empirical access implies that tools of observation and measurement must be addressed. This requires the description of engineering instruments and the possibility of unambiguous communication. If these tools are taken for granted, their states have to be conceived as ontic states. As a consequence, any epistemic description presupposing tools of empirical access must be understood relative to an ontic description of these tools. In this sense, *any epistemic description requires an ontological commitment*.

Accordingly, both epistemic and ontic elements are involved already at a fairly basic level of discussion. In order to avoid confusion, corresponding concepts of epistemic and ontic states need to be distinguished. If this is not taken care of, there is a risk of confounding issues belonging to categorially different domains of discussion. As Fetzer and Almeder emphasize [11, p. 101],

> "an ontic answer to an epistemic question (or vice versa)
> normally commits a category mistake."

For instance, it would be a category mistake to discuss an ontic description of an observation-independent reality in terms of measurements, and it would be a category mistake to discuss the epistemic description of the observation of facts in terms of theories not involving measurement as a key topic.

Drawing the distinction between epistemic and ontic descriptions does not imply, though, that the two categories are unrelated to each other. On the contrary, the crucial point is about the relationship between the two frameworks rather than the selection of one at the expense of the other. Based on an ontic description of a physical system, epistemic descriptions can be derived by additionally specifying *contexts* which distinguish relevant from irrelevant patterns of empirical reality. This distinction can be mathematically accomplished by introducing a *contextual topology*. One ontic description can give rise to numerous physically inequivalent epistemic descriptions.

3 Epistemic Descriptions of Quantum Systems

3.1 Every Experiment Requires a Boolean Context

In physics, epistemic descriptions deal with experiments and measurements. It is a basic fact of experimental physics that the registration of experimental results is described in terms of classical engineering science. This situation requires a *Boolean* domain of discourse, corresponding to Bohr's insight [5, p. 209]:

> "However far the phenomena transcend the scope
> of classical physical explanation, the account of
> all evidence must be expressed in classical terms."

From a modern point of view, "classical" is to be interpreted as "Boolean", not as macroscopic. Whether a system shows quantal or classical behavior is independent of its size. There are small molecular systems with classical properties, like the chirality of most biologically important molecules. Furthermore, measuring tools are not necessarily macroscopic. For example, molecular genetic coding realizes a highly reliable classical, irreversible and nonanticipating measuring instrument of molecular size.

3.2 Von Neumann's Statistical Formulation

The first mathematically rigorous codification of the ideas of the pioneers of quantum mechanics was given by von Neumann [27]. He *derived* the Hilbert-space representation for the states of quantum systems from the *uniqueness theorem* by Stone and von Neumann [26, 39]. This theorem says that for *finitely* many degrees of freedom there is (up to unitary equivalence) only *one* irreducible Hilbert-space representation of the fundamental canonical commutation relations. The restriction of von Neumann's codification to systems with only finitely many degrees of freedom is a severe limitation which excludes an explicit description of the environment (like the radiation field) or of measuring tools.

The Hilbert-space quantum theory introduced by von Neumann is a *statistical* theory which expresses the *expectation value* $\mathcal{E}_t(A)$ at time t of an observable A on a Kolmogorov probability space (Λ, Σ, μ_t) as $\mathcal{E}_t(A) = \int_\Lambda \lambda \, \mu_t(\mathrm{d}\lambda)$. Here A is a selfadjoint Hilbert-space operator with the spectral resolution $A = \int_\Lambda \lambda \, E(\mathrm{d}\lambda)$, where $\mathcal{B} \mapsto E(\mathcal{B})$ is a projection-valued set function on the Boolean σ-algebra Σ of Borel sets $\mathcal{B}$ in the spectrum Λ of A. The probability measure μ_t is given by $\mu_t(\mathcal{B}) = \mathrm{tr}\{D_t E(\mathcal{B})\}$, where the density operator D_t (a nonnegative trace-class operator with $\mathrm{tr}(D_t) = 1$) describes the statistical state at the instant t.

A statistical description refers to an *ensemble of experimental results as registered with laboratory instruments*. It may be tempting to interpret such an ensemble as a collection of individual systems. However, in quantum theory object systems are in general entangled with their environment. Therefore, a statistical description does not allow us to refer to properties of individual quantum objects. This does not mean that quantum systems do not "have" properties, but it means that the discussion of properties of individual quantum systems does not pertain to the domain of epistemic descriptions.

In von Neumann's approach, statistical states are represented as *linear functionals*, generalizing the idea that the expectation value of a sum of operators should equal the sum of the expectation values of individual operators (even if they do not commute).[5] This postulate is difficult to justify since the sum of non-commuting operators is not operationally explained. It is not at all trivial that statistical expectation values can be represented by *linear* functionals. This problem has been solved by Gleason's theorem [13], which implies that every σ-additive probability measure on the projection lattice on a Hilbert space $\mathcal{H}$ of dimension larger than two can be extended to a unique normalized positive normal linear functional $\mathfrak{B}(\mathcal{H}) \to \mathbb{C}$ on the algebra $\mathfrak{B}(\mathcal{H})$ of all bounded linear operators acting on $\mathcal{H}$. In quantum theory, such a normalized positive normal linear functional is called a *statistical state functional*. Gleason's theorem is of probabilistic nature and does not apply to individual descriptions.

In addition to the linearity of statistical state functionals, the evolution of statistical states is linear as well. In contrast to widespread misconceptions, this linearity is independent of the superposition principle. Physical statistical theories are based on abstract convex structures which can be defined without any reference to linear spaces [40]. A basic postulate for any statistical description is Mackey's axiom IX requiring the commutativity of the operation of convex linear combinations with the time evolution semigroup for a statistical ensemble [23, p. 81]. It is essentially this postulate which *implies* the linearity of the dynamics of any fundamental statistical theory.[6]

---

[5] Postulate E in [27], Chapt. IV.1, p. 164.
[6] The basic results are due to [16].

A well-known example is Koopman's Hilbert-space formalism which rephrases the *nonlinear* Hamiltonian equations of motion of *classical* mechanics in terms of *linear* equations of motion for classical statistical mechanics [19]. Another example is the equivalence of nonlinear stochastic differential equations in the sense of Itô with the linear Fokker–Planck equations.

3.3 The So-Called Measurement Problem

To include measurements in his statistical approach, von Neumann introduced the ad hoc postulate that a measurement of an observable A with a purely discrete nondegenerate spectrum and the spectral resolution $A = \sum_k a_k P_k$ ($P_k = P_k^2 = P_k^*$) transforms any initial density operator D into the eigenstate $D_j := P_j D P_j / \mathrm{tr}(D P_j)$ if the eigenvalue a_j is measured.[7] The nonlinear, non-deterministic, discontinuous, instantaneous change $D \to D_j$ formalizes the historical idea of "quantum jumps" and is usually referred to as the *projection postulate* (corresponding to the *reduction of the wave packet*). Since the density operator D always refers to a statistical ensemble rather than one particular outcome, von Neumann's discontinuous nonlinear map $D \to D_j$ describes a change from a statistical to an individual description.

In classical probability theory such a map can be expressed in terms of *Bayes' rule for updating probabilities*. That is, $\mathrm{tr}(D_j A)$ is the *conditional expectation* of A, under the condition that the event a_j has occurred. Since the projection postulate merely indicates a change from a statistical to an individual description, the interpretation of the nonlinear map $D \to D_j$ as a physical process is untenable. In the statistical description, the so-called *objectification problem* ("how can a pointer observable assume a definite value?") is ill-posed as well: a statistical description never leads to definite individual results, but to well-defined probabilistic predictions.

3.4 Contemporary Views on Measurement

A realistic theory of measurements has to be dynamical and, therefore, cannot be based on von Neumann's projection postulate. Moreover, the dated view that quantum measurements deal with the determination of eigenvalues of selfadjoint operators (the so-called "observables") has been superseded by modern response theory. It specifies a *classical output signal* of a quantum system in response to a *classical external stimulus*, say an electromagnetic field or a mechanical force. For example, no spectroscopic measurement is a direct measurement of energy. The response in a spectroscopic experiment is given in terms of a molecular electric or magnetic multipole moment. Only by using the intricate theory of line broadening and level shifts can approximate information about energy levels be *deduced*.

A mathematical model which relates the information about an input stimulus to the information contained in the output response is called an *information channel*. Thus, *information* is a central concept of any epistemic description. The output system can be characterized by a Kolmogorov probability space (Ω, Σ, ν), consisting of

---

[7] See [27], Sect. VI.1, pp. 222–225.

a set Ω (the sample space), a class Σ of subsets of Ω which is a σ-algebra (of experimentally decidable events), and a σ-additive probability measure ν on Σ. The classical statistical output states are given by probability densities, i.e. by positive normalized elements of the Banach space $L^1(\Omega, \Sigma, \nu)$ of equivalence classes of integrable complex-valued functions on Ω.

Within the traditional Hilbert-space framework, a quantum information channel can be described in terms of a Hilbert space $\mathcal{H}$ and an affine map $\Lambda : \mathfrak{B}(\mathcal{H})_* \to L^1(\Omega, \Sigma, \nu)$ which transfers a normal state functional $\rho \in \mathfrak{B}(\mathcal{H})_*$ on the W*-algebra $\mathfrak{B}(\mathcal{H})$ of all bounded operators on $\mathcal{H}$ to a probability density $f \in L^1(\Omega, \Sigma, \nu)$ of the classical output system, $\Lambda(\rho) = f$.[8] The statistical state functional ρ can also be represented by a density operator D_ρ acting on $\mathcal{H}$, $\rho(A) = \text{tr}(D_\rho A)$ for every $A \in \mathfrak{B}(\mathcal{H})$. Every affine mapping from the set of all density operators on the Hilbert space $\mathcal{H}$ to the set of probability measures on the measurable space (Ω, Σ) can be realized by a normalized positive operator-valued measure $F : \Sigma \to \mathfrak{B}(\mathcal{H})$, also called a *probability–operator measure*.[9] Every positive operator-valued measure $F : \Sigma \to \mathfrak{B}(\mathcal{H})$ defines by $\mu(\mathcal{B}) = \text{tr}\{D_\rho F(\mathcal{B})\}$ ($\mathcal{B} \in \Sigma$) a probability measure μ on the measurable space (Ω, Σ), and by $\mu(\mathcal{B}) := \rho\{F(\mathcal{B})\}$ and $\Lambda(\rho) := f := d\mu/d\nu$ a measurement channel $\Lambda(\rho)$.

Applying standard methods of statistical decision theory, information about the statistical state ρ of the object system can be inferred by observing the classical output system. Since the convex set of all statistical quantum states is in general not a simplex, *a statistical description of quantum systems does not determine an ensemble of individual systems.* That is, a description of a quantum system in terms of statistical, epistemic states does not allow a reconstruction of an individual description in terms of individual, ontic states. A statistical experiment yields information about the epistemic quantum state which determines the response behavior of the chosen experimental arrangement. Any statement about "intrinsic properties" of an individual object system is impossible in an epistemic description.

3.5 Full-Fledged Epistemic Descriptions

Essentially all verifications and applications of quantum theory are based on the engineering approach sketched in the preceding subsection. Nevertheless, this approach is not complete. Engineering quantum theory cannot not explain the existence of facts, and it does not specify under which experimental conditions measurable events can occur at all. It provides just the *conditional probability* for an event $\mathcal{B} \in \Sigma$ registered on a classical output device, *under the condition* that the measuring instrument has factually and irreversibly registered this event. To achieve a complete dynamical description of the measurement process, the tools necessary for a registration of the results of a measurement have to be included in the theoretical description.

Facts make particular propositions true and others false, so they are to be described in *Boolean* terms. This means that in quantum theory facts must be described by

---

[8] This fertile view was first proposed in [15].
[9] For details of the definition of positive operator-valued measures compare for example [3].

classical observables. They are defined via the *center* of the algebra of the full quantum theoretical description. The center $\mathfrak{Z}(\mathfrak{M})$ of an algebra $\mathfrak{M}$ is a commutative algebra, defined as the set of all elements of $\mathfrak{M}$ which commute with all elements of $\mathfrak{M}$. The selfadjoint elements of the center $\mathfrak{Z}(\mathfrak{M})$ of the algebra of observables of a quantum theoretical description are called *classical observables*.[10]

The emergence of classical observables can only be derived from quantum systems with infinitely many degrees of freedom, e.g. including the electromagnetic field. This is impossible under the condition of a locally compact phase space presupposed by the Stone–von Neumann theorem of traditional Hilbert-space quantum mechanics. If this condition is relaxed, the framework of *algebraic quantum theory* allows us to comprise classical observables.

A full statistical description for the measurement of an observable A with a purely discrete nondegenerate spectrum and the spectral resolution $A = \sum_k a_k P_k$ is given by the *linear* map $D \to \sum_k p_k D_k$, with $p_k := \text{tr}(DP_k)$. This map $D \to \sum_k p_k D_k$ is a statistical description of a measuring operation *if and only if* the final density operator $\sum_k p_k D_k$ represents a mixture of *classically disjoint states* with the weights p_k. Two quantum states described by two density operators D_r and D_s are called *disjoint*, if a classical observable Z exists such that $\text{tr}(ZD_r) \neq \text{tr}(ZD_s)$. Hence, disjoint states can be distinguished and classified operationally by the numerical value of an appropriate classical observable. While a nonpure reduced state of a quantum system entangled with its environment can be described by $p_1D_1 + p_2D_2 + \cdots p_nD_n$, a proper classical mixture is represented as a density operator by the expression $p_1\mathcal{D}_1 \oplus p_2\mathcal{D}_2 \oplus \cdots \oplus p_n\mathcal{D}_n$, where $D_1 := \mathcal{D}_1 \oplus 0 \oplus \cdots \oplus 0$, etc.

A quantum system can act as a measurement apparatus only if it can produce disjoint final states. If the fundamental dynamics is given by a one-parameter group of automorphisms, such a result cannot be completed in finite time. But Hepp proved that for appropriate quantum systems a one-parameter automorphism group exists such that for classically equivalent initial states their asymptotic limits exist for $t \to \infty$ and their final states are disjoint [14].

Hepp's crucial contribution did not solve the statistical measurement problem, though, since it leaves the possibility to reverse the measurement process at any finite time [2]. Since facts cannot be undone, a measuring process has to be governed by an *irreversible* dynamics. The decisive idea of combining the nonanticipative character of laboratory instruments with a dynamics generating asymptotically disjoint final states was worked out by Lockhart and Misra [22].[11] In the proposed dynamics the time-inversion symmetry is broken and given by a weakly contractive semigroup (based on a K-flow) which produces asymptotically disjoint final states in a strictly irreversible manner. The emergence of disjoint states occurs progressively over finite amounts of time.[12]

---

[10] By construction, classical observables depend on Planck's constant $\hbar$. The still widely held view that classical mechanics is given by the limit $\hbar \to 0$ is untenable.

[11] Compare also *Mathematical Reviews* 87k, #81006 (November 1987).

[12] It has been objected that a measurement process with asymptotically disjoint final states implies an infinite measurement time. This is a misunderstanding: *every measurement in engineering physics is of this type*. For details, compare [29, 31].

4 Ontic Descriptions of Quantum Systems

4.1 General Structure of Individual Descriptions

Quantum theory is well established only in its statistical epistemic formulations. Although Bohr defended the view that quantum theory refers to individual systems, he never indicated a mathematical formulation of his conviction. Most adherents of one or another version of the "Copenhagen interpretation" use von Neumann's statistical codification, which does not allow a description of individual systems or experiments.

Statistical descriptions of classical systems can be given in terms of Boolean algebras. In Hilbert-space quantum theory, the corresponding probabilistic structure is given by the *non-Boolean* lattice of closed subspaces of a Hilbert space. A key feature of this lattice is that it is *partially Boolean*. That is, it can be represented as a family of Boolean algebras pasted together such that their operations agree with each other wherever two or more Boolean algebras overlap.[13]

In the following, we adopt a similar structure for *individual* non-probabilistic descriptions. Here, classical systems are characterized by *commutative* algebras, while quantum systems are described by *non-commutative, but partially commutative* algebras.

4.2 Algebraic Framework for Individual Descriptions

The paradigmatic example for a non-probabilistic description of *individual systems* is classical point mechanics, where a state of an individual system at time t can be represented by a point $\gamma_t \in \Gamma$ of a locally compact phase space Γ. A fully equivalent algebraic description can be given in terms of commutative C*-algebras. The Gelfand representation of commutative Banach algebras implies that the commutative algebra $\mathcal{C}_0(\Gamma)$ of all continuous complex-valued functions on Γ vanishing at infinity is a commutative C*-algebra $\mathfrak{C}$, and every commutative C*-algebra can be represented as a function algebra $\mathcal{C}_0(\Gamma)$ where Γ is some locally compact space. In the algebraic description of classical systems with the underlying commutative C*-algebra $\mathfrak{C}$, every individual state is represented by a multiplicative linear functional υ_t on $\mathfrak{C}$.

For *individual quantum systems*, the role of the phase space in classical mechanics is taken over by a *unital simple non-commutative* C*-algebra $\mathfrak{A}$. The partially commutative structure of a quantum system is reflected by the requirement that every commutative C*-subalgebra $\mathfrak{C} \subset \mathfrak{A}$ represents an individual classical system. The singly generated (by the unit element **1** and a single selfadjoint element $A \in \mathfrak{A}$) commutative C*-subalgebras of $\mathfrak{A}$ represent the *intrinsic properties* of the system.

The *ontic state* of an individual system refers to *its mode of being at a given instant*. The required partially commutative structure implies that an ontic state has to be represented by a functional $\upsilon : \mathfrak{A} \to \mathbb{C}$ which is a linear positive functional on every commutative C*-subalgebra of $\mathfrak{A}$. Yet, for individual descriptions there is no

---

[13] For details, compare [17, 18].

reason to assume that states are represented by positive *linear* functionals on the full C*-algebra $\mathfrak{A}$. (Recall that in epistemic descriptions the linearity of state functionals is warranted by the probabilistic Gleason theorem.)

As a natural condition for ontic state functionals, Misra proposed *monotone positivity* [24]: A functional υ on a C*-algebra $\mathfrak{A}$ is called *monotone positive* if and only if $A \geq B$ implies $\upsilon(A) \geq \upsilon(B) \geq 0$ for all positive elements A, B in $\mathfrak{A}$. While every positive linear functional is monotone positive, the converse is not true. It is an open question whether nonlinear state functionals are necessary for individual descriptions of quantum systems, or what special role linear state functionals play.

4.3 Ontic Valuations of Intrinsic Properties

In classical point mechanics an ontic state is determined by the values which its associated intrinsic properties have at a given instant. This attribution of a dispersion-free numerical value to every physical property is an *ontic valuation* of the property by a real-valued point function $\upsilon : \mathfrak{A} \to \mathbb{R}$. This valuation is *continuous* in the sense that for a given ontic state small variations of any property lead to small variations of its valuation.

In the early days of quantum mechanics, Dirac proposed that a dynamical variable A *has* the value a if the state vector Ψ is the eigenvector of A with the eigenvalue a [8, p. 666]. This partial ontic valuation of properties by point functions, which is taken over from classical point mechanics, is inappropriate for quantum mechanics. The principal objection against Dirac's valuation is that it is not continuous: *Non-commuting elements cannot be valuated in a continuous manner by point functions*. For example, in traditional Hilbert-space quantum mechanics, for every projection $F = F^* = F^2$ and $\varepsilon > 0$ there exists another projection F_ε with $\|F - F_\varepsilon\| = \varepsilon$ which does not commute with F. If Ψ is an eigenvector of F with the eigenvalue 1, then F *has* the value 1 according to Dirac's proposal, while the projection F_ε cannot be valuated.

In 1935, Schrödinger proposed, in a qualitative way, a continuous ontic valuation of intrinsic properties in terms of a "blurring of all variables" and emphasized that this is a perfectly clear and consistent concept.[14] Since the valuation proposed by Schrödinger has nothing to do with missing information, uncertainty, inexactness, imprecision, indefiniteness, ambiguity, randomness, or probability, we refer to it simply as a *valuation by set functions*.

An intrinsic property of an individual quantum system is represented by a commutative C*-subalgebra $\mathfrak{C} \subset \mathfrak{A}$ generated by a single selfadjoint element $C \in \mathfrak{A}$ with the spectrum $\text{sp}\{C\} \subset \mathbb{R}$. Then $\mathfrak{C}$ is isometric *-isomorphic to the function algebra $\mathcal{C}(\text{sp}\{C\})$. An ontic valuation of the intrinsic property C can be characterized by the *continuous set function* $\mathcal{B} \mapsto \mu_t^{\mathfrak{C}}(\mathcal{B})$, where $\mathcal{B}$ is an element of the σ-algebra of

---

[14] See [38, pp. 811–812]. This idea of an "objective vagueness" has not become popular but is endorsed by some contemporary philosophers and physicists under terms like "Verschmiertheit", "properties lacking sharp values", "blurred variables", "ontic blurring", "objective fuzziness", "unsharp properties".

Borel sets of the spectrum sp{C} of $\mathfrak{C} \sim \mathcal{C}_0(\text{sp}\{C\})$. Here the measure $\mu_t^{\mathfrak{C}}$ is defined via the Bochner–Cramér representation of the characteristic function

$$s \mapsto \upsilon_t\{e^{isC}\} = \int_{\text{sp}\{C\}} e^{isx} \mu_t^{\mathfrak{C}}(dx), \quad s \in \mathbb{R},$$

where υ_t is the ontic state at time t. In both classical and quantum theories, *every* intrinsic property of an isolated or open system has a *uniquely* defined continuous ontic valuation with respect to *every* ontic state. Moreover, the ontic valuation of properties is *in no way probabilistic*. It refers to properties, which the system actually and objectively has, and not to potentialities, propensities, or dispositions. Therefore, ontic valuations should not be misinterpreted as fluctuations or statistical spreads of supposed valuations by point functions.

If an individual quantum system is entangled with its environment, then its ontic state cannot be extremal, and no intrinsic property has a pointwise valuation, $\upsilon(C^2) > \upsilon(C)^2$, or equivalently, $\int_{\text{sp}\{C\}} x^2 \mu_t^{\mathfrak{C}}(dx) > \left(\int_{\text{sp}\{C\}} x \mu_t^{\mathfrak{C}}(dx) \right)^2$.

4.4 Ontically Dependent Properties

The non-commutativity of the C*-algebra underlying an ontic, individual description of a quantum system exhibits a new kind of dependence of properties. Loosely speaking, two properties represented by the commutative C*-algebras $\mathfrak{C}_1$ and $\mathfrak{C}_2$ are called *ontically independent* if every possible ontic valuation of $\mathfrak{C}_1$ is compatible with every possible ontic valuation of $\mathfrak{C}_2$. A trivial requirement for the ontic independence of two properties is that they are *functionally independent*, i.e. that they cannot be commonly represented by a third one.

Accordingly, two properties represented by the commutative C*-algebras $\mathfrak{C}_1$ and $\mathfrak{C}_2$ are called *ontically dependent* in a nontrivial way, if they are functionally independent, $\mathfrak{C}_1 \cap \mathfrak{C}_2 = \mathbb{1}\mathbb{C}$, *and* if at least two ontic valuations υ_1 and υ_2 exist such that there is no ontic valuation υ extending both υ_1 and υ_2. Two intrinsic properties are called *incompatible* if they are functionally independent but ontically dependent.

Consider two selfadjoint elements C_1 and C_2 of a simple non-commutative C*-algebra $\mathfrak{A}$ with $C_1 C_2 \neq C_2 C_1$. There exist two pure linear functionals $\upsilon_1 \in \mathfrak{A}^*$ and $\upsilon_2 \in \mathfrak{A}^*$ such that $\upsilon_1(C_1^2) = \upsilon_1(C_1)^2$ and $\upsilon_2(C_2^2) = \upsilon_2(C_2)^2$. As proven by Misra, a simple C*-algebra $\mathfrak{A}$ admits a valuation by a monotone positive functional υ which is dispersion-free, $\upsilon(A^2) = \upsilon(A)^2$, for all selfadjoint elements $A \in \mathfrak{A}$, if and only if $\mathfrak{A}$ is commutative [24]. Consequently, there is no valuation υ such that $\upsilon(C_1^2) = \upsilon(C_1)^2$ and $\upsilon(C_2^2) = \upsilon(C_2)^2$ if $C_1 C_2 \neq C_2 C_1$. That is, *mutually non-commuting commutative C*-subalgebras of a C*-algebra represent incompatible properties*.

4.5 Superposition Principle and Ontic Entanglement

There are many physically equivalent representations of quantum theory. The traditional irreducible Hilbert-space representation has the important advantage that it allows a very simple mathematical formulation of the quantum mechanical superposition principle in terms of state vectors: Every linear combination of two state vectors is another state vector, called a coherent superposition of the two generating state vectors. Since pure states are represented by *rays* (and not by vectors), the underlying state space is the set of rays, i.e. *a nonlinear projective space*. As a consequence, the linearity of the traditional formulation of the quantum mechanical superposition principle is not an intrinsic feature, but a peculiarity of a cleverly chosen representation. From a conceptual point of view, a representation-independent formulation without linear structure is preferable: *Any two different pure states generate an uncountably infinite family of mutually different pure states different from the two generating pure states*.[15]

The meaning of a superposition state is not that a system "can be in two or more states at the same time", but that there is an *ontic holistic entanglement*, implying that the system considered is not composed of independently existing parts. Consider for example two subsystems described by mutually commuting C*-subalgebras $\mathfrak{B} \subset \mathfrak{A}$ and $\mathfrak{C} \subset \mathfrak{A}$ of an individual quantum system described by the C*-algebra $\mathfrak{A}$. Consider any linear state functional υ on $\mathfrak{A}$, two arbitrary elements $B_j \in \mathfrak{B}$ with $0 \leq B_j \leq 1$, and two arbitrary elements $C_k \in \mathfrak{C}$ with $0 \leq C_k \leq 1$, $B_j C_k = C_k B_j$, with $j, k = 1, 2$. A quantity describing the mutual ontic dependence of the two commuting subsystems is given by the functional $\kappa := (1/2)\,\upsilon(B_1 C_1 + B_1 C_2 + B_2 C_2 - B_2 C_1)$. Then Schwarz's inequality implies [20, 21].

$$|\kappa|^2 \leq 1 + \frac{1}{4} \upsilon\{(B_1 B_2 - B_2 B_1)(C_1 C_2 - C_2 C_1)\} \leq 2\,.$$

If $\mathfrak{B}$ or $\mathfrak{C}$ is commutative, we get $|\kappa| \leq 1$, a generalized Bell inequality. It is well known that in quantum theory Bell's inequality can be violated. If the generalized Bell inequality is violated, $|\kappa| > 1$, the state functional υ is *holistically entangled* with respect to the decomposition $(\mathfrak{B}, \mathfrak{C})$. The relation $|\kappa|^2 \leq 1 + (1/4)\upsilon\{(B_1 B_2 - B_2 B_1)(C_1 C_2 - C_2 C_1)\}$ shows that *ontic holistic entanglement between two subsystems is possible if and only if there exist incompatible properties in both subsystems*.

4.6 Dynamics of Individual Quantum Systems

The usual time-dependent Schrödinger equation is linear since it refers to a statistical description. This fact, however, does not imply the linearity of the state dynamics in individual descriptions. Since the quantum mechanical superposition principle is independent of any dynamical principle, *a nonlinear dynamics for individual quantum states is not in contradiction with the quantum mechanical superposition principle*. In particular, the quantum mechanical superposition principle does not require the invariance of superpositions under the time evolution.

According to current knowledge, the only way to *derive* a dynamics from fundamental first principles is to introduce, first, group-theoretical elementary systems

---

[15] For a representation-independent mathematical formulation of quantum mechanical superpositions, see [35].

and, then, interactions by gauge fields. Such a procedure is not satisfactory because elementary systems also represent the sources of the fields by which they interact.

Nevertheless, *bare elementary systems* are defined as ergodic representations of the presupposed kinematical group. Both the Lorentz and the Galilei group are *linear* Lie groups, implying that the dynamics of the corresponding *bare* elementary systems is governed by *linear* equations of motion. Yet, bare elementary systems are just auxiliary constructions. They are transformed to *dressed* systems by interactions with gauge fields (e.g. electromagnetism, gravitation). Dressing is a rather complicated procedure which may lead to nonlinear equations of motion, in particular due to self-interactions. Moreover, the inclusion of gauge fields enforces a discussion of quantum systems with infinitely many degrees of freedom which may have a classical part. An interaction of a quantum system with a classical subsystem entails *feedback effects* which necessarily generate nonlinearities in the dynamics.

Every quantum system with charged elementary systems (like electrons) inevitably interacts with the electromagnetic radiation field. The separation of an individual object system from its electromagnetic environment leads to an individual non-autonomous subdynamics for the object subsystem with a driving force which is chaotic in the individual sense of Wiener [46]. Since the electromagnetic environment acts as a K-system [43], Birkhoff's individual ergodic theorem applies [4], so that the external force acting on the object system can be regarded as an individual trajectory of a stochastic process. In this case, Birkhoff's individual ergodic theorem provides the crucial link between individual and statistical descriptions. It allows a transformation of the individual nonlinear chaotic dynamics into the linear dynamics of the statistical ensemble description.

On this basis an *individual* description of the measuring process is feasible. In simple models the reduced dynamics is then governed by a non-autonomous nonlinear equation which transforms pure ontic states into pure ontic states. Since the reduced object system is driven by a chaotic process (in the sense of Wiener), its behavior exhibits a sensitive dependence on the initial conditions of the environment. Epistemically indistinguishable but ontically different initial states of the environment can lead to asymptotically disjoint ontic final states. In the associated statistical description, results corresponding to the mentioned model by Lockhart and Misra can be achieved.

5 Relations Between Epistemic and Ontic Descriptions

5.1 From First Principles to Observed Phenomena

No fundamental theory does directly describe context-dependent phenomena of empirical science. To describe empirically accessible information in terms of a fundamental theory, the contexts introduced by measuring tools or pattern detection and recognition devices used by the experimentalist have to be taken into account explicitly. In mathematical terms, a context can be imposed on the fundamental theory by restricting its domain of validity and introducing a new, coarser topology compatible with the intrinsic topology of the underlying fundamental theory.

To get a mathematically complete and consistent theory, one has to close the derived theory in the new *contextual topology*. This results in a *qualitatively different* higher-level theory whose domain of validity may intersect nontrivially with the domain of validity of the fundamental theory. If neither one of the two domains is contained in the other, then any strong reductionist scheme fails. This situation usually leads to the emergence of *qualitatively new* properties.

There are many possibilities to introduce a contextual topology. A standard strategy uses singular asymptotic expansions which do not converge in the original topology of the ontic description. Examples include the emergence of phenomena of geometrical optics (such as "shadows") in the high-frequency limit, and the emergence of inductors, capacitors and resistors in the low-frequency limit of Maxwell's electrodynamics. Another instance is the emergence of the shape of molecules in the singular asymptotic Born–Oppenheimer expansion at the singular point of an infinite nuclear mass.[16] These and related examples show that for epistemic descriptions *the specification of a context is as important as the underlying ontic description*.

Another major issue in this context is the emergence of separate subsystems (parts) from an original entangled system as a whole. (Mereological emergence in this sense is different from mereological emergence in many philosophical discussions where a whole is considered to emerge from parts.) Any interaction with the system as a whole, e.g. by measurement, provides disentangled subsystems. The states of the subsystems can have properties which the entangled state before measurement did not have. If epistemic correlations between such properties violate Bell's inequality, they indicate the ontic entanglement of the original state.

A most powerful first principle in modern physics is *symmetry*. In the words of Weyl: "As far as I see, all a priori statements in physics have their origin in symmetry" [45, p. 126]. Yet, fundamental symmetries are not directly accessible by experiments. They can only be indirectly inferred by *symmetry breakings*.

A symmetry is spontaneously broken if the realized states show less symmetry than their associated equations of motion (or the Hamiltonian). Broken symmetries play a crucial role for the description of many physical phenomena. For example, ferromagnetism involves the spontaneous breakdown of the rotation symmetry, crystallization requires the spontaneous breakdown of the translation and rotation symmetry, superfluidity is related to the breakdown of the special Galilei symmetry, and superconductivity is connected with the spontaneously broken gauge symmetry.

At the molecular level, the parity symmetry is broken in the ground states of chiral molecules (such as L-amino acids or D-sugars). Moreover, every experiment presupposes irreversibility and requires *nonanticipative* measuring instruments, hence a distinction between past and future and the selection of a direction of time. Since fundamental physical laws do not distinguish between past and future, the time-inversion symmetry of the underlying fundamental ontic description has to be broken.

---

[16] See [30] for more details.

5.2 Emergence in Algebraic Quantum Theory

In algebraic quantum theory, a particular context can be introduced by imposing a contextually selected topology upon the state space of the C*-algebra $\mathfrak{A}$ of the underlying ontic description. This new topology has to be compatible with the algebraically determined, hence context-independent norm topology of $\mathfrak{A}$. It can be implemented by a particular reference state, given by a positive linear state functional ρ on the context-independent C*-algebra $\mathfrak{A}$. The so-called GNS-construction (according to Gel'fand, Naimark and Segal) then allows the construction of a context-dependent Hilbert space $\mathcal{H}_\rho$ and an associated faithful representation $\pi_\rho(\mathfrak{A})$ of the C*-algebra $\mathfrak{A}$ acting on $\mathcal{H}_\rho$.[17] The closure of $\pi_\rho(\mathfrak{A})$ in the weak topology of the algebra $\mathfrak{B}(\mathcal{H}_\rho)$ of all bounded operators acting on $\mathcal{H}_\rho$ is a context-dependent W*-algebra $\mathfrak{M}_\rho$, called the *algebra of contextual observables* with respect to the contextual topology generated by the reference state functional ρ. The relation $\mathfrak{A} \sim \pi_\rho(\mathfrak{A}) \subset \mathfrak{M}_\rho \subset \mathfrak{B}(\mathcal{H}_\rho)$ implies that all intrinsic properties appear also as contextual properties, but in addition there are new contextual properties which are not intrinsic.

Observables in the W*-algebra $\mathfrak{M}_\rho$ of contextual observables, which are outside the representation of the C*-algebra $\pi_\rho(\mathfrak{A})$, are called *emergent observables*. They represent properties which are novel in the sense that they are absent in the more fundamental context-independent C*-algebraic description. *The emergence of novelty in contextual descriptions is a compelling fact in algebraic quantum theory.* Simple examples of *emergent classical observables* are the temperature [41] and chemical potential [25] which arise in a most natural manner from a GNS-construction with respect to canonical KMS-states.

A basic common feature in all these cases of emergent behavior is the transition from an ontic to an epistemic description. If the interaction of an object system with its environment is not excluded, there are in general infinitely many *physically inequivalent* representations of the object system. Choosing one of these representations means to select a particular context for the epistemic description. *The simplicity of natural laws manifests itself only in the ontic description, while a representation of the richness of observable phenomena requires the multitude of inequivalent representations.*

5.3 Relative Onticity

Although a picture comprising both epistemic and ontic descriptions together with their relationship is very appealing, it is still too simplistic if epistemic and ontic elements must be considered in *one and the same* descriptive framework. For instance, such a situation is unavoidable if, as in Bohr's perspective, an epistemic description of a system is assumed to presuppose an ontic description of measuring tools. In such situations, a combination of ontic and epistemic elements is required in the desired descriptive framework. This difficulty can be resolved if it is realized that the distinction of epistemic and ontic descriptions can be applied to the entire hierarchy

---

[17] Compare for example [42], or [6].

of (perhaps partially overlapping) domains leading from fundamental particles in basic physics to chemistry and even to living systems in biology and psychology. Ontic and epistemic descriptions are then considered as *relative to* two (successive) domains in the hierarchy.

Let us briefly illustrate this by an example. From the fundamental viewpoint of quantum theory, atoms and molecules are objects with highly contextual properties, which can be described by interactions of electrons, nuclei, and their environments. However, from the viewpoint of a chemist one is usually not interested in these interactions, but in the shape, chirality, and similar features of molecules. In such a framework, it is reasonable to consider the properties of an atom or a molecule as intrinsic properties in an ontic description rather than as epistemic, contextual properties derived from more basic intrinsic properties of protons, neutrons, or even "more basic" constituents. While atoms and molecules are epistemically described within the domain of basic physics, they acquire ontic significance within the domain of chemistry.

Similarly, liquid water has properties which water molecules do not have, e.g. the property of wetness. It can be derived as an emergent, contextual property from an underlying ontic description. For a chemist, however, it would be absurd to dismiss the wetness of water as something which does not refer to an "independent reality" of water. Why should water only be wet if it is observed?

The central point of the concept of *relative onticity* is that states and properties of a system, which belong to an epistemic description in a particular domain, can be considered as ontic from the perspective of another domain. This idea resembles the (less formal) discussion of "ontological relativity" originally introduced by Quine [34]. Quine argues that if there is one ontology that fulfills a given theory, then there is more than one. He, thus, claims that it makes no sense to say what the objects of a theory are, beyond saying how to interpret or reinterpret that theory in another theory.

For Quine, any question as to the "quiddity" (the "whatness") of a thing is meaningless unless a conceptual scheme is specified relative to which it is discussed. He encourages "ontological commitment" in the sense that a most proper conceptual frame should be preferred for the interpretation of a theory. For Quine, the inscrutability of reference is the issue which causes the problems necessitating ontological relativity, not the unique assignment of referents as objects in the external world.[18]

Putnam has developed a related kind of ontological relativity, first called "internal realism", later sometimes modified to "pragmatic realism" [32, 33]. Ontological (sometimes conceptual) relativity in Putnam's internal realism differs from Quine's usage of the term in an important detail. While Quine's ontological relativity is due to the impossibility of a uniquely fixed relationship of our concepts to the totality of objects to which those concepts refer, Putnam's position is more radical insofar as he questions that we know what we mean when we speak of a totality of objects [7, p. 185]. This shift in emphasis is particularly interesting in view of the

---

[18] Compare [12].

holistic features of quantum systems which challenge the notion of an object as part of a system within an ontic description.

From this rough characterization,[19] a close relation between ontological relativity à la Putnam and the idea of relative onticity is apparent. Assuming that Putnam's notion of an object can be more precisely characterized by the states and properties of such an object, conceptual frames or schemes serve a purpose very similar to contextual representations in the framework of algebraic quantum theory. Both ontological relativity and relative onticity refer to a conception of realism where the states and properties of objects have to be described relative to a context, and they agree with regard to a basic assertion according to which there is a "real world as such". These basic issues can be successfully implemented in a sound formal manner, if epistemic and ontic descriptions are properly distinguished and related to each other.

6 Conclusions

Distinguishing epistemic and ontic descriptions of quantum systems is a key to avoid the category mistake of confounding concepts of observation-dependent and observation-independent quantum realities. Epistemic descriptions refer to a statistical description of ensembles of experimental outcomes. By contrast, ontic descriptions refer to individual systems without any respect to their observation or measurement. Epistemic and ontic descriptions require different mathematical codifications.

Although all applications of quantum theory are based on epistemic formulations, this does not imply that ontic descriptions are altogether pointless. Since they are free from particular contexts required for particular applications, ontic descriptions necessarily feature a higher degree of symmetry than epistemic descriptions. It is possible to derive epistemic descriptions from ontic descriptions, if the particular contexts can be implemented precisely enough. An ontic description of a system can give rise to many different epistemic descriptions mutually excluding each other.

An ultimately context-free ontic description or, equivalently, a universal context would be of perfect symmetry and cannot be formulated. Hence, any ontic description presupposes an "ontological commitment", an agreement about a fundamental domain of discourse based on some context which is as general as possible or reasonable. This entails a relativity of onticity which is crucial for the coherent discussion of emergent behavior and can be formalized by contextual symmetry breakings.

References

1. H. Atmanspacher and F. Kronz: Relative onticity. In: H. Atmanspacher, A. Amann and U. Müller-Herold (eds.): *On Quanta, Mind and Matter*. Kluwer, Dordrecht 1999, pp. 273–294
2. J.S. Bell: On wave packet reduction in the Coleman–Hepp model. *Helvetica Physica Acta* 48, 93–98 (1975)

---

[19] For more details see [1].

3. S.K. Berberian: *Notes on Spectral Theory*. Van Nostrand, Princeton 1966
4. G.D. Birkhoff: Proof of the ergodic theorem. *Proceedings of the National Academy of Sciences of the United States of America* 17, 656–660 (1931)
5. N. Bohr: Discussion with Einstein on epistemological problems in atomic physics. In: P.A. Schilpp (ed.): *Albert Einstein: Philosopher–Scientist*. Library of Living Philosophers, Evanston, Illinois 1949, pp. 199–241
6. O. Bratteli and D.W. Robinson: *Operator Algebras and Quantum Statistical Mechanics. I. C^*- and W^*-Algebras, Symmetry Groups, Decomposition of States*. 2nd edn. Springer, New York 1987
7. A. Burri: *Hilary Putnam*. Campus Verlag, Frankfurt 1994
8. P.A.M. Dirac: On the theory of quantum mechanics. *Proceedings of the Royal Society (London)* A112, 661–677 (1926)
9. A. Einstein: Autobiographical notes. In: P.A. Schilpp (ed.): *Albert Einstein: Philosopher–Scientist*. Library of Living Philosophers, Evanston, Illinois 1949, pp. 1–95
10. A. Einstein: Reply to criticism. In: P.A. Schilpp (ed.): *Albert Einstein: Philosopher–Scientist*. Library of Living Philosophers, Evanston, Illinois 1949, pp. 665–688
11. J.H. Fetzer and R.F. Almeder: *Glossary of Epistemology / Philosophy of Science*. Paragon House, New York 1993
12. R.F. Gibson: Quine, Willard Van Orman. In: J. Kim and E. Sosa (eds.): *A Companion to Metaphysics*, pp. 426–428. Blackwell, Oxford 1995
13. A.M. Gleason: Measures on the closed subspaces of a Hilbert space. *Journal of Mathematics and Mechanics* 6, 885–893 (1957)
14. K. Hepp: Quantum theory of measurement and macroscopic observables. *Helvetica Physica Acta* 45, 237–248 (1972)
15. A.S. Holevo: On the mathematical theory of quantum communication channels. *Problems of Information Transmission* 8, 47–54 (1972) [Russian original: *Problemy Peredači Informacii* 8, 63–71 (1972)]
16. R.V. Kadison: Transformations of states in operator theory and dynamics. *Topology* 3, Suppl. 2, 177–198 (1965)
17. S. Kochen and E.P. Specker: Logical structures arising in quantum theory. In: J. Addison, L. Henkin and A. Tarski (eds.): *The Theory of Models*. North Holland, Amsterdam 1965
18. S. Kochen and E.P. Specker: The problem of hidden variables in quantum mechanics. *Journal of Mathematics and Mechanics* 17, 59–88 (1967)
19. B.O. Koopman: Hamiltonian systems and transformations in Hilbert space. *Proceedings of the National Academy of Sciences of the United States of America* 17, 315–318 (1931)
20. L.J. Landau: On the violation of Bell's inequality in quantum theory. *Physics Letters A* 120, 54–56 (1987)
21. L.J. Landau: Experimental tests of general quantum theories. *Letters in Mathematical Physics* 14, 33–40 (1987)
22. C.M. Lockhart and B. Misra: Irreversibility and measurement in quantum mechanics. *Physica A* 136, 47–76 (1986)
23. G.W. Mackey: *The Mathematical Foundations of Quantum Mechanics*. Benjamin, New York 1963
24. B. Misra: When can hidden variables be excluded in quantum mechanics? *Nuovo Cimento* 47 A, 841–859 (1967)
25. U. Müller-Herold: Disjointness of β-KMS states with different chemical potential. *Letters in Mathematical Physics* 4, 45–88 (1980)
26. J. von Neumann: Die Eindeutigkeit der Schrödingerschen Operatoren. *Mathematische Annalen* 104, 570–578 (1931)

27. J. von Neumann: *Mathematische Grundlagen der Quantenmechanik*. Springer, Berlin 1932
28. A. Petersen: The philosophy of Niels Bohr. *Bulletin of the Atomic Scientist* 19, No. 7, 8–14 (1963)
29. H. Primas: The representation of facts in physical theories. In: H. Atmanspacher and E. Ruhnau (eds.): *Time, Temporality, Now*. Springer, Berlin 1997, pp. 241–263
30. H. Primas: Emergence in exact natural science. *Acta Polytechnica Scandinavica* Ma 91, 83–98 (1998)
31. H. Primas: Asymptotically disjoint quantum states. In: P. Blanchard, D. Giulini, E. Joos, C. Kiefer and I.-O. Stamatescu (eds.): *Decoherence: Theoretical, Experimental, and Conceptual Problems*. Springer, Berlin 2000, pp. 161–178
32. H. Putnam: *Reason, Truth and History*. Cambridge University Press, Cambridge 1981
33. H. Putnam: *The Many Faces of Realism*. Open Court, La Salle 1987
34. W.V. Quine: Ontological relativity. In: W.V. Quine (ed.): *Ontological Relativity and Other Essays*. Columbia University Press, New York 1969, pp. 26–68
35. J.E. Roberts and G. Roepstorff: Some basic concepts of algebraic quantum theory. *Communications in Mathematical Physics* 11, 321–338 (1969)
36. E. Scheibe: *Die kontingenten Aussagen in der Physik*. Athenäum Verlag, Frankfurt 1964
37. E. Scheibe: *The Logical Analysis of Quantum Mechanics*. Pergamon Press, Oxford 1973
38. E. Schrödinger: Die gegenwärtige Situation in der Quantenmechanik. *Naturwissenschaften* 23, 807–812, 823–828, 844–849 (1935)
39. M.H. Stone: Linear transformations in Hilbert space. III. Operational methods and group theory. *Proceedings of the National Academy of Sciences of the United States of America* 16, 172–175 (1930)
40. M.H. Stone: Postulates for the barycentric calculus. *Annali di Matematica Pura ed Applicata* 29, 25–30 (1949)
41. M. Takesaki: Disjointness of the KMS states of different temperatures. *Communications in Mathematical Physics* 17, 33–41 (1970)
42. M. Takesaki: *Theory of Operator Algebras I*. Springer, New York 1979
43. L.C. Thomas: A note on quantising Kolmogorov systems. *Annales de l'Institut Henri Poincaré, Physique théorique A* 21, 77–79 (1974)
44. C.F. von Weizsäcker: *Aufbau der Physik*. Hanser Verlag, München 1985
45. H. Weyl: *Symmetry*. Princeton University Press, Princeton 1952
46. N. Wiener: Generalized harmonic analysis. *Acta Mathematica* 55, 117–258 (1930)

Information and Fundamental Elements of the Structure of Quantum Theory

Časlav Brukner and Anton Zeilinger

According to Aage Petersen Niels Bohr liked to say: "There is no quantum world. There is only an abstract quantum physical description. It is wrong to think that the task of physics is to find out how Nature is. Physics concerns what we can say about Nature." In an analogous way, von Weizsäcker suggested that the notion of the elementary alternative, the "Ur", should play a pivotal role when constructing physics. Both approaches suggest that the concept of information should play an essential role in the foundations of any scientific description of Nature. We show that if, in our description of Nature, we use one definite proposition per elementary constituent of Nature, some of the essential characteristics of quantum physics, such as the irreducible randomness of individual events, quantum complementary and quantum entanglement, arise in a natural way. Then quantum physics is an elementary theory of information.

1 Introduction

All our description of objects is represented by propositions. The use of propositions is not a matter of our choice. In contrast, it is a necessity which is behind each of our attempts to learn something new about Nature and to communicate this knowledge with others. It is a necessity which we follow constantly and without any intention and it seems that there is no way to avoid it even if the phenomena to be described are highly counterintuitive and distinct from both our everyday experience and the classical world view. One may even say that there is no need to avoid it. The reason is that the only way we are able to understand any phenomena in Nature, including quantum phenomena, is exclusively through the epistemological structure of classical physics and everyday experience. Bohr [4] emphasized that "How far the [quantum] phenomena transcend the scope of classical physical explanation, the account of all evidence must be expressed in classical terms. The argument is simply that by the word 'experiment' we refer to a situation where we can tell others what we have done and what we have learned and that, therefore, the account of the experimental arrangement and the result of observation must be expressed in unambiguous language with suitable application of the terminology of classical physics." Von Weizsäcker [42] emphasized that the understanding of new physical theories will also be given in language: "This verbalized language must be the language spoken by those physicists who do not know yet the theory we are telling them. The language used in order to explain a theory which we propose in mathematical form is the language which has been existed before the theory. On the other hand, it is not self-evident that this language has a clear meaning at all, because if it had a completely clear meaning probable the new theory would not be needed. Thus it may happen that by applying our new formalism to experience – an application made possible by our existing language – we may tacitly or explicitly change the rules of this very language."

Rigorously speaking a system is nothing else than a construct based on a complete list of propositions together with their truth values. The propositions from the list could be (1) "The velocity of the object is v" or (2) "The position of the object is x" and could be associated both to classical and to quantum objects. Yet, there is an important difference between the two cases. From the theorems of Bell [2] and of Kochen and Specker [29] (1967) we know that for a quantum system one cannot assert definite (noncontextual) truth values to all conceivable propositions jointly. For example, if the proposition (1) above is a definite proposition, then the proposition (2) must necessarily be completely indefinite and vice versa. The two propositions are mutually exclusive. This is a specific case of quantum complementarity.

Therefore, in an attempt to describe quantum phenomena we are unavoidably put in the following situation. On one hand the epistemological structure applied has to be inherited from classical physics: the description of a quantum system has to be represented by the propositions which are used in the description of a classical system, and on the other hand, not all propositions cannot be assigned to a quantum system simultaneously. Now, a natural question arises: How to join these two, seemingly inconsistent, requirements? We suggest to use the concept of "knowledge" or "information". Then even in situations where we cannot assert simultaneously definite truth-values to mutually exclusive propositions we can assert measures of information about their truth values. The structure of the theory including the description of the time evolution can then be expressed in terms of measures of information[1]. To us this seems to be a change with the lowest possible "costs" in the epistemological structure of classical physics or even everyday experience. And since some costs are unavoidable anyway we believe that the information-theoretical formulation of quantum physics leads to the "easiest" understanding of the theory.

Also, from the point of view that the information content of a quantum system is fundamentally limited we will discuss precisely the empirical significance of the terms involved in formulating quantum theory, particularly the notion of a quantum state. However we are aware of the possibility that this might not carry the same degree of intuitive appeal for everyone. It is clear that it may be matter of taste whether one accepts the suggested concepts and principles as self-evident as we do

---

[1] Heisenberg [25] wrote: "The laws of nature which we formulate mathematically in quantum theory deal no longer with the particles themselves but with our knowledge of the elementary particles ... The conception of objective reality ... evaporated into the ... mathematics that represents no longer the behavior of elementary particles but rather *our knowledge of this behavior.*"

or not. If not, then one may turn the reasoning around and, following our approach in [6–8], argue for the validity of the statements given in the paper on the basis of known features of quantum physics.

The conceptual groundwork for the ideas presented here has been prepared most notably by von Bohr [5], Weizsäcker [41] and Wheeler [44]. In contrast to those other authors who look for deterministic mechanisms hidden behind the observed facts, these authors attempt to understand the structure of quantum theory as a necessity for extracting whatever meaning from the data of observations.

In recent years several ideas, different from our approach, were put forward suggesting that information can help us to learn more about the foundations of quantum physics. The foundations of quantum mechanics are interpreted in the light of quantum information [11, 12, 17, 18]. It was also suggested how to reduce quantum theory to few statements of physical significance by generalizing and extending classical probability theory [23, 24]. In another approach it was shown how certain elements of the structure of quantum theory emerge from looking for invariants of probabilistic observations assuming that any newly gained information shall lead to more accurate knowledge of these invariants [36–39].

2 Finiteness of Information, Ur, Elementary System

One of the most distinct features of quantum physics with respect to classical physics is that prediction with certainty of individual outcomes is only possible for a very limited class of experiments. Such a prediction is equivalent to saying that the corresponding propositions have definite truth values. For all other (complementary) propositions the truth values are necessarily indefinite. We suggest this to be a consequence of the feature that:[2]

The information content of a quantum system is finite.

With this we mean that a quantum system cannot carry enough information to provide definite answers to all questions that could be asked experimentally. Then, by necessity, the answer of the quantum system to some questions must contain an element of randomness. This kind of randomness must then be irreducible, that is, it cannot be reduced to "hidden" properties of the system. Otherwise the system would carry more information than what is available. Thus, without any additional physical structure assumed, we let the irreducible randomness of an individual event and complementarity, be a consequence of the finiteness of information.

How much information is available to a quantum system? If this information is limited than it is natural to assume that if we decompose a physical system, which may

---

[2] Feynman wrote: "It always bothers me that, according to the laws as we understand them today, it takes a computing machine an infinite number of logical operations to figure out what goes on in no matter how tiny a region of space and no matter how tiny a region of time, ... why should it take an infinite amount of logic to figure out what one tiny piece of space-time is going to do?" A closely related view was assumed by Landauer, who writes in his article "Information is Physical" [30]: " ... the laws of physics are ... limited by the range of information processing available.".

be represented by numerous propositions, into its constituents, each such constituent will be described by fewer propositions. This process of subdividing a system can go further until we reach a final limit when an individual system represents the truth value to one single proposition only. It is then suggestive to replace the above statement by a more precise one [50]:

The most elementary system represents the truth value of one proposition.

We call this the principle of quantization of information, a consequence of the countability of propositions. One may consider the above statement as a definition of what is the most elementary system. Note that the truth value of a proposition can be represented by one bit of information with "true" being identified with the bit value "1" and "false" being identified with the bit value "0". Thus, the principle becomes simply:

The most elementary system carries 1 bit of information.

We relate the notion of the most elementary system to that of the "Ur" introduced by von Weizsäcker. He was the first who introduced the concept of the most basic informational constituent of all objects ("Ur"). Von Weizsäcker wrote [42]: "It is certainly possible to decide any large alternative step by step in binary alternatives. This may tempt us to describe all objects as composite systems composed from the most simple possible objects. The simplest possible object is an object with a two-dimensional Hilbert space, the 'ur'. The word 'ur' is introduced to have an abstract term for something which can be described by quantum theory and has a two-dimensional Hilbert space, and nothing more."

How much information is contained in more complex systems consisting of N elementary systems? It is natural to assume that the information content of a complex system is proportional to the number of elementary constituents. The principle of quantization of information is thus generalized to [50]:

N elementary systems represent the truth values of N propositions,

or equivalently,

N elementary systems carry N bits.

Again one may consider these statements as definitions of what is a composite system consisting of N elementary systems. Note that the principle given above does not make any statement about how the information contained in N propositions (N bits) is distributed over the N systems. It can be represented by the N systems individually or, alternatively, it can be represented by N systems jointly. The latter is the feature of quantum entanglement, discussed in more detail below, for which Schrödinger [35] wrote: "If two separated bodies, each by itself known maximally, enter a situation in which they influence each other, and separate again, then there occurs regularly that which I have just called entanglement of our knowledge of the two bodies. The combined expectation-catalog consists initially of a logical sum of the individual catalogs; during the process it develops causally in accord with known law (there is no question whatever of measurement here). The knowledge remains maximal, but at its end, if the two bodies have again separated, it is not again split into a logical sum of knowledges about the individual bodies.

The fundamental statements given above seem to suggest that binary (yes-no) alternatives are representatives of basic information units of all systems. Consider the case of n-fold (i.e. ternary, quanternany etc.) alternatives generalizing the binary ones. Obviously, any $n=2^N$-fold alternative is decomposable into binary ones. Note that such an alternative can be realized in measurement of N elementary systems. Yet, it is not obvious how to decompose or how to consider a general n-fold alternative with $n \neq 2^N$? An interesting possibility would be to find the factorization of number n into its prime-number factors $p_1, p_2, \ldots$ and then to decompose the n-fold alternative into a sequential serial of p_1-fold alternatives, p_2-fold alternatives etc. Obviously such an approach would require to extend the notion of elementary system to all prime-number dimensional systems. Interestingly, as it will be shown later (see Sect. 6), only in these cases where the dimension of the quantum system is (a power of) a prime number, our definition of the total information content of the system is unambiguous. This might suggest that the notion of the elementary system should indeed be extended to all prime-number dimensions. However, in this manuscript we will mainly restrict our analysis to binary decomposable alternatives.

We would like to stress again that notions such as that a system "represents" the truth value of a proposition or that it "carries" one bit of information only implies a statement concerning what can be said about possible measurement results. For us a system is no more than a coherent representative of a compplete list of propositions.

3 Mutually Complementary Propositions

We consider an explicit example of an elementary system, the spin-1/2 particle, and the Stern–Gerlach experiment as depicted in Fig. 1 schematically. Depending on whether the upper or the lower detector plate is hit by a particle we call the outcome "yes" and "no" respectively, where "yes" and "no" represent the truth values of the proposition for the spin to be up along a chosen direction. The upper detector plate is hit with probability p. If it is not hit the other detector plate will be hit with probability $1 - p$. Therefore we consider a binary alternative. Different experimental situations are specified by the orientation θ of the magnet in the Stern–Gerlach apparatus as shown in Fig. 1.

Consider an elementary system specified by the true proposition "The spin along the z-axis is up" (or, equivalently, by the false proposition "The spin along the $-z$-axis is up"). This situation is described by the probabilities $p(0)=1$ and $p(\pi)=0$ for the "yes" outcome. Because, according to our principle, a spin can carry one bit of information only, each proposition: "The spin along the direction tilted at an angle θ $(0 < \theta < \pi)$ from the z-axes is up" has to be probabilistic (Fig. 2). How does the probability $p(\theta)$ of a "yes" count depend on the angle θ?

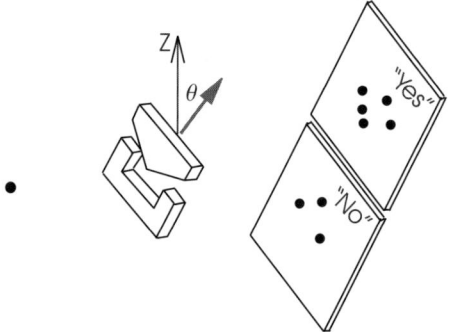

Fig. 1. Spin measurement of a spin-1/2 particle. The particle passes through the Stern–Gerlach magnet oriented at the angle θ, and then it hits one of the detector plates behind the Stern–Gerlach magnet. Depending on whether the upper or the lower detector plate is hit by a particle we call the outcome "yes" and "no", respectively

We assume that the mapping of θ to $p(\theta)$ is analytic[3] and monotonic. Then using the Cauchy theorem about continuous and monotonic functions one concludes that there has to be one and only one angle of orientation of the magnet in the Stern–Gerlach apparatus where the probabilities for a "yes" and for a "no" outcome are equal. Because of the symmetry of the problem this obviously has to be the angle $\pi/2$. For each direction **n** in the x-y plane (the green circle on the sphere in Figs. 2 and 3) the proposition "The spin along the **n**-axis is up" is completely indefinite, that is, we have absolutely no knowledge which outcome "yes" or "no" will be observed in a specific individual measurement.

Note, however, that in principle this equal number of yes-no outcomes could also be achieved by an ensemble of systems each giving a definite result for each direction such that the same number of "yes" or "no" results is obtained. Yet again this would imply that an individual system carries enough information to permit assignment of definite truth values to all possible propositions, in contradiction to our basic principle.

Consider now the state of a spin-1/2 particle specified by the proposition "The spin along the x-axis is up (down)". In this case we have complete knowledge which outcome will be observed when the Stern–Gerlach magnet is oriented along the $\pm x$-axis at the expense of the fact that we have absolutely no knowledge about the

---

[3] If, in contrast, $p(\theta)$ would only be sectionally analytic in θ then there would be points of nonanalyticity separating two regions in which the function $p(\theta)$ has different analytic forms. Thus the values of the function on a finite segment in the interior of a domain of analyticity would only determine, by the uniqueness theorem for analytic functions, the function up to the next point of nonanalyticity. Clearly, to describe such a system completely we would need catalogs both of functional values on finite segments in the interior of each domain of analyticity and of the positions of the points of nonanalyticity. Such a catalog would require large amount of information to describe the functional dependence and thus contradicts our desideratum of minimal information content of a quantum system.

Information and Structure of Quantum Theory 329

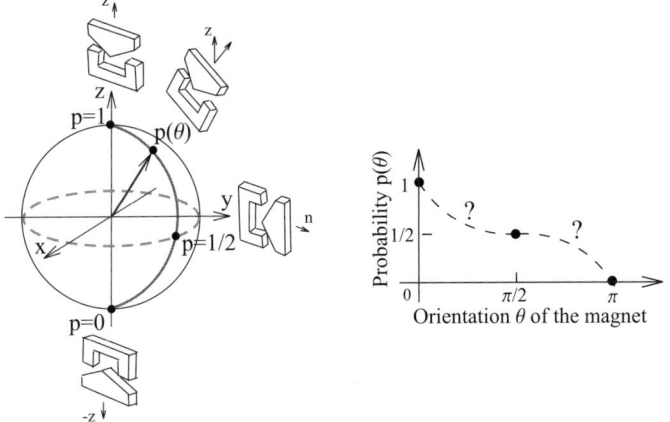

Fig. 2. The gradual change of the probability $p(\theta)$ of a "yes" ("spin up") count with a gradual change of the orientation θ of the magnet in the Stern–Gerlach apparatus. The measurement along the z-axis gives the result "yes" with certainty. Because of the symmetry of the problem the probabilities for a "yes" and for a "no" count in a measurement along any direction in the x-y plane (the dashed circle) are equal ($= 1/2$). How does the probability $p(\theta)$ of a "yes" count depend on θ explicitly?

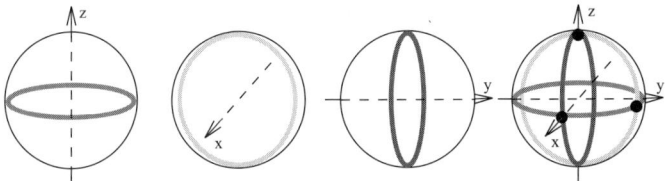

Fig. 3. The formation of mutually complementary propositions associated with orthogonal spin components. If measurement along the z-axis (x-axis) [y-axis] gives a definite result, measurement along any direction in the x-y plane, the equator (y-z plane, the first vertical circle) or [x-z plane, the second vertical circle] will be maximally random, respectively. There are altogether three mutually complementary spin measurements represented by three intersection points of the three circles

outcome for the orientation of the magnet along *any* direction in the y-z plane (the first vertical circle on the sphere in Fig. 3).

Finally, consider a spin-1/2 particle specified by the proposition "The spin along the y-axis is up (down)". In that case we know precisely the outcome of the experiment when the Stern–Gerlach magnet is oriented along the $\pm y$-axis at the expense of complete uncertainty about the outcome when it is oriented along any direction in the x-z plane (the second vertical circle on the sphere in Fig. 3).

There are, therefore, altogether *three mutually exclusive* or *complementary* propositions (represented by three intersection points of the three circles on the sphere in Fig. 3): "The spin along direction n_1 is up (down)", "The spin along direction n_2 is up (down)" and "The spin along direction n_3 is up (down)", where n_1, n_2 and n_3 are mutually orthogonal directions. These are propositions with a property of mutually exclusiveness: the total knowledge of one proposition is only possible at the cost of total ignorance about the other two complementary ones. In other words precise knowledge of the outcome of one experiment implies that all possible outcomes of complementary ones are equally probable.

Why are there exactly 3 mutually complementary propositions for the elementary system and not, e.g., 2 or 4? We do not understand that fully. However the discussion above indicates that there is a strong link between the number (3) of mutually complementary propositions and the (three-)dimensionality of ordinary space. We will come back once more to this question in the conclusions. But it is important to note that in any system with dichotomic (2-valued) observables there are always three complementary propositions even if these cannot be linked to the dimensionality of ordinary space.

4 Measure of Information in a Probabilistic Experiment

Consider a probabilistic experiment with n possible outcomes. Suppose that the experimenter plans to perform N trials of the experiment. All he knows before the trials are performed are the probabilities $p_1, \ldots, p_i, \ldots, p_n$ for all possible outcomes to occur: What kind of prediction can the experimenter make?

In general two cases are conceivable. The experimenter can ask: "What is the precise sequence of the N outcomes?" or "What is the number of occurrences of outcome i?". We will say that in answering the first question the experimenter makes a "deterministic" prediction and in answering the second one he makes a "probabilistic" prediction[4] (following discussion in [38]). The deterministic prediction can only make sense if different outcomes follow from an intrinsic difference between individuals of the ensemble measured – the situation which we have in classical measurements. Then the precise sequence of outcomes reveals which specific property which individual member of the ensemble carries. One can show that for sufficiently large N Shannon's measure of information [34]

$$H = -\sum_{i=1}^{n} p_i \log p_i \qquad (1)$$

---

[4] Summhammer [38] wrote: "I want to discard a deterministic link. The reason is that the amount of records available to the observer to form a conception of the world is always finite, so that many different sets of laws can be invented to account for them. Pinning down any one of these sets as *the* laws of nature is then purely speculative. On the other hand we have the probabilistic view, which is successfully used to interpret quantum observations. It seems that in this view we assign a minimum of information content to observed data. To see this, imagine the N trials of a probabilistic yes-no experiment, like tossing a coin, in which the outcome "yes" occurs L times. If we want to tell somebody else the result it is sufficient to state the values of N and of L. With the deterministic view, in which the precise sequence of outcomes is important, we would in general have to communicate many more details to enable the receiver to reconstruct this sequence.

is equal to the mean minimal number of yes-no questions (when the logarithm in (1) is taken to base 2) necessary to determine which particular sequence of outcomes occurs, divided by N. As we show in more detail elsewhere [8] this suggests that Shannon's measure of information is the adequate measure of the uncertainty in the case of deterministic prediction and thus in classical measurements.

In contrast to classical measurements, quantum measurements, in general, cannot be claimed to reveal an individual quantum system's property existing before the measurement is performed. This causes certain conceptual difficulties when we try to define the information gain in quantum measurements using the notion of Shannon's measure (for discussion of these points see [7–9, 21, 40]). Since outcomes of quantum-mechanical experiments are in general intrinsically probabilistic, there the experimenter can only make probabilistic predictions. If the experimenter decides to perform N experimental trails, all he can guess is the future number of occurrences of a specific outcome. Such a prediction will now be analyzed for the case of two possible outcomes "yes" and "no".

Because of the statistical fluctuations associated with any finite number of experimental trials, the number L of occurrences of the "yes" outcome in future repetitions of the experiment is not precisely predictable[5]. The random variable L is subject to a binomial distribution. Since it has a finite σ deviation, it fulfills Chebyshev's inequality [19]:

$$\text{Prob}\{|L - pN| > k\sigma\} \leq \frac{1}{k^2}, \qquad (2)$$

where the standard deviation σ is given by

$$\sigma = \sqrt{p(1-p)N} . \qquad (3)$$

This inequality means that the probability that the number L will deviate from the product pN by more than k standard deviations is less than or equal to $1/k^2$. In the case of small σ, large deviations in the number of occurrences of the "yes" outcome from

---

[5] Here, a very subtle and careful position was assumed by von Weizsäcker [42] who wrote: "It is most important to see that this [the fact that probability is not a prediction of the precise value of the relative frequency] is not a particular weakness of the objective empirical use of the concept of probability, but a feature of the objective empirical use of any quantitative concept. If you predict that some physical quantity, say a temperature, will have a certain value when measured, this prediction also means its expectation value within a statistical ensemble of measurements. The same statement applies to the empirical quantity called relative frequency. But here are two differences which are connected to each other. The first difference: In other empirical quantities the dispersion of the distribution is in most cases an independent empirical property of the distribution and can be altered by more precise measurements of other devices; in probability the dispersion is derived from the theory itself and depends on the absolute number of cases. The second difference: In other empirical quantities the discussion of their statistical distributions is done by another theory than the one to which they individually belong, namely by the general theory of probability; in probability this discussion evidently belongs to the theory of this quantity, namely of probability itself. The second difference explains the first one."

the mean value pN are improbable. In this case the experimenter knows the future number of occurrences with a high certainty. Conversely, a large σ indicates that not all highly probable values of L lie near the mean pN. In that case the experimenter knows much less about the future number of occurrences.

We suggest to identify the experimenter's uncertainty U with σ^2. Then it will be proportional to the number of trials. This important property guarantees that each individual performance of the experiment contributes the same amount of information, no matter how many times the experiment has already been performed. After each trial the experimenter's uncertainty about the specific outcome therefore decreases by

$$U = \frac{\sigma^2}{N} = p(1-p). \tag{4}$$

This is the lack of information about a specific outcome with respect to a single future experimental trial. If, instead of two outcomes, we have n of them with the probabilities $\boldsymbol{p} \equiv (p_1, p_2, \ldots p_n)$ for the individual occurrences, then we suggest to define the total lack of information regarding all n possible experimental outcomes as

$$U(\boldsymbol{p}) = \sum_{j=1}^{n} U(p_j) = \sum_{j=1}^{n} p_j(1-p_j) = 1 - \sum_{j=1}^{n} p_j^2. \tag{5}$$

The uncertainty is minimal if one probability is equal to one and it is maximal if all probabilities are equal.

This suggests that the knowledge, or information, with respect to a single future experimental trial an experimentalist possesses before the experiment is performed is somehow the complement of $U(\boldsymbol{p})$ and, furthermore, that it is a function of a sum of the squares of probabilities. A first ansatz therefore would be $I(\boldsymbol{p}) = 1 - U(\boldsymbol{p}) = \sum_{i=1}^{n} p_i^2$. Expressions of such a general type were studied in detail by Hardy, Littlewood and Pólya [22]. Notice that this expression can also be viewed as describing the length of the probability vector $\boldsymbol{p}$. Obviously, because of $\sum_i p_i = 1$, not all vectors in probability space are possible. Indeed, the minimum length of $\boldsymbol{p}$ is given when all p_i are equal ($p_i = 1/n$). This corresponds to the situation of complete lack of information in an experiment about its future outcome. Therefore we suggest to normalize the measure of information in an individual quantum measurement as obtaining finally

$$I(\boldsymbol{p}) = \mathcal{N} \sum_{i=1}^{n} \left(p_i - \frac{1}{n}\right)^2, \tag{6}$$

where $\mathcal{N}$ is the normalization[6]. Specifically, for a binary experiment the measure of information is given as

---

[6] In [6] only those cases were considered where maximally k bits of information can be encoded, i.e. $n = 2^k$. The normalization there is $\mathcal{N} = 2^k k/(2^k - 1)$. Then $I(\boldsymbol{p})$ results in k bits of information if one $p_i = 1$ and it results in 0 bits of information when all p_i are equal.

$$I(p_1, p_2) = 2\left(p_1 - \frac{1}{2}\right)^2 + 2\left(p_2 - \frac{1}{2}\right)^2 = (p_1 - p_2)^2. \tag{7}$$

It reaches its maximal value of 1 bit of information if one of the probabilities is one and it takes its minimal value of 0 bits of information if both probabilities are equal.

5 The Catalog of Knowledge of a Quantum System

Consider again a stationary experimental arrangement with two detectors, where only one detector fires in each experimental trial. The first detector, say, fires (we call this the "yes" outcome) with probability p_1. If it is does not fire the other detector fires with probability $p_2 = 1 - p_1$ (the "no" outcome).

Note that the experimenter's measure of information for the binary experiment as defined by (7) is invariant under permutation of the set of possible outcomes. In other words, it is a symmetrical function of p_1 and p_2. A permutation of the set of possible outcomes can be achieved in two manners, which may be called "active" and "passive". In the passive point of view the permutation is obtained by a simple relabelling of the possible outcomes and the property of invariance is self evident because relabelling obviously does not make an experiment more predictable.

From the active point of view, one retains the same labeling, and the permutation of the set of outcomes refers to a real change of the experimental set-up. For a spin measurement this would be a re-orientation of the Stern–Gerlach magnet. In that case the property of invariance states that the measure of information is indifferent under certain real physical changes of the experimental situation. This requirement is more stringent and may be precisely formulated as an invariance of the measure of information under interchange of the following two physical situations: (a) the probability for "yes" is p_1 and for "no" is p_2; and (b) the probability for "yes" is p_2 and for "no" is p_1. Yet these are different experimental situations.

In order to remove this ambiguity in the description of the experiment one can assign probabilities for occurrences or different numbers or other distinct labels to possible outcomes, the particular scheme is of no further relevance. For example, one can use the statement "the probability for the outcome 'yes' is 0.6, and for the outcome 'no' is 0.4", or the statement "the probability for the outcome 'yes' is 0.4, and for the outcome 'no' is 0.6" to distinguish between the situations (a) and (b) given above. Note that in both cases the measure of information as defined by (7) is $I = 0.04$.

Here we will remove this ambiguity by using a particular description which is based on the quantity

$$i = p_1 - p_2. \tag{8}$$

Then, on one hand, the sign of i differs between the two situations in (a) and (b), and on the other hand, the square of i is equal to the measure of information ($I = i^2$). Therefore i represents an economic and complete description of the experimental situation (equivalent to the assignment of specific probabilities for the two results)[7]

---

[7] We give another justification for introducing i. Our main goal in the next section will be to derive the functional dependence of probability $p_1(\theta)$ (recall $p_2(\theta) = 1 - p_1(\theta)$) on the

On the other hand, all the "quantum state" is meant to be is a representation of that catalog of our knowledge that is necessary to arrive at the set of, in general probabilistic, predictions for all possible future observations of the system. Such a view was assumed by Schrödinger [35] who wrote[8]: "Sie (die ψ-Funktion) ist jetzt das Instrument zur Voraussage der Wahrscheinlichkeit von Maßzahlen. In ihr ist die jeweils erreichte Summe theoretisch begründeter Zukunftserwartungen verkörpert, gleichsam wie in einem *Katalog* niedergelegt." The ψ function is characterized by a set of complex numbers which are very remote from our everyday experience. Yet, if the origin of the structure of quantum theory is to be sought in a theory of observations, of observers, and of meaning, then we should focus our attention not on complex numbers, but rather on real-value quantities which are directly observable[9]. Interestingly, quantum theory allows descriptions of quantum states in terms of real numbers. An example for this is the description of density operators in terms of the real coefficients in the decomposition into generators of SU(N) algebra (basis of generalized Pauli matrices as used in, e.g., [33]).

We will use a description of the state of an elementary system by a vector $\boldsymbol{i} = (i_1, i_2, i_3) = (p_x^+ - p_x^-, p_y^+ - p_y^-, p_z^+ - p_z^-)$, which is a catalog of knowledge about a set of three mutually complementary propositions and where, in the case of spin, p_x^+ is the probability to find the particle's spin up along x etc. The catalog $\boldsymbol{i}$ is a complete description of the system because its knowledge is sufficient to determine the probabilities for the outcomes of all possible future measurements.

Denote by θ an arbitrary direction within the y-z plane which is oriented at an angle θ with respect to the z-axis. Now, for all θ the propositions: $P_1(\theta)$: "The spin is up along the direction x", $P_2(\theta)$: "The spin is up along the direction θ," and $P_3(\theta)$: "The spin is up along the direction $\theta + 90°$" are mutually complementary. The different lists of the three mutually complementary propositions are labeled by a single experimental parameter θ as given in Fig. 4. They correspond to different representations $\boldsymbol{i}(\theta) = (i_1(\theta), i_2(\theta), i_3(\theta))$ of the catalog of our knowledge of the system as shown in Fig. 5.

6 Total Information Content of a Quantum System

The finiteness of the information covers not just the extreme cases of maximal knowledge of one proposition at the expense of complete ignorance of complementary

---

value of an experimental parameter θ. We will first derive the functional dependence $\boldsymbol{i}(\theta)$ and therefrom that of $p_1(\theta)$. Note that for this purpose one could not use $I(\theta)$ instead of $\boldsymbol{i}(\theta)$ because with any value $I(\theta)$ one can associate two physically non-equivalent situations (a) and (b) which also correspond to different values of the probabilities.

[8] Translated: "It (the ψ-function) is now the instrument for predicting the probability of measurement results. In it is embodied the respectively attained sum of theoretically grounded future expectations, somehow like laid down in a *catalogue*."

[9] As Peres put it: "After all, quantum phenomena do not occur in a Hilbert space. They occur in a laboratory."

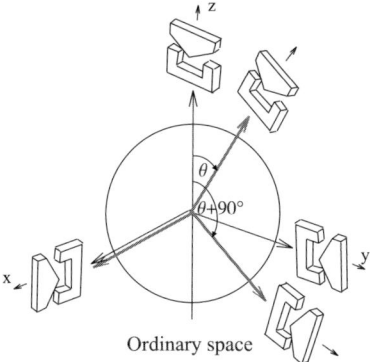

Fig. 4. Two sets of three mutually complementary Stern–Gerlach arrangements labeled by a single experimental parameter θ which specifies the orientations of the Stern–Gerlach magnets in the three experiments. The three experimental arrangements are associated to the mutually complementary propositions: $P_1(\theta)$: "The spin along the x-axis is up", $P_2(\theta)$: "The spin is up along the direction tilted at angle θ from the z-axes" and $P_3(\theta)$: "The spin is up along the direction tilted at angle $\theta + 90°$ from the z-axes"

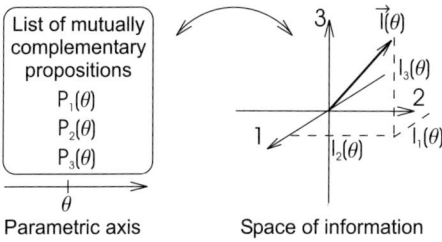

Fig. 5. Representation of the state of a quantum system by the information vector $\vec{i}(\theta)$. The components $(i_1(\theta), i_2(\theta), i_3(\theta))$ of the information vector are associated to the three mutually complementary propositions $P_1(\theta)$, $P_2(\theta)$ and $P_3(\theta)$

ones but it also applies to intermediate cases. For example, it has been pointed out that in the interference experiments one can obtain some partial knowledge about the particle's path and still observe an interference pattern of reduced contrast as compared to the ideal interference situation [15,48]. This suggests that the information content of the system can manifest itself as path information or as modulation of the interference pattern or partially in both to the extent defined by the finiteness of information [10]. Yet, how to define the total information content of a quantum system precisely?

Bohr [5] remarked that " ... phenomena under different experimental conditions must be termed complementary in the sense that each is well defined and that together they exhaust all definable knowledge about the object concerned". This suggests that the total information content of a quantum system is somehow contained in the full set of mutually complementary experiments. We define the total information (of 1 bit) of the elementary (or binary, or two-state) system as a sum of the individual measures of information over a complete set of three mutually complementary experiments

$$I_{\text{total}} = I_1 + I_2 + I_3 = 1 \,. \tag{9}$$

How to define the total information content of more complex systems? In an n-dimensional Hilbert space, one needs $n^2 - 1$ real parameters to specify a general density matrix ρ, which must be hermitean and have $\text{Tr}(\rho) = 1$. Since measurements within a particular basis set can yield only $n - 1$ independent probabilities (the sum of all probabilities for all possible outcomes in an individual experiment is one), one needs $n + 1$ distinct basis sets to provide the required total number of $n^2 - 1$ independent probabilities. Ivanović [27] showed that the required number $n + 1$ of unbiased basis sets (i.e. mutually complementary observables) indeed exists if n is a prime number, and Wootters and Fields [47] showed that it exists if n is any power of a prime number[10]. This suggests that the complete information represented by the density matrix is fully contained in a complete set of mutually complementary observables.

In the future discussion we take the number of mutually complementary observables in a complete set as given in the quantum theory, yet we expect that this number will also be derived in future from fundamental considerations.

Generalizing (9) we suggest to define the total information content of a n-dimensional quantum system as the sum of individual measures of information $I_i(\boldsymbol{p}^i)$ over a complete set of $n+1$ mutually complementary measurements

$$I_{\text{total}} = \sum_{i=1}^{n+1} I_i(\boldsymbol{p}^i) = \mathcal{N} \sum_{i=1}^{n+1} \sum_{j=1}^{n} \left(p_j^i - \frac{1}{n} \right)^2 \,. \tag{10}$$

Here $\boldsymbol{p}^i = (p_1^i, \ldots, p_n^i)$ are the probabilities for the outcomes in the i-th measurement. In the case of a system composed of N elementary systems and with appropriate normalization I_{total} results in N bits of information (for the system in a pure state).

The question whether or not one can find a complete set of mutually complementary observations in the general case of a Hilbert space of arbitrary dimensions is still open. If it should turn out to be the case, then the definition (10) can be applied to arbitrarily dimensional quantum systems. If, however, such sets only exist if the dimension is the power of a prime number, then we suggest to take this seriously, as implying that the prime number alternatives are the most basic informational constituents of all objects (see also introduction). On the basis of this assumption the information content of a complex system of general dimension could be defined as a sum of the information contents of its individual constituents (each with the dimension of a prime number) plus the information contained in the correlations between them.

For example, the system of dimension $n = p_1 \cdot p_2$ where p_1 and p_2 are the prime-number factors of n can be considered as a composite system consisting of

---

[10] A special case is a composite system consisting of N elementary systems, which has dimension $n = 2^N$.

two subsystems of dimensions p_1 and p_2 (The lowest dimension for which the existence of a complete set of mutually complementary observables has not been proven is $n = 6$. There $p_1 = 2$ and $p_3 = 3$.). One obtains $p_1^2 - 1$ independent numbers from a complete set of mutually complementary measurements of the first subsystem and $p_2^2 - 1$ independent numbers from such set of measurements of the second subsystems. Additional $(p_1^2 - 1)(p_1^2 - 1)$ numbers can be obtained from the correlations for measurements of the two subsystems. Therefore one obtains altogether $(p_1^2 - 1)(p_2^2 - 1) + (p_1^2 - 1) + (p_2^2 - 1) = p_1^2 p_2^2 - 1$ independent parameters, which is the number of independent parameters which completely define the density operators of our system of dimension $p_1 p_2$.

7 Malus' Law in Quantum Physics

Quantum theory predicts $p(\theta) = \cos^2(\theta/2)$ for the probability to find the spin up along the direction at an angle θ with respect to the direction along which the system gives spin up with certainty. From which deeper foundation emerges this law in quantum mechanics, originally formulated by Malus[11] for light? The most important contributions so far in that direction are those of Wootters [46], Summhammer [36,37] and Fivel [16]. In this section we argue that the most natural functional relation $p(\theta)$ consistent with the principle of quantization of information is indeed the sinusoidal dependence of Malus.

We wish to specify a mapping of θ onto $\boldsymbol{i}(\theta)$. It is of importance to note that we can invent this mapping freely. The reason for this is that θ will have functional relations to other physical parameters of the experiment. Then, the laws relating those parameters with the information vector $\boldsymbol{i}(\theta)$ can be seen as laws about relations between those parameters and θ plus a mapping of θ onto $\boldsymbol{i}(\theta)$. Which basic assumptions should we follow to obtain the mapping from θ to $\boldsymbol{i}(\theta)$ most appropriate for quantum mechanics?

There are two basic assumptions. The first one is the assumption of the *invariance* of the total information content under the change of representation of the catalog of our knowledge of the system. Or, in other words, it is the assumption that total information content must be independent of the particular choice of mutually complementary propositions considered (see Fig. 6). In the same spirit as choosing a coordinate system, one may choose any complete set of mutually complementary propositions to represent our knowledge of the system and the total information about the system must be invariant under that choice, i.e. for all θ

$$I_{\text{total}} = I_1(\theta) + I_2(\theta) + I_3(\theta) = i_1^2(\theta) + i_2^2(\theta) + i_3^2(\theta) = 1 . \tag{11}$$

In fact, this property of invariance is the reason why we may use the phrase "the total information content of the system" without explicitly specifying a particular reference set of mutually complementary propositions.

---

[11] Etienne Louis Malus (1775–1812), a French physicist, was almost entirely concerned with the study of light. He conducted experiments to verify Huygens' theory of light and rewrote the theory in analytical form. His discovery of the polarization of light by reflection was published in 1809 and his theory of double refraction of light in crystals in 1810.

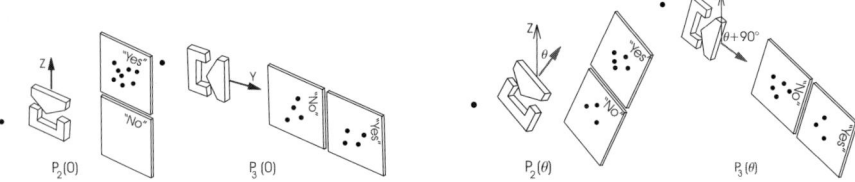

Fig. 6. Two different sets of mutually complementary spin measurements (the full sets include also the spin measurement along the x-axis which is not shown in the figures). They correspond to the following two sets of mutually complementary propositions: $\{P_1(0)$: "The spin along the x-axis is up", $P_2(0)$: "The spin along the y-axis is up", $P_3(0)$: "The spin along the z-axis is up"$\}$, and $\{P_1(\theta)$: "The spin along x-axis is up", $P_2(\theta)$: "The spin along the direction tilted at angle θ from the z-axes is up", $P_3(\theta)$: "The spin along the direction tilted at angle $\theta + 90°$ from the z-axes is up"$\}$. The total information carried by the spin is independent of the particular set of mutually complementary propositions considered, i.e. $I_{\text{total}} = I_1(0) + I_2(0) + I_3(0) = 0 + 0 + 1 = I_1(\theta) + I_2(\theta) + I_3(\theta)$ in the example shown

We suggest that only mappings where neighboring values of θ correspond to neighboring values of $i(\theta)$ are natural (compare Chap. 3). Thus, if we gradually change the orientation of the magnets in a set of Stern–Gerlach apparata defining a complete set of mutually complementary observables, then a continuous change of the information vector will result. The property of invariance defined by (11) implies that with a gradual change of the experimental parameter from θ_0 to θ_1 the information vector rotates in the space of information, i.e.,

$$i(\theta_1) = \hat{R}(\theta_1 - \theta_0, \theta_0) i(\theta_0) . \tag{12}$$

This transformation preserves the length of the information vector (Fig. 7). The rotation matrix $\hat{R}$ depends on two independent variables θ_0 and θ_1; here specific arguments $\theta_1 - \theta_0$ and θ_0 are chosen for convenience. Equation (12) expresses our expectation that the transformation law is linear[12], that is, independent of the actual information vector transformed. $\hat{R}(\theta_1 - \theta_0, \theta_0)$ is an orthonormal matrix

$$\hat{R}^{-1}(\theta_1 - \theta_0, \theta_0) = \hat{R}^T(\theta_1 - \theta_0, \theta_0).$$

---

[12] Precisely speaking, the invariance property only implies that the general transformation law $i(\theta_1) = f(\theta_1, \theta_0, i(\theta_0))$ is described by a mapping f which preserves the length of the information vector. Now, consider the situation where with probability w_A a system is prepared in state $i_A(\theta_0)$ and with probability w_B in $i_B(\theta_0)$. Then the information vector is given by $i(\theta_0) = w_A i_A(\theta_0) + w_B i_B(\theta_0)$. Let us now suppose that the experimental parameter in each of the three mutually complementary experiments is changed from the value θ_0 to θ_1. The individual information vectors $i_A(\theta_0)$ and $i_B(\theta_0)$ evolve independently, resulting in $w_A f(\theta_1, \theta_0, i_A(\theta_0)) + w_B f(\theta_1, \theta_0, i_B(\theta_0))$ for the total information vector at θ_1. This shows that the function f is linear: $f(\theta_1, \theta_0, w_A i_A(\theta_0) + w_B i_B(\theta_0)) = w_A f(\theta_1, \theta_0, i_A(\theta_0)) + w_B f(\theta_1, \theta_0, i_B(\theta_0))$ for convex sums over i_A and i_B. For an extension of the proof to arbitrary sums one can follow the idea from Appendix 1 of [23], which is there applied in a different context.

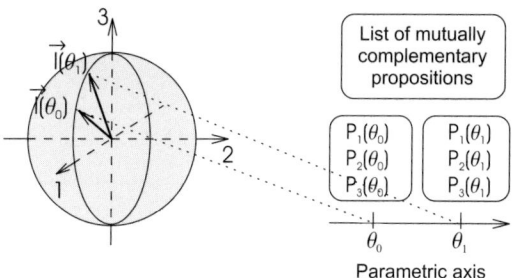

Fig. 7. A general rotation of the information vector from $i(\theta_0)$ to $i(\theta_1)$ due to a change of the physical parameter from θ_0 to θ_1

Notice that transformation matrices do not build up a group in general because of the explicit dependence on both the initial and final parametric value.

The second basic assumption in the derivation of the Malus' law in quantum physics is that no physical process *a priori* distinguishes one specific value of the physical parameter from others, that is, that the parametric θ-axis is *homogeneous*. In our example with the orientation of Stern–Gerlach magnets as the experimental parameter, the homogeneity of the parametric axis becomes equivalent to the *isotropy* of ordinary space. The homogeneity of the parametric axis precisely requires that if we transform the physical situations of the three complementary experiments together with the state of the system along the parametric axis for any real number b, we can never observe any effect. Using a more formal language this means the following. Suppose two lists each with three mutually complementary experimental arrangements are associated with a specific parametric value θ_0 and with some other value $\theta_0 + b$ ($-\infty < b < +\infty$) respectively. Furthermore, suppose the information vectors $i(\theta_0)$ and $i(\theta_0 + b)$ associated with the two lists are equal (i.e. all components of the two vectors are equal). The homogeneity of the parametric θ-axis then requires that if we change the physical parameter in each experiment by an equal interval of $\theta - \theta_0$ in the two lists of complementary experiments, the resulting information vectors will be equal again, as shown in Fig. 8. Mathematically, if $i(\theta_0) = i(\theta_0 + b)$ for all θ_0 implies $\hat{R}(\theta - \theta_0, \theta_0) i(\theta_0) = \hat{R}(\theta - \theta_0, \theta_0 + b) i(\theta_0 + b)$, then[13]

$$\hat{R}(\theta - \theta_0, \theta_0) = \hat{R}(\theta - \theta_0, \theta_0 + b). \tag{13}$$

The transformation matrix then depends only on the difference between the initial and final value of the experimental parameter, and not on the location of these values on the parametric θ-axis.

The orthogonality condition leads to the following general form of the transformation matrix

---

[13] We give another line of reasoning, that is to require the *same* functional dependence of the transformation law for each initial value θ_0 of the parameter. This can only be done with (13).

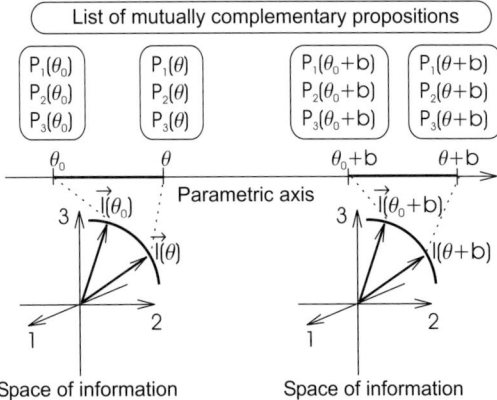

Fig. 8. The homogeneity of the parametric θ-axis

$$\hat{R}(\theta) = \begin{pmatrix} 1 & 0 & 0 \\ 0 & f(\theta) & -g(\theta) \\ 0 & g(\theta) & f(\theta) \end{pmatrix}, \quad (14)$$

where we take $\theta_0 = 0$ for simplicity and $f(\theta)$ and $g(\theta)$ are not yet specified but assumed to be analytical functions satisfying

$$f^2(\theta) + g^2(\theta) = 1, \; f(0) = 1 \text{ and } g(0) = 0 . \quad (15)$$

For $\theta = 0$ one has $\hat{R} = \hat{1}$ as there is no change of the physical situations of the complementary experiments.

We further require that a change of the experimental parameter in a set of mutually complementary arrangements from θ_0 to θ_1 and subsequently from θ_1 to θ_2 must have the same physical effect as a direct change of the parameter from θ_0 to θ_2. The resulting transformation will then be independent, whether we apply two consecutive transformations $\hat{R}(\theta_1 - \theta_0)$ and $\hat{R}(\theta_2 - \theta_1)$ or a single transformation $\hat{R}(\theta_2 - \theta_0)$

$$\hat{R}(\theta_2 - \theta_0) = \hat{R}(\theta_2 - \theta_1)\hat{R}(\theta_1 - \theta_0) . \quad (16)$$

This implies that transformation matrices build up the group of rotations SO(3), a connected subgroup of the group of orthogonal matrices O(3) which contains the identity transformation.

For the special case of infinitesimally small variation of the experimental conditions, (16) reads

$$\hat{R}(\theta + d\theta) = \hat{R}(\theta)\hat{R}(d\theta) . \quad (17)$$

Inserting the form (14) of the transformation matrix into the latter expression, one obtains

$$f(\theta + d\theta) = f(\theta)f(d\theta) - g(\theta)g(d\theta) . \quad (18)$$

Using conditions (15), we transform (18) into the differential equation

$$\frac{df(\theta)}{d\theta} = -n\sqrt{1 - f^2(\theta)},\qquad(19)$$

where

$$n = -g'(0)\qquad(20)$$

is a constant. The solution of the differential equation reads

$$f(\theta) = \cos n\theta,\qquad(21)$$

where we integrate between 0 and θ using the condition $f(0) = 1$ from (15). This finally leads to

$$\hat{R}(\theta) = \begin{pmatrix} 1 & 0 & 0 \\ 0 & \cos n\theta & -\sin n\theta \\ 0 & \sin n\theta & \cos n\theta \end{pmatrix}.\qquad(22)$$

This result directly gives the familiar expression

$$p = \cos^2 \frac{n\theta}{2}\qquad(23)$$

for probability in quantum theory.

Mathematically one could consider our result as a direct and immediate consequence of the theory of group representations; the cosine dependence follows from a particular representation of the rotation group. However from the physical perspective it is implied by two fundamental assumptions: (1) the total information of the system is invariant under the change of the representation of the catalog of our knowledge about the system and (2) the parametric space is homogeneous. If (1) and (2) are satisfied, then the probability must vary as $\cos^2 n\theta$, where n is a parameter not determined by the derivation. Quantum-mechanical probabilities are just of this form, with θ for a relative polarization angle and with $n = 1/2$ for electrons and neutrinos, or with $n = 1$ for photons, or with $n = 2$ for gravitons. The same functional dependence $\cos^2 \varphi$ undergoes also the probability to find a particle in a specific output beam in a Mach–Zehnder type interferometer with the phase shift φ between two paths inside the interferometer.

In the discussion so far we considered a change of a single experimental parameter and the rotation of the information vector within one plane only. This can be generalized. Let us define the orientations of three mutually orthogonal directions $\boldsymbol{n}_1(\alpha, \beta, \gamma)$, $\boldsymbol{n}_2(\alpha, \beta, \gamma)$ and $\boldsymbol{n}_3(\alpha, \beta, \gamma)$ in ordinary space by the Euler angles $0 \leq \alpha < 2\pi$, $0 \leq \beta \leq \pi$, and $0 \leq \gamma < 2\pi$. Then the mutually complementary propositions associated with measurements along the three directions can also be represented in terms of the Euler angles as $P_1(\alpha, \beta, \gamma)$: "The spin along the direction $\boldsymbol{n}_1(\alpha, \beta, \gamma)$ is up," $P_2(\alpha, \beta, \gamma)$: "The spin along the direction $\boldsymbol{n}_2(\alpha, \beta, \gamma)$ is up" and $P_3(\alpha, \beta, \gamma)$: "The spin along the direction $\boldsymbol{n}_3(\alpha, \beta, \gamma)$ is up".

Given a specific set of three orthogonal directions, all other sets of orthogonal directions can be obtained by rotating the reference set. Any general rotation for Euler's angles α, β, γ can be performed as a sequence of three rotations, the first around the z-axes by $0 \leq \gamma < 2\pi$, the second around the new y-axes by $0 \leq \beta \leq \pi$ and finally the third around the new z-axes by $0 \leq \alpha < 2\pi$.

A list of mutually complementary propositions associated with the spin measurements along directions obtained by the first rotation is $P_1(0, 0, \gamma)$, $P_2(0, 0, \gamma)$, and $P_3(0, 0, \gamma)$. Following the argumentation given above one obtains

$$\hat{R}(\gamma) = \begin{pmatrix} \cos\gamma & -\sin\gamma & 0 \\ \sin\gamma & \cos\gamma & 0 \\ 0 & 0 & 1 \end{pmatrix} \qquad (24)$$

for the corresponding transformation matrix in the space of information[14]. If we fix the angle of the first rotation at γ_0 and consider only propositions $P_1(0, \beta, \gamma_0)$, $P_2(0, \beta, \gamma_0)$ and $P_3(0, \beta, \gamma_0)$ about spins along directions obtained by the second rotation around the new y-axis for an angle $0 \leq \beta \leq \pi$, the corresponding transformation matrix reads

$$\hat{R}(\beta) = \begin{pmatrix} \cos\beta & 0 & \sin\beta \\ 0 & 1 & 0 \\ -\sin\beta & 0 & \cos\beta \end{pmatrix} . \qquad (25)$$

In the last step we fix both the angle γ_0 of the first rotation and the angle β_0 of the second rotation, and consider only sets of mutually complementary propositions $P_1(\alpha, \beta_0, \gamma_0)$, $P_2(\alpha, \beta_0, \gamma_0)$ and $P_3(\alpha, \beta_0, \gamma_0)$ about spins along directions obtained by the third rotation around the new z-axis for $0 \leq \alpha < 2\pi$. The corresponding transformation matrix is again of the form (24) with the angle α.

Finally, the transformation matrix for a general rotation in the space of information is given as

$$\hat{R}(\alpha, \beta, \gamma) = \hat{R}(\alpha)\hat{R}(\beta)\hat{R}(\gamma) . \qquad (26)$$

While these relations have an obvious meaningful for spin they hold equally for any elementary system. Specifically they also hold for a two-path interferometer [6].

---

[14] One should always keep in mind the difference between directions along which mutually complementary measurements are performed in ordinary space (such as the vertical direction and the direction at +45° along which a photon's polarization is measured, or three spatially orthogonal directions along which complementary spin components of a spin-1/2 particle are measured) and directions associated with mutually complementary propositions (components of an information vector) in the space of information. The latter always constitute an orthogonal coordinate system. These again have to be distinguished from the orthogonal directions in Hilbert space which do not correspond to complementary measurements.

8 Entanglement –
More Information in Joint Properties than in Individuals

Entanglement is the feature which distinguishes quantum physics most succinctly from classical physics as quantitatively expressed by the violation of Bell's inequalities [2, 13]). In 1964 John Bell obtained certain bounds (the Bell inequalities) on combinations of statistical correlations for measurements on two-particle systems if these correlations were to be understood within a realistic picture based on local properties of each individual particle. In a such a picture the measurement results are determined by properties the particles carry prior to and independent of observation (reality). Additionally the results obtained at one location are independent of any measurements or actions performed at space-like separation (locality). Quantum mechanics predicts violation of these constraints for certain statistical predictions for entangled systems. By today, the predictions of quantum physics have been confirmed in many experiments [1, 14, 31, 43].

In this section we will investigate how much information can be contained in the correlations between quantum systems in order to give an information-theoretic criterion of quantum entanglement. We suggest that a natural understanding of quantum entanglement results when one accepts that the information in a composite system can reside more in the correlations than in properties of individuals. The quantitative formulation of these ideas leads to a rather natural criterion of quantum entanglement[15].

The total information of a composite system can be distributed in various ways within the composite system. We will consider only that part of the total information of the system which is exclusively contained in correlations, or joint properties of its constituents. This is also the reason why, in contrast to the previous sections, now we will not consider complete sets of mutually complementary propositions for the composite system but just that subset of them which concerns joint properties of its constituents. As it is our final goal to compare that criterion with the one given by Bell-type inequalities where one considers correlations between spin measurements confined on each side within one plane we restrict our analysis to an x-y plane locally defined for each subsystem.

As an explicit example of a composite systems we consider a system consisting of two spin-1/2 particles. The propositions about their joint properties will be binary propositions, i.e. will be associated to experiments with two possible outcomes. They will be of the type: "The spin of particle 1 along x and the spin of particle 2 along y are the same", and its negation "The spin of particle 1 along x and the spin of particle 2 along y are different". Therefore the measure of information (7) for binary

---

[15] To this end we will follow Schrödinger's [35] view about entanglement: "Whenever one has a complete expectation-catalog – a maximum total knowledge – a psi-function – for two completely separated bodies, or, in better terms, for each of them singly, then one obviously has it also for the two bodies together, i.e., if one imagines that neither of them singly but rather the two of them together make up the object of interest, of our questions about the future. But the converse is not true. Maximal knowledge of a total system does not necessarily include total knowledge of all its parts, not even when these are fully separated from each other and at the moment are not influencing each other at all."

experiments can be applied. If we denote the probabilities for the two outcomes by p_{xy}^+ and p_{xy}^- respectively, then the information contained in proposition "The spin of particle 1 along x and the spin of particle 2 along y are the same (different)" is given by

$$I_{xy} = \left(p_{xy}^+ - p_{xy}^-\right)^2 . \tag{27}$$

We first consider a product state e.g. $|\psi\rangle = |+x\rangle_1|-x\rangle_2$. This is the case of a composite system composed of two elementary systems carrying therefore $N = 2$ bits of information, i.e. representing the truth value of two propositions. Here the state $|\psi\rangle$ represents the two-bit combination true-false of the truth values of the propositions about the spin of each particle along the x-axis: (1) "The spin of particle 1 is up along x" and (2) "The spin of particle 2 is up along x". Instead of the second proposition describing the spin of particle 2, we could alternatively choose a proposition which describes the result of a joint observation: (3) "The two spins are the same along x." Then the state $|\psi\rangle$ represents the two-bit combination true-false of the truth values of the propositions (1) and (3).

Evidently, for pure product states at most one proposition with definite truth-value can be made about joint properties because one proposition has to be used up to define a property of one of the two subsystems. In other words 1 bit of information defines the correlations. In our example where $|\psi\rangle = |+x\rangle_1|-x\rangle_2$ this 1 bit of information is fully contained in the correlations between spin x-measurements on the two sides, therefore

$$I_{xx} = 1 . \tag{28}$$

We call such states classically composed states.

Obviously, the choice of directions x and y within each of the planes of measurements on the two sides is arbitrary. It is physically not acceptable that the total information contained in correlations between spin measurements confined on each side within x–y planes depends on this choice. We therefore require that the total information contained in the correlations must be invariant upon the choice of general x and y measurement directions within the x–y planes on each side. Only with this requirement the statement "the total information contained in the correlations between measurements within the x–y planes" can have a meaning. This invariance property is guaranteed with our measure of information (27).

We define the total information contained in the correlations as the sum over the individual measures of information about a complete set of mutually complementary observations within the planes x–y on the two sides. The total information contained in the correlations is thus defined as the sum

$$I_{\text{corr}} = I_{xx} + I_{xy} + I_{yx} + I_{yy} \tag{29}$$

of the partial measures of information contained in the set of complementary observations within the x–y-planes. These observations are mutually complementary for product states and the set is complete as there exists no further complementary observation within the chosen x–y planes. By this we mean that for any product state complete knowledge contained in any proposition from the set: "The two spins are equal along x", "The spin of particle 1 along x and the spin of particle 2 along y are the same", "The spin of particle 1 along y and the spin of particle 2 along x are the same" and "The two spins are equal along y" excludes any knowledge about other three propositions.

Consider now a maximally entangled Bell state, e.g.

$$|\psi^-\rangle = \frac{1}{\sqrt{2}}(|+x\rangle_1|-x\rangle_2 - |-x\rangle_1|+x\rangle_2)$$
$$= \frac{1}{\sqrt{2}}(|+y\rangle_1|-y\rangle_2 - |-y\rangle_1|+y\rangle_2). \quad (30)$$

The two propositions here both are statements about results of joint observations [49], namely (1') "The two spins are equal along x" and (2') "The two spins are equal along y". Now the state represents the two-bit combination false-false of the two propositions. Note that here the 2 bits of information are all carried by the 2 elementary systems in a joint way, with no individual elementary system carrying any information on its own. In other words, as the two available bits of information are already exhausted in defining joint properties, no further possibility exists to also encode information in individuals. Therefore

$$I_{\text{corr}}^{\text{Bell}} = 2. \quad (31)$$

Note that in our example of Bell-state (30) $I_{xx} = I_{yy} = 1$ and $I_{xy} = I_{yx} = 0$. Also, note that the truth value for another proposition, namely, "The two spins are equal along z" must follow immediately from the truth values of the propositions (1') and (2'), as only 2 bits of information are available. Interestingly this is also a direct consequence of the formalism of quantum mechanics as the joint eigenstate of $\sigma_x^1 \sigma_x^2$ and of $\sigma_y^1 \sigma_y^2$ is also an eigenstate of $\sigma_z^1 \sigma_z^2 = -(\sigma_x^1 \sigma_x^2)(\sigma_y^1 \sigma_y^2)$.

In contrast to product states we suggest entanglement of two elementary systems to be defined in general such that *more than one bit* (of the two available ones) is used to define joint properties, i.e.

$$I_{\text{corr}}^{\text{entgl}} > 1 \quad (32)$$

for at least one choice of the x–y planes of measurements for the two elementary systems (or, equivalently, for that choice of the x–y planes (independently defined) on each side for which I_{corr} reaches its maximal value). Most importantly, this simple information-theoretic criterion of entanglement can be shown to be equivalent to a *necessary* and *sufficient* condition [26] for a violation of a Bell-type inequality for two-elementary systems [8]. A generalization of our information-theoretic criterion for entanglement to N elementary systems and its relation to the criteria for violation of Bell's inequalities can be found in [8].

9 Time Evolution of the Catalog of Knowledge

Any assignment of properties to an object is always a consequence of some observation. Using information obtained in previous observations we wish to make predictions about the future. Again our predictions might be formulated as, in general probabilistic, predictions about future properties of a system. Clearly, these predictions can be verified or falsified by performing measurements and checking whether the experimental results agree with our predictions. It is then important to connect past observations with future observations. Or, more precisely, to make, based on past observations, specific statements about possible results of future observations.

In quantum mechanics this connection between past observations and future observations exactly is achieved by the quantum-mechanical Liouville equation (for pure states it reduces to the Schrödinger equation)

$$i\hbar \frac{d\hat{\rho}(t)}{dt} = [\hat{H}(t), \hat{\rho}(t)] \,. \tag{33}$$

The initial state $\hat{\rho}(t_0)$ represents all our information as obtained by earlier observation. Using the quantum-mechanical Liouville equation we can derive a time evolved final state $\hat{\rho}(t)$ at some future time t which gives us predictions for any possible observation of the system at that time. In this section the dynamics of an elementary system is formulated as a time evolution of the catalog of our knowledge of the system. This is specified by the evolution of the information vector in the space of information. The Liouville equation will then be derived from the differential equation describing the motion of the information vector in the information space.

We will consider now the time evolution of an elementary system with no information exchange with an environment[16]. Suppose that the state of the system at some initial time t_0 is represented by the catalog $\boldsymbol{i}(t_0) = (i_1(t_0), i_2(t_0), i_3(t_0))$ of our knowledge. Now let the system evolve in time. Because there is no information exchange with an environment during the evolution, the total information of the system at some later time t must still be the same as at the initial time. This may be seen as an ultimate constant of the evolution of the system motion independent of the strength, time dependence or any other characteristic of the "external field" of the system. Therefore

$$I_{\text{total}}(t) = \sum_{n=1}^{3} i_n^2(t) = \sum_{n=1}^{3} i_n^2(t_0) = I_{\text{total}}(t_0) \,. \tag{34}$$

Mathematically, the conservation of the total information is equivalent to the conservation of the length of the information vector during its motion in the information space. This means that time evolution of an isolated quantum system is just a rotation of the information vector in the space of information (see footnote 12):

---

[16] If there is any information exchange between the system and the environment we cannot formulate system's evolution law independently of the environment, but we have to consider it as a subsystem of a larger system that contains both the system and the environment.

$$\boldsymbol{i}(t) = \hat{R}(t, t_0)\boldsymbol{i}(t_0) \,, \tag{35}$$

where again $\hat{R}(t, t_0)$ is a rotation matrix

$$\hat{R}^{-1}(t, t_0) = \hat{R}^T(t, t_0)$$

and $\hat{R}^T(t, t_0)$ is its transposed matrix.

The derivative of (35) with respect to time is

$$\frac{d\boldsymbol{i}(t)}{dt} = \frac{d\hat{R}(t, t_0)}{dt}\boldsymbol{i}(t_0) = \hat{K}(t, t_0)\boldsymbol{i}(t) \,, \tag{36}$$

where $\hat{K}(t, t_0) = (d\hat{R}(t, t_0)/dt)\hat{R}^T(t, t_0)$. We will now show that the operator $\hat{K}(t, t_0)$ is antisymmetric. We find

$$\hat{K}^T(t) = \hat{R}(t)\frac{d\hat{R}^T(t)}{dt} = \hat{R}(t) \lim_{\Delta t \to 0} \frac{\hat{R}^T(t + \Delta t) - \hat{R}^T(t)}{\Delta t}$$

$$= \hat{R}(t) \lim_{\Delta t \to 0} \hat{R}^T(t)\frac{\hat{R}(t) - \hat{R}(t + \Delta t)}{\Delta t}\hat{R}^T(t + \Delta t)$$

$$= \lim_{\Delta t \to 0} \frac{\hat{R}(t) - \hat{R}(t + \Delta t)}{\Delta t}\hat{R}^T(t) = -\hat{K}(t) \,,$$

where the initial time t_0 is identified with the time 0.

It is a well-known result of vector analysis that with every antisymmetric operator $\hat{K}$ one may uniquely associate the "vector of rotation" $\boldsymbol{u}$ by the relation[17]

$$\hat{K}\boldsymbol{y} = \boldsymbol{u} \times \boldsymbol{y} \qquad \text{for all } \boldsymbol{y} \,, \tag{37}$$

where "×" denotes the vector product. Using this result we now rewrite (36) as

$$\frac{d\boldsymbol{i}(t)}{dt} = \boldsymbol{u}(t, t_0) \times \boldsymbol{i}(t) \,. \tag{38}$$

Mathematically, this equation describes the rotation of the information vector around the axis $\boldsymbol{u}(t, t_0)$ which itself changes in the course of time. Physically, this is the formulation of the dynamical law for the evolution of the catalog of our knowledge.

---

[17] The operator $\hat{K}$ is represented by an antisymmetric matrix

$$\hat{K} = \begin{pmatrix} 0 & -k_{21} & -k_{31} \\ k_{21} & 0 & -k_{32} \\ k_{31} & k_{32} & 0 \end{pmatrix} \,.$$

From there we read out the components of the vector $\boldsymbol{u}$ of rotation as

$$u_1 = k_{32}, u_2 = -k_{31}, u_3 = k_{21} \,.$$

Based on the known features of the quantum formalism we will now argue for the validity of (38). Suppose that the quantum state of the system is described by the density matrix $\hat{\rho}$. We decompose the density matrix into the unit operator and the generators of SU(2) algebra (Pauli matrices)

$$\hat{\rho}(t) = \frac{1}{2}\hat{1} + \frac{1}{2}\sum_{j=1}^{3} i_j(t)\hat{\sigma}_j , \qquad (39)$$

where $\hat{\sigma}_j$ is Pauli spin operator for the direction $j = x, y, z$. Note that the component i_j of the information vector associated with the spin along the direction j is equal to the expectation value of spin along this direction, i.e. $i_j(t) = \text{Tr}(\hat{\rho}(t)\hat{\sigma}_j)$.

If we take a derivative of (39) in time we obtain

$$i\hbar \frac{d\hat{\rho}(t)}{dt} = \frac{1}{2}\sum_{j=1}^{3} \frac{i_j(t)}{dt}\hat{\sigma}_j . \qquad (40)$$

Inserting (38) on the right-hand side we find

$$i\hbar \frac{d\hat{\rho}(t)}{dt} = \frac{i}{2}\sum_{i,j,k=1}^{3} \epsilon_{ijk} u_i(t) i_j \hat{\sigma}_k . \qquad (41)$$

Since the Pauli matrices satisfy $[\hat{\sigma}_i, \hat{\sigma}_j] = 2i\sum_{k=1}^{3} \epsilon_{ijk}\hat{\sigma}_k$, we proceed with

$$i\hbar \frac{d\hat{\rho}(t)}{dt} = \frac{1}{4}\sum_{i,j=1}^{3} u_i(t) i_j (\hat{\sigma}_i\hat{\sigma}_j - \hat{\sigma}_j\hat{\sigma}_i) . \qquad (42)$$

Introducing the operator $\hat{H}(t)$ such that

$$u_i(t) := \text{Tr}(\hat{H}(t)\hat{\sigma}_i) , \qquad (43)$$

we finally obtain the quantum-mechanical Liouville equation

$$i\hbar \frac{d\hat{\rho}(t)}{dt} = [\hat{H}(t), \hat{\rho}(t)] . \qquad (44)$$

For the special case of a conservative system, the evolution of a quantum state in time is constrained by a higher constant of motion, namely our information about the energy of the system, apart from the ultimate one of the total information content of the system. In the space of information this corresponds to the rotation of the information vector around a fixed axis that is associated to our knowledge of energy of the system[18]. This is only possible if the axis u in (38) is fixed. This further implies

---

[18] Note that we consider elementary systems, that is, systems with two possible energy values. By information about the energy of the system we mean our knowledge about which of the two values will be observed in an appropriately designed experiment.

the existence of a minimal interval of time the information vector needs to make one complete rotation in the space of information (Fig. 9). After this time interval the values i for all propositions about the system take the same value. This time interval corresponds, for example, to Bloch oscillation period in quantum optics or to the de-Broglie wave-period for matter waves.

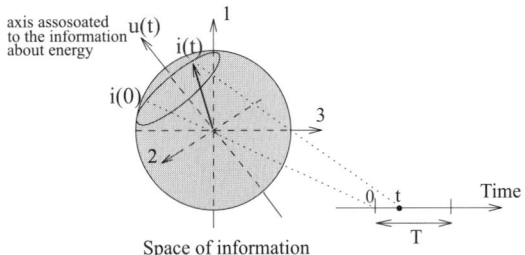

Fig. 9. One complete rotation of the information vector during a time elapse of the de-Broglie wave-period.

The result given above is just a very first and the most basic step toward an information-theoretical formulation of the quantum-mechanical evolution in time. Mathematically, it is an immediate consequence of the (nearly) isomorphism between SU(2) group of unitary rotations in a two-dimensional Hilbert space and SO(3) group of rotations in the real three-dimensional Euclidian space. Obviously the next step will be to consider more complex systems and to find position- and momentum-representations of the catalog of our knowledge in order to give an information-theoretical formulation of the Schrödinger equation.

10 Measurement – the Update of Information

In this section it will be argued that identifying the quantum state with the catalog of our knowledge leads to the resolution of many of the seemingly paradoxical features of quantum mechanics connected to the so-called measurement problem.

In a quantum measurement, we find the system to be in one of the eigenstates of the observable defined by the measurement apparatus. A specific example is the case when we are considering a wave packet as being composed of a superposition of plane waves. Such a wave packet is more or less well-localized, but we can always perform a position measurement on a wave packet which is better localized than the dimension of the packet itself. This, sometimes called "reduction of the wave packet" or "collapse of the wave function", can only be seen as a "measurement paradox" if one views this change of the quantum state as a real physical process. In the extreme case it is often even related to an instant collapse of some physical wave in space.

There is no basis for any such assumption. In contrast, there is never a paradox if we realize that the wave function is just an encoded mathematical representation of our knowledge of the system. When the state of a quantum system has a non-zero value at some position in space at some particular time, it does not mean that the system is physically present at that point, but only that our knowledge (or lack of knowledge) of the system allows the particle the *possibility* of being present at that point at that instant.

What can be more natural than to change the representation of our knowledge if we gain new knowledge from a measurement performed on the system? When a measurement is performed, our knowledge of the system changes, and therefore its representation, the quantum state, also changes. In agreement with the new knowledge, it instantaneously changes all its components, even those which describe our knowledge concerning regions of space quite distant from the site of the measurement. Then no need whatsoever arises to allude to notions like superluminal or instantaneous transmission of information.

Schrödinger [35] wrote[19]: "Bei jeder Messung ist man genötigt, der ψ-Funktion (= dem Voraussagenkatalog) eine eigenartige, etwas plötzliche Veränderung zuzuschreiben, die von der *gefundenen Maßzahl* abhängt und sich *nicht vorhersehen läßt*; woraus allein schon deutlich ist, daß diese zweite Art von Veränderung der ψ-Funktion mit ihrem regelmässigen Abrollen *zwischen* zwei Messungen nicht das mindeste zu tun hat. Die abrupte Veränderung durch die Messung ... ist der interessanteste Punkt der ganzen Theorie. Es ist genau *der* Punkt, der den Bruch mit dem naiven Realismus verlangt. Aus *diesem* Grund kann man die ψ-Funktion *nicht* direkt an die Stelle des Modells oder des Realdings setzen. Und zwar nicht etwa weil man einem Realding oder einem Modell nicht abrupte unvorhergesehene Änderung zumuten dürfte, sondern weil vom realistischen Standpunkt die Beobachtung ein Naturvorgang ist wie jeder andere und nicht per se eine Unterbrechung des regelmässigen Naturlaufs hervorrufen darf".

A closely related position was assumed also by Heisenberg, who wrote in a letter to Renninger dated February 2, 1960: "The act of recording, on the other hand, which leads to the reduction of the state, is not a physical, but rather, so to say, a mathematical process. With the sudden change of our knowledge also the mathematical presentation of our knowledge undergoes of course a sudden change.", as translated by Jammer [28].

With the only exception of the system being in an eigenstate of the measured observable, a quantum measurement changes the system into one of the possible

---

[19] Translated: "For each measurement one is required to ascribe to the ψ-function (= the prediction catalog) a characteristic, quite sudden change, which *depends on the measurement result obtained*, and so *cannot be foreseen*; from which alone it is already quite clear that this second kind of change of the ψ-function has nothing whatever in common with its orderly development *between* two measurements. The abrupt change by measurement ... is the most interesting point of the entire theory. It is precisely *the* point that demands the break with naive realism. For *this* reason one *cannot* put the ψ-function directly in place of the model or of the physical thing. And indeed not because one might never dare impute abrupt unforseen changes to a physical thing or to a model, but because in the realism point of view observation is a natural process like any other and cannot per se bring about an interruption of the orderly flow of natural events."

new states (these being defined by the measurement apparatus) in a fundamentally unpredictable[20] way. Thus it cannot be claimed to reveal a property existing before the measurement is performed. The reason for this is again the fact that a quantum system cannot, not even in principle, carry enough information to specify observation-independent properties corresponding to all possible measurements. In the measurement the state therefore must appear to be changed in accord with the new information, if any, *acquired* about the system together with unavoidable and irrecoverable *loss* of complementary information. Unlike a classical measurement, a quantum measurement thus does not just add (if any) some knowledge, it changes our knowledge in agreement with a fundamental finiteness of the total information content of the system. We will now bring the role of the observer in a quantum measurement to the center of our discussion. In classical physics we can assume that an observation reveals some property already existing in the outside world. For example, if we look at the moon, we just find out where it is and it is certainly safe to assume that the property of the moon to be there is independent of whether anyone looks or not. The situation is drastically different in quantum mechanics and it is just the very attitude of the Copenhagen interpretation giving a fundamental role to observation which is a major intellectual step forward over this naive classical realism.

We as observers have a significant role in the measurement process, because we can decide by choosing the measuring device which attribute will be realized in the actual measurement[21]. Since the information content of the system is limited, by choosing which measurement device to use we not only decide what particular knowledge will be gained, but simultaneously what complementary knowledge will be lost after the measurement is performed. Here, a very subtle position was assumed by Pauli [32] who writes: "The gain of knowledge by means of an observation has as a necessary and natural consequence, the loss of some other knowledge. The observer has however the free choice, corresponding to two mutually exclusive experimental arrangements, of determining *what* particular knowledge is gained and what other knowledge is lost (complementary pairs of opposites). Therefore every irrevocable interference by an observation about a system alters its state, and creates a new phenomenon in Bohr's sense."

11 Conclusions

The laws we discover about Nature do not already exist as "Laws of Nature" in the outside world. Rather "Laws of Nature" are necessities of the mind for any possibility

---

[20] Wheeler [45] stated that "... yes or no that is recorded constitutes an unsplittable bit of information".

[21] Wheeler explicates this by example of the well-known case of a quasar, of which we can see two pictures through the gravity lens action of a galaxy that lies between the quasar and ourselves. By choosing which instrument to use for observing the light coming from that quasar, we can decide here and now whether the quantum phenomenon in which the photons take part is interference of amplitudes passing on both sides of the galaxy or whether we determine the path the photon took on one or the other side of the galaxy.

to make sense whatsoever out of the data of experience. This epistemological structure is a necessity behind the form of all laws an observer can discover. As von Weizsäcker has put it, and Heisenberg quoted in [25]: "Nature is earlier than man, but man is earlier than natural science."

An observer is inescapably suspended in the situation of obtaining the data from observation, formatting concepts of Nature therefrom, and predicting the data of future observations. In observing she/he is able to distinguish only a finite number of results at each interval of time (compare [38, 39]). Therefore the experience of the ultimate experimenter is a stream of ("yes" or "no") answers to the questions posed to Nature. Any concept of an existing reality is then a mental construction based on these answers. Of course this does not imply that reality is no more than a pure subjective human construct. From our observations we are able to build up objects with a set of properties that do not change under variations of modes of observation or description. These are "invariants" with respect to these variations. Predictions based on any such specific invariants may then be checked by anyone, and as a result we may arrive at an intersubjective agreement about the model, thus lending a sense of independent reality to the mentally constructed objects.

In quantum experiments an observer may decide to measure different sets of complementary variables, thus gaining certainty about one or more variable at the expense of losing certainty about the other(s). Thus the measure of information in an individual experiment is not a candidate for the invariant from the previous paragraph as it depends on the specific experimental context. However the total uncertainty, or equivalently, the total information, is invariant under such transformation from one complete set of complementary variables to another. In classical physics a property of a system is a primary concept prior to and independent of observation and information is a secondary concept which measures our ignorance about properties of the system. In contrast in quantum physics the notion of the total information of the system emerges as a primary concept, independent of the particular complete set of complementary experimental procedures the observer might choose, and a property becomes a secondary concept, a specific representation of the information of the system that is created spontaneously in the measurement itself. Bohr (1934) wrote that "... a subsequent measurement to a certain degree deprives the information given by a previous measurement of its significance for predicting the future course of phenomena. Obviously, these facts not only set a limit to the *extent* of the information obtainable by measurement, but they also set a limit to the *meaning* which we may attribute to such information."

Theorems like those of Bell [2] and Greenberger–Horne–Zeilinger [20] state that randomness of an individual quantum event cannot be derived from local causes (local hidden variables). Quantum physics is not able to "explain why (specific) events happen" as pointed out by Bell [3]. It is beyond the scope of quantum physics to answer the question why events happen at all (why detectors click at all). Yet, if events happen, then they must happen randomly. The reason is the finiteness of the information. Any detailed description of reality that would be able to give an unambiguous answer to Bell's question, that is, any description that would be able to arrive at an accurate and detailed prediction of the particular process resulting in a particular event, will necessarily include the definition of a number of "hidden" properties of the system which would carry information as to which specific result will be observed for all possible future measurements. Therefore no answer can be given to Bell's question, because otherwise, quantum system would carry more information than what is available in principle.

It turns out that the lowest symmetry common to all elementary systems is the invariance of their total information content with respect to a rotation in a three-dimensional space. The dimensionality (3) of the information space is a consequence of the minimal number (3) of mutually exclusive experimental questions we may pose to an elementary system. This seems to justify the use of three-dimensional space as "the" space of the inferred world. Such a view was first suggested by von Weizsäcker [42]: "It [quantum theory of the simple alternative] contains a two-dimensional complex vector space with a unitary metric, a two-dimensional Hilbert space. This theory has a group of transformations which is surprisingly near-isomorphic with a group of rotations in the real three-dimensional Euclidian space. This has been known for a very long time. I propose to take this isomorphism seriously as being the real reason why ordinary space is three-dimensional."

We end with another quote of von Weizsäcker [42]: "But I feel these consideration make it plausible that quantum theory is not just one out of a thousand equally possible theories, and the one which happens to please God so much that he chose to create a world in which it would be true. I rather think, if we had understood quantum theory just a little bit better than we understand it so far it would turn out to be a fairly good approximation towards the formulation of a theory which contains nothing but the rules under which we speak about future events if we can speak about them in an empirically testable way at all."

It has not escaped our attention that our considerations presented here may be viewed as providing the necessary justification for von Weizsäcker's point of view as expressed in these quotations.

Acknowledgement. We acknowledge discussions with Terry Rudolph, Christoph Simon, Johann Summhammer and Marek Żukowski. This work is supported by the Austrian FWF project F1506, and by the QuComm project of the European Union, No: IST-1999-1003 "Long Distance Photonic Quantum Communication".

References

1. A. Aspect, P. Grangier and G. Roger: *Phys. Rev. Lett.* 47, 460–463 (1981)
2. J.S. Bell: *Physics* 1, 195–200 (1964); reprinted: J.S. Bell: *Speakable and Unspeakable in Quantum Mechanics.* Cambridge Univ. Press, 1987
3. J.S. Bell: *Physics World* (August 1990)
4. N. Bohr: In: P.A. Schillp (ed.): *Albert Einstein: Philosopher-Scientist.* (The Library of Living Philosophers Evanston, IL) 1949, p. 200. A copy can be found at the web site (http://www.emr.hibu.no/lars/eng/schlipp/Default.html)
5. N. Bohr: *Atomic Physics and Human Knowledge.* Wiley, New York 1958
6. Č. Brukner and A. Zeilinger: *Phys. Rev. Lett.* 83, 3354–3357 (1999)

7. Č. Brukner and A. Zeilinger: e-print quant-ph/0008091, 2000
8. Č. Brukner, M. Żukowski and A. Zeilinger: e-print quant-ph/0106119, 2001
9. Č. Brukner and A. Zeilinger: *Phys. Rev. A* 63, 022 113 1–10 (2001)
10. Č. Brukner and A. Zeilinger: *Phil. Trans. R. Soc. Lond. A* 360, 1061 (2002)
11. C.M. Caves, C.A. Fuchs and R. Schack: *J. Math. Phys.* 43, 4537 (2002)
12. C.M. Caves, C.A. Fuchs and R. Schack: *Phys. Rev. A* 022 305 (2001)
13. J. Clauser, M. Horne, A. Shimony and R. Holt: *Phys. Rev. Lett.* 23, 880–884 (1969)
14. S.J. Freedman and J.S. Clauser: *Phys. Rev. Lett.* 28, 938–941 (1972)
15. B.G. Englert: *Phys. Rev. Lett.* 77, 2154 (1996)
16. D.I. Fivel: *Phys. Rev. A* 59, 2108 (1994)
17. C.A. Fuchs: e-print quant-ph/0205039, 2002
18. C.A. Fuchs: e-print quant-ph/0106166, 2001
19. B.V. Gnedenko: *The Theory of Probability*. Mir Publishers, Moscow 1976
20. D.M. Greenberger, M. Horne, A. Shimony and A. Zeilinger: *Am. J. Phys.* 58, 1131–1143 (1990)
21. M.J.W. Hall: e-print quant-ph/0007116, 2000
22. G. Hardy, J.E. Littlewood and G. Pólya: *Inequalities*. Cambridge University Press, 1952
23. L. Hardy: e-print quant-ph/0101012, 2001
24. L. Hardy: e-print quant-ph/0111068, 2001
25. W. Heisenberg: *Daedalus* 87, 95 (1958)
26. R. Horodecki, P. Horodecki and M. Horodecki: *Phys. Lett. A* 200, 340–344 (1995)
27. I. Ivanović: *J. Phys. A* 14, 3241 (1981)
28. M. Jammer: *The Philosophy of Quantum Mechanics*. J. Wiley & Sons, New York 1974
29. S. Kochen and E.P. Specker: *J. Math. and Mech.* 17, 59 (1967)
30. R. Landauer: *Physics Today* 23, May 1991
31. J.W. Pan, D. Bouwmeester, H. Weinfurter and A. Zeilinger: *Nature* 403, 515–518 (2000)
32. W. Pauli: In: C.P. Enz and K. von Meyenn (eds.): *Writings on Philosophy and Physics*. Translated by Robert Schlapp. Springer Verlag, Berlin 1955
33. J. Schlienz and G. Mahler: *Phys. Rev. A* 52, 4396–4404 (1995)
34. C.E. Shannon: *Bell Syst. Tech. J.* 27, 379 (1948). A copy can be found at http://cm.bell-labs.com/cm/ms/what/shannonday/paper.html
35. E. Schrödinger: *Naturwissenschaften* 23, 807–812; 823–828; 844–849 (1935). Translation published in *Proc. Am. Phil. Soc.* 124, 323–338 and in: J.A. Wheeler and W.H. Zurek (eds.): *Quantum Theory and Measurement*. Princeton University Press, Princeton, pp. 152–167. A copy can be found at www.emr.hibu.no/lars/eng/cat
36. J. Summhammer: *Found. Phys. Lett.* 1, 123 (1988)
37. J. Summhammer: *Int. J. Theor. Phys.* 33, 171 (1994)
38. J. Summhammer: e-print quant-ph/0008098, 2000. To appear in: C. Calude (ed.): *The Third Millenium*
39. J. Summhammer: e-print quant-ph/0102099, 2001
40. C.G. Timpson: e-print quant-ph/0112178, 2001
41. C.F. von Weizsäcker: *Aufbau der Physik*. Carl Hanser, München 1958
42. C.F. von Weizsäcker: In: L. Castell, M. Drieschner and C.F. von Weizsäcker (eds.): *Quantum Theory and the Structures of Time and Space*. Hanser, München 1975. Papers presented at a conference held in Feldafing, July 1974
43. G. Weihs, T. Jennewein, C. Simon, H. Weinfurter and A. Zeilinger: *Phys. Rev. Lett.* 81, 5039–5043 (1998)
44. J.A. Wheeler: Law without Law. In: J.A. Wheeler and W.H. Zurek (eds.) *Quantum Theory and Measurement*. Princeton University Press, Princeton 1983, p. 182

45. J.A. Wheeler: *Proc. 3rd Int. Symp. Foundations of Quantum Mechanics.* 354 (1989), Tokyo
46. W.K. Wootters: *Phys. Rev D* 23, 357 (1981)
47. W.K. Wootters and B.D. Fields: *Ann. Phys.* 191, 363 (1989)
48. W.K. Wootters and W.H. Zurek: *Phys. Rev. D* 19, 473 (1979)
49. A. Zeilinger: *Phil. Trans. Roy. Soc. Lond.* 1733, 2401–2404 (1997)
50. A. Zeilinger: *Found. Phys.* 29, 631–643 (1999)

Multiple Quantization in Fock Space

Dirk Graudenz

1 Introduction

A recurring theme in Carl Friedrich von Weizsäcker's synthesis of physics and philosophy is that at the core of quantum theory there is a non-classical logic. Application of this logic to its own statements leads to the concept of *multiple quantization* [1]. The reconstruction of quantum theory as done in [2,3] is an explicit implementation of this idea, starting with the quantization of an abstract classical alternative. The result of this first step is interpreted as three-dimensional momentum space. The second quantization step yields the Weyl, Dirac and Maxwell equations. The third quantization step would then lead to the familiar quantum field theories of matter and gauge fields.

The full program has not been carried out, partly due to the conceptual problem of interactions (here parabose statistics has been a promising candidate, cf. [1]). Moreover, a useful calculational framework to study the iterated quantization of a system is lacking.

- The standard approach to quantum mechanics is canonical quantization based on a Schrödinger wave function. The formulation of multiple quantization in this language is possible in principle, but eventually yields intractable wave functions on a space of wave functions, themselves the result of a former quantization step.
- Quantum field theory can be developed using the method of "second quantization". Despite its name, second quantization really is "first" quantization of a *classical* field or, stated differently, the construction of a multi-particle theory in Fock space from the quantum mechanics of a single particle.
- Path-integral quantization, nowadays the standard tool for quantization, offers no obvious clue to the problem at hand.

None of these quantization procedures is directly suitable for a systematic study of multiple quantization.

This paper presents a first step towards a framework for multiple quantization based on a simple generalization of the Fock space approach. The organization is as follows. Sect. 2 briefly reviews canonical field quantization both in the Schrödinger and in the Fock space representation, giving explicit formulas to connect the two approaches in the case of a scalar field. In Sect. 3, the framework for multiple quantization is introduced, modeled after the Fock space representation from Sect. 2. The paper closes with a summary and some open questions.

2 Setting the Stage: Quantization of a Free Scalar Field

Canonical quantization of a scalar field in Fock space and in the Schrödinger representation starts with a classical theory expressed in terms of the field $\Phi(x)$ and conjugated momenta $\pi(x)$.[1] In analogy to quantum mechanics, $\Phi(x)$ and $\pi(x)$ are promoted to operators acting on a Hilbert space and fulfilling canonical commutation relations[2]

$$[\Phi(x), \pi(y)] = i\,\delta(x-y)\,. \tag{1}$$

Following [4], a Fourier decomposition leads to

$$\Phi(x) = \int d\tilde{k}\,\left[a(k)\,e^{-ikx} + a^\dagger(k)\,e^{ikx}\right]$$

$$\pi(x) = -i\int d\tilde{k}\,\omega_k\,\left[a(k)\,e^{-ikx} - a^\dagger(k)\,e^{ikx}\right]\,. \tag{2}$$

Here $d\tilde{k} = dk/[(2\pi)^3\,2\omega_k]$ is an invariant measure, and $\omega_k = \sqrt{k^2+m^2}$ is the energy of a particle with momentum k and mass m. The $a^\dagger(k)$ and $a(k)$ are creation and destruction operators for particles with momentum k:

$$\left[a(k), a^\dagger(l)\right] = (2\pi)^3\,2\omega_k\,\delta(k-l)\,. \tag{3}$$

Anticipating things to come, it is useful to introduce the operators

$$a(x) = \int d\tilde{k}\,\sqrt{2\omega_k}\,a(l)\,e^{ilx}\,, \quad a^\dagger(x) = \int d\tilde{k}\,\sqrt{2\omega_k}\,a^\dagger(l)\,e^{-ilx}\,, \tag{4}$$

which can be shown to be destruction and creation operators, respectively, for a particle at position x. They fulfill the commutation relation

$$\left[a(x), a^\dagger(y)\right] = \delta(x-y)\,. \tag{5}$$

In terms of $\Phi(x)$ and $\pi(x)$ they can be represented as[3]

$$a(x) = \frac{1}{\sqrt{2}}\left[\sqrt[4]{-\Delta_x+m^2}\,\Phi(x) + i\,\frac{1}{\sqrt[4]{-\Delta_x+m^2}}\,\pi(x)\right]$$

$$a^\dagger(x) = \frac{1}{\sqrt{2}}\left[\sqrt[4]{-\Delta_x+m^2}\,\Phi(x) - i\,\frac{1}{\sqrt[4]{-\Delta_x+m^2}}\,\pi(x)\right]\,. \tag{6}$$

---

[1] The field argument x, contrary to standard notation, stands for a spatial 3-vector.
[2] In units $\hbar = 1$, $c = 1$.
[3] Please note that the destruction and creation operators for a particle at x involve arbitrary powers of the Laplacian Δ_x and are thus non-local in $\Phi(x)$ and $\pi(x)$.

The vacuum (ground state) Ω, not containing any particles, is annihilated by the destruction operators: $a(x)\Omega = 0$.

2.1 Canonical Quantization in Fock Space

The Hilbert space $\mathcal{H}$ can be constructed as a Fock space:

$$\mathcal{H} = \bigoplus_{n=0}^{\infty} \mathcal{H}_n = \bigoplus_{n=0}^{\infty} \mathrm{Sym}\left(\bigotimes^n \mathcal{H}_1\right) . \tag{7}$$

$\mathrm{Sym}\left(\bigotimes^n \mathcal{H}_1\right)$ is the symmetrized tensor product of n copies of $\mathcal{H}_1$. $\mathcal{H}_n$ is spanned by basis vectors of the form

$$|x_1 x_2 \ldots x_n\rangle = a^\dagger(x_1) a^\dagger(x_2) \ldots a^\dagger(x_n) \Omega , \tag{8}$$

representing a state with particles at $x_1, x_2, \ldots, x_n$. In particular, $\mathcal{H}_1$ is the one-particle subspace. The unit operator can be written as

$$\mathbf{1} = \sum_n \frac{1}{n!} \int \mathrm{d}x_1 \ldots \int \mathrm{d}x_n \, |x_1 x_2 \ldots x_n\rangle\langle x_1 x_2 \ldots x_n| , \tag{9}$$

and a state Ψ can be decomposed as

$$\Psi = \sum_n \frac{1}{n!} \int \mathrm{d}x_1 \ldots \int \mathrm{d}x_n \, \Psi(x_1, x_2, \ldots, x_n) |x_1 x_2 \ldots x_n\rangle . \tag{10}$$

The complex number

$$\Psi(x_1, x_2, \ldots, x_n) = \langle x_1 x_2 \ldots x_n | \Psi \rangle \tag{11}$$

is the probability amplitude to find particles at $x_1, x_2, \ldots, x_n$ in the state Ψ.

2.2 Canonical Quantization Based on a Schrödinger Wave Function

The Schrödinger wave function approach represents the algebra (1) on functionals $\Psi(\varphi)$ of classical fields $\varphi(x)$ by means of

$$\hat{\Phi}(x) = \varphi(x), \quad \hat{\pi}(x) = -\mathrm{i}\frac{\delta}{\delta \varphi(x)} . \tag{12}$$

The symbol $\hat{}$ is used to distinguish the functional from the operator representation of the preceeding section. $\hat{\Phi}(x)$ is a multiplication, $\hat{\pi}(x)$ is a functional differentiation operator. The quantity $\Psi(\varphi)$ is interpreted as the probability amplitude to find the classical field configuration φ. The inner product of two states χ and Ψ is given by a functional integral over all possible classical field configurations:

$$\langle \chi | \Psi \rangle = \int \mathcal{D}\varphi \, \overline{\chi(\varphi)} \, \Psi(\varphi) . \tag{13}$$

The state $|\varphi_0\rangle$ representing a classical field φ_0 with probability one has the wave function

$$\Psi_{\varphi_0}(\varphi) = \delta(\varphi - \varphi_0) = \prod_x \delta[\varphi(x) - \varphi_0(x)] \ . \tag{14}$$

An arbitrary state Ψ therefore has the wave function $\Psi(\varphi) = \langle\varphi|\Psi\rangle$. Up to normalization, the vacuum state is

$$\Omega(\varphi) = \exp\left[-\frac{1}{2}\int dx\, \varphi(x)\sqrt{-\Delta_x + m^2}\,\varphi(x)\right] \tag{15}$$

and satisfies $\hat{a}(x)\,\Omega(\varphi) = 0$. Explicit expressions for the functional representation $\Psi_{x_1 x_2 \ldots x_n}(\varphi) = \langle\varphi|x_1 x_2 \ldots x_n\rangle$ of Fock space states can be derived successively by the following identities:

$$\hat{a}^\dagger(x) = \frac{1}{\sqrt{2}}[F(x) - D(x)]$$

$$F(x) = \sqrt[4]{-\Delta_x + m^2}\,\varphi(x)$$

$$D(x) = \frac{1}{\sqrt[4]{-\Delta_x + m^2}}\frac{\delta}{\delta\varphi(x)}$$

$$D(x)(fg) = (D(x)f)\,g + f D(x)g$$

$$D(x)\,F(y) = \delta(x - y)$$

$$D(x)\,\Omega(\varphi) = -F(x) \tag{16}$$

The first two non-trivial terms are

$$\langle\varphi|x\rangle = \frac{1}{\sqrt{2}}2F(x)\,\Omega(\varphi)$$

$$\langle\varphi|x\,y\rangle = \frac{1}{\sqrt{2}^2}[4F(x)F(y) - 2\delta(x-y)]\,\Omega(\varphi) \ . \tag{17}$$

The polynomials in F are generalizations of the Hermite polynomials and allow the calculation of the particle content of a classical field configuration φ. For instance, the amplitude to find exactly two particles at x and y is given by

$$\langle x\, y|\varphi\rangle = \left(2\sqrt[4]{-\Delta_x + m^2}\,\varphi(x)\sqrt[4]{-\Delta_y + m^2}\,\varphi(y) - \delta(x-y)\right)\Omega(\varphi) \ . \tag{18}$$

2.3 Comparison of the Two Approaches

Using the unit operator from (9), a general probability amplitude can be written as

$$\Psi(\varphi) = \sum_n \frac{1}{n!}\int dx_1 \ldots \int dx_n\, \Psi(x_1, x_2, \ldots, x_n)\,\langle\varphi|x_1 x_2 \ldots x_n\rangle \ . \tag{19}$$

Here $\Psi(\varphi)$ is expressed in a Volterra-type series[4], the coefficients being the n-particle amplitudes. This equation is to be compared with (10).

What is the general structure of the two quantization procedures? The wave function approach starts with a classical field $\varphi : \mathbb{R}^3 \to \mathbb{R}$. The configuration space of the system forms a real vector space V, and Schrödinger wave functions are given by $\Psi : V \to \mathbb{C}$. The building block of the Fock space approach is $\mathcal{H}_1$, which is a vector space of complex one-particle wave functions $\rho : \mathbb{R}^3 \to \mathbb{C}$. Despite different physical interpretations, the two spaces have, up to a factor of two (real vs. complex), the same number of degrees of freedom, so one may identify

$$\mathcal{H}_1 = \mathbb{C} \otimes V . \tag{20}$$

Thus, given a classical system with (real) configuration space V, the corresponding Fock space can be built according to (7) on the basis of $\mathcal{H}_1 = \mathbb{C} \otimes V$.

3 A Framework for Multiple Quantization

The final remarks in the preceeding section permit the construction of the state space for the case of multiple quantization of a system. Let $\mathcal{K}$ be the Hilbert space of a quantum system. Quantization of the quantum system in the Schrödinger representation is done by forming wave functions $\Psi : \mathcal{K} \to \mathbb{C}$. Since $\mathcal{K}$ is a complex vector space, it can be decomposed into the direct sum of two real vector spaces $\mathcal{K}_1 \oplus \mathcal{K}_2$.[5] If the "classical" configuration space V from the preceeding section is identified with $\mathcal{K}_1 \oplus \mathcal{K}_2$, the state space of wave functions $\Psi(\varphi) = \Psi(\varphi_1 \oplus \varphi_2)$ is equal to the Fock space constructed from

$$\mathcal{H}_1 = \mathbb{C} \otimes V = \mathbb{C} \otimes (\mathcal{K}_1 \oplus \mathcal{K}_2) \cong (\mathbb{C} \otimes \mathcal{K}_1) \oplus (\mathbb{C} \otimes \mathcal{K}_2) \cong \mathcal{K} \oplus \mathcal{K} . \tag{21}$$

The full Fock space is therefore

$$\mathcal{H} = \bigoplus_{n=0}^{\infty} \operatorname{Sym} \left[\bigotimes^n (\mathcal{K} \oplus \mathcal{K}) \right] . \tag{22}$$

As an example, consider the case of $\mathcal{K} = \mathbb{C}^2$, the quantum version of a classical alternative, with basis $\{|0\rangle, |1\rangle\}$. $\mathcal{K} \oplus \mathcal{K}$ will have a basis $\{|0\rangle, |1\rangle, |2\rangle, |3\rangle\}$, and thus $\mathcal{H}$ is spanned by states $|u_1 u_2 \ldots u_n\rangle$ with $u_i \in \{1, 2, 3, 4\}$.[6]

Starting with a Hilbert space $\mathcal{H}^{(0)}$, the full hierarchy of Hilbert spaces of multiple quantization can be built according to

$$\mathcal{H}^{(N+1)} = \bigoplus_{n=0}^{\infty} \operatorname{Sym} \left[\bigotimes^n (\mathcal{H}^{(N)} \oplus \mathcal{H}^{(N)}) \right] . \tag{23}$$

---

[4] $\langle \varphi | x_1 x_2 \ldots x_n \rangle$ contains the operator $\sqrt[4]{-\Delta_x + m^2}$ and the ground-state exponential $\Omega(\varphi)$.
[5] If $\mathcal{K}$ is a space of complex-valued wave functions, $\mathcal{K}_1$ and $\mathcal{K}_2$ could represent the real and imaginary parts.
[6] It is peculiar that the "doubling" to describe "anti-urs" introduced *ad hoc* in [1] appears here as a natural consequence of the quantization step.

4 Summary and Open Questions

This paper has presented a Fock space-type construction of the state space of a multiply quantized system, modeled according to a comparison of canonical quantization in the Fock and in the Schrödinger representation. The construction given in Sect. 3 is just a first step towards a full understanding of multiple quantization. To mention a few open questions:

- (22) is based on the assumption of Bose statistics – what is the interpretation of a similar construction employing Fermi statistics?
- As mentioned in the introduction, multiple quantization follows as a consequence from the application of logic to itself. What is the physical interpretation of this approach (besides giving us three–dimensional space and quantum fields) – is there anything that can be falsified experimentally?
- Is there a relation to the theory of the measurement process? A measurement device gains information about the measured system, and measurements can be iterated (a measurement performed on the measurement device itself, etc., with an observer eventually taking note of the measurement result).
- What about the dynamics of a multiply quantized system? Standard Fock-space quantization defines a multi-particle picture, with interactions of the particles put in either by hand or derived from the field-theory Hamiltonian. Is it conceivable that a similar mechanism is possible here? If so, what is the interpretation of this kind of interaction?
- What is the role of gauge symmetry? Assuming that the system has a local gauge symmetry, how does this translate into a symmetry for the multiply quantized system?

References

1. C.F. von Weizsäcker: *Aufbau der Physik*. Carl Hanser Verlag, München 1985
2. C.F. von Weizsäcker: *Zeitschrift für Naturforschung* 13a, 245 (1958)
3. C.F. von Weizsäcker, E. Scheibe and G. Süssmann: *Zeitschrift für Naturforschung* 13a, 705 (1958)
4. C. Itzykson and J.-B. Zuber: *Quantum Field Theory*. McGraw Hill, Singapore 1980

Part V

The Ur-Hypothesis and Its Implications for Particle Physics and Cosmology

The Ur-Hypothesis

Lutz Castell

"Den lieb ich der Unmögliches begehrt."
Goethe, Faust II

The very large cosmological number

$$N = \frac{c^3}{\hbar u \kappa^2} \approx 10^{120} \tag{1}$$

plays a central role in the following three hypotheses concerning the observable universe; u is the energy density and κ the gravitational constant $\kappa = 8\pi G/c^2$. C. F. von Weizsäcker observed in 1974 [1] that

$$N = \frac{U}{\hbar c/R} = R\left(\frac{U}{\hbar c}\right), \tag{2}$$

where U equals the total energy $U \approx u R^3$ of the observable universe, and R is the scale constant. Equation (2) is obtained from (1) by Einstein's energy density relation

$$u \approx \frac{c^2}{\kappa R^2} \tag{3}$$

eliminating κ from equation (1). Consequently C.F. von Weizsäcker interpreted $\hbar c/R$ as the energy of an "ur", the lowest excitation in a spherical universe of radius R (Einstein universe), or in an anti-de-Sitter universe of curvature $-1/R^2$. N is the number of decidable alternatives in a finite universe.

Eliminating u, equation (1) can be written as

$$N = R^2 \left(\frac{c}{\hbar \kappa}\right), \tag{4}$$

where $(\hbar \kappa/c)^{1/2}$ is the Planck length l_P. This form is the basis of the "holographic principle" of cosmology, first stated by G. 't Hooft and L. Susskind in the 90th [2].

The surface of the observable universe is divided into pixels of the size of the Planck area. All information inside the observable universe can be written on its surface, and N is the number of binary digits, which describe the observable universe today.

There is a third way to interpret equation (1). In 1920 A. Eddington [3] observed that the 3rd power of the classical electron radius can be written as a product of R and the Planck area. Using instead of the classical electron radius the Compton wave length of the proton, we can write

$$\lambda_p^3 \approx R \frac{\hbar \kappa}{c}.$$

Now equation (4) can be put in the form

$$N = R^3 \left(\frac{1}{\lambda_p}\right)^3. \tag{5}$$

So the Planck pixels on the surface correspond to the 3-dimensional nuclear cells of the size λ_p^3 inside the observable universe. We assume that this is not just a coincidence at the present epoch, but that this law holds in general. Equation (5) was interpreted by C.F. von Weizsäcker that you need at least $3R/\lambda_p = 3N^{1/3} \approx 10^{40}$ urs to specify the states of a massive elementary particle.

Equations (4), (5), (1) can be written as

$$\begin{aligned} R &= l_P N^{1/2} \\ \lambda_p &= l_P N^{1/6} \\ U &= m_P c^2 N^{1/2} \\ u &= \frac{\hbar c}{l_P^4} N^{-1}, \end{aligned} \tag{6}$$

where m_P is the Planck mass $m_P = (\hbar/c\kappa)^{1/2}$. These formulas suggest that if the physical constants $\hbar$, c, κ do not vary in time, i.e. the size of the Planck pixels on the surface is constant, the total information will vary. In [4] it was calculated that the expectation value of N varies with Minkowski time as

$$N(t) = \frac{c}{\hbar} \frac{Ut^2}{R} = \frac{t^2}{t_P^2}$$

where $t_P = l_P/c$ is the Planck time. That means that at the big bang the universe started in fact with one qubit of information $N(t) = 1$ at the time $t = t_P$, with the highest possible energy density

$$u = \frac{\hbar c}{l_P^4}.$$

(Note that Heisenberg's uncertainty principle suggests $N > 1/2$.)

The formulas (6) become

$$R = ct$$
$$\lambda_p = l_P^{2/3}(ct)^{1/3}$$
$$U = \frac{\hbar t}{t_P^2}.$$

The number of protons in the observable universe equals $N^{2/3}$ and will increase; the mass of the proton, however, will decrease with time. The above formulae are consistent with General Relativity. From (6) we obtain for the pressure $p = u/3$. The time dependence leads to the spatial curvature $k = 1$ or 0. The model has only 6 metrical (spatial) automorphisms.

References

1. C.F. von Weizsäcker: In *Quantum Theory and the Structures of Time and Space*. Vol. 1. Hanser Verlag, München 1975, p. 213
2. G.'t Hooft: *Nucl. Phys.* B342, 471 (1990); L. Susskind: *J. Math. Phys.* 36, 6377 (1995); M. Srednicki: *Phys. Rev. Lett.* 71, 666 (1993)
3. A. Eddington: *Space, Time and Gravitation*. Cambridge University Press, 1920, p. 179
4. L. Castell: In L. Castell, M. Drieschner, C.F. von Weizsäcker (eds.): *Quantum Theory and the Structures of Time and Space*. Vol. 2, Hanser Verlag München 1977 p. 130

The Momentum Eigenstates and the Lorentz-Invariant State

Lutz Castell

I have some difficulties about which topic in C.F. von Weizsäcker's ur theory I should write. There are fascinating new developments in this area, which lead to a quantized form of time and space, but which may be too lengthy for this volume. So I want to report on an elegant derivation of a formula, in which C.F. von Weizsäcker has been most interested. This is the following: The eigenstates of the momentum operator of a mass 0, spin 0 particle (in the simplest case) can be interpreted as superpositions of many-ur and anti-ur states in a quite natural way [1].

In this physical interpretation an ur, or binary alternative, is a superposition of two neutrino states (left-handed, positive energy), two positive energy solutions Ψ' and Ψ'' of the mass zero Weyl equation spread over the whole universe:

$$\left(\sigma_i \frac{\partial}{\partial y_i} + \frac{\partial}{\partial y_4}\right) \Psi = 0 ,$$

$$\Psi' = R^{3/2} \frac{y_i \sigma_i + (y_4 - iR)}{[y_i^2 - (y_4 - iR)^2]^2} \begin{pmatrix} 1 \\ 0 \end{pmatrix} ,$$

$$\Psi'' = R^{3/2} \frac{y_i \sigma_i + (y_4 - iR)}{[y_i^2 - (y_4 - iR)^2]^2} \begin{pmatrix} 0 \\ 1 \end{pmatrix} ,$$

where $R > 0$ is the scale constant of the universe [2].

We start with the unitary representations of the Poincaré group of positive energy, mass 0, and spin 0. The Lie algebra is defind by the self-adjoint operators in the 3-dimensional momentum space,

$$M_{ij} = -i\left(p_i \frac{\partial}{\partial p_j} - p_j \frac{\partial}{\partial p_i}\right), i, j, k = 1, 2, 3 ,$$

$$M_{i4} = -ip \frac{\partial}{\partial p_i} ,$$

$$P_i = p_i, \; P_4 = p = +\sqrt{p_i p_i} \; .$$

These operators are defined in a dense domain of the Hilbert space

$$\int a(p_i)^* a(p_i) p^{-1} d^3 p < \infty$$

of square integable functions. In the case of mass 0 representations the Poincaré group can be extended to the full conformal group $SO_0(4,2)$ on the same Hilbert space. The additional generators are the generators of the dilatations

$$D = i \left(p_k \frac{\partial}{\partial p_k} + 1 \right)$$

and of the special conformal transformations

$$K_i = p_i \nabla^2 - 2 \left(p_k \frac{\partial}{\partial p_k} + 1 \right) \frac{\partial}{\partial p_i} \; ,$$

$$K_4 = p_i \nabla^2, \quad \text{where} \quad \nabla^2 = \frac{\partial}{\partial p_k} \frac{\partial}{\partial p_k} \; .$$

The maximal compact sub-group of this 15-dimensional symmetry group is $SO(4) \times SO(2)$, whose generators are the M_{ij}, the $M_{i5} = -1/2(P_i + K_i), i = 1, 2, 3$, and the $M_{46} = 1/2(P_4 - K_4)$. It is this fact, which makes a physical interpretation of the ur theory possible. The operators M_{12}, $M_{12}^2 + M_{23}^2 + M_{31}^2 = M^2$, and M_{46} represent a set of 3 <u>discrete</u> quantum numbers, l_3, $(l) \times (l+1)$, and E_n, instead of the 3-momentum p_i. We shall use the corresponding eigenfunctions to represent the unit-operator $p\delta^3(p-p')$. The orthonormal set of eigenfunctions is given by the eigenfunctions of the angular momentum operators, and

$$M_{46} \varphi = \frac{p}{2}(1 - \nabla^2)\varphi = E_n \varphi \; ,$$

$$\varphi_{l,l_3,n} = N_{l,n} Y_{l,l_3}(\varphi, \theta) L_n^{2l+1}(2p) e^{-p} p^l \; .$$

The Y_{l,l_3} are the normalized spherical harmonics, and the L_n^α the Laguerre polynomials. The normalisation factor $N_{l,n}$ is given by

$$N_{l,n} = \left[\frac{n! 2^{2l+2}}{(n+2l+1)!} \right]^{1/2} \; .$$

The angular momentum takes the values $l = 0, 1, 2, 3, \ldots$, $-l \le l_3 \le l$, and the eigenvalues E_n are given by $E_n = n + l + 1$, $n = 0, 1, 2, 3, \ldots$, see Fig. 1.

The state $a(p)$ can be expanded into

$$a(p) = \sum b_{l,l_3,n} \varphi_{l,l_3,n}(p) \; .$$

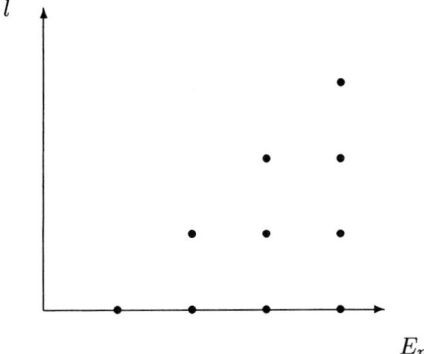

Fig. 1. SO(4)×SO(2) covariant states for the mass 0 representation

The unit operator can be expanded correspondingly (completeness relation)

$$p\delta^3(p-p') = \sum_{l,l_3,n} \varphi^*_{l,l_3,n}(p')\varphi_{l,l_3,n}(p) . \tag{1}$$

It is the generalized eigenfunction of P_i for the eigenvalue p'_i. In the discrete basis, we can interpret the mass 0 states in terms of urs: Every state $|\varphi_{l,l_3,n}\rangle$, $n > 0$ can be represented as a bound state of n symmetrical urs, and n symmetrical anti-urs [1]. The functions $\varphi^*_{l,l_3,n}(p'_i)$ represent the weight of every state in the expansion of the momentum eigenfunctions into the discrete basis.

The Lorentz-Invariant State

The Klein-Gordon equation $\Box\varphi = 0$ has as Lorentz-invariant solution $\varphi =$ constant. The positive frequency component $a(p_i)$ of $\varphi(y)$ is defined by

$$\varphi(y) = \frac{1}{(2\pi)^{3/2}} \int a(p_i) e^{ipy} \frac{d^3p}{p} ,$$

The invariant norm is given by

$$\frac{i}{2}\int \varphi^* \frac{\overleftrightarrow{\partial}}{\partial y_4} \varphi d^3y = \int a^*a \frac{d^3p}{p} .$$

For $\varphi = C$ we obtain

$$a(p) = (2\pi)^{3/2} C\delta^3(p)p ,$$

which equals

$$(2\pi)^{3/2} C \frac{\delta(p)}{p} \frac{1}{2\pi} ,$$

If we put $p' = 0$ in equation (1) we obtain

$$p\delta^3(p) = \sum \varphi_{00n}^*(0)\varphi_{00n}(p),$$

$$\varphi_{00n}^*(0) = \sqrt{\frac{4}{n+1}}\sqrt{\frac{1}{4\pi}}L_n^1(0) = \frac{1}{\sqrt{\pi}}\sqrt{n+1},$$

$$b_{00n} = C \cdot 2\sqrt{2\pi}\sqrt{n+1}.$$

There is an other interesting calculation, which leads to the same result. The normalized coherent states of positive energy of the Klein-Gordon equation are given by [3]

$$\varphi = \frac{1}{\sqrt{2\pi}}\frac{\sqrt{(\xi-\xi^*)^2}}{(y-\xi)^2}, \quad (\operatorname{Im}\xi_\mu)^2 < 0, \quad \operatorname{Im}\xi_4 > 0,$$

where ξ_μ is a complex parameter defined in the forward cone. If we choose $\xi_\mu = (0,0,0,i\lambda)$ we obtain

$$\varphi = \frac{\sqrt{2}}{\pi}\frac{\lambda}{y^2 + 2i\lambda y_4 + \lambda^2},$$

which in the limit $\lambda \to \infty$ approaches the constant $1 : \pi/\sqrt{2}\lambda\varphi \to 1$.

On the other hand, we can transform the Klein-Gordon equation $\Box\varphi(y) = 0$ from Minkowski-space into the Einstein-space $S_3 \times S_1$, and obtain

$$\left[\frac{1}{\sin^2\rho}\frac{\partial}{\partial\rho}\sin^2\rho\frac{\partial}{\partial\rho} + \frac{1}{\sin^2\rho}\left(\frac{1}{\sin\theta}\frac{\partial}{\partial\theta}\sin\theta\frac{\partial}{\partial\theta} + \frac{1}{\sin^2\theta}\frac{\partial}{\partial\varphi}\frac{\partial}{\partial\varphi}\right)\right.$$

$$\left. -\frac{\partial}{\partial\alpha}\frac{\partial}{\partial\alpha} - 1\right]\Phi(\varphi,\theta,\rho,\alpha) = 0,$$

$$\Phi = \frac{1}{2}\left[y_i^2 - (y_4 - i)^2\right]^{1/2}\left[y_i^2 - (y_4 + i)^2\right]^{1/2}\varphi(y),$$

where (φ, θ, ρ) and (α) are the polar coordinates on S_3 and S_1, respectively. The norm is given by

$$\frac{i}{2}\int \Phi^* \overset{\leftrightarrow}{\frac{\partial}{\partial\alpha}} \Phi d^3\omega.$$

The constant solution is transformed into the singular solution

$$\Phi = \frac{C}{\cos\rho + \cos\alpha}. \tag{2}$$

The normalized, positive energy solutions on $S_3 \times S_1$ with angular momentum 0 are given by

$$\frac{1}{\pi}\frac{1}{\sqrt{2}}\frac{1}{\sqrt{n+1}}C_n^1(\cos\rho)e^{-i(n+1)\alpha}, \quad E_n = n+1.$$

The $C_n^1(\cos\rho)$ are the Gegenbauer polynomials. The positive frequency coefficients of solution (2) are now again

$$\sqrt{n+1} \cdot 2C\pi\sqrt{2}.$$

References

1. L. Castell: In L. Castell, M. Drieschner and C.F. von Weizsäcker (eds.): *Quantum Theory and the Structures of Time and Space*. Vol. 1, C. Hanser Verlag, München 1975, p. 147
2. L. Castell: In L. Castell, M. Drieschner and C.F. von Weizsäcker (eds.): *Quantum Theory and the Structures of Time and Space*. Vol. 2, C. Hanser Verlag, München 1977, p. 130
3. L. Castell: *Phys. Rev.* D6, 563 (1972)

C.F. von Weizsäcker's Reconstruction of Physics: Yesterday, Today, Tomorrow

Holger Lyre

Heisenberg ... hatte seine nichtlineare Spinorfeldtheorie auf die Voraussetzung der allgemeinen ... Quantentheorie und die Forderung der Poincaré-Gruppe und der Isospingruppe gegründet. Als ich ihn fragte, warum genau diese Gruppen, sagte er etwa: 'Das kann ich nicht mehr begründen. Mit einer Forderung muß man eben anfangen.' Ich fragte: 'Warum überhaupt Gruppen?' Er, etwa: 'Wenn man mit einer Forderung anfängt, dann ist Symmetrie der beste Anfang. Symmetrie ist schön, das wußte schon Platon. Darin drückt sich die zentrale Ordnung aus.' ...
Ich aber blieb unbefriedigt. Mir schien, man solle eine rationale Behauptung wie die Geltung einer Symmetrie womöglich noch rational begründen. Meine Vermutung war: Symmetrie bedeutet die Trennbarkeit des jeweils untersuchten Gegenstandes vom Rest der Welt ... Die Frage ist, ob man diesem Gedanken noch eine strengere Fassung geben kann.

C.F. von Weizsäcker ([18, p. 910])

1 The Origin: The Philosophy of Physics

Carl Friedrich von Weizsäcker's thinking has always crossed the borders between physics and philosophy. Being a physicist by training he still feels at home in the physics community, as a philosopher by passion, however, his mind cannot stop thinking at the limits of physics. His physical ideas are based on the general conceptual and methodological preconditions of physical theories. The above quotation does indicate this: Heisenberg's hint on symmetry as a natural starting point for a physicist – even though already philosophically motivated by Plato – left Weizsäcker "dissatisfied." He even wanted to explore the reasons for the symmetries themselves.

Such a line of reasoning about the foundations of physics has brought Weizsäcker into an abstract program of a possible reconstruction of physics in terms of yes-no-alternatives, which he called "ur theory." I shall start this paper with a review of the basic ideas of ur theory: the definition of an ur and the connection between ur-spinors and spacetime (Sect. 2: "Yesterday"). I then go over to some of ur theory's present borders: the construction of quantized spacetime tetrads and the difficulties to incorporate gravity and gauge theories (Sect. 3: "Today"). It goes without saying that my brief review will be far from being complete – see in particular the developments presented in [4] In the last section (Sect. 4: "Tomorrow") I shall discuss the possible prospects of ur theory – partly with a view to modern quantum gravity approaches, but mainly in connection with its philosophical implications. Here, one of the crucial questions is, whether form, or, modern, information is an entity *per se* and what particular consequences this may have.

2 Yesterday: Urs, Spinors and Spacetime

Weizsäcker's first book on the philosophy of physics, *Zum Weltbild der Physik*, appeared in 1943 in its first edition – when his professional main concern still was 'real physics'. But already at this time his 'real interest' had shifted to the foundations of quantum mechanics. About ten years later in 1954, in a spa, Bad Wildungen, Weizsäcker had the crucial idea: the quantum theory of a binary alternative *is* the theory of objects in a three-dimensional space. Mathematically, he had just stumbled about the fact that $SU(2)$, the quantum theoretical symmetry group of a binary alternative, is locally isomorphic to the three-dimensional rotation group $SO(3)$ in Euclidean space. Philosophically, however, this was more "satisfying" than just to start with a certain symmetry, since now the symmetry itself was distinguished by the fact that it is the symmetry of the simplest logical object – a yes-no-alternative. Couldn't this be a reason for the three-dimensionality of position space?

Weizsäcker's approach rests on two main 'ingredients': firstly, the idea of reducing physics to binary alternatives and, secondly, the connection between spin structures and the structures of time and space. This second motive has caused David Finkelstein [5] to subsume Weizsäcker's approach under the heading of "spinorism" and this will be my main concern in this and the following section. For a moment, however, we shall consider the first mentioned, rather philosophical ingredient. Here the idea is that – in its core – physics reduces to predicting measurement outcomes. Measurement outcomes may be restated in terms of empirically decidable n-fold alternatives, and, trivially, any n-fold alternative can be embedded into a (Cartesian) product of binary alternatives. Binary alternatives, in turn, can be considered as bits of information. Thus physics reduces to information or, more precisely, *potential* information.

Weizsäcker has called the empirically decidable binary alternatives in his reconstruction *ur-alternatives* (from the German prefix 'Ur-': 'original,' 'elementary,' 'pre-'). A bit whimsically, he later called them *urs*. Urs may be considered the fundamental objects in physics. As a matter of principle, any physical object can be built out of them. The reader may note that at this point an interesting conceptual shift has occurred. We started from the notion of information, a subjective notion in the first place, and applied it to the objective notion of a physical object. This is another way of saying that objects are reduced or even 'made out of' information. It seems that via this shift information has been gained the status of a substance. In the last section I shall indeed come back to this challenging formula, as our present working hypothesis, however, we will take an information-theoretic reductionist view of physics in the sense that physical objects are entirely characterized by the information which can be gained from them.

In quantum theory in particular, this view has a lot of plausibility. Quantum objects are represented in terms of their Hilbert state spaces, their quantum states correspond to empirically decidable alternatives. Any quantum object may further be de-composed or embedded into the tensor product of two-dimensional objects, nowadays called quantum bits or qubits. Urs, therefore, are in fact nothing but qubits (cf. Sect. 4).

Weizsäcker's vision of the structure of physics as a quantum theory of ur-alternatives has its roots in the above mentioned fact that the essential symmetry group of urs, which is $SU(2)$, is the double covering of the rotational symmetry group of three-dimensional position space. The guiding line is that, if the idea of urs as fundamental entities is true, the symmetry of such urs should play an essential role in the reconstruction of physics and in the phenomenology of our empirical world. And, of course, position space is probably the most essential feature of the empirical world!

But this only gives rise to spinorism in general, the mathematical motive of deriving the spacetime structure from a primarily given spin structure.[1] A more precise ur-theoretic ansatz was chosen in Weizsäcker's 1958 papers *Die Quantentheorie der einfachen Alternative (Komplementarität und Logik II)* – internally called "KL II" – and the "Dreimännerarbeit" KL III together with Erhard Scheibe and Georg Süßmann [15, 16]. KL II starts from the well-known and deep connection between tensors and spinors: there always exists a mapping from spinor representations to tensor representations of the various rotation groups. A lightlike four-vector, in particular, can be written in terms of the Pauli matrices σ_μ as

$$k_\mu = \sigma^\mu_{\dot{A} B} u^{\dot{A}} u^B , \qquad (1)$$

where u^A is a spinor and dotted indices denote complex conjugate components. This relation highlights the link between the homogeneous Lorentz group $SO(1, 3)$ and $SL(2, \mathbb{C})$, the unimodular group in spinor space. But instead of elaborating the thus defined spin structure of Minkowski space, the focus of KL III lies on Weizsäcker's idea of "multiple quantization." Generally speaking, the procedure of quantization

---

[1] To the best of my knowledge, Weizsäcker presented the first hint on spinorism in a short note from 1951 on the occasion of Werner Heisenberg's 50th birthday, where he drew attention to the fact that both position space and Hilbert space are provided with a quadratic metric and that in order to give a deeper reason for this one presumably needs a property common to both spaces (which indeed could be the spin structure). He closes that perhaps the question of the structure of the state space is fundamental, whereas the structure of position space is a dependent and derived one – a rather bold and visionary remark in 1951 (and, of course, still today)! If we now take the publication date of this note [14] then this Festschrift can simultaneously be considered as celebrating 50 years of Weizsäcker's idea of spinorism. However, the rather mathematical hint from 1951/52, the second ingredient of ur theory, was only filled with conceptual life two years later at the above mentioned spa. This little episode also sheds light on the crucial interplay between philosophy and physics in Weizsäcker's thinking.

can be split into two steps: (i) the transition from a (classically) discrete number of degrees of freedom to infinitely many degrees, and (ii) the transition to operator-valuedness and, hence, appropriate commutation relations. Starting from a simple, classical yes-no-alternative a_A, step (i) means the construction of a wavefunction $\varphi(a_A)$, i.e. a spinor $u_A \equiv \varphi(a_A)$. Relation (1) allows to transform $u_A \Leftrightarrow k_\mu$, then (ii) implies the transition $k_\mu \to \hat{k}_\mu$. This is the 'first quantization' of a binary alternative. On the level of 'second quantization' wavefunctions $\varphi(k_\mu)$ occur, where operators $\hat{k}_\mu$ act upon. Interpreting k_μ as energy-momentum four-vector, we get quantum mechanical wavefunctions $\psi(x_\mu)$ in Minkowski space after Fourier transformation. The next, third level of quantizing $\psi(x_\mu)$ corresponds to usual quantum field theory.

Weizsäcker, Scheibe and Süßmann derived in their paper the Weyl and – after doubling the state space of urs and working with bi-spinors – the Dirac and Klein-Gordon equation and also the homogeneous Maxwell equations as algebraic identities![2] Since for a nullvector one obtains $k_\mu k^\mu = 0$, the massless Klein-Gordon equation follows for instance from

$$\hat{k}_\mu \hat{k}^\mu \, \varphi(k_\mu) = 0 \quad \overset{FT}{\Longleftrightarrow} \quad \Box \, \psi(x_\mu) = 0 \, . \tag{2}$$

Let $D(1/2, 0)$ be the fundamental representation of $SL(2, \mathbb{C})$ and $D(0, 1/2)$ its complex-conjugate, then the transition to bi-spinors effectively means to work with $D(1/2, 0) \oplus D(0, 1/2)$. I shall show in the next section how this combines with a proper usage of the full spin structure of urs.

We have so far presented ur-theoretic arguments to introduce Minkowski spacetime as a local spacetime model, we shall now go over to global spacetime considerations. We start again with a quantum theory of binary alternatives, but this time with usual quantum bits obeying a unitary norm. The symmetry group of urs then contains $SU(2)$, $U(1)$ and the complex conjugation. As a Lie group manifold, this yields

$$SU(2) \times U(1) \,=\, \mathbb{S}^3 \times \mathbb{S}^1 \, . \tag{3}$$

Weizsäcker has made the far-reaching assumption – for which the above portrayed spinorism is just the motive – that $\mathbb{S}^3$ itself should be considered a model of global cosmic space [17]. It follows that urs are wavefunctions on $\mathbb{S}^3$. As Thomas Görnitz [6] has shown, this leads to a remarkably fresh look at the large numbers in physics on the basis of the multiplicity of the regular representations of $SU(2)$, in accordance with the Planck scale and Weizsäcker's earlier calculations. This will be explained in a moment. Moreover, for the appropriate treatment of global cosmic time and time in general, the distinction between past and future plays a major role in Weizsäcker's philosophy of physics. He considers it a very precondition of empirical science, on which the concept of separable alternatives has to be built. In my opinion it is therefore highly implausible to take $U(1)$ as a model of cosmic time, which would be cyclic then. Obviously, time demands a special treatment (see, however, [3] for an opposite view).

---

[2] Details of the derivation must be left out due to lack of space, the reader may consult KL III or [7, Sect. 2.4].

We now seek to calculate the number of urs, starting from Weizsäcker's early estimations in the 70's. As a wavefunction on global space, one ur can be thought of as the alternative of being in the 'one half' or the 'other half' of the universe. Suppose now we want to localize a nucleon in the universe. The Compton wavelength can be understood as a natural measure of localizing a certain particle. For a nucleon, the ratio between the cosmic radius R and the Compton wavelength λ is about 10^{40}. We therefore have to decide 10^{40} alternatives in each spatial direction to localize that nucleon. Hence, in ur theory the assumption is made that a nucleon *is* 10^{40} urs (up to two or three orders of magnitude). Accordingly, an electron is about 10^{37} urs.

The Compton wavelength λ of a nucleon is actually a distinguished measure of length. If we take the whole energy content of the universe to subdivide space into equal intervals, then λ drops out. In this sense it is indeed an elementary length. We may of course measure up smaller regions of space, but then, as a matter of principle, do we loose the possibility to perform measurements of other regions at the same time with the same accuracy. In this sense the number of elementary spatial cells, which is $N = (R/\lambda)^3 = 10^{120}$, is the total number of elementary "events" in the world and, thus, the total number of urs (the one bit decision whether, for instance, a particular cell is occupied by a nucleon). Accordingly, the dimension of the Hilbert space of urs is $2^{10^{120}}$. As a *result*, not as an input, Weizsäcker got the number of nucleons as $10^{120}/10^{40} = 10^{80}$ – in accordance with empirical results!

As already mentioned, Görnitz has refined these estimations and put them on a more solid, group-theoretical basis. He considered the regular representation of $SU(2)$, i.e. the representation in terms of square-integrable functions of the Hilbert space $\mathcal{L}^2(\mathbb{S}^3)$ on $SU(2)$ itself as a homogeneous space $\mathbb{S}^3$. The multiplicity of the irreducible representations of the reduced tensor product of spin-1/2-representations of $SU(2)$ shows a characteristic cut-off at functions with a wavelength of the order $l_o = R/\sqrt{N}$, where R is the radius of $\mathbb{S}^3$ and N the number of ur-functions (cf. [6] for details). Obviously, l_o is now the lower bound of spatial distances which can be measured in a cosmos with N urs at maximum. l_o is indeed a fundamental elementary length in the sense that it represents the smallest spatial resolution physically possible. Remarkably, from the above $N = 10^{120}$ we get $l_o = 10^{-60} R$, i.e. the Planck length! Conversely, if we already know about the Planck scale from other considerations (as we usually do), we get $N = 10^{120}$ as a result – in accordance with Weizsäcker's original, more hand-waving estimations.

3 Today: Tetrads, Gravity and Gauges

In the preceding section we have discussed the possibility of introducing spacetime from an ur-hypothesis and also deriving the form of the basic equations for matter fields (mainly the Dirac equation) and the simplest free interaction field (the homogeneous Maxwell equations). From a broader perspective two questions arise: Shouldn't the ur-theoretic ansatz, a way of deriving spacetime itself, directly lead to gravity (or, perhaps, even quantum gravity)? Shouldn't we be able to derive interaction gauge theories in general? Ambitious questions, indeed, but it is perhaps not astonishing that in abstract accounts like ur theory questions like these arise at the very beginning. And despite of their ambitious character and also far from really answering them I try to discuss in the following some possible ways to tackle these questions.

As a working hypothesis, let us, again, take $SU(2) = \mathbb{S}^3$ as a model of global position space. Urs can be understood as non-local functions on $SU(2)$ and do naturally represent a spinor dyad (with spinors u^A, v^A satisfying $u_A v^A = -v_A u^A = 1$). It is now well-known that a spinor dyad is equivalent to a null-vector vierbein or tetrad, where the four null-vectors have the form (1), but consist of mixed combinations of u_A and v_A (the details of this and the following paragraph can be found in [8]). By considering suitable linear combinations of the null-vectors, such a null-tetrad can generally be written in real-valued form $\theta_\mu^\alpha = (t_\mu, x_\mu, y_\mu, z_\mu)$, where the space-like vectors x^μ, y^μ, z^μ represent a spatial dreibein tangent to $\mathbb{S}^3$ together with an orthogonal time-like vector t^μ. The interesting point is that, since the tetrad is written in terms of ur-spinor components, a (first) quantization of urs also induces a quantization of the tetrad. Such a quantized ur-tetrad, however, means nothing but quantized coordinates in our spacetime model.

As a remarkable feature of the quantized ur-tetrad it turns out that the time operator $\hat{t}_\mu$ is just the ur-number operator $\hat{n} = 1/1 \sum_r \{\hat{a}_r^+, \hat{a}_r\}$ in the Fock space of Bose urs. This is consistent with Weizsäcker's assumption that the growth of the total number of urs is a measure of temporal cosmic evolution. Now, since the number operator has indeed a lower bound at zero, the global spacetime model $\mathbb{S}^3 \times \mathbb{R}^+$ with a time parameter manifold $\mathbb{R}^+$ seems justified (thus avoiding globally closed time-like curves as mentioned in Sect. 2).

Can all this give us a hint how to describe (quantum) gravity in ur theory? We may think of θ_μ^α as representing four vector bosons, i.e. massless 'gravitons' with spin 1 and get a corresponding wave equation $\Box \theta_\mu^\alpha(x) = 0$ analogous to (2). This could perhaps describe the gravitational field in non-standard form (i.e. not as a spin-2 field) in the linearized limit. However, what we are really looking for is a recipe to make this field dynamical and couple it to matter. This was actually a main disadvantage of KL III. The theory of the free Maxwell field followed from the fact that $D(1/2, 0) \otimes D(1/2, 0) = D(1/2, 1/2) \oplus D(0, 0)$ consists of a spin-1 triplett and a spin-0 singulett, where the former can be written in terms of an anti-symmetric tensor with the algebraic properties of Maxwell's field strength tensor. Even if we grant this as a 'derivation' of Maxwell's free theory, and even if we have also obtained Dirac's equation as a free matter field theory, the *dynamical coupling* of both fields is still missing.

Today, the usual recipe of coupling matter and interaction fields is the gauge principle. The idea is to start from a global symmetry of the free matter field theory and then to postulate the corresponding local symmetry as well. The geometrical picture is that of a fiber bundle over spacetime, where the fibers are represented by the local symmetry group. It thus turns out that the inhomogeneous term in the covariant derivative (i.e. covariant under local gauge symmetry transformations) has the interpretation of a bundle connection. A word of caution, however: most textbooks present the connection coupling term as a genuine way to introduce a real physical field. But this is certainly an overestimation of a mere symmetry requirement! The appearance of a connection only adjusts the postulate of local gauge symmetry, the connection itself must still be considered flat, i.e. with vanishing bundle curvature. A true physical interaction field requires non-vanishing curvature and hence non-flat connections, but these are clearly not *enforced* by the logic of the gauge principle. Local gauge transformations must be understood as a mere change in the position representation of the wavefunction and are thus physically vacuous. The situation is analogous to choosing curvilinear coordinates in flat space (which leads to Christoffel symbols, i.e. connections, but no gravitational field, i.e. non-vanishing curvature). Therefore, one needs a true physical input to generally justify the connection term as non-flat. In general relativity this input is the equivalence principle. It has been argued that one possible way to establish the connection term as a real coupling term could be a *generalization* of the equivalence principle by introducing generalized 'inertial' and 'field charges' [9].

Now, a gauge theory is characterized by a certain gauge group. One nice starting point for an ur-theoretic gauge theory of gravity could be the fact that the quantized ur-tetrad generates a group which itself could be used as a proper gauge group. It can be shown that the Lie algebra of the ur-tetrad operators is 12-dimensional and that the corresponding Lie group is isomorphic to $SL(2,\mathbb{C}) \times SL(2,\mathbb{C})$. Whether and how this group can be used for a proper gauge approach is still an open question. One interesting point to be mentioned is that the operators of this algebra "live" in flat Minkowski spacetime (in ur theory this is reflected by unimodular groups rather than unitary ones). As in many other gauge approaches of gravity we would therefore describe the gravitational field in a flat space – and perhaps in this sense it is indeed just a local field as other Yang-Mills fields. But the point is that in ur theory it will nevertheless be possible to have a global spherical model of the cosmos, since we got global curvature right from the beginning by taking $\mathbb{S}^3$ as the cosmic model.

4 Tomorrow: Qubits, Holographic Principle and the Ontology of Information

In this last section I shall come back to the more philosophical issues of ur theory. In Sect. 2 we already noticed that, ontologically speaking, information in ur theory seems to gain the status of a modern notion of substance. Two questions are simply unavoidable then: Does information exist without a carrier? Does information exist without an observer or information-gathering system?

Again, the very idea of ur theory is to characterize physical objects entirely by the information which can be gained from them. The further, novel feature is that even space or spacetime is reconstructed here as a mere device to represent information. By the time Weizsäcker proposed it, this was a revolutionary new perspective. Nowadays, in modern quantum gravity, there is a strikingly similar discussion about the deeper connections between space and information, which has its roots in the considerations of black hole entropy. In the early days, black holes were thought of as characterized by three quantities only: mass, angular momentum and charge ("no hair"-property).

But this in turn means that we could use them as 'entropy graves' by simply feeding them with the products of high entropy processes. This caused Jakob Bekenstein to attribute thermodynamical properties to black holes [1]. In particluar, the surface of the event horizon A in Planck units turns out as a suitable measure of the entropy content

$$S = \frac{1}{4} A, \qquad (4)$$

thus leading to a generalized second law of thermodynamics.

Formula (4) is in many respects quite remarkable. As Gerard 't Hooft [13] has pointed out, it does characterize physical objects on the most fundamental level not by the three-dimensional volume they occupy in space, but by a 'projection' of their degrees of freedom on a two-dimensional area much like a hologram. He thus calls this idea the *holographic principle*. From an ur-theoretic point of view, it is rather the characterization of physical objects in terms of pure information, which becomes evident here. Indeed, if we calculate the entropy and, hence, information content of the whole universe in Planck lengths, we get $S_u \approx (10^{60})^2 = 10^{120}$ bits. In the same way we get $S_n \approx (\lambda/l_o)^2 = 10^{40}$ bits for a nucleon – hence, the ur-numbers Weizsäcker already found in the 70's! For an electron, however, (4) leads to $S_e \approx (\lambda_e/l_o)^2 = 10^{46}$ in contrast to the 10^{37} urs stated in Sect. 2.

Another calculation may suffice (cf. [6]). Suppose spherical symmetry $A = 4\pi R^2$ and the Schwarzschild radius $R = 2M$, then (4) transforms to

$$S = 4\pi M^2 . \qquad (5)$$

Again, for the whole universe $M_u = 10^{60} m_o$ we get $S_u = 10^{120}$ bits as above. The information content of a particle with mass m then actually is the entropy difference of a universe with or without such a particle, and hence

$$\Delta S = 4\pi \left((M_u + m)^2 - M_u^2 \right) \approx 8\pi M_u m. \qquad (6)$$

We now get $S_n \approx 10^{40}$ and $S_e \approx 10^{37}$ in accordance with the results from Sect. 2. It seems therefore that ur theory does not directly support the view of the holographic principle – and perhaps this could be a helpful insight for other programs as well.

Indeed, the relevance of the Bekenstein entropy has been acknowledged and partially explained in modern quantum gravity programs – in string theory as well as in the quantum loop approach [12]. Weizsäcker's explanation – historically the first – is a further alternative. The Ashtekar-inspired, canonical quantum loop approach has indeed some core similarities with ur theory. Like Penrose's twistor approach, quantum loop gravity has its most suitable representation in terms of spin networks. It is certainly the most powerful among the spinoristic programs today, whose key feature is that they are background-free. Of course, all of the programs mentioned are mathematically by far more elaborated than ur theory, which merely is an outline or perhaps a raw framework of a spinoristic theory. The true advantage of Weizsäcker's approach is rather its philosophical underpinning – and this may also serve other programs.[3]

To explore the implications of a true ontology of information in physics, consider equation (5) again. It indeed highlights the ur-theoretic view that energy-matter *is* information. In a future quantum gravity, formula (5) might obtain the same status as Einstein's $E = mc^2$, which indicates the ontological equivalence of energy and matter. However, would this suffice to consider information as a substance? Surely, the age-old distinction between matter and form lurks behind this question. Aristotle considered a (physical) object a *synholon* that is a composition of form (*eidos*) and matter (*hyle*). Form comes into matter, but none of these can exist independently, as far as physical objects are concerned. For Aristotle the essence of a thing is its form, the collection of all the attributes characterizing the object completely. In modern terms: the total information which can be gained from it. Nevertheless Aristotle insisted in the necessity of matter. He claimed the existence of a *prote hyle*, a "first matter" as a kind of a neutral, universal substance. The *prote hyle* is nothing we can specify any further, any form is stripped off from it. It does therefore not belong to the realm of physics, but can be considered the *hypokeimenon* of form – that on which form is based. Such a concept is clearly a pure metaphysical one and whether we like to make demands on it lies *per definitionem* beyond physics. What can empirically be known about an object, is *per se* reducible to information and in this sense physics indeed reduces to information. Perhaps, ur theory is just the most rigorous anccount to take this insight seriously.

So far, information has been treated in an absolute sense. However, from a conceptual point of view the notion of information is inherently a context-related, relative concept presupposing semantics information does only exist in relation to the difference of two semantical levels. E.g. the sign H on a sheet of paper has its 'meaning' as the eighth letter of the Latin capital alphabet or the seventh of the Greek; but it may also mean just a collection of ink molecules. To specify a particular amount of information, a certain semantics must be presupposed, absolute numbers of information are 'meaningless'.

From this point of view the above derived ur and, hence, bit numbers appear as highly questionable. Under which "context" do they fall? How could one specify the semantics of physical objects? Actually, the idea of (context-) "relatedness" in physics is not at all a new one, relativity theory heavily relies on it. In ur theory, the generators of the Poincaré group ([3] and [17, p. 407]) create and annihilate urs. Quite naturally, then, the information content of a physical state becomes a frame-dependent statement. This even applies to the information content of the universe. We may think of the 10^{120} bits of the world or its age of 10^{60} Planck times in relation to observers in the rest-frame of the cosmos only. Proponents of certain quantum gravity approaches, as for instance Carlo Rovelli [11], advocate that only an extreme relationalist view of space and time holds – as far as general covariance in general relativity and its consequences in quantum gravity are concerned. Ur-theory partially supports this view. On the other hand, there is no other system left outside the universe. Thus,

---

[3] For a recent dialogue between philosophy of science and quantum gravity research see [2].

the information content of the universe has an 'absolute' meaning in an operational sense. It is therefore not clear to me whether it makes sense to tackle cosmology in a rigorous generally covariant manner, where there is no distinguished observer left. Operationally we *are* distinguished as inner observers of the universe with respect to its rest-frame. In ur theory the cosmic model is therefore not derived as a solution of a certain set of basic equations, but as the epitome of the existence of observers and physical objects in spacetime itself: the representation of one ur. Global cosmic spacetime should perhaps be treated significantly different than local considerations.

Heisenberg's saying that there is no physics left, if we leave aside the possibility of any observers at all, does apply here, too. It was originally intended as a comment on the quantum measurement problem, not on problems about global spacetime. Ur theory combines both meanings: urs are essentially quantum wavefunctions of the cosmos – "cosmic qubits" so to speak. The relevance of the observer as a 'device' whose factual observation 'creates' the observed thing is in ur theory both supported from a rigorous quantum as well as information theoretic basis. As we already indicated in Sect. 2, the ur as a qubit represents potential information. Measurements must be understood as transitions from potential to actual information. The upshot is that indeed the world only comes into being for measuring observers – not for them as particular individuals, but for the idea and possibility of such observers in general. Kant would have called this the *transcendental subject*. In my view it is the unavoidable epistemological consequence of a fundamental physics of quantum information, and, eventually, the philosophy of physics presumably has to deal with questions of this sort.

Our speculations have come to a certain end now. Perhaps I was able to support a bit the view that ur theory is an account which allows for challenging considerations on fundamental questions of physics and the philosophy of physics, questions which are not even touched upon within the mainstream. As a physical program, ur theory surely suffers from mathematical elaboration as compared to other programs and does therefore not even earn the name "theory" in a strict sense. But it is still highly impressing that Weizsäcker had his main starting ideas already at a time, when other programs were far from being born. The main ingredients of ur "theory" – information and spinorism – do exist as central motives in modern approaches as well (as pointed out in the above). Perhaps the future development may lead to a mutual stimulation between other programs and the ur-hypothesis – at any rate this is what I wish Carl Friedrich von Weizsäcker from the heart.

References

1. J.D. Bekenstein: Black holes and entropy. *Physical Review D* 7, 2333–2346 (1973)
2. C. Callender and N. Huggett (eds.): *Physics meets Philosophy at the Planck Scale* Cambridge University Press, Cambridge 2001
3. L. Castell: In: Castell, Drieschner and Weizsäcker (eds.): *Quantum Theory of Simple Alternatives*. Volume 1, 1975
4. L. Castell, M. Drieschner and C.F. von Weizsäcker (eds.) *Quantum Theory and the Structures of Time and Space* (6 volumes). Hanser, Munich 1975–1986

5. D. Finkelstein: Finite Physics. In: R. Herken (ed.): *The Universal Turing Machine – A Half-Century Survey*. Springer, Vienna 1994
6. T. Görnitz: Abstract quantum theory and space-time-structure. I. Ur theory and Bekenstein-Hawking entropy. *International Journal of Theoretical Physics* 27(5), 527–542 (1988)
7. H. Lyre: *Quantentheorie der Information*. Springer, Vienna 1998
8. H. Lyre: Quantum space-time and tetrads. *International Journal of Theoretical Physics* 37(1), 393–400 (1998) (E-print quant-ph/9703028)
9. H. Lyre: A generalized equivalence principle. *International Journal of Modern Physics D* 9(6), 633–647 (2000) (E-print gr-qc/0004054)
10. H. Lyre: *Informationstheorie. Eine philosophisch-naturwissenschaftliche Einführung* UTB 2289. Fink, Munich 2002.
11. C. Rovelli: Halfway through the woods: Contemporary research on space and time. In: J. Earman and J. Norton (eds.): *The Cosmos of Science*. University of Pittsburgh Press/Universitätsverlag Konstanz 1997
12. C. Rovelli: Strings, loops and the others: A critical survey on the present approaches to quantum gravity. In: N. Dadhich and J. Narlikar (eds.): *Gravitation and Relativity: At the Turn of the Millennium*. Poona University Press, Poona 1998 (E-print gr-qc/9803024)
13. G. 't Hooft: Obstacles on the way towards the quantisation of space, time and matter – and possible resolutions. *Studies in History and Philosophy of Modern Physics* 32, 157–180 (2001)
14. C.F. von Weizsäcker: Eine Frage über die Rolle der quadratischen Metrik in der Physik. *Zeitschrift für Naturforschung* 7, a:141 (1952)
15. C.F. von Weizsäcker: Die Quantentheorie der einfachen Alternative (Komplementarität und Logik II). *Zeitschrift für Naturforschung* 13, a:245–253 (1958)
16. C.F. von Weizsäcker, E. Scheibe and G. Süssmann: Komplementarität und Logik, III. Mehrfache Quantelung. *Zeitschrift für Naturforschung* 13, a:705–721 (1958)
17. C.F. von Weizsäcker: *Aufbau der Physik*. Hanser, Munich 1985
18. C.F. von Weizsäcker: *Zeit und Wissen*. Hanser, Munich 1992

The Operational Structure of Spacetime

H. Saller

1 Ontic or Praxic Spacetime

For Newton, time and space were like fixed unchangeable boxes wherein the dynamics of mass points takes place – like fleas jumping around. By Leibniz, space and also time were considered as relational concepts characterizing a dynamics. Although even Newton himself disliked the independence of space, time and matter, his axioms and the building erected thereon, were extremely sucessful – first in experiments, but also for theories. For about two hundred years, it determined the way physicists incorporated the spacetime aspects in their theories. Kant's epistemological analysis to understand space and time as preconditions for our expericences was not connected enough with experiments to leave more than philosophical traces in physics.

With Einstein space and time came into closer contact. In contrast to much folklore, they stayed cleanly separated – special relativity tells us exactly which translations are timelike and which are spacelike. These are invariant concepts. However, their distinction by the sign of the indefinite Lorentz metric is no longer linear – e.g. the linear combination of two spacelike translations can be timelike and vice versa. Timelike translations, i.e. the double cone with the causally influential past and the causally influentiable future, on the one side, and spacelike translations as the one-shell hyperboloid of a causally unrelated generalized presence, on the other side, constitute no vector subspaces of the real 4-dimensional Minkowski spacetime translations. In addition to the old concepts of time and space, the causal skin of the timelike double cone, i.e. the lightlike translations where each space travel becomes a time travel, were found to be a consequence of a highest speed of action which induces the indefinite metrical, only partially ordered relativistic spacetime structure. A nontrivial causal structure for spacetime translations, mathematically a reflexive, transitive and antisymmetric order relation, requires an indefinite orthogonal group, established by the Lorentz transformations. Definite Euclidean orthogonal transformations on spacetime translations are compatible only with a trivial order relation wherewith each point exhausts its own past and future.

Special relativity is still a theory with independent spacetime and matter whereas in general relativity spacetime does not look like a box for fleas, but like part of a snail where the body forms the shell and vice versa – spacetime geometry and energy-momenta condition each other.

Quantum theory makes the decisive step from an ontic interpretation and description of physics to a praxic one: In quantum mechanics, at least position comes as an operator which, according to the experimental setup, can be measured in different projections which each have an ontic interpretation by its own, e.g. as an orbit of an electron. Position and momentum are operators, time remains a real parameter whose complex representation, especially in unitary groups, induces the parametrization of probabilities by the quantum characteristic complex probability amplitudes. In contrast to much folklore, there is no genuine energy-time uncertainty; what is used under this name is Heisenberg's position-momentum uncertainty relation in a disguised form.

The naive expectation that a special relativistic extension of quantum mechanics – to incorporate the Faraday-Maxwell field concepts – could lead to an operational structure also for time did not materialize. In relativistic quantum field theory, e.g. in quantum electrodynamics, space is de-operationalized again – both time and space are treated as real parameters, the operational structure is implemented in the value space of the fields, i.e. where we measure in units like Volt or Coulomb. In this respect, quantum theory on, of or for non-quantum spacetime is unsatisfactory, but once more – as experienced in the basically different approaches of Newton and Leibniz – extremely sucessful. The coincidence between experimentally measured and theoretically calculated numbers, e.g. for the anomalous magnetic moment of the electron in quantum electrodynamics with its complicated and sophisticated renormalization-regularization machinery and concepts, is overwhelming.

2 Operations and Symmetries

If the human prejudices which are anchored in our macroscopic existence and evolution, break down – which concepts remain for an explanation of the physical phenomena? What effects the laws of nature – their regularity, their constancy in time and space, e.g. their – apparent – equality in other galaxies, and, by no way least, the 'unreasonable effectiveness of mathematics' (Eugene Wigner) for their formulation? Can it be that physics is applied mathematics which itself can be considered with Bertrand Russell as an extension of logic? Or, perhaps, in a more appropriate hierarchical order: Is mathematics and, finally, even logic a condensation of our physical experiences – in contrast to its characterization as an everywhere and always necessary unique form of thinking?

Such a reversion of the order of abstraction – logic not as the precondition, but as the last consequence of experiences and, therewith, a highly experimental science – is suggested by the quantum structure induced new description of a logic. In contrast to a classical subset oriented distributive logical lattice, as introduced by George Boole, quantum logic, as formulated by Garrett Birkhoff and John von Neumann, is complex linear and constituted by a lattice of Hilbert subspaces. It can be operationalized by characterizing the Hilbert subspaces by projectors, a special kind of operators, embedded, so to say, as diagonals in the whole nonabelian set of physical operations. Operators characterize physical phenomena – they can be concatenated, one after the other, there is a trivial operator and – in some cases – they can even be inverted.

Quantum operators as monoids, sometimes even groups, acting on complex vector spaces have invariants which allows the concepts of symmetry. Therefore from the beginning, the description of basic structures in physics tried to invoke symmetry as an ultimate explanation. It started with the five regular Platonic solids – tetra-, hexa-, octa- and icosahedron for fire, earth, air and water, all embedded in the quintessential dodecahedron. It continued with Johannes Kepler – trying in his 'Mysterium Cosmographicum' to understand the six planets, known in his time – five godfathers of the weekdays (without Sunday and Monday) and the earth – as regularly circling on the simultaneous in- and out-spheres between the five Platonic solids, nested in each other. The differential geometric formulation of Lie groups acting on manifolds as used in the Erlanger Programm of Felix Klein to characterize geometries was taken up by Einstein to formulate special relativity as a pseudo-orthogonal geometry with the Lorentz group acting on 4-dimensional spacetime translations and, even more, for general relativity as theory for the invariants with respect to the huge diffeomorphism group of the 4-dimensional spacetime manifold. To study the formulation with and the consequences of acting groups became a basic tenet in physics, as exemplified by the identification of conservation laws with operational invariants as done by Emmy Nöther.

With Hermann Weyl, group theory entered massively quantum theory: In the operational quantum structures, objects, formalized by bound or scattering states and, in special relativity, by elementary particles, can be characterized as eigenvectors with respect to the action of spacetime translations and rotations. Physical properties are eigenvalues or invariants of these operations.

A bound state in quantum mechanics, e.g. for the nonrelativistic hydrogen atom, is characterizable by a finite set of rational numbers as eigenvalues and invariants of compact groups, e.g. angular momentum and perihel conservation number for the Coulomb potential. Starting from this algebraic point of view, Schrödinger wave functions are one of many possible representations only, their space dependence characterizes one action group with rational discrete invariants for the bound states.

Wigner connects flat spacetime and free particles: He classifies elementary particles as infinite dimensional vector spaces acted upon with unitary irreducible representations of the Poincaré group whose rank 2 gives rise to two invariant properties for particles: mass from a continuous spectrum for the noncompact spacetime translation group and discrete spin or polarization for the compact rotation groups. Together with the integer valued electromagnetic charge number, these are the characteristic properties of particles. Additional operations like isospin or color are used to define and structurize interactions, e.g. the gauge interactions of the standard model. In the particle projection, these internal (charge related) operations are broken as symmetries, the hypercharge-isospin by postulating an asymmetric ground state, or confined as proposed for color with the hadronization of the strong interaction parametrizing gluon and quark fields.

3 How Simple Can Be Simple?

If all physical objects and properties are characterizable with operations, the race is on to find the ultimate action group. If particles can be defined with spacetime action groups, an operational structure for spacetime may be a satisfactory step in the physical explanation chain for nature.

And from the mathematical structural point of view: Are there simplest action groups which – like the primes for the natural numbers – can build all physically relevant groups? The everyday word 'simple' has a definite meaning for Lie groups[1] and their infinitesimalization by Lie algebras: A simple group has no proper invariant subgroup, a simple Lie algebra is defined to be nonabelian without a proper ideal – i.e. the simple operational structures are like the prime numbers, they cannot be decomposed. The simple finite dimensional Lie algebras were completely classified by Elie Cartan. All relevant Lie algebras, e.g. as used in the standard model for the electroweak and strong interactions of elementary particles, are either abelian like the Lie algebra of the spacetime translations and the phase group $U(1)$ for hypercharge or simple like the Lie algebras of the special unitary group in two complex dimensions $SU(2)$ for isospin, of $SU(3)$ for color and of the Lorentz group $SO(1, 3)$.

In addition to their classification, Cartan gave also all finite dimensional representations of all simple Lie algebras. Therewith, the finite dimensional mathematics of simple operational structures is completely clarified. After that, 'only' the picture, i.e. the 'physical Tizian' in this beautiful mathematical frame, is missing: Which operational Lie algebras and groups are used in nature for interactions and in which representations? And why those?

In Cartan's classification of the simple operation structures also Platon and Kepler could find their favorites: The regular triangle, square and pentagon used for the five 3-dimensional Platonic solids can be refound in the four main series and the five exeptional Lie algebras. The 2-dimensional regular triangle and square can be related to the rank 2 origins of the main Lie algebra series for special linear, symplectic and orthogonal operations. The origin of all simple Lie algebras, which is simultaneously orthogonal, symplectic and special, is the 'simplest simple' operational structure with rank 1, the Lie algebra of the special complex group $SL(2, \mathbb{C})$ with its compact form, the unitary group $SU(2)$. This group is of paramount importance for physics. For instance, it is used in faithful representations with even dimension for the spin of fermions and in $SO(3)$-representations with odd dimensions for the spin of bosons. The odd dimensionality three of our space characterizes an irreducible representation of the action group $SU(2)$ in the form of rotations $SO(3)$. The $SU(2)$-complexification as real 6-dimensional Lie structure $SL(2, \mathbb{C})$ with two real invariants is the twofold cover of the orthochronous Lorentz group for spacetime.

The 'mother of all simple compact symmetry operations' $SU(2)$, used directly for spin as a spacetime related external operation group and for isospin as internal operation group acting on chargelike degrees of freedom, can be related to the 'Platonic distance' $[-1/2, 1/2]$ with the two end points the (iso)spin eigenvalues

---

[1] From now on, more and more mathematical experience will be required.

for eigenvectors in the defining Pauli representation. Any simple Lie algebra contains a more refined multiple **SU**(2)-structure: It is combined by more and more exemplars of this simplest Lie algebra where the diagonalizable Cartan operators are 'glued' together. For instance, the 8-dimensional rank 2 Lie algebra of the group **SU**(3), used for strong interaction parametrizing color transformations, connects with each other three 3-dimensional rank 1 **SU**(2)-Lie algebras where only two combinations of the three Cartan operators remain independent. This buildup principle can be seen in analogy to the construction of regular 2-dimensional polygons where several equally long 1-dimensional distances are glued together, e.g. a triangle and square with three and four distances as sides respectively, then of regular 3-dimensional polyhedrons connecting regular 2-dimensional polygons, e.g. an octahedron eight triangles or an hexahedron six squares, etc. The rank of a simple Lie algebra, which gives the number of the independent invariants with respect to the related operations, has to be seen in analogy to the dimensionality of a regular polytop.

A last example of some relevance in particle physics and relativity: The roots (nontrivial adjoint eigenvalues) of the Lie algebra of the special group in four complex dimensions **SL**(4, $\mathbb{C}$) with rank 3 and dimension 15 occupy the twelve ($12 = 15 - 3$) corners of the 3-dimensional Archimedean cuboctahedron which is limited by eight regular triangles and six regular squares and has $24 = 8 \times 3 = 6 \times 4$ edges. It can be seen as the intersection of the Platonic octa- and hexahedron which are dual in the following sense: Four triangles make an octahedron corner, three squares make an hexahedron corner, $(4, 3) \leftrightarrow (3, 4)$. At each corner in the cuboctahedron, there meet two triangles and two squares.

The compact form of the **SL**(4, $\mathbb{C}$)-Lie algebra is the Lie algebra of **SU**(4) which is used in unification approaches with the four complex dimensions related to the three color degrees of freedom of quark fields and one leptonic degree of freedom. One noncompact real form of the **SL**(4, $\mathbb{C}$)-Lie algebra is the Lie algebra of **SU**(2, 2) which plays a role in the context of particle-antiparticle structures with spin 1/2. It is isomorphic to the Lie algebra of the conformal group **SO**(2, 4). Another noncompact real form is the Lie algebra of the group **SU**(1, 3) used for a complex embedding of the Lorentz group **SO**(1, 3) to define the state space metric for the four components of the electromagnetic quantum field where, because of the indefinite unitary structure, only two components – the left and right circularly polarized photon – have a particle interpretation. Finally, another noncompact real form of the **SL**(4, $\mathbb{C}$)-Lie algebra characterizes the Lie algebra of **SL**(4, $\mathbb{R}$) or, isomorphic, of **SO**(3, 3) which is used as special part of the full linear group **GL**(4, $\mathbb{R}$) in the tangent spacetimes of a curved real 4-dimensional spacetime manifold.

I mention this rich example also in memory of a discussion with Carl Friedrich von Weizsäcker long ago where our topic were the pseudo-orthogonal features in general relativity, as seen in the local isomorphy **SO**(3, 3) $\sim$ **SL**(4, $\mathbb{R}$) with **SO**(3, 3) the invariance group for the Killing metric of the Lorentz group **SO**(1, 3) and **SL**(4, $\mathbb{R}$) arising in **GL**(4, $\mathbb{R}$)/**SO**(1, 3), the real 10-dimensional tetrad manifold of the spacetime translations.

This example may also illustrate that it is almost impossible not to be tempted by one of these beautiful Lie symmetries and to exploit the chosen favorite for an explanation of fundamental physical structures. This was the case with Platon and Kepler and, also today, the literature is full of such 'explanations' – not all of them can be true, to say the least.

4 The Binary Alternative, Spin and Space

If the operational quantum aspect is fundamental, then it's a good bet to start the nonabelian structures with the simplest simple Lie algebra whose compact form is the (iso)spin Lie algebra.

A strong motivation for Weizsäcker was a possible connection of quantum operations to a logical interpretation – to what he calls the basic alternatives or 'die Ure'. In a mathematical formalization, the basic alternative (bit, up-down, closed-open etc.) comes as a set $A = \{\bullet, \circ\}$ with two different elements which can be created and annihilated or related to each other by operations on the alternative, formalized as follows

$$\emptyset \leftrightarrow \bullet, \quad \emptyset \leftrightarrow \circ, \quad \bullet \leftrightarrow \circ.$$

In contrast to classical deterministic physics with a valuation of the alternatives by true-false with the numbers $\{0, 1\}$ or to thermostatistics with real probabilities $[0, 1] \subset \mathbf{R}$, a quantum valuation uses the complex numbers $\mathbf{C}$ for what is called probability amplitudes.

The field of the complex numbers $\mathbf{C}$ – in contrast to the real ones – has, with the canonical conjugation $\alpha \leftrightarrow \bar{\alpha}$, a nontrivial automorphism which allows – as first proposed by Wigner – the nontrivial implementation of time reversal and a distinction of past and future, with Weizsäcker characterized by facts and open possibilities respectively. To clear up and to understand from the philosophical, physical and mathematical point of view how causal order, mathematically a binary reflexive, transitive and antisymmetric relation, and direction of time, the probability amplitudes of quantum structure and the complex numbers are related to each other was one of the main efforts of Weizsäcker's work in physics.

Having the complex numbers at hand, the alternative allows the concept of linear superpositions – the coexistence of undecided alternatives. This is mathematically formalized by a free vector space $\mathbf{A}^1$ of two complex dimensions, constituted by the mappings of the alternative $A \longrightarrow \mathbf{C}$ giving it a complex valuation, i.e. $\bullet \longmapsto \alpha$ and $\circ \longmapsto \beta$

$$\mathbf{A}^1 = \{\alpha u^1 + \beta u^2 \mid \alpha, \beta \in \mathbf{C}\} \cong \mathbf{C}^2$$

$$\text{with } \bullet \cong u^1 = \begin{pmatrix} 1 \\ 0 \end{pmatrix}, \; \circ \cong u^2 = \begin{pmatrix} 0 \\ 1 \end{pmatrix}, \; \emptyset \cong \begin{pmatrix} 0 \\ 0 \end{pmatrix}.$$

The operations on the alternative in the form of the ur – with the nice notation u^r, $r = 1, 2$

$$\sigma^+ = \begin{pmatrix} 0 & 1 \\ 0 & 0 \end{pmatrix}, \; \sigma^- = \begin{pmatrix} 0 & 0 \\ 1 & 0 \end{pmatrix} \Rightarrow \begin{cases} \sigma^+.u^1 = 0, & \sigma^+.u^2 = u^1 \\ \sigma^-.u^1 = u^2, & \sigma^-.u^2 = 0 \end{cases}$$

span – together with their commutator, the Cartan eigenvalue operator – the Lie algebra of **SU**(2)

$$[\sigma^+, \sigma^-] = \sigma^0 = \begin{pmatrix} 1 & 0 \\ 0 & -1 \end{pmatrix}, \quad [\sigma^0, \sigma^\pm] = \pm 2\sigma^\pm .$$

The canonical number conjugation induces a scalar product of the ur-space and endows it with a Hilbert space structure with probability amplitudes and expectation values

$$\langle u^r | u^s \rangle = \delta^{rs} \Rightarrow \mathbf{A}^1 \text{ is a Hilbert space} .$$

Binary alternatives $\mathbf{A}^1 \cong \mathbb{C}^2$ with the action of **SU**(2) can be lumped together to multiple alternatives $\mathbf{A}^n \cong \mathbb{C}^{1+n}$ or vice versa – all multiple alternatives can be resolved into binary ones. The related Hilbert spaces are acted upon with irreducible **SU**(2)-representations. They arise as the totally symmetric tensor products of the fundamental ur-space

$$\text{Hilbert space } \mathbf{A}^n \cong \bigvee^n \mathbf{A}^1 \cong \mathbb{C}^{1+n} \text{ with } \langle u^a | u^b \rangle = \delta^{ab} .$$

Here one has to discuss the statistics of the ur-alternative – Bose, Fermi or parastatistics? The introduction of the anti-ur u_r^* requires the discussion of the related conjugation operation.

In a spin interpretation of **SU**(2) the multiplicity of an alternative is related to spin by $n = 2j$ with the eigenvectors having spin eigenvalues $\{-j, -j+1, \ldots, j-1, j\}$ – sharp values of decided alternatives.

Therewith, a Birkhoff–von Neumann logical lattice with finite dimensional Hilbert spaces can be interpreted as a collection of alternatives arising from basic binary ones.

The Lie algebra for the binary alternative itself can be interpreted as a Hilbert space for a doubled ur-alternative

$$\mathbb{C}^3 \cong \mathbf{A}^2 = \mathbf{A}^1 \vee \mathbf{A}^1 \ni u^* \sigma u \quad \text{with} \quad \alpha_+ \sigma^+ + \alpha_0 \sigma^0 + \alpha_- \sigma^- = \begin{pmatrix} \alpha_0 & \alpha_+ \\ \alpha_- & -\alpha_0 \end{pmatrix} .$$

It was Weizsäcker's proposal to interpret the group **SU**(2), related to this adjoint structure as our 3-dimensional position space. Therewith the tangent space translations are given by the **SU**(2)-Lie algebra, formalized by the real compact form

$$\begin{pmatrix} \alpha_0 & \alpha_+ \\ \alpha_- & -\alpha_0 \end{pmatrix} \Rightarrow i \begin{pmatrix} x_3 & x_1 - ix_2 \\ x_1 + ix_2 & -x_3 \end{pmatrix} = i\boldsymbol{\sigma} x, \quad x_a \in \mathbb{R}^3 .$$

The determinant as unique invariant gives the Euclidean metric with the definite length of space translations

$$\det i\boldsymbol{\sigma} x = x^2 .$$

The rotations of space vectors around the axis φ (angle and direction) are described by the adjoint $\mathbf{SU}(2)$-action – infinitesimally

$$i\sigma x \longmapsto [\frac{i}{2}\sigma\varphi, i\sigma x] = i\sigma(\varphi \times x).$$

It is a very suggestive, simple and direct approach to introduce position space as the space with three real valuation parameters for the binary alternative. The arising global (nonlinear) model

$$\exp i\sigma x = \cos x + i\frac{\sigma x}{x}\sin x \in \mathbf{SU}(2) \text{ with } x^2 = x^2$$

gives the group $\mathbf{SU}(2)$ as compact position space which has the finite volume parametrization of a full 3-dimensional sphere. According to the Peter-Weyl theorem, the binary alternative and their products are the irreducible components of the regular representation of $\mathbf{SU}(2)$, i.e. of the Hilbert space with the square integrable functions $\{\mathbf{SU}(2) \longrightarrow \mathbb{C}\}$ on global position space.

The discussion of this proposal was the topic of many long and controversial conversations between Weizsäcker and myself, I enjoyed very much. To start with a basic importance of the operational $\mathbf{SU}(2)$-structures, we could agree. But to take the daring and fascinating step, to switch from there to the interpretation of position space as being directly this group with the related cosmological consequences, we disagreed. I will come in the following to this – of course, very deep – disagreement.

5 The External-Internal Dichotomy

Weyl proposed a unification of electromagnetic with gravitational interactions by the freedom of a spacetime varying scale transformation – therefrom the general name 'gauge' transformations, counteracted with a Lorentz vector field which he identified with the electromagnetic potential. Einstein disagreed – the wave length and the spectrum of the light of the stars would be path dependent. Together with Fritz London, the idea of a spacetime dependent real 1-dimensional transformation group for electrodynamics was modified from the noncompact dilatation group $\mathbf{D}(1) = \exp \mathbb{R}$ to the compact phase group $\mathbf{U}(1) = \exp i\mathbb{R}$. The internal symmetries were born – last, but not least under the influence of the complex formulation of quantum mechanics and the relevance of the group $\mathbf{U}(1)$ for the probability amplitudes therein.

The internal operations have proliferated – electromagnetic $\mathbf{U}(1)$-transformations come embededd in hypercharge-isospin $\mathbf{U}(2)$-transformations for electroweak interactions, accompanied in the standard model by color $\mathbf{SU}(3)$-transformations for strong interactions. Up today, the twofoldness with external operations, spacetime translations and rotations on the one side, and unitary internal ones, hyperisospin-color on the other side, remains a mystery. In one of his later unification attempts, Einstein also complexified the metrical spacetime tensor to include electrodynamics – without convincing success.

Coming back to Weizsäcker's ur theory: The basic $\mathbf{SU}(2)$ for the binary alternative – has it to be considered for external spin or for internal isospin rotations, or is it responsible for both? How to describe the branching and disentanglement of the basically logical concepts for the ur-alternative into external and internal physical operations? At least for me, the attempts in this direction seem not very concrete, the internal compact structure remains a stepchild. Why the external-internal dichotomy? What induces the gauge structures of the internal interactions? Where is a way to understand interaction coupling constants? Those structures are so fundamental that I do not expect that, without a qualitative pre-understanding, only detailed and possibly complicated calculations can reveal them. And what, if internal transformations are operationally deeper and spacetime operations can be understood only as a background for them – as a screen for their experimental projection? An understanding of spacetime without a simultaneous understanding of internal transformations seems incomplete – I suspect, even impossible.

All these problems should possibly be considered not in the context of position space alone, but by adressing full spacetime. If $\mathbf{SU}(2)$ is the operational model for global space – what is the operational model for global (nonlinear) spacetime?

With the basic importance of spin $\mathbf{SU}(2)$ as the twofold universal cover of the space rotation group $\mathbf{SO}(3)$ there is connected the covering of the orthochronous spacetime Lorentz group $\mathbf{SO}_0(1,3)$ by the special linear group $\mathbf{SL}(2,\mathbb{C})$. With such an embedding, the Pauli representation of space translations is embedded into the Cartan representation of spacetime translations

$$\sigma x = \begin{pmatrix} x_3 & x_1 - ix_2 \\ x_1 + ix_2 & -x_3 \end{pmatrix} \hookrightarrow \begin{pmatrix} x_0 + x_3 & x_1 - ix_2 \\ x_1 + ix_2 & x_0 - x_3 \end{pmatrix} = \mathbf{x} \ .$$

The time translations and the invariant Lorentz length are related to the coefficients of the characteristic polynomials $\det(\mathbf{x} - \lambda)$ which characterizes the causal spacetime structure by the positivity of its solutions

$$\frac{1}{2} \operatorname{tr} \mathbf{x} = x_0, \quad \det \mathbf{x} = x_0^2 - \mathbf{x}^2$$
$$\mathbf{x} \text{ futurelike} \iff \operatorname{spec} \mathbf{x} \geq 0 \ .$$

In Cartan's representation, the spacetime translations constitute the hermitian vector subspace of the C*-algebra of all 2-dimensional complex linear transformations with the canonical complex number induced order relation. If nonlinear spacetime comes as exponential, it is embedded in the regular group $\mathbf{GL}(2,\mathbb{C})$. By simply exponentiating $i\mathbf{x}$ which embeds Weizsäcker's proposal $i\sigma x$ for space translations, the nonlinear space accompaning global time would be modelled by the cyclic finite compact group $\mathbf{U}(1)$

$$\exp i\mathbf{x} = e^{ix_0} e^{i\sigma x} \in \mathbf{U}(1) \circ \mathbf{SU}(2) = \mathbf{U}(2) \ .$$

A global spacetime $\mathbf{U}(2)$ with cyclic time $\mathbf{U}(1)$ requires a new discussion of an appropriate transitive causality concept.

One could use different conjugation properties for time and space, e.g. the spacetime translations in the form of Hamilton's quaternions – one real and three imaginary coordinates $x_0 + i\sigma x$ in contrast to four imaginary ones $ix_0 + i\sigma x$. With such an ansatz global time would be noncompact infinite in contrast to compact finite global space as seen in a naive exponentiation

$$e^{x_0} e^{i\sigma x} \in \mathbf{D}(1) \times \mathbf{SU}(2) .$$

In this approach, the causal structure for time is as usual. However, the Lorentz transformations of tangent spacetime implemented for spacetime $\mathbf{x}$ or $i\mathbf{x}$ as adjoint action

$$\mathbf{x} \longmapsto s\mathbf{x}s^* = \Lambda.\mathbf{x} \quad \text{with } s \in \mathbf{SL}(2, \mathbb{C}), \ \Lambda \in \mathbf{SO}_0(1, 3)$$

would require a new interpretation.

Or the whole concept for the connection of space with time is too simple: An embedding of Pauli's space representation into Cartan's spacetime representation does not make sense at all – time, on the one side, and space, on the other side, enter the operational structure of spacetime completely differently. How can one dare to formalize the deep structure and meaning of time by such a simple procedure?

There is another interpretation of operational spacetime which is my favorite and which is shortly sketched in the end: The complex 2×2-structure of spacetime allows nonlinear spacetime as the exponent of hermitian tangent translations – similar to the boost parametrization by momenta

$$e^{\xi_0} e^{\sigma \xi} = x_0 + \sigma x \in \mathbf{D}(2) \Rightarrow \begin{cases} x_0 = e^{\xi_0} \cosh |\xi| \\ |x| = e^{\xi_0} \sinh |\xi| \\ \dfrac{x}{|x|} = \dfrac{\xi}{|\xi|} \end{cases}.$$

Nonlinear spacetime $\mathbf{D}(2)$ can be parametrized by the positive cone of tangent Minkowski spacetime – there is no need to change the causality concepts and the Lorentz transformation properties. Also in this case global space is finite, its radius changing with time.

This spacetime model incorporates from the beginning a nonabelian internal transformation group. Being isomorphic to the symmetric space of the compact unitary automorphisms in all complex linear ones

$$\exp \mathbb{R}^4 = \mathbf{D}(2) \cong \mathbf{GL}(2, \mathbb{C})/\mathbf{U}(2) \cong \mathbf{D}(1) \times \mathbf{SO}_0(1, 3)/\mathbf{SO}(3)$$

it displays local hyperisospin $\mathbf{U}(2)$-transformations as arising in the electroweak interactions of the standard model. The symmetric space $\mathbf{GL}(2, \mathbb{C})/\mathbf{U}(2)$ as operation classes can be implemented – in analogy to the real tetrad manifold $\mathbf{GL}(4, \mathbb{R})/\mathbf{SO}(1, 3)$ of tangent spacetime in general relativity – by a spacetime dependent complex dyad which has an interpretation as a 'hinge' field transforming the class parametrizing external spacetime degrees of freedom to the internal compact unitary ones. It is conceivable that the product representations of dyad operationalized spacetime yields – via its invariants – also the properties of elementary particles which, then, can be apostrophized as spectrum of the nonabelian and noncompact nonlinear spacetime operations.

Obviously, this proposal is different from Weizsäcker's model for global space. It reflects another prejudice and temptation by operational Lie structures – spacetime not as unitary group, but as the orientation manifold of a unitary group. Beauty for physical theories may be necessary, it is not sufficient. The proof of all these puddings – if considered as physical theories – can be only in the experimental eating, e.g. in qualitatively and numerically testable consequences for mass spectra, for interactions and their coupling constants, for cross sections etc.

Ur Theory and Space-Time Structure

David Ritz Finkelstein

1 The ur

Ur theory is the attempt by C.F. von Weizsäcker to build quantum theory up from the bottom out of one elementary entity, a binary quantum variable, the ur, with L. Castell, T. Görnitz, and other co-workers.

The ur program makes the world something like a pure quantum computer. The ur plays the role for cosmology that the entity called a qubit plays for modern quantum computation theory.

The ur program bases itself first on quantum kinematics, which allows us to represent all possible operations on a quantum entity by square matrices of a fixed size. And second, on special relativity, A quantum binary variable has the most basic symmetry of special relativity, that of the Lorentz group. The two-component mode vector of the ur supports the Lorentz group in the same way as the two-component wave-function of the left-handed fermions. This makes a binary quantum variable a natural unit from which to build a space-time.

Such observations motivated other physicists in the mid-twentieth century to make similar conceptual explorations of atomistic quantum space-time structure, including R. Feynman, R. Penrose, and myself. A program of a quantum space-time is now also being advanced by A. Connes and co-workers.

Early on Weizsäcker recognized that urs could not have any of the most familiar statistics, Maxwell, Fermi, or Bose. Qubits that have been considered (sometimes implicitly) for the space-time code include

- Dirac spins with Maxwell-Boltzmann statistics (Feynman),
- Weyl spins with various parastatistics (Weizsäcker et al.),
- Pauli spins with Bose statistics (Penrose),
- Weyl spins with Bose statistics (Finkelstein),
- Multiple quantification with Fermi–Dirac statistics (Finkelstein), and
- Multiple quantification with Clifford statistics.

For decades none of these efforts reached the level of making new physical predictions. Only the intrinsic beauty and unity of the concept of quantum space-time kept the study alive. Now the situation looks different.

2 The quantum universe

Any quantum theory of the universe raises a fundamental philosophical problem. Quantum theory was originally supposed to describe the operations that an experimenter can do. Any actual experimenter must ignore much more of the universe than he can measure sharply. How can quantum cosmology be given an operational meaning? Bohr expressed serious doubts about the meaningfdulness of this concept at first.

Weizsäcker's response to this question was that in principle there is no reason to think that the concept of universe is meaningful in quantum theory, since even if it existed we would be a small part of it able to observe precisely only even smaller parts of it.

In practice Weizsäcker treated this question of principle in much the same way as most quantum physicists, including Heisenberg, since Dirac first formulated quantum field theory: He ignored it and went blithely ahead with wave-functions for the universe.

I hesitated over the construct of the quantum universe for years, but recently came to a limited acceptance. I think now that what Weizsäcker and most physicists do is probably right.

To be sure, their practice violates Heisenberg's original principle of operationalism. And it's about time, too. Eventually operationalism too becomes a fetish. an absolute that must be relativized.

Bohr too ultimately relented and endorsed broader extensions of quantum theory that take into account the atomic structure of the experimenter and the relativity of all constructs.

Basing quantum theory on a fixed actual experimenter outside the system was helpful for communicating the quantum theory to a community of classical physicists. But making an absolute split between the experimenter and the system under experimental study is just as bad as making an absolute split between time and space. Indeed, the two splits are one. In elementary mechanics the time coordinate is read from the experimenter's clock while the space coordinate is measured from the system. Ultimately, if one wishes to reconcile the warring viewpoints of all possible experimenters in one framework, one must unify system and experimenter as well as space and time. That quantum unity serves as a quantum cosmos.

Our quantum strategy corresponds satisfactorily to Laplace's classical strategy when $\hbar \to 0$. Where Laplace imagined an omniscient intellect, we imagine a Cosmic Experimenter; not merely a passive observer but a potent doer as well, who prepared the entire universe in the past and will make quantum measurements on it on judgment day. We extract physical predictions from this legend and construct an operational theory by computing what the CE would find for the variables we observe if the CE ignored the variables that we must ignore. The theory of the CE is still a quantum theory. We can assign an absolute state to the universe no more than we can to a hydrogen atom. Our CE is as subject to complementarity and indeterminacy as we are.

3 Quantum logic

Von Weizsäcker was impressed in the early 1940's by how the quantum theory of Heisenberg and Bohr changed the effective logic of physical systems. When Von Neumann set up the predicate algebra of his quantum logic he shook all physics, which had been built on the classical logic of Aristotle, Boole, Frege, and Whitehead and Russell. The classical logic deals with acts of perception or selection whose order of execution is supposed not to matter. This commutativity makes it possible to accumulate complete knowledge of the system, summed up in its state, later perceptions simply adding to the store of information created by earlier ones. In the non-commutative quantum logic complete knowledge is impossible, because later perceptions invalidate earlier ones in an uncontrollable way. The concept of state becomes inappropriate.

These early papers on quantum logic led von Weizsäcker to undertake the reconstruction of physics on the basis of quantum logic, and specifically quantum binary decisions, or qubits. He proposed to build the structures of physics from one quantum binary element, the ur, because he was struck by the fact that a qubit is isomorphic in its algebra and group to a quantum spin 1/2, from which one can in principle reconstruct space-time.

I undertook a similar project later, also stimulated by the work of Von Neumann. As a student I had learned that computers more than any other system can be formulated in purely logical terms, so I set out to model the universe as a quantum logical engine, a computer running on quantum logic.

The Von Neumann logic is wrong for a cosmic computer, however, though better than Boole's.

There is not much reason to doubt that the world is a microscopically reversible process. A computer representing it should be reversible, like the classical computers designed by Bennett and Fredkin [2, 6], as well as quantum and relativistic. The quantum logic it incorporates should be reversible.

But the acts of selection or filters considered by Boole and Von Neumann are irreversible. They are idempotent, in the sense that filtering twice gives the same result as filtering once: $e^2 = e$. One root of this equation is $e = 0$, which has no inverse. Filters cannot be the basis of a reversible logic.

A reversible logic has to be be founded on reversible processes like beam-splitting and phase-shifting rather than filtration. Half-wave plates obey $e^2 = 1$ instead of $e^2 = e$. The logic is then a group instead of a partially ordered set.

The classical logical group operation is the exclusive disjunction or XOR, which is a group operation on predicates. A reversible quantum logic can be based on a quantum XOR. Until recently I worked with the quantum form of the disjoint union or partial disjunction POR, which has no inverse.

A reversible logic requires an additional concept to distinguish between the two roots $+1$ and -1 of the equation $e^2 = 1$, representing true and false. Grassmann's concept of degree does this. A predicate of degree D is the product of D orthogonal unit predicates. Degree distinguishes TRUE ⊤, with maximum degree, from FALSE ⊥, with degree 0.

The quantum version of this reversible logic can be represented within an algebra consisting of polynomials in unit classes or qubits $g_1, \ldots, g_N$ subject to the same relations relations as the classical reversible XOR logic, up to sign. Instead of commuting these qubits anticommute. These relations define a Clifford algebra. The elements of the reversible quantum logic are represented relative to one frame by the 2^n linearly independent products that can be formed from the N generators g_n.

The quantum analogue of the classical transition from a set X to its power set 2^X is the transition from a Clifford algebra **X** to the power Clifford algebra $2^{\mathbf{X}}$. Writing V for the vector space spanned by the units g_n, we therefore write $C = 2^V$ for the Clifford algebra generated by those units.

Since C is also a vector space, we can iterate the exponential, forming 2^C. This process leads to a master Clifford algebra C^∞ corresponding to the classical theory of (hierarchically finite) sets.

This is the Clifford algebra generated from $\mathbb{R}$ by one free linear operator $\iota : C^\infty \to C^\infty$, the *unifier* that binds any generators $g_1, \ldots, g_N$ one chooses into one new generator $\iota(g_1 \ldots g_N)$ also written $\{g_1 \ldots g_N\}$.

Nowadays, we use this Clifford algebra as the basic reversible quantum-logical predicate algebra for the cosmic computer rather than the logic of idempotents suggested by Von Neumann. It has the structure of a higher-order set theory, not just a predicate algebra. We call the quantum ensembles it describes not sets but *squads*, to signal their quantum nature.

Von Weizsäcker and Thomas Görnitz recognized that none of the familiar particle statistics would work for urs, and experimented with parastatistics. Our choice of the Clifford logic also determines a novel statistics, one of the class of double-valued representations of the permutation group catalogued long ago by Wiman and Schur and first applied in physics by Wilczek. Our qubits anti-commute when they are disjoint. Parastatistics usually considers only single-valued representations.

At the start, the chronon and the ur differed in their locality. Initially I assumed that the chronon was submicroscopic, to assure infinitesimal locality in the continuum limit. Weizsäcker considered the ur as a binary decision that might choose between two halves of the universe.

Since then, the criterion of group simplicity has compelled me to assign to the chronon complementary qualities akin to position and momentum. This makes them more like urs. They are approximately local when we determine their position-like variables, but reach out across the universe when we determine their momentum-like variables.

4 Non-associative logic

The Von Neumann logic, like Boole's, is based on an associative algebra.

It seems self-evident a priori that the composition of physical processes must obey the associative law, and I am not aware of experimental violations of associativity so far.

The non-associativity that I consider here is not the one entertained by Pascual Jordan in the 1920's. He turned attention within the usual quantum theory from the non-commutative composition product xy of operators to the commutative non-associative product $xy+yx$ of variables in the same theory. The Jordan non-associative product does not have the meaning "y after x"; but is rather "the square of the sum minus the sum of the squares of x and y". He does not discuss the composition product "y after x".

I still assume that in nature the composition "y after x" is the fundamental product, but I also assume that it is not exactly associative for several reasons, especially for Segal stability. We are concerned just with the product that represents physical process-composition or concatenation: doing one thing after another; not the Jordan product. Other products with other meanings than composition will always exist, of course. Even if the composition product is associative, the Jordan and Lie products that it defines will generally be non-associative.

Many have considered non-associative quantum theories, for reasons outlined below. What I present here is not so much the results of research on this topic by me as the process of deciding whether such research is indicated in our present situation.

4.1 Non-associativity and the Strong Force

The most famous non-associative algebra is the octonions $\mathbb{O}$. What first led physicists to consider octonionic physics seriously was the strong force [8]. The symmetry group of the strong interactions can be more simply expressed in octonionic terms than complex or quaternionic.

Since then it has been noticed that the algebra $T = \mathbb{C} \otimes \mathbb{Q} \otimes \mathbb{O}$ composed of the four division algebras (the algebra $\mathbb{R}$ need not be written, since it is the identity of this product) well-describes internal degrees of freedom of the quarks [4]. Indeed, the connection between forces and division algebras seems to exist for all the forces but gravity:

1. The electromagnetic gauge algebra u_1 is the Lie algebra generated by one imaginary unit in the division algebra $\mathbb{C}$.
2. The electroweak gauge algebra su_2 is the Lie algebra generated by two anticommuting imaginary units in the division algebra $\mathbb{Q}$.
3. The strong gauge algebra su_3 is the Lie algebra generated by (left-multiplication by) three anticommuting imaginary units in the division algebra $\mathbb{O}$, *restricted to the centralizer of one imaginary unit*.

There are no more division algebras, and no more internal charges.

The italicized clause "*restricted to the centralizer of one imaginary unit*" in paragraph 3 above spoils the uniformity of the paragraphs. Without this clause one would construct the 14-dimensional group G_2 instead of its 8-dimensional subgroup SU_3. So one has to tinker a bit to fit the forces to the division algebras. One might simply insert the offending clause in the first two paragraphs as well, since it is tautologous there.

4.2 Non-associativity and Measurement

When we make the evolutionary leap from classical to quantum physics we renounce complete representation of the physical system and learn to represent instead the processes we carry out upon the system. This enables us to cope with predicative processes that do not commute: $ab \neq ba$. To represent the system would misrepresent the process.

It seems inevitable that a further evolution will occur as we deal with first processes further from direct control, such as those in the extreme microcosm, where the atomic nature of our instruments must be taken into account. When we are forced to use second instruments to manipulate first instruments, we learn about first processes through second processes. Then we might have to renounce complete representation of the first process and learn to represent the second process instead. To represent the first process might misrepresent the second process.

Indeed, associativity is but a higher kind of commutativity, namely the commutativity of all left multiplications with all right multiplications.

To understand non-associativity better we should look at its practical or operational meaning.

The experimental sign of non-associative composition is the following.

Consider three alleged processes, carried out by "black" boxes A, B, C, each with one inlet and one outlet. We write BA for the process that is carried out by a composite box consisting of B and A connected in series: the outlet of A is physically connected to the inlet of B so that systems emerging from A are channeled to B. When we write $X = BA$ we mean that the black box X can replace BA in every experiment without the change being statistically detectable by a judge whose only access to the boxes is through the initial inlet and the final outlet [5]. The judge is disqualified if he peeks at the system between A and B.

Likewise we define a black box $Y = CB$.

Non-associativity means that even though $X = BA$ and $Y = CB$, the judge finds $CX \neq YA$.

In such cases the resultant process of three parts A, B, C depends anomalously on the order of the connections, $(CB)A$ or $C(BA)$, keeping the order CBA of the elements fixed; as in quantum theory the resultant process of two parts A, B depends anomalously on the order in which its parts are performed, AB or BA. In this ultraquantum theory, the second process changes the first process, just as in quantum theory the first process changes the system.

For macroscopic processes associativity is the familiar case. The most familiar non-associative product is chemical combination. For example we may think of H_2SO_4 as a product $H_2(SO_4)$ that is very different from $(H_2S)(O_4)$, a metastable mixture of two gases. Here the order of association matters because chemical binding can saturate. In general, non-associative composition describes and results from a kind of binding.

Binding can happen when the surfaces of the processes are not too small compared to their volumes. Small black boxes are, like molecules, "all surface", and interconnecting them can transform them much as chemical combination does.

It would therefore not be surprising if the second quantum evolution has a fundamental scale-time like an ur, that sets the scale of non-associativity in the way that the fundamental action $\hbar$ sets the scale of non-commutativity.

For small operations, close to the identity, non-commutativity is a second-order effect while non-associativity is a third-order effect; it is reasonable to expect physical constants of different units to measure their strengths.

In the above consideration we did not write the multiplication signs that connect processes in series. This builds in the presupposition that connecting does nothing essential to what we connect. Let us be more explicit. We write $(La)b$ for the product ab as a function of b, and $(Rb)a$ for the same product as a function of a. Here La and Rb represent second processes including the second process of connection. Associativity then takes the form

$$La\ Rb = Rb\ La\ . \tag{1}$$

This indeed asserts that two second processes commute.

Non-associativity of processes thus replays the most familiar quantum phenomenon on a higher level: second processes that were expected to commute do not.

Such evolutions seem inevitable because we cannot do literally the very same process twice. Each instance of a process is intrinsically unique and non-repeatable. Every equation connecting processes, such as the commutative or associative laws, is a comparison that we make between two entities that are actually distinct, remote from each other, and therefore unequal. Underlying each such equation is a higher-level process that transforms one side of the equation to the other. The equation asserts that the totality of these higher-level processes is integrable in the sense that they can all be represented by equality together without inconsistency.

Since all representations of processes by complex matrices are associative, when we lose associativity we indeed give up representation of the first process in the usual sense.

4.3 Non-associative Algebra

The mathematical theory of non-associative algebras has been extensively studied [3]. A general "product" $\times : L \otimes L \to L$, possibly non-associative, with an identity element $I : I \times x = x = x \times I$ and an inverse x^{-1} is called a loop (on L). If V is a vector space, a bilinear product $\times : V \otimes V \to V$ is called an algebra. An algebra with identity element $I \in V$ is called unital. A unital algebra whose elements $x \neq 0$ have inverse x^{-1} is called a division algebra. A division algebra is a bilinear loop.

A norm on a division algebra is a positive-definite *real-valued* multiplicative quadratic form $\|x\|$; it has

$$\begin{aligned} \|x\| &\geq 0, \\ \|x\| = 0 &:\equiv: x = 0, \\ \|y\|\|x\| &= \|yx\|, \end{aligned} \tag{2}$$

for all $x, y \in A_n$. A positive-definite norm is not a necessity of quantum theory today.

The only normed division algebras (up to isomorphism) are the $A_N = \mathbb{R}, \mathbb{C}, \mathbb{Q}, \mathbb{O}$ ($N = 0, 1, 2, 3$) generated by N pairwise-anticommuting square roots of -1 whose $2^N - 1$ independent products other than 1 are also pairwise-anticommuting square roots of -1. The only non-associative algebra of these is the octonion algebra $A_3 = \mathbb{O}$.

One builds up Clifford algebras Cliff(N) too by adjoining n independent anti-commuting imaginary units, but preserving associativity. Then some products of generators commute, and some are square roots of $+1$.

One builds up the Cayley–Dickson algebra D_N by adjoining N independent anti-commuting imaginary units, requiring their products also to be anti-commuting imaginary units. The Cayley–Dickson algebras are just Clifford algebras with some signs in their multiplication tables changed, to make the elements all mutually anti-commute.

The two lines of algebras both agree on the reals, the complex numbers, and the quaternions, but then they part company: Clifford goes next to an 8-dimensional matrix algebra while Cayley–Dickson produces the 8-dimensional octonions.

4.4 Non-associativity and Stability

Stability seems to require non-associativity. It is already known that commutativity, as a linear constraint on the structure tensor of a theory, is an unstable condition [9]. Associativity is a bilinear constraint on the structure tensor: the associator $a(bc) - (ab)c$ is required to vanish identically. Presumably this too is an unstable condition.

The deformation of algebras from associative to non-associative is among the processes intensively studied in the operad theory of Peter May, the theory of associahedrons or Stasheff polytopes, and the theory of deformation quantisation of Flato and Sternheimer among others; I am grateful to Z. Oziewicz for calling my attention to this work.

4.5 Non-associativity and the Chronon

Infinitesimal geodesics compose non-commutatively, due to curvature, but associatively, to second order. Finite geodesics, however, compose non-associatively as well as non-commutatively; I thank Z. Oziewicz for pointing this out. In this product, the geodesics start at the origin, so one geodesic has to be transported along another before they can be composed. This is an instance of one process acting on another during their composition. This composition of finite geodesics is a prototype quasigroup and the origin of much of their interest [3].

If there is truly a quantum of time, the closest approximant in nature to an infinitesimal geodesic may be a finite one of length χ, the chronon. Since finite geodesics are non-associative, we might expect finite non-associativity for chronons, with an associator $\sim O(\chi^3)$. This suggests that χ measures the intrinsic non-associativity of elementary processes in the way that $\hbar$ measures the intrinsic non-commutativity.

4.6 Non-associativity and Reflexivity

Reflexivity is the newest leg of the platform of fact from which we launch our speculations on non-associativity. Irrreflexive processes, processes that do not act on themselves, combine associatively. It is easy to see from elementary computer models that processes that act on or refer to themselves do not necessarily compose associatively. If the dynamical laws are elements of the very cosmos that they propel, the dynamical processes they represent might be non-associative.

4.7 Non-associativity and Set Theory

When we work with a non-associative algebra we must remember not to shift association symbols such as parentheses and brackets as freely as we do in associative algebras. Immovable association symbols were introduced by Cantor to describe set formation. The sets $\{\{a, b\}, c\}$, $\{a, \{b, c\}\}$, and $\{a, b, c\}$ are all different. Set theory may thus be regarded as a non-associative algebra. It has an associative product and an associator $\iota = \{\ldots\}$ that defines a non-associative product $\{a, b, c, \ldots\}$ of any number of factors. Its quantum correspondent is the cosmic Clifford algebra C^∞ already introduced.

This algebra is still more non-associative than the Cayley–Dickson algebras in the following sense. If g_1, g_2, g_3 are three orthogonal generators (chronons or urs) then in the Cayley–Dickson algebras $((g_1 g_2) g_3)$ is still linearly dependent on $(g_1 (g_2 g_3))$, as in an associative algebra. Only the sign is strange. In set theory and C^∞, however, $\{\{g_1 g_2\} g_3\}$ and $\{g_1 \{g_2 g_3\}\}$ are algebraically independent. It is routine to model any of the Cayley–Dickson algebras in a quotient algebra of C^∞ with respect to the ideal generated by the appropriate linear dependency relations.

4.8 Non-associativity and Symmetry

The line of Cayley–Dickson algebras and the line of Clifford algebras differ outstandingly in their symmetries.

The groups of the Clifford algebras form a line of orthogonal groups of rapidly growing dimension. They contain the unitary groups as centralizer subgroups of an imaginary element.

The Cayley–Dickson algebras agree with the Clifford algebras up to the quaternions, but the group G_2 (dimension 14) of the octonions is already smaller than the corresponding Clifford group SO_8 (dimension 28). Higher Cayley–Dickson algebras have still fewer continuous symmetries compared to the orthogonal and unitary series now in use.

5 Summation

Facing the personal problem of whether to develop associative quantum logics as a physical language, or or more general non-associative ones, I find the correspondence between forces and division algebras too startling too ignore. I have not done it full justice here.

I also take the stability argument seriously. However that leads just as naturally to C^∞, as to the Cayley–Dickson algebras. The Clifford algebra is on the one hand associative in the usual sense, and on the other – I realize now – as non-associative as one could like, because of the bracer ι.

There are other reasons that lead me to prefer C^∞ over the Cayley–Dickson series of logic algebras.

First, gravity does not fit this pattern. Its gauge group has been taken to be the space-time translation group $\mathbb{R}^4$.

Another reason to follow the Clifford line rather than the Cayle-Dickson is that quantum theory works for very many dimensions and seems associative in its higher-dimensional forms. I do not consider seriously the possibility of a sub-lunary sphere that separates the logic of subparticle processes from that for particle processes.

If quantum theory followed the Cayley–Dickson line, it seems likely that the hydrogen atom would be very different than the one we know, which fits beautifully into the line of associative matrix algebras. We would be foolhardy to discard all the abundantly verified symmetries along the matrix line to account more easily for the one SU_3 symmetry.

One well-known problem with non-associative algebras like the octonions and the higher Cayley–Dickson algebras is that the concepts of linear dependence and dimension misbehave. The second law of thermodynamics depends on the uniqueness and additivity of dimension, and these break down, so presumably the second law of thermodynamics fails in the statistical thermodynamics of non-associative systems. The Clifford series does not have this problem, due to its associative product.

One needs to see whether the law of large numbers can wash out non-associativity in the way that it can wash out non-commutativity in a suitable limit. There should be a correspondence principle from the non-associative logic to the associative, just as there is from the associative, non-commutative to the commutative.

At present it seems to me more productive, if less radical, to use a Clifford-algebraic logic, with both associative and non-associative products, for present physics. Perhaps the most plausible way to SU_3 is still that of Grand Unified Theory, which presents SU_3 as a broken SO_{10}.

6 Acknowledgements

I thank Zbigniew Oziewicz for instructive discussions of non-associative algebra.

References

1. J. Baez: The octonions. *Bull. Amer. Math. Soc.* 39, 145–205 (2002)
2. C.H. Bennett: Logical reversibility of computation. *IBM Journal of Research and Development* 6, 525–532 (1973)
3. R.H. Bruck: *A Survey of Binary Systems*. Springer, New York 1946
4. G.M. Dixon: *Division Algebras: Octonions, Quaternions, Complex Numbers and the Algebraic Design of Physics*. Kluwer, Dordrecht 1994

5. D. Finkelstein: *Quantum Relativity*. Springer, Heidelberg 1996
6. E. Fredkin and T. Toffoli: Conservative Logic. *International Journal of Theoretical Physics* 21, 219–253 (1982)
7. Th. Görnitz, and C.F. v Weizsäcker: Quantum interpretations. *International Journal of Theoretical Physics* 26, 921 (1987)
8. A. Pais: Phys. Rev. Lett. 7, 291–293 (1961)
9. I.E. Segal: A class of operator algebras which are determined by groups. *Duke Mathematics Journal* 18, 221 (1951)
10. T. Smith: quant-ph/9503009, LANL
11. C.F. von Weizsäcker: Komplementarität und Logik. *Naturwissenschaften* 20, 545–555 (1955)
12. F. Wilczek and A. Zee: Families from spinors. *Physical Review* D25, 553 (1982)
13. F. Wilczek: *Projective statistics and spinors in Hilbert space.* hep-th/9806228 [LANL]

Weizsäcker's Ur Theory – A Cosmological Point of View

Thomas Görnitz

1 Introduction

Among the physical innovations of the 20th century which have not yet found general recognition one of the most interesting is C.F. von Weizsäcker's ur theory. The basic ideas of this theory were epistemological and quantum logical considerations [25,26] dealing with the foundations of quantum theory. In the early papers on ur theory, Weizsäcker investigated, later also with Scheibe and Süssmann [24], in which way it could be possible to reduce quantum theory to quantum logic.

With the concept of multiple quantization a fundamental structure of quantum theory was used. It allows deriving the quantum mechanical particles from a pure logical substructure. A particle can be composed from abstract quantized alternatives, the ur-alternatives or urs, in an analogue way a quantum field can be composed of quantized particles. The so-called Fock-Space-construction enables to build up a quantum field as an infinite sum over all possible numbers of particles and in the same way a particle can be composed from combinations of urs. Originally, Weizsäcker operated with the method of the so-called "naïve quantization". Later, Castell [7] used the orthodox method of second quantization with creation- and annihilation operators. These operators are subject of Bose and, as also proposed later, of Para-Bose statistics.

For the quantized alternative the role of its fundamental symmetry group plays $SU(2)$, and beside this $U(1)$, the phase group, and the complex conjugation K, which refers to the essential complex character of any quantum theory. The group $SU(2)$ is locally isomorphic to $O(3)$, the group of rotations in an Euclidean space of three dimensions. Weizsäcker has stressed that from this isomorphism one may therefore deduce the three-dimensionality of position space from quantum logic. As Drieschner worked out later [9], it is a consequence of ur theory that for every physical object a representation exists in three-dimensional position space if it is constructed from ur-alternatives.

However, real physical particles live in four-dimensional space-time and will be represented by a unitary representation of a relativistic group. With modified creation- and annihilation-operators for the urs, Görnitz and Weizsäcker could show that this method permits to construct particles in a de Sitter world [11]. This is a space-time with closed position space and open time.

The creation- and annihilation-operators of Bose type lead to an unphysical representation with a closed cosmological time. These operators which set all urs as beeing identical are unable to create particles in Minkowski space and neither particles with a restmass. In the context of conventional physics one looks for particle representations in Minkowski space. Therefore it is necessary to take back step by step the indistinguishability of the ur-alternatives, which was postulated at the starting point of the theory.

The first step is to introduce a distinguishability of urs in respect to the complex conjugation K and the symmetry group $U(1)$. The later one is correlated with the time development of the free ur. By postulating a distinction between urs and anti-urs which posses a positive respectively negative time behaviour it becomes possible to represent the complex conjugation K linearly.

The "fourfold-urs", defined in this way, enable the representation of all massless particles in Minkowski space by Bose-quantization (Castell [6]). In a next step, urs are subject to para-statistics, represented by Green indices. This allows a distinction between the different objects that the urs constitute as Heidenreich [20] has postulated. By this differentiation the urs can obey Para-Bose-Statistics [20–22]. By means of urs and anti-urs this in principle enables the construction of not only the massless but also of all the massive elementary particles with any spin value. Görnitz, Graudenz and Weizsäcker [15] made concrete calculations. Massless particles with every spin state were explicitly constructed, later also states of massive particles with spins zero and 1/2 were assembled in co-operation with Schomäcker.

In this approach to particle physics, two serious problems remain unsolved:

The first one consists of the question in which way in the continuous spectrum of possible masses one can give reasons for the existence of only few sharp rest masses of the real particles, like electrons or protons.

The second one is the definition of charge in the frame of ur theory.

Charge is invariant under all operations of the Poincaré group. Therefore a hypothetical charge operator has to commute with all generators of this group. Together with Schomäcker in Braunschweig all operators up to the 6th order in creation and annihilation operators that commute with the Poincaré group have been constructed. Because of insufficient computing power these first calculations have not given conclusive results [16].

A solution of the first problem, the problem of the sharp rest masses, can only be expected in the framework of a realistic cosmology. In my opinion, a theory in Minkowski space is unable to supply mathematical relations among basic physical quantities. Weizsäcker made a first contribution [27] through his estimate of the number of urs that constitute a nucleon. At the time of its publication, its value, 10^{40} bits, was so far from any imaginable physical value that its acceptance was marginal, and, I am afraid to say, it remains so untill now.

A second possible reason for the minor public interest in the ur-concept could be the incompatibility of the cosmological model, presented in the *Aufbau der Physik* [28], with Einstein's theory of general relativity. The model not only requires a cosmological constant, but also leads to a gravitational theory with a varying gravitational constant. Since the work of Dirac, Jordan and Ludwig such theories came out of fashion in physics. Further on, in the 80s, the world models with cosmological terms became unpopular. The effort to explain its value by quantum field theoretical methods leads to results which differ from any realistic value, which is comparable to the empirical data, by a factor of order of 10^{120}. Therefore in the *Aufbau der Physik,* the chapter on general relativity closes with the sentence: "Wir lassen diese Fragen und damit die urtheoretische Beschreibung der Kosmologie offen" (We leave these questions open and, in connection with this, the ur theoretical description of cosmology).

2 The Way to a Connection Between Ur Theory and Cosmology

In the 70s, there were efforts by Becker, Castell [2] and by Roman [23] to connect ur theory with cosmology. In these papers, urs were treated in a more conventional way, as constituents of mass zero particles. I have aimed to connect the ur theory - as the consequent quantum theory - with the gravitational theory. Therefore, I started with the black holes. This is the point in the framework of gravitational theory where quantum effects can not be ignored.

During the 70s and 80s many physicist looked at black holes as imaginative ideas of some theoreticians, the foundation was laid for its integration into the general structure of physics by Bekenstein and Hawking. This was achieved by a definition of entropy for black holes. As a consequence, the theory of black holes became linked to thermodynamics.

As I could show, the entropy of the black holes is the central link between the ur-theoretical concepts and the conventional part of physics, e.g. general relativity and quantum mechanics.

Entropy is that part of information on a system that cannot be used for its description. But its value can be computed by means of models of its internal structure. By its horizon a black hole does not allow any information on its interior to get outside. Solely three numbers, the mass, spin and charge, are available to characterise objects containing millions of solar masses. If by physical law valid outside the horizon no other information can be grasped, the entropy of this object must exceed any scale that was imaginable until then in physics.

For a long time the notion of entropy of black holes could not be integrated into normal physics. The usual field theoretical methods give values that are distant from the right ones by many orders of magnitude. But estimates of realistic values could easily be linked to ur-theoretical considerations. This opened the possibility of obtaining results on ur theory without the explicit use of its postulates.

The present paper will start with ur theory and will then show which cosmological conclusions can be drawn from it.

3 The Introduction of Space

The states of an ur, of a qubit in modern language, constitute a two-dimensional complex space $\mathbb{C}^2$. It is a representation space for a spin-1/2-representation $^2D_{1/2}$ of $SU(2)$, which is an irreducible component of the regular representation of this group in the Hilbert space $\mathbb{L}^2(SU(2)) = \mathbb{L}^2(\mathbb{S}^3)$. Therefore all states of an ur can be represented by functions on $\mathbb{S}^3$, possessing only one node plane. In ur theory the sphere $\mathbb{S}^3$ is interpreted as the position space and an ur thus becomes an "object" extended over the whole cosmological space. If there are more urs present, they constitute higher dimensional representations of its symmetry group $SU(2)$ by tensor products. Each irreducible component ^nD in it can again be represented by functions over $\mathbb{S}^3$ possessing different frequencies dependent on n. The wave functions of an irreducible representation ^nD have wavelengths of the order R/n. So with it is possible to get higher spatial resolutions because sharper wave packets can be formed. Let R denote the radius of curvature of the cosmic space and let N be the number of urs in it at a given time. N urs constitute a representation $(^2D_{1/2})^{\otimes N}$ of $SU(2)$. Its decomposition into irreducible components is given by the Clebsch–Gordan-series (we define $|N/2| = k$ for $N = 2k$ or $N = 2k+1$):

$$(^2D_{1/2})^{\otimes N} = \bigoplus_{j=0}^{|N/2|} \frac{N!(N+I-2j)}{(N+I-j)!j!} {}^{2|N/2|-2j+1}D_{|N/2|-j} \ .$$

The distribution of wavelengths that can be found in a cosmos containing N urs is obtained from the factor of multiplicity

$$f(j) = \frac{N!(N+1-2j)}{(N+1-j)!j!} \ .$$

The representations with the largest wavelength, $^1D_0$ or $^2D_{1/2}$ respectively, correspond to $j = |N/2|$ and have multiplicity $f(N/2) = O(2^{N/2}N^{-3/2})$.

The maximum of the multiplicity $f(j)$ is at

$$j_{\max} = (1/2)\left[N - (N+2)^{1/2}\right]$$

or for $N \gg 1$, at

$$j_{\max} \approx (1/2)\left[N - N^{1/2}\right]$$

and belongs to the representations $^{2\sqrt{N}}D_{\sqrt{N}}$.

From D_0 to $D_{\sqrt{N}}$ the multiplicity $f(j)$ rises from

$$f(N/2) = 1 \cdot O\left(2^{N/2}N^{-3/2}\right)$$

to

$$f\left[(1/2)\left(N - \sqrt{N}\right)\right]k = O\left(2^{N/2}N^{-1}\right) = \sqrt{N} \cdot O\left(2^{N/2}N^{-3/2}\right) \ .$$

After the maximum, we have between $j = (1/2)(N - \sqrt{N})$ and $j = 0$ an exponential decrease of $f(j)$ from the order of $2^{N/2}$ to 1. It can be approximated by

$$f(j_{\max} - (1/2)a) = 2N \exp\left[-\left(1 + a/\sqrt{N}\right)^2\right] .$$

So the multiplicities in the Clebsch–Gordan decomposition of the tensor product are large for representations $^k D$ with $k = 2\sqrt{N}$. After this there is an exponential decrease for the multiplicities and the corresponding representations could be neglected. This is so because representations with localizations sharper than $R/2\sqrt{N}$ will almost never occur.

Therefore in a cosmos of curvature radius R and containing N urs the shortest possible physically realisable length, which later will be identified with the Planck length ($\lambda_0 = 1$), is of the order

$$\frac{R}{\sqrt{N}} \sim \lambda_0 = 1 .$$

For the cosmic radius we get

$$R \sim \lambda_0 \sqrt{N} \sim \sqrt{N} .$$

A single ur has a wavelength R, thus an energy of order $1/R$ corresponds to it:

$$U_{\mathrm{ur}} \sim 1/R .$$

If there are N urs in the universe, the order of the total energy of all the urs is

$$U_{\mathrm{universe}} \sim N U_{\mathrm{ur}} \sim N/R \sim N(1/\sqrt{N}) \sim \sqrt{N} \sim R$$

and the energy density behaves in its order like

$$\mu \sim U/R^3 \sim 1/R^2 .$$

In a first approximation we assume - as almost always in cosmology - that the urs behave like an ideal fluid. The finite cosmos is an *isolated* system in the sense of thermodynamics (perhaps the only one) and the first law of thermodynamics will hold for it. (This assumption is less restrictive than the $dU = 0$ normally used.)

$$dU + p\, dV = 0 \tag{1}$$

or

$$\left(1 + p \cdot 3R^2\right) dR = 0 . \tag{2}$$

In a realistic cosmological model R cannot be a constant, so we have to take the solution

$$dR \neq 0$$

and for the pressure as a function of R it follows from (1,2) that
$$p = -1/(3R^2) = -\mu/3 \tag{3}$$
A negative pressure is always a hint for instability – which an evolving cosmos will indeed show.

Usually, for cosmological models one argues that "it is reasonable to assume p is non-negative" [17]. But the restrictions on energy and pressure, which in general relativity are reasonable indeed, do not restrict p to positive values [17]: All the relativistic conditions on energy,

the *weak energy condition*, $\mu = 0$ and $\mu + p = 0$;
the *strong energy condition*, $\mu + 3p = p$ and $\mu + p = 0$;
and the *dominant energy condition*, $\mu = 0$ and $\mu = p = -\mu$

which implies that matter cannot travel faster than light, are fulfilled by the ur-theoretic "ideal fluid".

There is no reason that urs must behave like ordinary matter and radiation.

For an isotropic fluid with an energy-momentum tensor given by equation (3), i.e.,
$$T_i^k = \mathrm{diag}(\mu, -p, -p, -p) = \mathrm{diag}(\mu, \mu/3, \mu/3, \mu/3).$$

Einstein's equation (κ, Einstein's gravitational constant, is equal to 1 in Planck units) is
$$G_i^k = -\kappa T_i^k$$

For a perfect fluid this reduces to
$$\kappa\mu = 3\left[1 + (dR/dt)^2\right]/R^2$$
$$\kappa p = \left[1 + (dR/dt)^2 + 2R\left(dR^2/dt^2\right)\right]/R^2$$

which in the present case is equivalent to
$$dR^2/dt^2 = 0$$

and gives as solution a Robertson–Walker cosmos with the structure $\boldsymbol{R}^+ \oplus \boldsymbol{S}^3$ and with the metric ($k = +1$ and $\Lambda = 0$)
$$ds^2 = dt^2 - R^2(t)\left[(1-r^2)^{-1} dr^2 + r^2 d\Omega^2\right].$$

With
$$R(t) = R(0) + vt$$

$\mu(t)$ and $p(t)$ behave like
$$\mu(t) = 3\kappa^{-1}\left(1 + v^2\right)/[R(0) + vt]^2 = 3\left(1 + v^2\right)/R^2$$
$$p(t) = -\kappa^{-1}\left(1 + v^2\right)/[R(0) + vt]^2 = -\left(1 + v^2\right)/R^2$$

The constant expansion velocity v appears as a universal and fundamental constant of this comic model. This gives a good reason to let it become equal to the one fundamental velocity in physics, the velocity of light, i.e. to 1 in Planck units.

4 An Effective Energy-Momentum Tensor

We can decompose $_{(ur)}T_i^k$ into a sum of energy-momentum tensors for matter, light, and vacuum:

$$_{(ur)}T_i^k = {}_{(matter)}T_i^k + {}_{(light)}T_i^k + {}_{(vacuum)}T_i^k$$

or (m = matter, l = light)

$$\begin{pmatrix} \mu & & & \\ & \mu/3 & & \\ & & \mu/3 & \\ & & & \mu/3 \end{pmatrix} = \begin{pmatrix} \mu_m & & & \\ & 0 & & \\ & & 0 & \\ & & & 0 \end{pmatrix} + \begin{pmatrix} \mu_l & & & \\ & -\mu_l/3 & & \\ & & -\mu_l/3 & \\ & & & -\mu_l/3 \end{pmatrix} + \begin{pmatrix} \lambda & & & \\ & \lambda & & \\ & & \lambda & \\ & & & \lambda \end{pmatrix} \quad (4)$$

The vacuum term looks like the cosmological term that Einstein had introduced to allow an unchanging cosmological model. In the above model there are changes in spacelike extension. By the rules of quantum theory the vacuum has to react on these as opposed to a cosmological constant.

Calling ω the energy density ratio between matter and light to vacuum

$$(\mu_m + \mu_l)/\lambda = \omega$$

then (4) results in

$$\mu = (\omega + 1)\lambda$$

and

$$(\omega + 1)\lambda/3 = -\mu_l/3 + \lambda$$

so

$$\mu_l = (2 - \omega)\lambda$$

because of $\mu_l \geq 0$ it follows $2 \geq \omega \geq 0$ and

$$\mu_m = (2\omega - 2)\lambda$$

from $\mu \geq 0$ we get $2 \geq \omega \geq 1$.
 In terms of μ we have

$$\lambda = \mu/(\omega + 1)$$
$$\mu_l = \mu(2 - \omega)/(\omega + 1)$$
$$\mu_m = 2\mu(\omega - 1)/(\omega + 1)$$

As special cases we have

$$\mu_m = \mu_l = \mu \cdot 2/7, \quad \lambda = \mu \cdot 3/7$$
$$\mu_m = 0 \Rightarrow \mu_l = \mu/2, \quad \lambda = \mu/2$$
$$\mu_l = 0 \Rightarrow \mu_m = \mu \cdot 2/3, \quad \lambda = \mu/3$$

If we introduce an "effective" cosmological constant which is constant only around a cosmic time t_0

$$_{\text{eff}}\Lambda = \kappa\lambda(t_0)$$

we get an "effective" Einstein equation

$$_{\text{eff}}G_i^k + {}_{\text{eff}}\Lambda\delta_i^k = -\kappa\left[{}_{\text{matter}}T_i^k + {}_{\text{light}}T_i^k\right]$$

which is valid for cosmological times around t_0.

The cosmological constant Λ is usually understood as the vacuum density of the universe. Of course, this interpretation leads to strange properties of the "vacuum" such as a pressure that is as large as the negative energy density for each component.

In our units the observational value is of the order of $\Lambda \sim 1/R^2$. The smallness of this value may be one of the most serious problems in modern cosmology [19]. If Λ is actually a constant its fine-tuning will be hard to explain. On the other hand, in the model given above, the right order of magnitude for Λ follows directly without use of doubtful inventions such as the anthropic principle.

5 An Estimation for the Entropy of Particles

Now we will compute the entropy related to simple models of particles that in some sense are localisable in their Compton volume.

Let n_i urs with $n_i < \sqrt{N}$ form a particle. Then it will have energy of the order of

$$m_i = n_i/R$$

and from its Compton wavelength λ_i its elementary volume is of the order of

$$v_i = \lambda_i^3 = (R/n_i)^3.$$

Then the number z_i of the ith type of particles can be as large as $z_i = N/n_i$, and the number of places P_i for it is equal to

$$P_i = V/v_i = R^3/(R/n_i)^3 = n_i^3.$$

If the particles are of bosonic type then the thermodynamic probability of putting z_i of them in P_i places is

$$W_{Bi} = (P_i + z_i)!/(P_i!z_i!)$$

For a freely chosen n_i we get

$$S_{Bi} = \ln W_{Bi}$$
$$= \left(n_i^3 + N/n_i\right) \ln \left(n_i^3 + N/n_i\right) - n_i^3 \ln n_i^3 - N/n_i \ln(N/n_i)$$
$$= n_i^3 \left[\ln \left(1 + N/n_i^4\right) - N/n_i^4 \ln \left(1 + n_i^4/N\right)\right] .$$

With $\alpha_i = n_i^4/N$, and $N^{-1} \ll \alpha_i \ll N$, it follows

$$S_{Bi} = N^{3/4}\alpha_i^{3/4} \left[\ln \left(1 + \alpha_i^{-1}\right) - \alpha_i^{-1} \ln (1 + \alpha_i)\right] = N^{3/4} f_B(\alpha_i)$$

The entropy S_{Bi} has its maximal value at $\alpha_i = 17,5$ or $n_i \approx 2N^{1/4}$. With $f_B(17,5) = 1,902$ we get

$$S_{B\max} \approx 2N^{3/4} .$$

If we decide to look for particles of a fermionic type then at most one particle can be at one place. The thermodynamic probability to put z_i fermionic particles on P_i places is

$$W_{Fi} = (P_i)!/\left[(P_i - z_i)!(z_i)!\right] .$$

For $n_i = N^{1/4}$ it is $W_{Fi} = 1$, and for $n_i = N^{1/2}$ we have

$$\ln W_{Fi} = N^{3/2} \ln \left(N^{3/2}\right) - \left(N^{3/2} - N^{1/2}\right) \ln \left(N^{3/2} - N^{1/2}\right) - N^{1/2} \ln \left(N^{1/2}\right)$$
$$\approx N^{1/2}(1 + \ln N) .$$

With $n_i^4 = \alpha_i N$, and $1 \leq \alpha_i \leq N$, one has

$$S_{Fi} = N^{3/4}\alpha_i^{3/4} \left[-\ln \left(1 - \alpha_i^{-1}\right) + \alpha_i^{-1} \ln (\alpha_i - 1)\right] = N^{3/4} f_F(\alpha_i)$$

The entropy S_{Fi} has its maximal value at $a_i = 22,5$ resp. at $n_i \approx 2N^{1/4}$; with $f_F(22,5) = 1,8782$ we get, not unlike the Bose case,

$$S_{F\max} \approx 2N^{3/4} .$$

6 Ur Number and Bekenstein–Hawking–Entropy for Particles

Let us now have a more intense look at the black holes. By assigning entropy to the black hole one must also assign temperature to it; i.e. the black hole has to radiate. The resolution of this difficulty was given by Hawking [18], showing that quantum effects cause black holes to create and emit particles, i.e., application of quantum theory led to reasonable results for the exterior region of the black hole.

But *quantum-theoretic considerations should also be valid for the interior solution.*

Any quantum system has a ground state and its energy depends on the system's extension. The horizon of a black hole contained a finite volume; it is an ideal box. The ground state of such a box has to depend on the radius of the event horizon, but usually this necessity is not recognised in general relativity.

If the ur-theoretical considerations are taken into account, it can be shown that the interior of a Schwarzschild black hole can be described by a completely stationary Robertson–Walker space-time, i.e., replacing the Schwarzschild singularity by a Friedmann singularity [14].

The largest black hole imaginable would be one with a mass equal to the mass of the universe. In this case the Schwarzschild radius of the black hole would become of the same order of magnitude as the Friedmann-radius of the contemporary universe.

In case of one single nucleon falling into this largest possible black hole, the entropy of the black hole would raise by the amount of $10^{40.6}$ bit. This is almost exactly the value Weizsäcker gave in 1972.

7 Conclusions

If we take as empirical input the number of urs for a nucleon with $10^{40.6}$ then the number of urs in the universe is 10^{122}.

The present age of the universe appears as 10^{61} Planck times, the radius of our cosmos is of the order of 10^{61} Planck lengths and the present energy density is of order 10^{-122} Planck masses and is is of the same order as the present cosmological term.

If we identify our models of particles with the photons and neutrinos respectively that create the background radiation, their most likely energy will be about $10^{-10.2} m_{\text{nucleon}} = 10^{-2.2}$ eV, which is not very far from the empirically 3 Kelvin of the background radiation.

The maximal entropy S_{max} in the Bose as well in the Fermi case is equal to 10^{90}. That is the value of the particle-related entropy in the present universe. For a long time in the history of physics this value was seen as an unexplainable one.

References

1. A.O. Barut, and H.-D. Doebner (eds.): *Conformal Groups and Related Symmetries, Physical Results and Mathematical Background.* Springer-Verlag, Berlin 1986
2. J. Becker and L. Castell: *Act. Phys. Austriaca, Suppl.* XVIII, 885 (1977)
3. J.D. Bekenstein: *Physical Review* D 7, 2333–2346 (1973)
4. J.D. Bekenstein: *Physical Review* D 9, 3292–3300 (1974)
5. H.J. Blome, J. Hoell, W. Priester: *Bergmann-Schaefer, Lehrbuch der Experimentalphysik, Bd. 8, Sterne und Weltraum.* de Gruyter, Berlin 2002
6. L. Castell: *Nucl. Phys.* B13, 231 (1969)
7. L. Castell: in: *Quantum Theory and the Structures of Time and Space.* Vol. I, Hanser, München and Nuovo Cimento 29A, 1975, p. 445

8. L. Castell, M. Drieschner and C.F. von Weizsäcker (eds.): *Quantum Theory and the Structures of Space and Time.* Vol. I–V, München, Hanser 1975, 1977, 1979, 1981, 1983
9. M. Drieschner: *Voraussage Wahrscheinlichkeit Objekt.* Lecture Notes in Physics 99, Springer, Berlin 1979
10. M. Drieschner, Th. Görnitz, and C.F. von Weizsäcker: *International Journal of theoretical Physics* 27, 289–306 (1987)
11. Th. Görnitz and C.F. von Weizsäcker: in: *Conformal Groups and Related Symmetries, Physical Results and Mathematical Background.* A.O. Barut and H.-D. Doebner (eds.), Springer-Verlag, Berlin 1986
12. Th. Görnitz: *International Journal of Theoretical Physics* 27, 527–542 (1988)
13. Th. Görnitz: *International Journal of Theoretical Physics* 27, 659–666 (1988)
14. Th. Görnitz and E. Ruhnau: *Intern. Journ. Theoret. Phys.* 28, 651–657 (1989)
15. Th. Görnitz, D. Graudenz and C.F. von Weizsäcker(1992): *Intern. J. Theoret. Phys.* 31, 1929–1959 (1992)
16. Th. Görnitz and U. Schomäcker: *Group theoretical aspects of a charge operator in an ur-theoretical framework.* talk given at: GROUP 21, Applications and Mathematical Aspects of Geometry, Groups, and Algebras, Goslar, 1997
17. S.W. Hawking and G.F.R. Ellis: *The Large Scale Structure of the Universe.* University Press, Cambridge 1973
18. S.W. Hawking: *Communications in Mathematical Physics* 43, 199–220 (1975)
19. S.W. Hawking: in: *Quantum Structure of Space and Time.* M.J. Duff and C.J. Isham (eds.): University Press, Cambridge 1982
20. W. Heidenreich: Thesis. Technical University of Munich, 1981
21. P. Jacob: Thesis. Technical University of Munich, 1981
22. Th. Kühnemund: Thesis. Technical University of Munich, 1982
23. P. Roman: in: Castell, Drieschner, Weizsäcker (eds.): *Quantum Theory an the Sructures of Space and Time.* Vol. II, IV, V. Hanser, München 1977, 1981, 1983
24. E. Scheibe, G. Suessmann and C.F. von Weizsäcker: Mehrfache Quantelung, Komplementarität und Logik III. *Zeitschrift für Naturforschung* 13a, 705 (1958)
25. C.F. von Weizsäcker: Komplementarität und Logik I. *Naturwissenschaften* 42, 521–529, 545–555 (1955)
26. C.F. von Weizsäcker: Komplementarität und Logik II. *Zeitschrift für Naturforschung* 13a, 245 (1958)
27. C.F. von Weizsäcker: in: *Quantum Theory and the Structures of Time and Space.* Vol. I, Hanser, München 1975
28. C.F. von Weizsäcker: *Aufbau der Physik,* Hanser, Munich 1985

Ur Theory and Cosmological Phase Transition

Jörg D. Becker and Lutz Castell

1 Do Large Numbers Have a Meaning?

Following the Pythagorean tradition of number mysticism, Weyl (1919), Eddington (1920), Dirac (1937), and others speculated about the significance of "large numbers" which arise from the natural constants and from cosmology. For instance, the ratio of the electromagnetic force and the gravitational force between electrons amounts to $\sim 10^{42}$. From the velocity of light c, Planck's constant $\hbar$, the gravitational constant κ, and the matter density ρ of the universe, we may form the dimensionless quantity $X = c/(\hbar\rho\kappa^2)$ which amounts to $\sim 10^{120}$. If there was any connection between the Planck mass $m_P = \sqrt{\hbar(c/\kappa)}$, $\kappa = 8\pi G/c^2$, the rest mass m_0 of baryons and X, we could assume $m_0 = m_{Pl} \cdot f(X)$ where f is some function of X [3]. Taking $f(X) = X^{1/6}$ we get into the range of the meson masses. Is this just an entertaining game, or is there a possible explanation and interpretation? We shall present a cosmic scenario in which the Bose–Einstein condensation of ur systems gives a precise meaning to these quantities.

2 From Ur Space to Minkowski Space

Democritus conjectured that all that exists was made up from "atoms". During the 19th century scientists were very happy when they discovered that matter was indeed made up from small building blocks which they consequently called "atoms", too, until they discovered that atoms were made up from still smaller objects which were thought to be truly elementary, and hence called "elementary particles". Since 1964 (Gell-Mann, Zweig) we know that elementary particles are made up from still smaller constituents which have been named "quarks".

How should we now deal with this problem? Several suggestions have been made. First of all one might say that quarks are truly the smallest constituents of matter, but can we trust such an argument any longer? Heisenberg has suggested that the basic object of physics is a nonlinear spinor field which is not directly connected to a single particle but generates all types of particles in a dynamical process. Wilson has proposed that the process of zooming into an object and discovering its constituents goes on forever; this idea led him to the concept of renormalization group. Whitehead has argued that one should consider elementary processes as building bocks of reality.

Weizsäcker has started to look at the problem from the other end: Which object could be the primary building block of matter, having enough structure for this purpose but being in principle undividable itself? His conclusion was that a *binary alternative* (called *ur*) could be such an object. Since to his opinion quantum physics is the basic physical theory he postulated that these urs were not classical but quantum physical objects, being subject to superposition and revealing their properties only after a measurement. (In fact they are nothing else but q-bits of quantum computation.) Mathematically the urs can be represented by elements of the space $\mathbb{C}^2$.

In spirit this is similar to Spencer Brown's idea [6] of *distinctions* being the primary building blocks of mathematics, even if he has realized it in a different mathematical representation. Both approaches however suggest that *information* is the primary notion to be considered.

Urs are supposed to live on their own, and not only particles and fields but even space and time are thought to be phenomena which are generated by urs. (The number of urs which make up our universe has been estimated to be of the order of 10^{120} – a first hint that our magic number X may have some real connotation.) Hence a mechanism is necessary which performs this task. Castell [2] has suggested such a formalism which follows basically group theoretical arguments, and wich is displayed in Table 1.

Consider the natural invariance group of an ur, i.e. the transformations which leave norm and transition probabilities invariant. It is composed from the group U_2 of unitary 2×2 matrices and the group K of complex conjugation. This group cannot be represented linearly on the space $\mathbb{C}^2$, but a linear representation is possible on the space $\mathbb{C}^4$. This construction leads to the "doubling of the components of the ur", but since this is only a matter of representation the physical content of the ur has not

$\mathbb{C}^2$	→ linear represent-ation	$\mathbb{C}^4$		$S_3 \times S_1$
↓ symmetry group		↓ symmetry group		↑ represent-ation on
$SU_2 \times K$		$SU(2,2)$	→ local isomorphism	$SO_0(4,2)$

Table 1. From Ur space to Einstein space

changed. Or has it? The new space now allows for a larger symmetry group – it is the group SU(2,2). Since this group is locally isomorphic to the group SO_0 (4,2) we may represent it on a space $S_3 \times S_1$. Such a space is homeomorphic to an Einstein universe.

This group theoretical approach may look somewhat artificial at the first sight, but no longer if we remember that all "classical" equations of motion, such as Maxwell's or Dirac's equation, specify irreducible representations of the inhomogeneous Lorentz group and hence reflect the symmetry properties of the underlying physical space (in this case the Minkowski space).

3 Photon Condensation in an Einstein Universe

Massless particles (such as neutrinos and photons) arise in this space in a natural way as irreducible representations of the SO_0 (4,2). In particular, photons may be considered as being bound states of urs u and antiurs u*, where the lowest energy state consists just of one symmetric pair of urs (uu), the next state is of the form (uuuu*), and so forth. Because the quantum numbers of the maximally compact subgroup SO(4) × SO(2) specify the states, there is a discrete energy basis, and the lowest energy level is greater than zero. The energy spectrum is given by (Castell [2])

$$\epsilon_n = \frac{\hbar c}{R} \cdot (n + |s| + 1),$$

with the degeneracies

$$g_n = (n + 2|s| + 1)(n + 1),$$

R is the radius of the Einstein space, $n = (0, 1, 2, \ldots)$ is the principle quantum number, and s is the helicity of the particle; for photons, $s = \pm 1$. The question how matter, i.e. massive particles, arise in this context remains to be discussed. Einstein himself expressed the opinion that matter arises from the condensation of the electromagnetic field, i.e. of photons. Also C.F. von Weizsäcker has advocated the idea that the phenomenon of mass was somehow connected to a phase transition. For constructing a model along this line of thought we have to deal with statistics of photons.

The first approach to a statistics of ur systems was presented by Paul Roman [5]. Then J. Becker and L. Castell [1] considered an Einstein universe filled with photons. Because the lowest energy state is greater than 0, and Bose–Einstein condensation takes place, one may identify the condensate with the matter of the universe.

To summarize their arguments, let us start with the grand canonical ensemble,

$$J = -\sum_{n=0}^{\infty} g_n \ln\left(1 - \Lambda' e^{-\beta \varepsilon_n}\right)$$

where $\Lambda = \Lambda' \cdot \exp(-\beta \epsilon_0)$ is the absolute activity to be determined from the total number N of photons in the universe,

$$N = \Lambda' \cdot \partial J/\partial \Lambda'.$$

For a large number N of photons, and for $T \gg T_R = \hbar c/(kR)$ (which for our universe is $3 \cdot 10^{-29}$K; k is the Boltzmann constant), all thermodynamic quantities can be calculated. They depend crucially on the degeneracies g_n. The absolute activity stays close to $\Lambda = 1$ up to a critical temperature T_c (where the deviation from 1 is of the order of $1/\sqrt{N}$); it then has a turning point and decreases like T^{-3}. Hence we may define the critical temperature T_c as the turning point of $\Lambda(T)$. Below T_c a large proportion of all photons populate the ground state, i.e. the Bose condensate. Practically the specific heat exhibits a jump (more precisely: a very rapid variation, because we must not take the thermodynamic limit $V \to \infty$) at T_c of $9kN \cdot \zeta(3)/\zeta(2)$ where ζ is the Riemann zeta function, thus classifying the phase transition as second order. In general, the following relation between N and T_c holds:

$$\left(\frac{T_c}{T_R}\right)^3 = \frac{N}{4\zeta(3)}.$$

If we now identify the photon condensate with the matter in our universe, we may determine N and T_c. We assume a condensate dominated universe. According to standard cosmology, we have $R^2 = (3/2)/(\kappa\rho)$; R is the cosmic scale factor and ρ the matter density of the universe; κ is the gravitational constant. We get (for $s = \pm 1$; $u = \rho c^2$ is the energy density)

$$N = \frac{u \cdot \pi^2 \cdot R^4}{\hbar c} = \frac{3}{2} \cdot \frac{\pi^2 c}{\kappa \hbar} R^2$$

$$T_c^3 = \frac{\pi^2 \hbar^2 c^2}{4\zeta(3) k^3} \cdot uR = \frac{3}{8} \cdot \frac{\pi^2}{\zeta(3)} \cdot \frac{\hbar^2 c^4}{k^3} \cdot \frac{1}{\kappa R}.$$

From this formula we see that for infinite volume, $R \to \infty$, T_c would vanish, and so would the phenomenon of photon condensation. (These results follow also from the work of Landsberg and Dunning-Davies [4] concerning the Bose condensation of relativistic particles in a finite volume if we take the limit $m \to 0$.) Inserting all values we get

$$N = \frac{9}{4}\pi^2 X = 2.9 \cdot 10^{120}; \quad T_c = 3 \cdot 10^{11} K$$

Since photons in the ground state consist just of one ur pair (helicity $+1$) or one antiur pair (helicity -1), 2N is essentially the number of urs in the universe. The obtained numerical value comes close to other estimates of this number. The critical temperature corresponds to an energy $E_c = kT_c$ which is of the order of the hadron masses. We don't think this is purely incidental. One might say that by combining the natural constants in an arbitrary way one always has a chance of arriving at these numbers. On the other hand the degeneracies g_n of the photon states enter crucially into the formulae of the model, and they are certainly not incidental.

4 Conclusion

We have shown that a radiation dominated Einstein universe exhibits Bose–Einstein condensation of photons, and we have identified the liquid phase with the baryons. Even if this does not represent the full physical reality it shares certain features with reality; so some of the conclusions may also be valid in more realistic models to be developed.

References

1. J.D. Becker, L. Castell: Photon Condensation in an Einstein Universe. *Acta Physica Austriaca Suppl* XVIII, 885 (1977)
2. L. Castell: *Nucl. Phys.* B13, 231 (1969)
3. L. Castell: An example for the possible role of cosmology in elementary particle physics. In: L. Castell, C.F. von Weizsäcker (eds.): *Quantum theory and the structures of time and space,* Vol. 3. Hanser, München 1979
4. P.T. Landsberg, J. Dunning-Davies: *Phys.Rev.* 138A, 109 (1965)
5. P. Roman: Statistics of Ur Systems. In: L. Castell, M. Drieschner and C. F. von Weizsäcker (eds.): *Quantum theory and the structures of time and space,* Vol. 2. Hanser, München 1977
6. G. Spencer Brown: *Laws of Form.* George Allen and Unwin, London 1971

Phase Transitions in an Expanding Universe: Paul Roman's Models and Some Remarks on Entropy

Jörg D. Becker

The question of how photon condensation could emerge in an expanding universe remains open. Assuming that the formula for the transition temperature T_c which has been derived in the preceding contribution [1],

$$(kT_c)^3 = \frac{3}{8} \cdot \frac{\pi^2}{\zeta(3)} \cdot \frac{\hbar^2 c^4}{\kappa R}$$

was valid at any time we would have the following choices:

1. kT_c varies with R (even if moderately) which means that in the past of our universe the transition temperature was higher than it is now. If there was a correlation between T_c and the hadron masses these masses would have changed in time.
2. kT_c is constant which means that at least one natural constant, c, $\hbar$, or κ, changes in the course of the history of the universe.
3. In a more general theory (L. Castell, private communication) $1/R$ is replaced by $(\dot{R} + c^2)/R$. For $R \approx t^2/4 + c^2 k$, T_c is constant.

It is not sure whether observational evidence would support such variations in time.

1 Photon Condensation in an Expanding Universe

It was Paul Roman [5] who first studied this issue. He started with a universe which is radiation dominated. The scenario is that in the very early phase of the universe (say for a cosmic time $t < 10^{-7}s$) there are only photons, and the temperature is above T_c. When the temperature has fallen to this value condensation occurs, and droplets of the photon liquid are formed which may be interpreted as "baryons". Roman noticed that all arguments about the Bose-Einstein condensation of photons remain valid, just that now we have a different relation between ρ and R:

$$\rho = \frac{3}{4} \cdot \frac{c^2}{\kappa} \cdot \frac{\tau^2}{R^4}$$

where τ is the (hypothetial) life time a pure radiation universe would experience without condensation. However, after condensation the universe goes over to the usual Friedman type.

Roman arrives at the following conclusions:

$$T(t) = \frac{b\sqrt{\tau}}{cR(t)}$$

$$b = \left(\frac{15h^3c^3}{8\pi^5 k^4}\right)^{1/4} \left(\frac{3}{4\kappa}\right)^{1/4}$$

$$\frac{T(t)}{T_R(t)} = \frac{bk\sqrt{\tau}}{\hbar}$$

$$N = \frac{3}{4} \frac{\pi^2 c^3}{\hbar\kappa} \tau^2$$

$$T_c = \frac{90}{4\pi} \frac{c^2\hbar^3}{k\kappa^4} \frac{1}{b^3} \frac{\sqrt{\tau}}{R_c} \;.$$

It is interesting to note that in this model T/T_R and N are independent of cosmic time. From standard cosmological formulae one may conclude that $\tau \approx 10^{16}$s; hence $N \approx 10^{118}$. T_c depends on the radius at the time of the phase transition. Roman uses the typical baryon masses to determine the temperature, and thus the radius, at which condensation occurs. He gets a critical temperature of $T_c \approx 10^{13}$K.

2 Bubble Formation and Entropy

So far there is no evidence for relating T_c to the masses of elementary particles. Roman has approached this question by estimating the work needed for forming bubbles. For this work ("latent heat") Roman derives the expression

$$Q = \frac{4\pi^4}{45} \frac{k^4}{\hbar^3} b^3 \tau^{3/2} T_c \;.$$

Taking the observational value of the number of bubbles, i.e. the number of baryons in the universe N_B, he arrives at the conclusion that the entropy of the background radiation per baryon is of the order of $10^9 k$ which is in the range of the observed value. This quantity is highly significant because "our universe would be unrecognizable indeed", as Roman put it, if this value would be different. It is believed that it is so large because of some irreversible process in the early universe, and the formation of bubbles could be such a process.

3 Linking Ur Cosmology to More Conventional Approaches

In a subsequent paper Roman [6] presented a considerably enlarged scenario. He now starts with a pure ur universe. There is no matter and no radiation, but a large cosmological constant Λ. Taking into account the ground state energy of the urs, Λ can be expressed by the number of urs N_o in the ground state,

$$\Lambda = \frac{3\hbar G}{\pi c^5 R^4} N_o \, .$$

Thus we have an empty LeMaitre universe. It expandslike

$$R(t) = \sqrt{\frac{3}{\Lambda}} \cosh\left(\sqrt{\frac{3}{\Lambda}} t\right)$$

such that there is no singularity at t=0; instead we have

$$R(0) = \sqrt{\frac{3}{\Lambda}} = \sqrt{\frac{\hbar G}{\pi c^5} N_o} \, .$$

Since the urs themselves obey Bose statistics there is a first phase transition from the ur world to the GUT world, i.e. a world containing gauge vector bosons, two-component fermions, and Higgs scalars. According to the scenario, this phase transition should take place at $t = 7.3 \cdot 10^{-38}$s. Since this is larger than the Planck time, $t_{Pl} = \sqrt{G\hbar/c^5} \sim 5.4 \cdot 10^{-44}$s, quantum gravity does not enter in the GUT phase. Only shortly afterwords, at $t = 9 \cdot 10^{-38}$s, a second phase transition takes place, leading to the matter world. Supercooling due to the delayed formation of the new phase takes place, followed by a re-heating which finally takes us to the usual Friedmann universe. Again the number of urs in the universe, as well as the entropy per baryon, are comperable to the values given for the previously discussed scenarioes.

In this scenario the number of urs is not constant. At $t = 0$, this number is $\sim 2.8 \cdot 10^{12}$. What is its significance? Since there is no singularity at $t = 0$, and $R(t)$ is time reversal invariant (arount $t = 0$), we have a kind of Hindu universe, which is born again, and the urs present at $t = 0$ may express the karma of its previous life cycle. By now the number of urs has increased to 10^{119}, mainly during the phase transition to a matter dominated universe. The problem is that in this scenario there are many assumptions which enter, and thus the interpretation of the results remains uncertain.

4 Conclusions and Some Remarks on Entropy

We have presented several model universes which exhibit Bose-Einstein condensation of photons and other ur systems, and we have identified the liquid phase with the baryons. Even if the models do not represent physical reality they share certain features with reality; so some of the conclusions may also be valid in more realistic models to be developed.

An interesting question is how the number N of urs varies with cosmic time t. In Roman's first scenario [5] the number of urs is constant. If this would be the case in reality we would have to distinguish between ur theory as a theory of objects in the space $\mathbb{C}^2$ and the hypothetical interpretation as binary quantum alternatives; the world would be made up from urs, but they would all have been there from the very beginning in a prestabilized harmoy – somehow like the Leibniz monads: each one lives isolated from the others, yet contains the potential information of the whole universe which is generated from aggregates of monads. (The difference is that the urs would all be identical.)

In his second scenario [6] the number of urs starts with about 10^{12} at $t = 0$ and rises only moderately. Then, during the transition from a GUT to a Friedmann universe, i.e. in the stage of supercooling and bubble formation, there is a rapid rise to about 10^{119}. Once that the phase transition is completed the number of urs practically stays constant in t. Such a picture would suggest that entropy production is tightly correlated with ur production.

In the scenario which Lutz Castell has contributed to this volume [4] the universe starts just with *one single* ur, and the number of urs rises continuously $\sim t^2$ with time to the present value. The Minkowski time-translation it the reason for the production of urs [3].

All scenarioes are "phenomenological" in the sense that the process of ur production itself remains hidden. Is it ur production which drives the cosmic development? Or is ur production a consequence of the latter? To answer this question it would be necessary to have a microscopic description of ur production. This would first of all require an interaction between urs, but since urs form bound states a corresponding "Hamiltonian" may already be hidden in the models.

Furthermore it is an open question whether ur production is a purely physical phenomenon, or whether also structure formation (atoms, galaxies, stars, amino acids, plants etc.) or pragmatic information (see for instance [2]) can be related to it.

We may conclude that ur theory is not a closed chapter in the history of science but a real challenge to future investigations.

References

1. J.D. Becker, L. Castell: *Ur theory and cosmological phase transition.* This volume
2. J.D. Becker: Learning as gain of information. In: G. Dalenoort (ed.): *The paradigm of self-organization II.* Gordon and Breach, London 1994;
   J.D. Becker, I. Antoniou: Bergsonian time and system maturation. In: R. Buccheri (ed.): *Studies on the structure of time.* Plenum Press, New York 2000
3. L. Castell: Ur-Theory for Physicists. In: L. Castell, M. Drieschner and C.F. von Weizsäcker (eds.): *Quantum Theory and the Structures of Tima and Space,* Vol. 2. Hanser, Munich 1977
4. L. Castell: *The Ur-Hypothesis.* This volume
5. P. Roman: The world before the era of baryons, or: What happened when the clock read 10^{-7} seconds? In: L. Castell, C.F. von Weizsäcker (eds.): *Quantum theory and the structures of time and space,* Vol. 4. Hanser, München 1981
6. P. Roman: From the ur-world through the GUT-world to the matter-world. In: L. Castell, C.F. von Weizsäcker (eds.): *Quantum theory and the structures of time and space,* Vol. 5. Hanser, München 1983

Appendix

C. F. von Weizsäcker:
Biographical Data
and Selected Bibliography in Physics

1912	Born in Kiel (June 28)
1924	First meeting with his life long friend Georg Picht
1927	First meeting with Werner Heisenberg in Copenhagen
1929	Student of physics and mathematics at the Universities of Berlin (summer semester 1929), Leipzig (winter 1929/30–winter 1930/31), Göttingen (summer 1931), Leipzig (winter 1931/32–summer 1933)
1932	First meeting with Niels Bohr
1933	Doctorate in physics under W. Heisenberg, University of Leipzig
1933/34	Visiting scientist in Bohr's Institute for Theoretical Physics at the University of Copenhagen
1934–36	Assistant of Werner Heisenberg in Leipzig, habilitation: "On the spin dependence of nuclear forces" (June 1936)
1936–42	Assistant to Lise Meitner at the Kaiser Wilhelm Institute for Chemistry in Berlin-Dahlem (6 months). After Habilitation, member of the Kaiser Wilhelm Institute for Physics in Berlin-Dahlem (1936), Privatdozent for theoretical physics at the University of Berlin (1937)
1937	Married to Gundalena Wille
1939–45	Member of the *Uranverein*

1942–44   Associate professor of theoretical physics at the University of Strasbourg

1944–45   Return to the Kaiser Wilhelm Institute for Physics in Berlin, transfer of the work to Hechingen and Haigerloch in January 1945

1945–46   Interned from April 1945 to January 1946 (from July 3, 1945 at Farm Hall, England together with nine other German scientists)

1946–57   Head of the department for theoretical physics of the Max Planck Institute for Physics in Göttingen and honorary professor at the University of Göttingen

1957   Göttingen Declaration on the dangers of nuclear war together with 17 other German nuclear scientists (April 13)

1957–70   Chair of philosophy at the University of Hamburg

1961   Member of the Order pour le Mérite for Sciences and Arts

1969–74   Chairman of the Board of Directors of the German Volunteer Service (DED)

1970–80   Director of the Max Planck Institute zur Erforschung der Lebensbedingen der wissenschaftlich-technischen Welt, Starnberg

Since 1970   Honorary professor at the Ludwig Maximilian University, Munich

Honorary Doctorships and Prizes

Carl Friedrich von Weizsäcker is honorary doctor of the Faculty for Catholic Theology, University of Tübingen (1977); the Karl Marx University of Leipzig (1987); and of the Faculty for Communication and Historical Sciences of the Technical University of Berlin (1987).

C.F. von Weizsäcker has received the Max Planck Medal of the German Physical Society (1957), the Goethe Prize of the City of Frankfurt (1958), the Peace Prize of the German Book Trade (1963), the Arnold-Reymond Prize for Physics (1964), the Wilhelm Bölsche Medal (1965), the Erasmus Prize by the Erasmus Prize Foundation Rotterdam (1969), the Theodor Heuss Prize (1978), the Ernst Hellmut Vits Prize of the University of Münster (1981) and the Templeton Prize for Progress in Religion (1989).

Festschrifts and Books on C. F. von Weizsäcker

Erhard Scheibe and Georg Suessman (eds.): *Einheit und Vielheit. Festschrift für Carl Friedrich von Weizsäcker zum 60. Geburtstag.* Vandenhoeck & Ruprecht, Göttingen 1973

Klaus Michael Meyer-Abich (ed.): *Physik, Philosophie und Politik. Festschrift für Carl Friedrich von Weizsäcker zum 70. Geburtstag.* Hanser, München und Wien 1982

C.F. von Weizsäcker's scientific work and his political engagement are discussed in the following books:

Mathias Schüz: *Die Einheit des Wirklichen. Carl Friedrich von Weizsäcker's Denkweg.* Neske, Pfullingen 1986

Peter Ackermann, Wolfgang Eisenberg, Helge Herwig and Karlheinz Kannegießer (eds.): *Erfahrung des Denkens – Wahrnehmung des Ganzen.* Akademie Verlag, Berlin 1989

Michael Drieschner: *Carl Friedrich von Weizsäcker zur Einführung.* Junius, Hamburg 1992

Thomas Görnitz: *Carl Friedrich von Weizsäcker. Ein Denker an der Schwelle zum neuen Jahrtausend.* Herder, Freiburg 1992

Ulrich Bartosch: *Weltinnenpolitik.* Duncker und Humblot, Berlin 1995

Elisabeth Kraus: *Von der Uranspaltung zur Göttinger Erklärung. Otto Hahn, Werner Heisenberg, Carl Friedrich von Weizsäcker und die Verantwortung des Wissenschaftlers.* Würzburg 2001

Carl Friedrich von Weizsäcker: Lieber Freund, lieber Gegner! Briefe aus fünf Jahrzehnten. Ausgewählt und mit Anmerkungen versehen von Eginhard Hora. Hanser, München und Wien 2002

C. F. von Weizsäcker's Publications in Physics

A bibliography of C.F. von Weizsäcker's publications up to the beginning of 1973 is contained in the Festschrift editied by Erhard Scheibe and Georg Suessmann, pp. 292–304. In 1977 at the Max Planck Institute in Starnberg, R. Skottke has collected a bibliography of 39 pages on the occasion of C.F. von Weizsäcker's 65th birthday. A bibliography up to the beginning of 1988 is contained in: P. Ackermann et al. (eds.): *Erfahrung des Denkens – Wahrnehmung des Ganzen*, Akademie Verlag, Berlin 1989, pp. 211–246. A bibliography is also available from the archives of the Max Planck Society in Berlin.

The following selected bibliography is focussed on publications in physics. Beyond physics, Carl Friedrich von Weizsäcker has published several books and many papers on philosophy, religion, politics, on the prevention and abolition of war, and on the civilian use of nuclear energy. Many, if not most, of these publications have links to physics and science in general. There is, therefore, a measure of arbitrariness which of these publications are included in this selected bibliography.

1931

Ortsbestimmung eines Elektrons durch ein Mikroskop. *Zeitschrift für Physik* 70, 114–130 (1931)

1932

Grenzfragen der Philosophie und der modernen Physik. *Himmelswelt. Zeitschrift für Astronomie und ihre Grenzgebiete* 42, 129–138 (1932)

1933

Durchgang schneller Korpuskularstrahlen durch ein Ferromagnetikum. *Annalen der Physik* 17, 869–896 (1933)

1934

Ausstrahlung bei Stößen sehr schneller Elektronen. *Zeitschrift für Physik* 88, 612–625 (1934)

Zur Kritik der Ungenauigkeitsrelationen. *Naturwissenschaften* 22, 48, 807–808 (1934)

1935

Zur Theorie der Kernmassen. *Zeitschrift für Physik* 96, 431–458 (1935)
Die für den Bau der Atomkerne maßgebenden Kräfte. *Physikalische Zeitschrift* 36, 779–785 (1935)
Die für den Bau der Atomkerne maßgebenden Kräfte. *Zeitschrift für technische Physik* 16, 385–390 (1935)

1936

Über die Spinabhängigkeit der Kernkräfte. *Zeitschrift für Physik* 102, 572–602 (1936)

Metastabile Zustände der Atomkerne. *Die Naturwissenschaften* 24, 813–814 (1936)

Fortschritte in der Theorie des Atomkerns. *Forschungen und Fortschritte* 12, 171–172 (1936)

1937

Über die Möglichkeit eines dualen β-Zerfalls von Kalium. *Physikalische Zeitschrift* 38, 623–624 (1937)

Über Elementumwandlungen im Innern der Sterne. *Physikalische Zeitschrift* 38(6), 176–191 (1937)

Die Atomkerne – Grundlagen und Anwendung ihrer Theorie. Akademische Verlagsgesellschaft, Leipzig 1937, 215 p.

Ist in der Gegenwart eine systematische Einheit der Wissenschaft möglich? *Die Tatwelt* 13(2), 68–78 (1937)

1938

Über Elementumwandlungen im Innern der Sterne II. *Physikalische Zeitschrift* 38(17/18), 633–646 (1938)

Neuere Modellvorstellungen über den Bau der Atomkerne. *Die Naturwissenschaften* 26, 209–217 (1938); 225–230

1939

Umwandlung chemischer Elemente im Inneren der Sterne. *Forschungen und Fortschritte* 15, 159 (1939)

Kernumwandlung als Quelle der Sternenergie. *Verhandlungen der deutschen physikalischen Gesellschaft* 20(3), 2–4 (1939)

Der zweite Hauptsatz und der Unterschied von Vergangenheit und Zukunft. *Annalen der Physik* 5, 36, 3 and 4, 275–283 (1939). Reprinted in: *Die Einheit der Natur.* Hanser, München und Wien 1971

Zum Wefelmeierschen Modell der Transurane. *Die Naturwissenschaften* 27, 133 (1939)

1941

Zur Deutung der Quantenmechanik. *Zeitschrift für Physik* 118, 489–509 (1941)

Die theoretische Deutung der Spaltung von Atomkernen. *Forschungen und Fortschritte* 17, 10–11 (1941)

Die Physik der Gegenwart und das physikalische Weltbild. *Naturwissenschaften* 29(13), 185–194 (1941)

Das Verhältnis der Quantenmechanik zur Philosophie Kants. *Die Tatwelt* 17(3), 66–98 (1941); 18(2), 105–109 (1942)

1942

Atomtheorie und Philosophie. *Die Chemie* 55, 99–104 (1942), 121–126

Die Atomlehre der modernen Physik. *Volk im Werden* 10(4/5), 95–105 (1942)

Die Auswirkungen des Satzes von der Erhaltung der Energie in der Physik. In: *Robert Mayer und das Energieprinzip*. VDI, Berlin 1942, pp. 149–176

1943

Theorie des Mesons. In: W. Heisenberg (ed.): *Kosmische Strahlung*. Berlin 1943, pp. 90–110

Zum Weltbild der Physik. Hirzel, Leipzig 1943, 378 p. English: *The World View of Physics*. Routledge & Kegan, London 1952, 219 p. (12 editions: 1943, 1944, 1945, 1949, 1951, 1958, 160, 1963, 1970, 1976). French: *Le monde vu par la physique*. Flammarion, Paris 1956. Dutch: *Het wereldbeeld in de fysica*. Het spectrum, Utrecht, Antwerpen 1959. Danish: 1962. Italian: *L'immagine fisica del mondo*, 1962. Spanish: *La imagen física del mundo*. La Editorial Católica 1974

Deutung einer Auswahlregel für Neutronen- und Protonenemission aus Kernen ungerader Ladung. *Naturwissenschaften* 31, 207–208 (1943)

Über die Entstehung des Planetensystems. Zum 75. Geburtstag von Arnold Sommerfeld. *Zeitschrift für Astrophysik* 22, 319–355 (1944)

1944

Die Unendlichkeit der Welt. Eine Studie über das Symbolische in der Naturwissenschaft. *Die Chemie* 57(1/2), 1–6 (1944); 57(3/4), 17–22

1946

Das Kontinuitätsprinzip in der heutigen Naturwissenschaft. In: *Gottfried Wilhelm Leibniz. Vorträge aus Anlaß seines 300. Geburtstags abgehaltenen Tagung*. Hanseatische Gilden, Hamburg 1946, pp. 283–301

1947

Die Entstehung des Planetensystems. *Naturwissenschaften*, 21/23, 7–9, 71 (1947)

Zur Kosmogonie. *Zeitschrift für Astrophysik* 24, 1/2, 181–206 (1947)

Zur statistischen Theorie der Turbulenz. *Angewandte Chemie* A 59, 60 (1947)

Physikalische Grundlagen der Isotopen-Chemie. *Angewandte Chemie* A 59, 105–108 (1947)

1948

Die Geschichte der Natur. Zwölf Vorlesungen. Hirzel, Stuttgart 1948, 138 p. (8 editions: 1948, 1954, 1956, 1962, 1970, 1979). English: *The History of Nature*, London

1951, 179 p. and Univ. of Chicago Press, Chicago IL 1949. Japanese: Horitsu Bunkasha, Kyotoi, 1968. Norwegian: Land og Kirke, Oslo 1955. Swedish: Natur och Kultur, Stockholm 1955. Spanish: Ed. Rialp, Madrid 1962. Czech: 1969

Das Spektrum der Turbulenz bei großen Reynoldsschen Zahlen. *Zeitschrift für Physik* 124, 7/12, 614–627 (1948)

Kosmogonie. In: P. Ten Bruggenkate (ed.): *Naturforschung und Medizin in Deutschland 1939–1946*, Vol. 20: Astronomie, Astrophysik und Kosmogonie. Dieterich, Wiesbaden 1948, pp. 413–426

Die obere Grenze der Spiralnebel. *Naturwissenschaften* 35(6), 188–189 (1948)

Die Rotation kosmischer Gasmassen. *Zeitschrift für Naturforschung* 3a, 524–539 (1948)

With W. Heisenberg: Die Gestalt der Spiralnebel. *Zeitschrift für Physik* 125, 4–6, 290–292 (1948)

Die Unendlichkeit der Welt. Eine Studie über das Symbolische in der Naturwissenschaft. In: *Vision. Deutsche Beiträge zum geistigen Bestand* 1, 3, 246–263 (1948)

Der begriffliche Aufbau der theoretischen Physik. Mimeographed lecture notes by R. Skottke. Göttingen 1948, 233 p.

Wilfried Wefelmeier (Nachruf). *Zeitschrift für Naturforschung*, 3a, 370 (1978)

1949

Eine Bemerkung über die Grundlagen der Mechanik. Max von Laue zum 70. Geburtstag gewidmet. *Annalen der Physik* 6(6), 67–68 (1949)

Turbulence in Interstellar Matter. In: *Proc. Symp. on the Motion of Gaseous Masses of Cosmic Dimensions*, p. 200. Paris 1949

Beziehungen der theoretischen Physik zum Denken Heideggers. In: *Martin Heideggers Einfluß auf die Wissenschaften*. A. Franke AG, Bern 1949, pp. 172–174

1950

Zum Problem des Wärmetods. *Mitteilungsblatt für mathematische Statistik* 2(3), 224–225 (1950)

The experiment. Its nature and its limits. *Measure* 2(2), 35–47 (1951)

1951

Anwendungen der Hydrodynamik auf Probleme der Kosmogonie. In: *Festschrift zur Feier des 200-jährigen Bestehens der Akademie der Wissenschaften in Göttingen*. Springer, Berlin 1951, pp. 86–122

The Evolution of Galaxies and Stars. *The Astrophysical Journal* 114(2), 165–186 (1951)

Kontinuität und Möglichkeit (Werner Heisenberg zum 50. Geburtstag am 5. Dezember 1951 gewidmet). *Naturwissenschaften* 38, 533–543 (1951)

Einstein and the Philosophy of Physics. *Measure* 2(2), 231–240 (1951). German: Einstein und Bohr. Der Streit um den Realitätsbegriff des Physikers. *Wort und Wahr-*

heit 7, 119–126 (1952). Reprinted in: *Zum Weltbild der Physik*, 4th edn. Hirzel, Stuttgart

Das neue Bild vom Weltall. *Merkur* 5(9), 805–819 (1951). English: The new picture of the universe. *The Listener* 49, 1245 (1953)

1952

Eine Frage über die Rolle der quadratischen Metrik in der Physik. Werner Heisenberg zum 50. Geburtstag. *Zeitschrift für Naturforschung* 7a(1), 141 (1952)

With J. Juilfs: *Physik der Gegenwart*. Athenäum, Bonn 1952, 166 p., 2nd edn. Vandenhoeck & Ruprecht, Göttingen 1957, 134 p. English: *Contemporary physics*. Hutchinson, London 1957, 150 p.; Braziller, New York 1957, 150 p. Spanish: *La física actual*, Ed. Rialp, Madrid 1964

1953

Klassische Feldtheorie der Mesonen. In: W. Heisenberg (ed.): *Kosmische Strahlung*. Springer, Berlin 1953, pp. 535–546

1954

Die Geschichte der Natur. Vandenhoeck & Ruprecht, Göttingen 1954. Swedish edition: *Natures historia*, Stockholm 1955. Spanish edition: *Historia de la naturaleza*, Ed. Rialp, Madrid 1962

Genäherte Darstellung starker instationärer Stoßwellen durch Homologie-Lösungen. *Zeitschrift für Naturforschung* 9a(4), 269–275 (1954)

Eine Bemerkung über die Gestalt der kugelförmigen Sternhaufen. *Zeitschrift für Astrophysik* 35B, 252–254 (1954/55)

1955

Komplementarität und Logik. Niels Bohr zum 70. Geburtstag am 7.10.1955 gewidmet. *Naturwissenschaften* 42(19), 521–529 (1955)

Komplementarität und Logik (Schluß). *Naturwissenschaften* 42(20), 545–555 (1955)

Über einige Begriffe in der Naturwissenschaft Goethes. Nachwort in Bd. 13: *Naturwissenschaftliche Schriften der Hamburger Goethe-Ausgabe* (1955); reprinted in: *Festschrift für Robert Boehringer*, ed. by E. Boehringer and W. Hoffmann. Mohr, Tübingen 1957, pp. 697–771

Problems connected with the origin of the spiral arms. In: *Gas Dynamics of Cosmic Clouds. A symposium held at Cambridge, England*. Amsterdam 1955, pp. 167–171

1956

Gestaltkreis und Komplementarität, In: Paul Vogel: *Viktor von Weizsäcker. Arzt im Irrsal der Zeit*. Vandenhoek & Ruprecht, Göttingen 1956, pp. 21–53

Die Illusion deutscher Atombomben. In: Carl Seelig: *Helle Zeit – dunkle Zeit. In memoriam Albert Einstein.* Europa Verlag, Zürich 1956, pp. 130–133

1957

Atomenergie und Atomzeitalter: Zwölf Vorlesungen. Fischer, Frankfurt a.M. 1957, 165 p.

Operative Logik und Mathematik. *Naturwissenschaften* 44(18), 482–485 (1957)

Erklärung der 18 Atomwissenschaftler, 1957. English translation: Declaration of the German nuclear physicists. *Bulletin of the Atomic Scientists* 13(6), 228 (1957)

1958

Die Quantentheorie der einfachen Alternative (Komplementarität und Logik II). *Zeitschrift für Naturforschung* 13a, 245–253 (1958)

With E. Scheibe and G. Suessmann: Komplementarität und Logik III. Mehrfache Quantelung. *Zeitschrift für Naturforschung* 13a, 705–721 (1958)

Über Sternentstehung. Vortrag zur Verleihung der Max-Planck-Medallie der Deutschen Physikalischen Gesellschaft am 28.9.1957 in Heidelberg. In: E. Brüche and W. Wessel: *Physikertagung Heidelberg 1957.* Physik Verlag, Mosbach 1958, pp. 44–60

Philosophische Fragen der Naturwissenschaft. Vortrag gehalten am 24. Oktober 1957 in Berlin auf der Jahrestagung der Deutschen Forschungsgemeinschaft. *Merkur* 12(9), 801–820 (1958)

Goethe und die Natur. Rede gehalten in der Paulskirche bei der Verleihung des Goethe-Preises der Stadt Frankfurt a.M., 31th August 1958, City of Frankfurt a.M., Frankfurt a.M., 1958, 13–17

Die gegenwärtigen Aussichten einer Begrenzung der Gefahr eines Atomkrieges, part 1–3. In: DIE ZEIT, Hamburg, 15th May 1958, 4 p.; 22nd May 1958, 3 p.; 29nd May, 3 p.; 5th June 1958, 3 p. Special print by DIE ZEIT under the title: *Mit der Bombe leben* (1958), 24 p. Reprinted in: *Der bedrohte Friede,* 1981

Do we want to save ourselves? *Bulletin of Atomic Scientists* 14(5), 180–184 (1958)

Descartes und die neuzeitliche Naturwissenschaft. Akademische Rede gehalten am 13. November 1957 an der Universität Hamburg, Univ. of Hamburg, 1958, 29 p. (Hamburger Universitätsreden)

Ich-Du und Ich-Es in der heutigen Naturwissenschaft. *Merkur* 12(2), 124–128 (1958)

1959

Erinnerungen an Wolfgang Pauli, 25.4.1900–15.12.1958. *Zeitschrift für Naturforschung* 14a(5), 435–440 (1959)

1961

Die Einheit der Physik. In: F. Bopp: *Werner Heisenberg und die Physik unserer Zeit.* Vieweg, Braunschweig 1961, pp. 23–46

Werner Heisenberg zu seinem 60. Geburtstag am 5.12.1961. *Physikalische Verhandlungen* 1(12), 343–346 (1961)

1962

Kopernikus, Kepler, Galilei. In: K. Oehler and R. Schaeffler: *Einsichten. Gerhard Krüger zum 60. Geburtstag (1962)*. Klostermann, Frankfurt a.M. 1962, pp. 376–394

Eine Anwendung der Theorie der Kristallgitter auf Fragen der Bevölkerungsstatistik. *Physikalische Blätter* 18, 553 (1962)

1964

Die Tragweite der Wissenschaft. Schöpfung und Weltentstehung. Die Geschichte zweier Begriffe. Hirzel, Stuttgart 1964, 243 p., new printing 1990. English: *The Relevance of Science. Creation and Cosmogony. Gifford Lectures 1959–1960*. Harper & Row, New York 1964 and Collins, London 1964

Kants "Erste Analogie der Erfahrung" und die Erhaltungssätze der Physik. In: H. Delius und G. Patzig: *Argumentationen. Festschrift für Josef König*. Vandenhoeck & Ruprecht, Göttingen 1964, pp. 256–275

1965

La physique – une unité à construire. Discours prononcé à l'occasion de la collation du Prix Arnold Raymond. Publications de l'Université de Lausanne 28, 19–43 (1965); German: Die Einheit der Physik als konstruktive Aufgabe. *Philosophia Naturalis* 9(3), 247–265 (1966)

1966

Kants Theorie der Naturwissenschaft nach P. Plass. *Kant-Studien* 56(3–4), 528–544 (1966)

1967

Die Einheit der Physik. Festvortrag bei der Eröffnung der Physikertagung in München 1966. *Physikalische Blätter* 23(1), 4–14 (1967). English: The Unity of Physics. In: R.S. Cohen and M.W. Wartofsky (eds.): *Boston Studies in the Philosophy of Science*, Vol. V. Proceedings of the Boston Colloquium for the Philosophy of Science 1966/68. Reidel, Dordrecht 1969, pp. 460–473

Möglichkeit und Bewegung. Eine Notiz zur aristotelischen Physik. In: *Festschrift für Josef Klein zum 70. Geburtstag*. Vandenhoek & Ruprecht, Göttingen 1967, pp. 73–84

Das Problem der Zeit als philosophisches Problem. Wichern, Berlin 1967, 35 p. (Erkenntnis und Glaube. Schriften der Evangelischen Forschungsakademie, Ilsenburg, Vol. 28)

Das Weltbild des Atomwissenschaftlers unserer Zeit. *Universitas* 22(8), 785–790 (1967)

1971

Die Einheit der Natur, Hanser, München und Wien 1971, 491 p. (5 editions: 4 (1972), 5 (1979), paperback edn.: Deutscher Taschenbuch-Verlag, 1974). English: *The unity of nature*, transl. by F.J. Zucker, Farrar, Straus & Giroux, New York 1980, 406 p. Polish: Panstwory Institut Wydawniczy, Warzawa 1978

The Copenhagen Interpretation. In: T. Bastian (ed.): *Quantum Theory and Beyond*. Cambridge University Press, Cambridge, UK 1971, pp. 25–31

The Unity of Physics. In: T. Bastian (ed.): *Quantum Theory and Beyond*. Cambridge University Press, Cambridge, UK 1971, pp. 229–265

Notizen über die philosophische Bedeutung der Heisenbergschen Physik. In: H.P. Dürr (ed.): *Quanten und Felder. Physikalische und philosophische Betrachtungen zum 70. Geburtstag von Werner Heisenberg*. Vieweg, Braunschweig 1971, pp. 11–26

Kant and modern science, *Synthese* 23, 75–95 (1971)

1972

Voraussetzungen des naturwissenschaftlichen Denkens. Herder, Freiburg 1972, 141 p.

Der Zusammenhang der Quantentheorie elementarer Felder mit der Kosmognie. In: J.-H. Scharf (ed.): *Grundfragen der Quanten- und Relativitätstheorie*. Symposium der Deutschen Akademie der Naturforscher Leopoldina, Halle/Saale 1974, pp. 61–80 (*Nova Acta Leopoldina* NF 212, 39)

Evolution und Entropiewachstum. *Nova Acta Leopoldina* NF 37, 206, 515–530 (1972)

Die philosophische Interpretation der modernen Physik. Zwei Vorlesungen. Barth, Leipzig 1972 (*Nova Acta Leopoldina* NF 37, 207, 7–39 (1972))

Platonic natural science in the course of history. *Main Currents in Modern Thought* 29, 3–13 (1972)

1973

Die Aktualität der Tradition: Platons Logik. Vortrag gehalten vor der Schelling-Kommission der Bayerischen Akademie der Wissenschaften am 19.3.1973 in München. In: *Philosophisches Jahrbuch* 80(2), 221–242 (1973)

Comment on Dirac's Paper. In: J. Mehra (ed.): *The physicist's conception of nature*. Reidel, Dordrecht 1973, pp. 55–59

Classical and quantum description. In: J. Mehra (ed.): *The physicist's conception of nature*. Reidel, Dordrecht 1973, pp. 635–667

Physics and Philosophy. In: J. Mehra (ed.): *The physicist's conception of nature*. Reidel, Dordrecht 1973, pp. 736–746

Probability and quantum mechanics. *British Journal for the Philosophy of Science* 24, 321–337 (1973)

Der Gedanke der Einheit der Natur. *Almanach der Österreichischen Akademie der Wissenschaften*. Wien 1973, pp. 40–59

Kants 'Erste Analogie der Erfahrung' und die Erhaltungssätze der Physik. In: G. Prauss (ed.): *Kant. Zur Deutung seiner Theorie von Erkennen und Handeln.* Kiepenheuer & Witsch, Köln 1973, pp. 151–166

1974

Geometrie und Physik. In: C.P. Enz and J. Mehra (eds.): *Physical reality and mathematical description. Dedicated to Josef Maria Jauch on the occasion of his 60th birthday.* Reidel, Dordrecht 1974, pp. 48–90

Evolution und Entropiewachstum. In: E.U. von Weizsäcker (ed.): *Offene Systeme.* Klett, Stuttgart 1974, pp. 200–221

Der Zusammenhang der Quantentheorie elementarer Felder mit der Kosmogonie. Vortrag gehalten anlässlich des Symposiums "Grundfragen der Quanten- und Relativitätstheorie" der Deutschen Akademie der Naturforscher Leopoldina in Eisenach 10.–15.4.1972. In: *Nova Acta Leopoldina NF* 39, 312, 61–80 (1974)

1975

Who is the subject in physics? In: T.M.P. Mahadevan (ed.): *Spiritual Perspectives. Essays in Mysticism and Metaphysics.* Arnold-Heinemann, New Delhi

Editor with L. Castell and M. Drieschner: *Quantum theory and the structures of time and space.* Papers presented at a conference held in Tutzing, July 1974. Hanser, München und Wien, 1975, 252 p.

The philosophy of alternatives. In: L. Castell et al. (eds.): *Quantum theory and the structures of time and space.* Papers presented at a conference held in Tutzing, July 1974, Hanser, München und Wien 1975, pp. 213–230

Werner Heisenberg, 5. Dezember 1901–1. Februar 1976. Eine Gedenkrede, *Universitas* 31(7), 681–691 (1976)

1977

Der Garten des Menschlichen. Beiträge zur geschichtlichen Anthropologie. Hanser, München und Wien 1977, 612 p. English: *The Ambivalence of Progress. Essays on Historical Anthropology*, Paragon House, New York 1988, 304 p.

With B.L. v. d. Waerden: *Werner Heisenberg.* Hanser, München und Wien 1977, 104 p.

Werner Heisenberg in memoriam/Heisenbergs Entwicklung seit 1927/Heisenberg – Physik und Philosophie/Notizen über die Bedeutung der Heisenbergschen Physik. In: C.F. von Weizsäcker et al. (eds.): *Werner Heisenberg.* Hanser, München und Wien 1977, pp. 7–14, 25–40, 41–50, 51–83

Heidegger und die Naturwissenschaft. In: Marx, W. (ed.): *Heidegger.* Freiburger Universitätsvorträge zu seinem Gedenken. Alber, Freiburg 1977, pp. 63–86

Editor with L. Castell and M. Drieschner: *Quantum theory and the structures of time and space*, Vol. 2. In Memoriam Werner Heisenberg. Hanser, München und Wien 1977, 252 p.

Heisenberg's conception of physics. In: L. Castell et al. (eds.): *Quantum theory and the structures of time and space*, Vol. 2. In Memoriam Werner Heisenberg. Hanser, München und Wien 1977, pp. 9–19. German: *Zeit und Wissen*, 796–808 (1992)

Binary alternatives and space-time structure, In: L. Castell et al. (eds.): *Quantum theory and the structures of time and space*, Vol 2. In Memoriam Werner Heisenberg. Hanser, München und Wien 1977, pp. 86–112

Deskriptive zeitliche Logik. *Philosophische Rundschau* (1977/8)

1978

Quantentheorie elementarer Objekte. Deutsche Akademie der Naturforscher Leopoldina, Barth, Leipzig, 19 p. (*Nova Acta Leopoldina* NF 230, 49, 1978)

1979

Editor with L. Castell and M. Drieschner: *Quantum theory and the structures of time and space*, Vol. 3: Papers presented at a conference held in Tutzing, July 1978. Hanser, München und Wien 1979, 200 p.

A reconstruction of quantum theory. In: L. Castell et al. (eds.): *Quantum theory and the structures of time and space*, Vol. 3: Papers presented at a conference held in Tutzing, July 1978. Hanser, München und Wien 1979, pp. 7–35

1981

Ein Blick auf Platon: Ideenlehre, Logik und Physik. Reclam, Stuttgart 1981, 140 p. (Universalbibliothek; 7731)

Editor with L. Castell and M. Drieschner: *Quantum theory and the structures of time and space*, Vol. 4: Papers presented at a conference held in Tutzing, July 1980. Hanser, München und Wien 1981, 250 p.

The Present Status of the theory of Urs. In: L. Castell et al. (eds.): *Quantum theory and the structures of time and space*, Vol. 4: Papers presented at a conference held in Tutzing, July 1980. Hanser, München und Wien 1981, pp. 7–26

Der Bedrohte Friede. Politische Aufsätze. Hanser, München und Wien 1981. Paperback edn. 1983

1983

Wahrnehmung der Neuzeit. Hanser, München und Wien 1983, 440 p.

Editor with L. Castell: *Quantum theory and the structures of time and space*, Vol. 5: Papers presented at a conference held in Tutzing, July 1982. Hanser, München und Wien 1983, 195 p.

Urs, particles, fields. In: L. Castell et al. (eds.): *Quantum theory and the structures of time and space*, Vol. 5: Papers presented at a conference held in Tutzing, July 1982. Hanser, München und Wien 1983, pp. 34–53

Geometrie und Physik. In: D. Mayr and G. Suessmann (eds.): *Space, Time and Mechanics: basic structures of a physical theory*. Reidel, Dordrecht 1983, pp. 39–86

1984

The unity of nature. In: R.S. Cohen et al. (eds.): *Physical sciences and history of physics*. Reidel, Dordrecht 1984, pp. 239–254

Paul Adrien Maurice Dirac. DIE ZEIT, Hamburg, November 1984, reprinted in: *Zeit und Wissen*, 809–813 (1992)

Zeit, Physik, Metaphysik. In: Ch. Kink (ed.): *Die Erfahrung der Zeit. Gedenkschrift für Georg Picht*. Klett-Cotta, Stuttgart 1984, pp. 17–34

1985

Aufbau der Physik. Hanser, München und Wien 1985, 662 p. Paperback edn.: Deutscher Taschenbuch-Verlag, München 1988

Reminiscence from 1932. In: French and Kennedy (eds.): *Niels Bohr. A centenary volume*. Cambridge, MA 1985, pp. 183–190

Zeit, Physik, Metaphysik. In: K. Michalski (ed.): *Der Mensch in den modernen Wissenschaften*. Castelgandolfo-Gespräche 1983. Klett-Cotta, Stuttgart 1985

Quantum theory and space-time. In: P. Lathi (ed.): *Symposium on the foundations of modern physics 1985: 50 years of the Einstein–Podolsky–Rosen Gedankenexperiment*, Joensuu, Finland 16–20 June 1985. World Scientific, Singapore 1985, pp. 223–237

Werner Heisenberg. In: L. Gall (ed.): *Die großen Deutschen unserer Epoche*. Propyläen, Berlin 1985

"Der Sokrates unter den Physikern" – Niels Bohr – Philosoph der Quantentheorie vor hundert Jahren geboren. Frankfurter Allgemeine Zeitung, 5.10.1985

1986

Editor with L. Castell: *Quantum theory and the structures of time and space*, Vol. 6: Papers presented at a conference held in Tutzing, July 1984. Hanser, München und Wien 1986, 175 p.

Reconstruction of quantum theory. In: Lutz Castell et al. (eds.): *Quantum theory and the structures of time and space*, Vol. 6: Papers presented at a conference held in Tutzing, July 1984. Hanser, München und Wien 1986, pp. 7–41

Die Logik zeitlicher Aussagen und die Grundlagen der Physik. *Information Philosophie* 14(3), 7–22 (1986)

With Th. Görnitz: De-Sitter representations and the particle concept in an urtheoretical cosmological background. In: A.O. Barut et al. (eds.): *Conformal groups and related symmetries. Physical results and mathematical background*. Proceedings of a symposium held at the Arnold Sommerfeld Institute for Mathematics (ASI), Technical University of Clausthal, Germany, August 12–14, 1985. Springer, Berlin 1986 (Lecture notes in physics, 261)

Evolution und Entropiewachstum. In: E.U. von Weizsäcker (ed.): *Offene Systeme I: Beiträge zur Zeitstruktur von Information*. 2nd revised edition. Klett-Cotta, Stuttgart

1986, pp. 200–221 (Forschungen und Berichte der Evangelischen Studiengemeinschaft, 30)

1987

With Th. Görnitz: Remarks on S. Kochen's interpretation of quantum mechanics. In: P. Lahti (ed.): *The Copenhagen interpretation 60 years after the Como lecture.* Symposium on the foundations of modern physics 1987, Joensuu, Finland, 6–8 August 1987. World Scientific, Singapore 1987

With Th. Görnitz: Quantum Interpretations. *Intern. Journ. of Theor. Physics* 26, 921 (1987)

1988

With Th. Görnitz: Copenhagen and Transactional Interpretations. *Intern. Journ. of Theor. Physics* 27, 237–250 (1988)

With M. Drieschner and Th. Görnitz: Reconstruction of abstract quantum theory. *Intern. Jour. of Theor. Physics* 27, 289–306 (1988)

With E. Rudolph: *Zeit und Logik bei Leibniz,* Stuttgart 1989

1991

Der Mensch in seiner Geschichte. Hanser, München und Wien 1991, 244 p.

With Th. Görnitz: *Quantum theory as a theory of human knowledge.* In: Symposium on the foundations of modern physics 1990: Quantum theory of measurement and related philosophical problems, Joensuu, Finland 13–17 August 1990, ed. by P. Lathi. World Scientific, Singapore 1991

With Th. Görnitz: Quantum-realistic interpretation. *Foundation of Physics* 21, 311–321 (1991)

With Th. Görnitz: Steps in the Philosophy of Quantum Theory. In: J. Hennig, W. Lücke and J. Tolar (eds.): *Differential Geometry, Group Representations and Quantizations* Springer, Heidelberg 1991 (Lect. Notes in Physics 379)

1992

Editor with H.-P. Dürr, E. Feinberg and B.L. van der Waerden: *Werner Heisenberg.* München 1992

Nachruf auf W. Heisenberg. In: H.-P. Dürr et. al. (eds.): *Werner Heisenberg.* München 1992

Zeit und Wissen. Hanser, München und Wien 1992, 1184 p. Paperback edn.: Deutscher Taschenbuch-Verlag, München 1995

With D. Graudenz and Th. Görnitz: Quantum field theory of binary alternatives. *Intern. Journ. of Theor. Physics* 31, 11 (1992)

With Th. Görnitz and E. Ruhnau: Temporal asymmetry as precondition of experience: The foundation of the arrow of time. *Intern. Journ. of Theor. Physics* 31(11), 37–46 (1992)

1993

With Th. Görnitz: *Events in Quantum Theory*. In: P. Busch, P. Lathi and P. Mittelstaedt (eds.): Symposium on the Foundations of Modern Physics, Köln. World Scientific, Singapore 1993

1997

Time – Empirical Mathematics – Quantum Theory. In: H. Atmanspacher and E. Ruhnau (eds.): *Time, Temporality, Now*. Proceedings of the International Workshop at Ringberg, February 26–March 1, 1996. Springer, Berlin 1997

2000

Hypothese über Ure und Strings. In: G. Pospiech (ed.): *Atome und Quanten im Unterricht – Erfahrungen und Perspektiven*. Institut für Didaktik der Physik der J.W. Goethe-Universität Frankfurt a.M. 2000, ISBN 3-9806144-1-7

The Authors

H. Atmanspacher

Harald Atmanspacher received a Ph.D. in physics for a thesis on multimode laser systems and he completed a habilitation in the field of nonlinear dynamics and complex systems at the University of Potsdam in 1995. From 1986 to 1988, Dr. Atmanspacher was a research scientist at the Max Planck Institute für extraterrestrische Physik in Garching. He is now head of the Theory and Data Analysis Department at the Institut für Grenzgebiete der Psychologie und Psychohygiene (IGPP) in Freiburg, Germany. His fields of work and interest are nonlinear dynamics, complex systems, mind-matter problems and selected topics in the history and philosophy of science.

H. Bethe

Hans A. Bethe was born in Strasbourg in 1906. He studied physics in Frankfurt and worked at the University of Tübingen from where he was dismissed in 1933. He briefly worked with Arnold Sommerfeld in Munich and with Niels Bohr in Copenhagen, from where he joined Cornell University, Ithaca NY in 1935. From 1942 he took part in the Manhattan Project, in which he became head of the theory division of the Los Alamos laboratory. After the was, he returned to Cornell University, where his principal field of research was elementary particle physics. In 1967 he was awarded the Nobel Prize in Physics for his work on fusion cycles in stars.

Č. Brukner

Časlav Brukner studied Physics at the Universities of Belgrade and Vienna. In 1999, he obtained his doctorate in Physics at Vienna with a thesis on 'Information in Individual Quantum Systems'. Dr. Brukner collaborates with A. Zeilinger on the conflict between quantum mechanics and local realism and for the estabishment of protocols in quantum communication and quantum computation.

L. Castell

Lutz Castell studied under A. Salam, F. Bopp and J. Hamilton, and was habilitated 1969 by W. Heisenberg. He is professor for theoretical physics at the Technical University of Munich. From 1970 to 1984 he was a member of the physics group at the Max Planck Institute zur Erforschung der Lebensbedingungen der Wissenschaftlich-Technischen Welt in Starnberg which was headed by C.F. von Weizsäcker.

M. Drieschner

Michael Drieschner is professor for Natural Philosophy at the University of Bochum, Germany. He has received his doctorate in physics from the University of Hamburg where he was a student of C.F. von Weizsäcker. From 1970 to 1978, he was member of the Max Planck Institute zur Erforschung der Lebensbedingungen der Wissenschaftlich-Technischen Welt in Starnberg, where he worked with C.F. von Weizsäcker on the philosophical and mathematical foundations of quantum theory.

G. Emch

Gérard G. Emch is professor for mathematics at the University of Florida, Gainsville, USA. He works on mathematical physics, analysis (in particular operator algebras and their use in quantum theory), quantum probability and non-commutative geometry and on the history and philosophy of sciences.

D.R. Finkelstein

David Ritz Finkelstein is professor of physics at the Georgia Institute of Technology, Atlanta, USA. His research activity centers on reconciling the fundamental concepts and principles of quantum theory and space-time theory.

Th. Görnitz

Thomas Görnitz is professor at the Institute for Didactics of Physics at the University of Frankfurt. He works on didactic and philosophical questions of Quantum Theory and Cosmology, didactical aspects of the links between natural sciences and humanities. He is chairman of the association for the Carl Friedrich von Weizsäcker Foundation. Thomas Görnitz collaborates, since 1984, with C.F. von Weizsäcker in the development of Ur Theory.

K. Gottstein

Klaus Gottstein is experimental physicist. He has been the head of the elementary particle physics group from 1958 to 1971 under Werner Heisenberg at the Max Planck Institute for Physics and Astrophysics in Munich. From 1971 to 1974 he served as Attaché for Science at the German embassy in Washington, USA. He became a member of the Max Planck Institute zur Erforschung der Lebensbedingungen der Wissenschaftlich-Technischen Welt in Starnberg working with Carl Friedrich von Weizsäcker (1975–1980). Prof. Gottstein has directed a research unit of the Max Planck Society until 1992.

D. Graudenz

Dirk Graudenz has studied physics in Hamburg. He held a postdoctoral fellowship in Aachen, at the Lawrence Berkeley Laboratory and was a fellow at CERN from 1994 to 1997. He obtained his habilitation in physics in 1996. He was Senior Research Associate at the Paul Scherrer Institut in Vilingen, Switzerland from 1997 to 2000 and lecturer at ETH Zurich. His areas of research concerned deep inelastic electron-proton collisions and the physics of Higgs bosons. He has worked with C.F. von Weizsäcker on questions of gauge theories and quantum field theory from 1988 to 1997. He works as a business consultant in Hamburg.

R. Grosse

Ruth Grosse has studied psychology. She has been for 25 years secretary of Carl Friedrich von Weizsäcker. She works as a graphologist.

S. Großmann

Siegfried Großmann is professor emeritus for theoretical physics at the Department of Physics of the University of Marburg, Germany. His areas of research are theoretical and mathematical physics, nonlinear (chaotic) dynamics, structure formation in non-equilibrium systems, hydrodynamic turbulence, stochastic processes, laser theory, and phase transitions. From 1990 to 1993, Prof. Großmann has served as chairman of the Commission for Fundamental Research of the Federal Ministry of Science and Technology. He is a member of the Berlin-Brandenburgische Akademie der Wissenschaften, of the Deutsche Akademie der Naturforscher Leopoldina, and of the European Academy of Sciences and Arts. In 1995 he received the Max Planck Medal, in 1997 the Karl Kuepfmueller Ring.

J. Habermas

Jürgen Habermas, born in 1929, is philosopher and sociologist. From 1964 he was professor of philosophy at the University of Frankfurt, where he was a member of the Frankfurt School of Critical Theory. From 1971 to 1980 he was director of the Max Planck Istitute zur Erforschung der Lebensbedingungen in der wissenschaftlich-technischen Welt in Starnberg, together with Carl Friedrich von Weizsäcker. He remained in Starnberg until 1982 as director of the Max Planck Institute für Sozialwissenschaften from where he returned to the University of Frankfurt as professor professor for philosophy. From 1983, Prof. Habermas was external member of the Max Planck Institute für Psychologie in Munic.

Th. Henning

Thomas Henning is director of the Max-Planck-Institut für Astronomie in Heidelberg and of the Astrophysics Institute of Friedrich Schiller University in Jena. His research concentrates on the interstellar medium, star formation, accretion disks and radiation hydrodynamics. Thomas Henning is scientific coordinator of the German Center for Infrared and Optical Interferometry whose aim is to coordinate efforts by German institutions in obtaining and interpreting astronomical interferometric data from optical to mid-infrared wavelengths.

R. Hilfer

Rudolf Hilfer is associate professor for physics at the University of Mainz and the University of Stuttgart. He has doctorates in political science from the University of Göttingen and in physics from the University of Munich. His fields of research is the correlated transport in disordered systems, dynamic percolation in superionic conductors, pattern formation and nonlinear phenomena, spinglasses, microstructure and transport in porous materials, foundations of the theory of critical phenomena, computer simulation of soft condensed matter, theory of non equilibrium phase transitions, finite size scaling theory, computer simulation of lipid monolayers, fractional dynamics.

O. Ischebeck

Otfried Ischebeck is physicist. He has taught at the University of Algiers (1978–81) and at the Institut Supérieur Pédagogique Bukavu, Congo (1981–85). He was a member of the research group of Horst Afheldt in the Max Planck Society (1986–89), at the Institute for Experimental Physics of the University of Hamburg and the Institute for Peace Research and Security Policy at the University of Hamburg (1990–95). He is currently working at the Bavarian Center for Applied Energy Research in Munich.

C. Kiefer

Claus Kiefer is professor of physics at the University of Cologne. His principle field of research is decoherence and the appearance of a classical world in Quantum Theory.

R. Lüst

Reimar Lüst is astrophysicist. In 1960, he became a member of the Max Planck Institute for Physics and Astrophysics in Munich and in 1963 he became director of the Max Planck Institute for Extraterrestrial Physics in Garching/Munich. At the same time he was professor of both universities of Munich. From 1969–1972, Prof. Lüst was president of the German Research Council. From 1972–1984, he was president of the Max Planck Society. He was director and later on vice president of the European Space Organization and, from 1984 to 1990, General Director of the European Space Organization. He served as President of the Alexander von Humboldt Foundation from 1989 to 1999.

H. Lyre

Holger Lyre is assistant professor at the Philosophy Department of the University of Bonn. His research concerns the philosophy of science, epistemology and neurophilosophy.

K. v. Meyenn

Karl v. Meyenn is physicist with special interest in the historical development of quantum theory, nuclear and elementary particle physics. He is editor of the correspondence of Wolfgang Pauli.

K.M. Meyer-Abich

Klaus Michael Meyer-Abich is professor for natural philosophy at the University of Essen, Germany and at the Institute for Cultural Sciences of the Science Center of Rhineland-Westphalia. From 1970 to 1972, he was a member of the Max Planck Institute zur Erforschung der Lebensbedingungen der Wissenschaftlich-Technischen Welt in Starnberg. He was Senator for Science of the City of Hamburg (1984–1987) and he has served as member of the Commissions of the German Bundestag for Future Energy Policy (1979–1982) and for the Protection of the Earth's Atmosphere (1987–1992).

P. Mittelstaedt

Peter Mittelstaedt has obtained his doctorate under Werner Heisenberg in Göttingen. After research at CERN in Geneva (1958), at the Max-Planck-Institut für Physik und Astrophysik in Munich (1959–62, 1963–65), and at MIT in Cambridge MA (1962–63), he became Lecturer in Physics in Munich and, in 1965, Professor for Theoretical Physics of the University of Cologne. His principle fields of research concern the foundations of quantum theory, relativity theory, the philosophy of natural sciences, the theory of sciences and logic. From 1970 to 1973, Prof. Mittelstaedt has served as rector and vice rector of the University of Cologne and as Vice Rector for Research from 1991 to 1994.

G. Neuneck

Götz Neuneck is research fellow at the Institute for Peace Research and Security Policy at the University of Hamburg. He works on arms control, the analysis of weapons technologies and on mathematical modeling of processes in security policy. Dr. Neuneck is representative of the Pugwash Conferences in Germany.

H. Primas

Hans Primas is professor for physical and theoretical chemistry at ETH Zurich. His research areas range from nuclear magnetic resonance spectroscopy, theoretical chemistry, theory reduction, the measurement problem, holism and realism in quantum theory, to the dialogue between Wolfgang Pauli and Carl Gustav Jung.

H. Rechenberg

Helmut Rechenberg is physicist at the Max Planck Institute of Physics (Werner Heisenberg Institute) in Munich. His main field of interest is the historical development of quantum theory and of elementary particle physics. He is the curator of the Werner-Heisenberg-Archive.

E. Ruhnau

Eva Ruhnau studied physics, mathematics and philosophy in Germany and Canada. She received her Ph.D. in mathematics from the Technical University of Munich. She was teaching at the Universities of Edmonton (Canada), Munich, Jena, Hamburg and Berlin. She worked as a researcher in Edmonton, Munich, Princeton, San Diego, Jülich and Tokyo. Since 1997 she is Director of the Human Science Center of the University of Munich. She has worked and published in the fields of differential geometry, the concept of time in physics and philosophy and its mathematical modeling, the foundations of quantum theory, and in the neurosciences.

H. Saller

Heinrich Saller is theoretical physicist at the Max Planck Institute of Physics (Werner Heisenberg Institute) in Munich and he teaches at the Ludwig Maximilian University in Munich. His research concerns symmetry principles of elementary particles, quantum field theory and the structures of time and space.

G. Süssmann

Georg Süssmann is professor emeritus of the University of Munich. He has received his doctorate under C.F. von Weizsäcker in Göttingen and has been a member of the research group on hadron physics and nuclear physics in Munich.

G. Wolschin

Georg Wolschin is professor for physics at the Institute of Theoretical Physics of the University of Heidelberg. His principle research are is the physics of quarks and gluons.

E. Teller

Edward Teller was born in Hungary. He studied physics at the University of Munich (1928), he has worked with Werner Heisenberg in Leipzig and obtained his doctorate from the University of Göttingen. From this time he has kept a close relationship with C.F. von Weizsäcker. In 1934, he was a member of the Institute for Theoretical Physics of Niels Bohr in Copenhagen from where he emigrated to the United States in 1935. After work with Enrico Fermi in Chicago and Robert Oppenheimer in Berkely, he joined the Manhattan Project in 1941. Promoting the establishment of Lawrence Livermore National Laboratory, he served as director of Lawrence Livermore from 1958 to 1960, to become Professor of Physics at Large for the University of California. Since 1975, Prof. Teller is a member of the Hoover Institution on War, Revolution and Peace at the University of Stanford, California.

A. Zeilinger

Anton Zeilinger in professor for Experimental Physics at the University of Vienna, Austria. His research concerns the basic structures of physical systems and their quantum theoretic analysis. He has been professor at the Technical University of Munich 1990–1999, at the University of Innsbruck 1995, Visiting Professor at the Collège de France, Paris 1996–1998, and Visiting Research Fellow, Merton College, Oxford University 1999. In 1998, he served as President to the Austrian Physical Society.

Printing: Saladruck Berlin
Binding: Stürtz AG, Würzburg